CRC Handbook of Free Radicals and Antioxidants in Biomedicine

Volume II

Editor

Jaime Miquel, Ph.D.
Research Associate
Linus Pauling Institute of Science and Medicine
Palo Alto, California
and
Professor of Neurogerontology
Department of Neurochemistry
University School of Medicine
Alicante, Spain

Associate Editors

Alexandre T. Quintanilha, Ph.D., D.Sc.
Director
Center for Atmospheric and Biospheric Effects of Technology
Applied Science Division
Lawrence Berkeley Laboratory
and
Adjunct Professor
Department of Physiology-Anatomy
University of California
Berkeley, California

Hans Weber, Ph.D.
Medical Writer
Syntex Research
Palo Alto, California

CRC Press, Inc.
Boca Raton, Florida

Library of Congress Cataloging-in-Publication Data

Handbook of free radicals and antioxidants in biomedicine.

Includes bibliographies and index.
1. Free radicals (Chemistry) 2. Antioxidants.
3. Membrane lipids--Peroxidation. I. Miquel, Jaime.
II. Quintanilha, Alexandre T. III. Weber, Hans,
1936- [DNLM: 1. Antioxidants. 2. Biology.
3. Free Radicals. 4. Medicine. QD 471 H236]
RB170.H36 1989 616.07'1 87-35535
ISBN-0-8493-3268-0 (Volume I)
ISBN-0-8493-3269-9 (Volume II)
ISBN-0-8493-3270-2 (Volume III)

This book represents information obtained from authentic and highly regarded sources. Reprinted material is quoted with permission, and sources are indicated. A wide variety of references are listed. Every reasonable effort has been made to give reliable data and information, but the author and the publisher cannot assume responsibility for the validity of all materials or for the consequences of their use.

Direct all inquiries to CRC Press, Inc., 2000 Corporate Blvd., N.W., Boca Raton, Florida, 33431.

International Standard Book Number 0-8493-3268-0 (Volume I)
International Standard Book Number 0-8493-3269-9 (Volume II)
International Standard Book Number 0-8493-3270-2 (Volume III)

Library of Congress Card Number 87-35535
Printed in the United States

PREFACE

Only about two centuries have elapsed since the discovery of oxygen, in spite of the fact that most life, including human life, is oxygen-dependent. It was only a few decades ago that the importance of oxygen radicals in biology and medicine began to be realized.

In the first half of our century, radical research was a theoretical pursuit, more appealing to physicists and chemists than to biologists and clinicians. This situation changed completely with the advent of nuclear fission and the increased interest in radiation research. This brought intensive work on radical reactions in irradiated tissue, and simultaneous technical breakthroughs made it possible to detect these highly reactive and unstable molecular fragments.

In the mid-1950s, because of concepts originated at the Radiation Laboratory of the University of California at Berkeley, interest in free radical biology spread from radiation research to other areas of biology and medicine. This was largely due to Rebeca Gerschman, who proposed that oxygen radicals play a role in both oxygen toxicity and aging, and to Denham Harman, originator of the concept that free radicals are involved not only in aging but also in age-related conditions such as amyloidosis, atherosclerosis, and cancer. These concepts are still controversial but stand on much firmer ground than when first enunciated. It has been shown by now that free radicals are indeed released in living cells, and the evidence suggests that they cause (or modulate) aging, inflammation, and carcinogen-induced mutagenesis.

In addition to the above, it has been established that free radicals are involved in cellular respiration, microsomal detoxification, and membrane turnover, as well as in the cytopathological reactions to ionizing radiation, anoxia hyperoxia, and stress.

The last decades have also witnessed great strides in the field of the free radical ''antidotes'', i.e., the free radical scavengers and antioxidants provided by evolution to ward off the injurious effects of free radicals. In the short period from 1966 to 1969, De Duve clarified the detoxifying role of peroxysomes, and Fridovich discovered a key line of enzyme defense against oxygen radicals, namely the superoxide dismutases. These achievements, in addition to the finding that prostaglandin biosynthesis requires the formation of lipoperoxides, awoke interest in the whole range of subjects which are covered in this Handbook, from the physics and detection of free radicals to the stimulation of antioxidant defenses as a novel therapeutic approach.

Free radical research has become fashionable, as reflected in the abundance of symposia and publications dealing with the subject. However, there is a scarcity of textbooks which deal systematically and in a concise and simple fashion with both the fundamental and the clinically relevant aspects of free radicals and antioxidants. This Handbook is a step in that direction. In the introduction the early scientific discoveries and some theoretical concepts which are seldom acknowledged, despite their importance in paving the way to the contemporary accomplishments, are reviewed. This is followed by a critical review of specific issues in the fields of free radicals and antioxidants, and by a final selection of models and methods for radical-related experimental and clinical research.

It is the hope of the authors that the information gathered in this Handbook will be useful not only to students, researchers, and teachers, but also to health professionals who wish to get acquainted with this intriguing offshoot of the biomedical sciences.

J. Miquel
A. Quintanilha
H. Weber

EDITOR

Jaime Miquel, Ph.D., obtained his Ph.D. in Pharmacology at the University of Madrid, Spain. He has been active in biomedical research at the Cajal Institute in Spain, The National Institutes of Health in Bethesda, Maryland, and from 1961 to 1983 served as Chief of the Experimental Pathology Branch and Research Scientist of the NASA Ames Research Center in Moffett Field, California. Presently, Dr. Miquel holds the positions of Research Associate at the Linus Pauling Institute of Science and Medicine in Palo Alto, California, and Professor of Neurogerontology at the University School of Medicine in Alicante, Spain. He has received numerous NASA awards for research on the biological effects of space flight, including the first study ever performed (as part of the collaborative Soviet-American "Cosmos" project) on the aging process of weightless animals maintained on board space biosatellites. Dr. Miquel is a Fellow of the Gerontological Society of America and a member of the editorial boards of seven international journals dealing with gerontology and geriatrics. His current research interests are the cytopathological and molecular mechanisms of aging, with focus on the role played by oxygen radicals in physiological and pathological processes.

ASSOCIATE EDITORS

Alexandre T. Quintanilha, Ph.D., D.Sc., is Director of the Center for Atmospheric and Biospheric Effects of Technology and Assistant Head of the Applied Science Division at the Lawrence Berkeley Laboratory. He is also an Adjunct Professor of Physiology at the University of California, Berkeley and a Professor of Biophysics at the *Instituto de Ciencias Biomedicas* "Abel Salazar" in Porto, Portugal. Dr. Quintanilha was born in Maputo, Mozambique, obtained his B.Sc.(Hon.) at the University of Witwatersrand, Johannesburg, in 1968 in Theoretical Physics, and his Ph.D. in Solid State Physics in 1972 in the area of superconductivity. In 1981 he obtained his *Agregation* from the University of Porto. Since 1974 he has been at the University of California, Berkeley working in the areas of free radicals and oxygen toxicity in biology. He has spent extended periods both at the University of Paris, Orsay, and at the Biomedical Institute in Porto. His more recent interests include spectroscopic techniques for biological studies and environmental mechanisms of oxidative damage and adaptation.

Dr. Quintanilha is a member of the Society for Free Radical Research, American Society for Biochemistry and Molecular Biology, Biophysical Society, *Sociedade Portuguesa de Bioquimica,* and a Fellow of the International Cell Research Organization. He is an editor for several international journals.

Among other awards, he has received Fellowships from the Gulbenkian Foundation and the National Academy of Science, National Research Council. He has been a visiting professor at many universities, and has presented more than 30 invited lectures at international and national meetings. He has published more than 60 research papers in peer-reviewed journals. He has edited and coedited three other books on issues relating to oxidative stress and the environment.

Hans U. Weber, Ph.D., studied life sciences and chemistry at the Federal Institute of Technology, Zürich, Switzerland, where he received an M.S. in Chemistry and a Ph.D. for his thesis in plant biochemistry (1964). He moved to California and, as biochemist for the Foundation for Nutrition and Stress Research, Redwood City, studied human nutrition in relation to disease and aging. In 1979, he joined the management of the Alexander Medical Foundation, which sponsored a program of research on superoxide dismutase. At the same time he was a contributing editor to a newsletter on health. Other writing projects followed, and in 1987 Dr. Weber joined the Medical Writing Group of Syntex Research, Palo Alto, California. He has been particularly interested in the biology and mechanisms of aging and is a member of the Society for Free Radical Research, the American Aging Association, the American Medical Writers Association, and the Society for Technical Communication.

CONTRIBUTORS

Volume II

Soraya Abbasi, M.D.
Assistant Professor
Department of Pediatrics
University of Pennsylvania
Philadelphia, Pennsylvania

Daniel Acosta, Ph.D.
Professor
Department of Pharmacology and Toxicology
College of Pharmacy
University of Texas
Austin, Texas

Maria Jose Alcaraz, Ph.D.
Assistant Professor
Department of Pharmacology
Faculty of Pharmacy
Valencia, Spain

Gyorgy Balla, Ph.D.
Assistant Professor
Department of Pediatrics
University Medical School
Debrecen, Hungary

Joseph J. Barboriak, Sc.D.
Professor of Pharmacology
Department of Pharmacology
Medical College of Wisconsin
Milwaukee, Wisconsin

Michael E. Begin, Ph.D.
Efamol Research Institute
Kentville, Nova Scotia, Canada

Adrianne Bendich, Ph.D.
Senior Scientist
Department of Vitamins and Clinical Nutrition
Hoffmann-La Roche, Inc.
Nutley, New Jersey

Carlo Bertoni-Freddari
Center for Surgical Research
Gerontology Research Department
I.N.R.C.A.
Ancona, Italy

Robert E. Beyer, Ph.D.
Professor of Biological Sciences
Department of Biology
University of Michigan
Ann Arbor, Michigan

Graham W. Burton, Ph.D.
Senior Research Officer
Division of Chemistry
National Research Council of Canada
Ottawa, Ontario, Canada

Ciro Costagliola, M.D.
Eye Clinic
First School of Medicine
University of Naples
Naples, Italy

Mary J. De Long, Ph.D.
Department of Environmental Health Sciences
Johns Hopkins University
Baltimore, Maryland

J. H. Fellman, Ph.D.
Professor
Department of Biochemistry
Oregon Health Sciences University
Portland, Oregon

Emil Ginter, Ph.D.
Department of Biochemistry
Institute of Nutrition
Bratislava, Czechoslovakia

Hannu Heiskala, M.D.
Pediatrician
Children's Hospital
University of Helsinki
Helsinki, Finland

Helen Mintz Hittner, M.D.
Clinical Professor of Ophthalmology and Pediatrics
Baylor College of Medicine
Houston, Texas

Peter J. Hornsby, Ph.D.
Associate Professor
Department of Cell and Molecular Biology
Medical College of Georgia
Augusta, Georgia

David F. Horrobin, D.Phil.
Efamol Research Institute
Kentville, Nova Scotia, Canada

Keith U. Ingold, Ph.D.
Associate Director
Division of Chemistry
National Research Council of Canada
Ottawa, Ontario, Canada

Lois H. Johnson, M.D.
Clinical Associate Professor
Department of Pediatrics
University of Pennsylvania
Philadelphia, Pennsylvania

John H. Kalbfleisch, Ph.D.
Assistant Professor
Department of Anesthesiology
Medical College of Wisconsin
Milwaukee, Wisconsin

Laszlo Karmazsin, Ph.D.
Professor and Head of Pediatric Clinic
Pediatric Clinic
University Medical School
Debrecen, Hungary

Martin L. Katz, Ph.D.
Research Assistant Professor
Department of Ophthalmology
University of Missouri School of Medicine
Columbia, Missouri

Masayoshi Kibata,, M.D.
Director
Department of Internal Medicine
National Sanitorium
Minamiokayama Hospital
Okayama, Japan

Heinz Koehler, M.D.Sc.
Professor
Department of Internal Medicine
Karl Marx University
Leipzig, East Germany

Prayad Komaratat, Ph.D.
Associate Professor
Department of Biochemistry
Mahidol University
Bangkok, Thailand

Motoharu Kondo, Ph.D.
Director
First Department of Medicine
Kyoto Prefectural University of Medicine
Kyoto, Japan

Frank L. Kretzer, Ph.D.
Associate Professor of Ophthalmology and Cell Biology
Baylor College of Medicine
Houston, Texas

Jaime Miquel, Ph.D.
Research Associate
Linus Pauling Institute of Science and Medicine
Palo Alto, California, and
Professor of Neurogerontology
Department of Neurochemistry
University School of Medicine
Alicante, Spain

Nobuhiro Morisaki, M.D.
Associate Professor
Second Department of Internal Medicine
College of Medicine
Chiba University
Chiba, Japan

James S. Nelson, M.D.
Division Head, Neuropathology
Department of Pathology
Henry Ford Hospital
Detroit, Michigan

Herminia Pasantes-Morales, Ph.D.
Professor
Department of Neurosciences
Institute of Cellular Physiology
Universidad Nacional Autónoma de México
Mexico City, Mexico

Hans-Joachim Peters, Dr.rer.nat.
Department of Internal Medicine
Karl Marx University
Leipzig, East Germany

P. Puig-Parellada, Ph.D.
Associate Professor
Department of Pharmacology
Faculty of Medicine
University of Barcelona
Barcelona, Spain

Graham E. Quinn, M.D.
Assistant Professor
Department of Ophthalmology
University of Pennsylvania
Philadelphia, Pennsylvania

Alexandre T. Quintanilha, Ph.D., D.Sc.
Director
Atmospheric and Biospheric Effects of Technology
Applied Science Division
Lawrence Berkeley Laboratory, and
Adjunct Professor
Department of Physiology-Anatomy
University of California, Berkeley
Berkeley, California

Kenneth Ramos, Ph.D.
Assistant Professor
Department of Pharmacology and Toxicology
Philadelphia College of Pharmacy and Science
Philadelphia, Pennsylvania

Ernesto Rinaldi, M.D.
Eye Clinic
First School of Medicine
University of Naples
Naples, Italy

W. Gerald Robison, Jr., Ph.D.
Chief
Section of Pathophysiology
National Eye Institute
National Institutes of Health
Bethesda, Maryland

Guillermo T. Sáez, M.D.
Professor
Department of Biochemistry and Molecular Biology
University of Valencia
Valencia, Spain

Yasushi Saito, M.D.
Professor
Second Department of Internal Medicine
College of Medicine
Chiba University
Chiba, Japan

Pirkko Santavuori, M.D.
Children's Hospital
University of Helsinki
Helsinki, Finland

Nongnuch Settasatian, M.Sc.
Lecturer
Department of Clinical Chemistry
Khon Kaen University
Khon Kaen, Thailand

Dorota Soltysiak-Pawluczuk, Ph.D.
Project Investigator
Department of Experimental Pediatrics
University of Texas System Cancer Center
Houston, Texas

Angel Villar, Ph.D.
Professor and Head
Department of Pharmacology
Faculty of Pharmacy
Madrid, Spain

Jose Viña, M.D.
Professor and Chairman
Department of Physiology
Faculty of Medicine
University of Valencia
Valencia, Spain

Juan R. Viña, M.D.
Professor
Department of Biochemistry and Molecular Biology
Faculty of Medicine
University of Valencia
Valencia, Spain

Alexander Yeu-Ming Wang, Ph.D.
Associate Professor and Deputy Chairman
Department of Experimental Pediatrics
University of Texas System Cancer Center
Houston, Texas

Hans U. Weber, Ph.D.
Medical Writer
Syntex Research
Palo Alto, California

Tuomas Walter Westermarck, M.D.
Vice-Superintendent
Helsinki Central Institution for the Mentally Retarded
Killinmäki Kirkkonummi
University of Helsinki
Helsinki, Finland

Prapon Wilairat, Ph.D.
Associate Professor
Department of Biochemistry
Mahidol University
Bangkok, Thailand

Toshikazu Yoshikawa, M.D., Ph.D.
Assistant Professor
First Department of Medicine
Kyoto Prefectural University of Medicine
Kyoto, Japan

Imre Zs.-Nagy, M.D.
Professor and Head
V.I.L.E.G., Hungarian Section
University Medical School
Debrecen, Hungary

CRC HANDBOOK OF FREE RADICALS AND ANTIOXIDANTS IN BIOMEDICINE

Volume I

Introduction

Historical Introduction to Free Radical and Antioxidant Biomedical Research

Physicochemical and Physiological Aspects

Chemical and Biochemical Aspects of Activated Oxygen: Singlet Oxygen, Superoxide Anion, and Related Species

Formation of Electronically Excited States in Biological Systems: Contribution of Singlet Oxygen and Excited Carbonyl Compounds to Biological Chemiluminescence

Role of Metals in Oxygen Radical Reactions and Oxidative Stress

Peroxides: Physiological Function and Role in Drug Toxicity and Carcinogen Activation

Pro-Oxidant Action of Antioxidants

The Role of Free Radicals in the Cytotoxicity of Cysteine

Generation of Oxygen Radicals by the Anticancer Quinone Drugs Adriamycin, Aclacinomycin A, Carbazilquinone, and Mitomycin C in NADPH-Cytochrome P-450 Reductase and Xanthine Oxidase Systems

Thyroid Hormone, Active Oxygen, and Lipid Peroxidation

Effects of Oxygen Radicals and Antioxidants on Calcium Ion Transport across Lipid Bilayers and Cellular Membranes

Leukotrienes and Related Substances: Inflammatory and Pharmacological Effects

Free Radicals from Catecholamine Hormones, Neuromelanins, and Neurotoxins

Membrane Phospholipids Modulate Synaptic GABA Receptor Binding through the Generation of Active Oxygen Radicals

Free Radicals, Peroxides, Aging, and Disease

Free Radicals and Lipid Peroxidation in Cell Pathology

Free Radical-Induced Lipoperoxidation and Aging

Functional Consequences of Free Radical Damage to Cell Membranes

Free Radicals and Human Disease

Serum Lipid Peroxide Levels in Subjects Having Certain Diseases and in Aging

Neutrophil-Induced Erythrocyte Injury

Plasma Lipid Peroxides in Patients with Vascular Disease and in Middle-Aged Normal Subjects with a High Risk of Atherosclerosis

Neutrophil Infiltration of the Myocardium: A Potential Mechanism for Free Radical Necrosis

Xanthine Oxidase, Free Radicals, and Myocardial Ischemia

The Role of Oxygen-Derived Free Radicals in the Pathogenesis of Myocardial Ischemia and Reperfusion-Induced Injury

The Role of Free Radicals in Drug-Induced Myocardial Effects

Involvement of Oxygen Radicals in Pulmonary Edema

Free Radical Lipid Peroxidation in Rheumatoid Arthritis

Volume II

Antioxidants and Free Radical Scavengers: Biochemical Mechanisms and Clinical Applications

The Role of Antioxidants in the Function of the Adrenal Cortex
Essential Fatty Acids: Metabolism, Interaction with Free Radicals, and Clinical Use
Mechanisms of Antioxidant Action: Preventive and Chain-Breaking Antioxidants
The Role of Coenzyme Q in Free Radical Formation and Antioxidation
Induction by Substituted Phenols of Enzymes that Inactivate Electrophilic Compounds
Anti-inflammatory Effects of Natural Phenolic Compounds
Centrophenoxine as OH· Free Radical Scavenger
Interactions between Vitamins C and E and Cytochrome P-450
Taurine and Hypotaurine and Membrane Lipid Peroxidation
The Physiological Functions of Glutathione
Effects of Exercise-Induced Oxidative Stress on Liver Glutathione
Vitamin E and Red Blood Cell Glutathione
Interaction between Antioxidant Vitamins C and E and Its Effect on Immune Responses
Effect of Vitamin E Status on Hepatic Cholesterol 7α-Hydroxylase
Free Radical Lipid Peroxidation and Vitamin E in Liver Injury
Cytotoxic Actions of Isoproterenol in Cardiac Cells: Protective Effects of Antioxidants
Chronic Vitamin C Deficiency, Hypercholesterolemia, and Atherosclerosis
The Action of Free Radical Scavengers and Anti-inflammatory Drugs on Synovial Fluid Degradation Induced by Free Radicals
α-Tocopherol and Plasma Lipids in Man
Effect of Vitamin E on Low-Density Lipoprotein Metabolism in the Arterial Wall
Serum α-Tocopherol, Coenzyme Q, and TBA-Reactive Substance in Acute Myocardial Damage and Stroke
Selenium and Lipid Peroxidation in Human and Experimental Cardiovascular Disease
The Neuropathology of Chronic Vitamin E Deficiency in Humans and Other Mammals
Synaptic Plasticity in the Cerebellar Glomeruli of Rats: Effects of Aging and Vitamin E Deficiency
α-Tocopherol as a Potential Modifier of the Toxicity of Anthracycline Antibiotics
Antioxidant Therapy in Neuronal Ceroid Lipofuscinosis (NCL)
Consequences of Impaired Antioxidant Protection of the Retina
Vitamin E and Retinopathy of Prematurity: Historical and Clinical Aspects
Pathogenesis and Antioxidant Suppression of Retinopathy of Prematurity
Pharmacological Protection against Free Radical-Induced Retrolental Fibroplasia

Volume III

Selected Models and Methods for Biological and Clinical Research
The Carbon Tetrachloride Hepatotoxicity Model: Free Radicals and Calcium Homeostasis
Models of Brain Lipoperoxidation: Inhibitory Effects of Mannitol, Vitamin E, and Glucocorticoids
Role of Oxidative Stress in Tumor Promotion
Effects of Vitamin C on Experimental Malignant Tumors
Fatty Acid Composition of Phospholipids in Heart Muscle Modified by Age, Diet, and Stress
Lesions of Aortic Intima Caused by Linoleic Acid Hydroperoxide
Peroxide-Induced Membrane Damage in Red Blood Cells
Oxidized Fatty Acids and Phosphatidic Acid Promote Ca^{2+} Translocation in Model Membranes: Studies with Arsenazo III-Containing Liposomes
Studies of Phospholipid Peroxidation in Liposomes
Spectroscopic Techniques in the Study of Lipid Oxidation (except ESR Techniques)
Chemiluminescence Methods for Assays of Lipoxygenase, 1,4-Pentadiene Fatty Acids, and Oxygen Concentrations
Measurement of Intracellular Oxygen
Chemical Determination of Oxidative Damage to DNA
Electron Spin Resonance in Biology
General Mechanisms of Spin Trapping in Vitro and in Vivo
Spin-Trapping Studies of Xenobiotic-Mediated Toxicity
Application of Electron Spin Resonance Techniques to the Detection of Free Radicals in Muscle Tissue
A Simple Fluorometric Assay for Lipid Peroxides in Blood Serum or Plasma
Chromogenic Determination of Lipid Hydroperoxides by Sesamol Dimer
Measurement of Individual Antioxidants and Radical-Trapping Activity
Analysis of Ascorbic Acid in Body Fluids and Tissues
Analysis of Tocopherol in Blood and Liver
Methods for the Analysis of Vitamin E in Animal Tissue
Determination of Coenzyme Q (Ubiquinone) in Tissues of Experimental Animals
Determination of Reduced and Oxidized Ubiquinones
HPLC Analysis of Thiols Using Monobromobimane Fluorescent Labeling: Application to Radioprotective Drugs
Selenium in Human Urine: Determination and Clinical Implications
Determination of Glutathione Peroxidase
Convenient Assays for Superoxide Dismutase

To my wife Margarita

TABLE OF CONTENTS

Volume II

ANTIOXIDANTS AND FREE RADICAL SCAVENGERS: BIOCHEMICAL MECHANISMS AND CLINICAL APPLICATIONS

The Role of Antioxidants in the Function of the Adrenal Cortex 3
Peter J. Hornsby

Essential Fatty Acids: Metabolism, Interaction with Free Radicals, and Clinical Use ... 17
David F. Horrobin and Michael E. Begin

Mechanisms of Antioxidant Action: Preventive and Chain-Breaking Antioxidants 29
Graham W. Burton and Keith U. Ingold

The Role of Coenzyme Q in Free Radical Formation and Antioxidation 45
Robert E. Beyer

Induction by Substituted Phenols of Enzymes That Inactivate Electrophilic Compounds 63
Mary J. De Long

Anti-inflammatory Effects of Natural Phenolic Compounds 77
Maria Jose Alcaraz and Angel Villar

Centrophenoxine as OH· Free Radical Scavenger.. 87
I. Zs.-Nagy

Interactions between Vitamins C and E and Cytochrome P-450 95
Emil Ginter

Taurine and Hypotaurine and Membrane Lipid Peroxidation 105
H. Pasantes-Morales and J. H. Fellman

The Physiological Functions of Glutathione 121
Juan R. Viña, Guillermo T. Sáez, and Jose Viña

Effects of Exercise-Induced Oxidative Stress on Liver Glutathione..................... 133
Alexandre T. Quintanilha

Vitamin E and Red Blood Cell Glutathione 145
Ciro Costagliola and Ernesto Rinaldi

Interaction between Antioxidant Vitamins C and E and Its Effect on Immune Responses 153
Adrianne Bendich

Effect of Vitamin E Status on Hepatic Cholesterol 7 α-Hydroxylase 161
Nongnuch Settasatian, Prapon Wilairat, and Prayad Komaratat

Free Radical Lipid Peroxidation and Vitamin E in Liver Injury 167
Toshikazu Yoshikawa and Motoharu Kondo

Cytotoxic Actions of Isoproterenol in Cardiac Cells: Protective Effects of Antioxidants 177
Kenneth Ramos and Daniel Acosta

Chronic Vitamin C Deficiency, Hypercholesterolemia, and Atherosclerosis 187
Emil Ginter

The Action of Free Radical Scavengers and Anti-inflammatory Drugs on Synovial Fluid Degradation Induced by Free Radicals 197
P. Puig-Parellada

α-Tocopherol and Plasma Lipids in Man 207
John H. Kalbfleisch and Joseph J. Barboriak

Effect of Vitamin E on Low-Density Lipoprotein Metabolism in the Arterial Wall..... 215
Yasushi Saito and Nobuhiro Morisaki

Serum α-Tocopherol, Coenzyme Q, and TBA-Reactive Substance in Acute Myocardial Damage and Stroke.. 227
Masayoshi Kibata

Selenium and Lipid Peroxidation in Human and Experimental Cardiovascular Disease 237
Hans-Joachim Peters and Heinz Koehler

The Neuropathology of Chronic Vitamin E Deficiency in Humans and Other Mammals 243
James S. Nelson

Synaptic Plasticity in the Cerebellar Glomeruli of Rats: Effects of Aging and Vitamin E Deficiency.. 255
Carlo Bertoni-Freddari

α-Tocopherol as a Potential Modifier of the Toxicity of Anthracycline Antibiotics..... 269
Dorota Soltysiak-Pawluczuk and A. Yeu-Ming Wang

Antioxidant Therapy in Neuronal Ceroid Lipofuscinosis (NCL) 281
Tuomas Westermarck, Pirkko Santavuori, and Hannu Heiskala

Consequences of Impaired Antioxidant Protection of the Retina........................ 289
Martin L. Katz and W. Gerald Robison, Jr.

Vitamin E and Retinopathy of Prematurity 303
Lois Johnson, Soraya Abbasi, and Graham Quinn

Pathogenesis and Antioxidant Suppression of Retinopathy of Prematurity 315
Frank L. Kretzer and Helen M. Hittner

Pharmacological Protection against Free Radical-Induced Retrolental Fibroplasia 333
Laszlo Karmazsin and Gyorgy Balla

Index 341

Antioxidants and Free Radical Scavengers: Biochemical Mechanisms and Clinical Applications

THE ROLE OF ANTIOXIDANTS IN THE FUNCTION OF THE ADRENAL CORTEX

Peter J. Hornsby

INTRODUCTION

The adrenal cortex is the site of the synthesis of the steroid hormones such as the glucocorticoid cortisol and the mineralocorticoid aldosterone. The pathway of biosynthesis of these steroids from cholesterol involves a sequence of transformations using cytochrome P-450 enzymes. The high content of cytochrome P450 enzymes in the adrenal cortex presents special problems of oxygen radical toxicity in this tissue. The evidence that the high content of antioxidant compounds in the adrenal cortex, principally ascorbate, may serve to protect cytochrome P450 enzymes from the damaging effects of oxygen radical species formed during biosynthesis of the steroid hormones and the significance of the loss of enzymatic activity of some of the enzymes by this mechanism in the regulation of the zonation of aldosterone and cortisol synthesis in the adrenal cortex in pathological or normal conditions are reviewed below.

PSEUDOSUBSTRATE EFFECTS ON CYTOCHROME P450

Cytochrome P450/Pseudosubstrate Effects in Loss of 11β-Hydroxylase Activity in Senescence

Our work on the effects of oxygen radicals and antioxidants on adrenocortical cytochrome P450 enzymes, particularly 11β-hydroxylase, originated in an investigation of the causes of the poor expression of some of these enzymes during senescence of bovine adrenocortical cells in culture.[1,2] Both glucocorticoid (cortisol) and mineralocorticoid (aldosterone) synthesis were affected. Aldosterone secretion results from the presence of a cytochrome P450 species with corticosterone methyl oxidase (CMO) activity. It has been thought that CMO results from the activity of a separate cytochrome P450 ($P450_{CMO}$), but there is evidence that CMO is actually an activity of the mitochondrial cytochrome P450 with 11β-hydroxylase activity, cytochrome $P450_{11\beta}$.[3,4] The presence of a particular lipid or other cofactor may be required to confer CMO activity on cytochrome $P450_{11\beta}$. In the absence of this factor, cytochrome $P450_{11\beta}$ synthesizes corticosterone, 18-hydroxydeoxycorticosterone, and 18-hydroxycorticosterone from deoxycorticosterone, but not aldosterone.[5,6] CMO activity is increased on exposure of adrenocortical zona glomerulosa cells to elevated medium (K^+), whereas 11β-hydroxylase activity, but not CMO, is increased after incubation with ACTH or other agents which raise intracellular cyclic AMP.[7] Loss of cytochrome $P450_{11\beta}$ (11β-hydroxylase and CMO) activity in bovine adrenocortical cells in culture was associated with accumulation in the culture medium of steroids synthesized by the cells, such as cortisol. These steroids appeared to act as pseudosubstrates for this enzyme.[8-11] The interaction of steroid pseudosubstrates with cytochrome $P450_{11\beta}$ is a major cause of the lack of proper expression of 11β-hydroxylase and CMO in adrenocortical cells in senescence in culture.

Decrease in Halflife of Cytochrome $P450_{11\beta}$ with Pseudosubstrate Steroids

When bovine adrenocortical cells are incubated with pseudosubstrate steroids, e.g., cortisol or several other adrenal steroids, the activity of 11β-hydroxylase decreases.[8] For example, the halflife of the enzyme activity was reduced from 40 to 9 hr in the presence of 30 μM cortisol.[8] In cultured rat and bovine adrenocortical cells, CMO activity is also decreased when pseudosubstrate steroids are added to the cells.[7,11] Steroids were used at concentrations

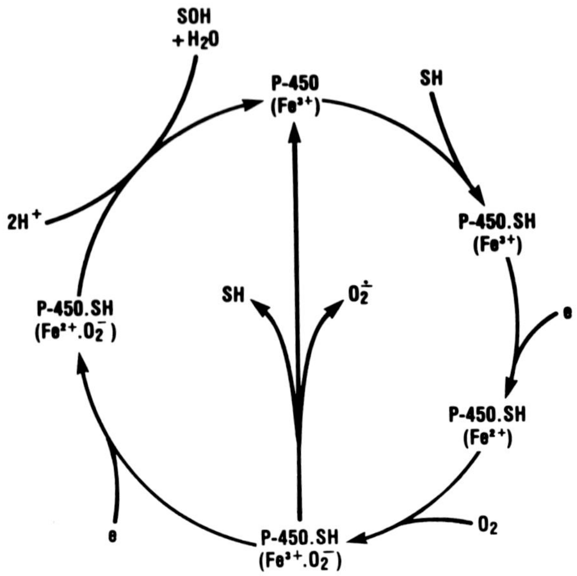

FIGURE 1. Proposed mechanism of oxygen radical production by adrenocortical cytochrome $P450_{11\beta}$. Ferric cytochrome P450 binds substrates or pseudosubstrates (products) (SH) and forms a P450· pseudosubstrate· O_2 complex. With substrates, the cycle is completed normally with production of product (SOH) and water. With pseudosubstrates, the complex breaks down to form ferric P450 and superoxide ($O_2^{\overline{\cdot}}$). (From Hornsby, P. J. and Crivello, J. F., *Mol. Cell. Endocrinol.*, 30, 123, 1983. With permission from Elsevier Scientific Publishers.)

likely to exist in the adrenal cortex in vivo.[12] In bovine glomerulosa cell cultures, cortisol and other steroids cause losses of both 11β-hydroxylase and CMO activities.[11] However, in rat glomerulosa cell cultures, added cortisol and other steroids cause loss of CMO activity, as evidenced by loss of aldosterone and 18-hydroxycorticosterone production, without affecting 11β-hydroxylase activity, as evidenced by continued formation of corticosterone and 18-hydroxydeoxycorticosterone;[7] similar findings have also been reported in explant cultures of fetal rat and fetal human adrenals.[13]

Pseudosubstrates Appear to Interact at the Enzyme Substrate Site

Steroids active in causing loss of 11β-hydroxylase and CMO include cortisol, corticosterone, and androstenedione, but not aldosterone.[7,8,11] All the steroids found to decrease the halflife of 11β-hydroxylase have an 11β-hydroxy group or no substituent at position 11 of the steroid nucleus.[8] Steroids with an 11α-hydroxy group or an 11-keto group were ineffective. The steroid structure for loss of CMO activity was similar to that required for loss of 11β-hydroxylase activity.[11] The requirement for the carbon at position 11 either to have a β-hydroxy group or to be able to be hydroxylated suggested that the products of the enzyme interact with the enzyme substrate site, i.e., enzyme products may act as pseudosubstrates.[8,9]

Hypothesis for Loss of 11β-Hydroxylase Activity with Pseudosubstrates

Destruction or inactivation of cytochrome P450 enzymes during metabolism of certain substrates or interaction with pseudosubstrates may be a widespread phenomenon.[14] The loss of activity may be initiated by the release of superoxide from the enzyme when attempted metabolism of pseudosubstrates occurs and normal hydroxylation cannot take place (Figure 1). If superoxide initiates lipid peroxidation, damage to cytochrome P450 enzymes may result from peroxidase activity of cytochrome P450. Peroxidase activity is a general property of hemeproteins.[15-17] When homolytic cleavage of the peroxide bond occurs on interaction of hemeproteins with peroxides, radicals are formed which give rise to a variety of phenomena, such as quasi-lipoxygenase activity,[18,19] facilitation of lipid peroxidation,[20,21] co-oxygenation of certain substrates,[22-24] and attack of the radicals on the heme moiety with formation of a distinctive class of damage products and loss of activity of the hemeprotein.[25-27] These phenomena are established for hemoglobin, peroxidases, and prostaglandin synthetase, as well as cytochrome P450. Cytochrome $P450_{11\beta}$ has been shown to be inactivated on incubation with peroxides.[28,29] The evidence that 11β-hydroxylase and CMO are activities of the same cytochrome suggests that their differential sensitivity to loss of activity with pseudosubstrates must result from differences in the associated lipids or other factors[3,4] which confer these different activities on cytochrome $P450_{11\beta}$.

PROTECTION OF CYTOCHROME $P450_{11\beta}$ BY ANTIOXIDANTS, INHIBITORS, AND LOWERED OXYGEN

In our initial investigations,[8] various substances, particularly BHA (butylated hydroxyanisole, 2(3)-*tert*-butyl-4-methoxyphenol), dimethyl sulfoxide (Me_2SO), and ascorbic acid, were found to protect adrenocortical cells against loss of 11β-hydroxylase activity in the presence of steroid pseudosubstrates. The loss of 11β-hydroxylase activity was also reduced when the concentration of oxygen was lowered.[8] Similarly BHA, Me_2SO, and lowered oxygen were found to prevent loss of CMO activity.[10,11]

We suggested that BHA acted to inhibit lipid peroxidation (i.e., reacted with lipoperoxy radicals) and that Me_2SO scavenged hydroxyl radicals, formed as a secondary product of $O_2^{\overline{\cdot}}$ released from cytochrome $P450_{11\beta}$.[8,9,14]

In subsequent studies, many other compounds were found to prevent the pseudosubstrate-mediated loss of activity. In particular, inhibitors of cytochrome P450 protected adrenocortical cells against the loss of 11β-hydroxylase and CMO activities in the presence of pseudosubstrates. For example, metyrapone is both a potent inhibitor and protector for 11β-hydroxylase and CMO.[10,11,30] In fact, CMO activity was so unstable in the presence of endogenous steroids and air that basal CMO activity could only be preserved when cells were cultured in the continuous presence of the inhibitor.[10]

Comparison of Protection by Phenols, Sulfoxides, and Metyrapone

We compared the mode of action of these various protectors of 11β-hydroxylase in the presence of pseudosubstrates. In addition to BHA, several other phenols were active at micromolar concentrations (Figure 2).[31] Me_2SO provided a greater absolute level of protection than the phenols, but at much higher concentrations. Various aryl and alkyl sulfoxides tested were found to be protective, at concentrations considerably lower than those required for Me_2SO (Figure 3).[32] In particular, dibenzyl sulfoxide completely protected 11β-hydroxylase at a concentration of 1 m*M*. The ED_{50} for protection by this compound (160 μ*M*) is less than 1/200 of that for Me_2SO (34 m*M*). Metyrapone was active at much lower concentrations, with complete protection of 11β-hydroxylase at a concentration of 1 μ*M*.[31,32]

Phenols and Sulfoxides More Effective at Lowered Oxygen

When cultures were incubated with phenols or sulfoxides together with 50 μ*M* cortisol at three different oxygen concentrations (2 and 19% as well as the standard 5%), each compound being used at its optimal concentration, we found that lowering the oxygen concentration itself resulted in a small protective action on 11β-hydroxylase activity and that all of the tested substances were more effective in protection of 11β-hydroxylase at low oxygen and less effective in air.[31,32]

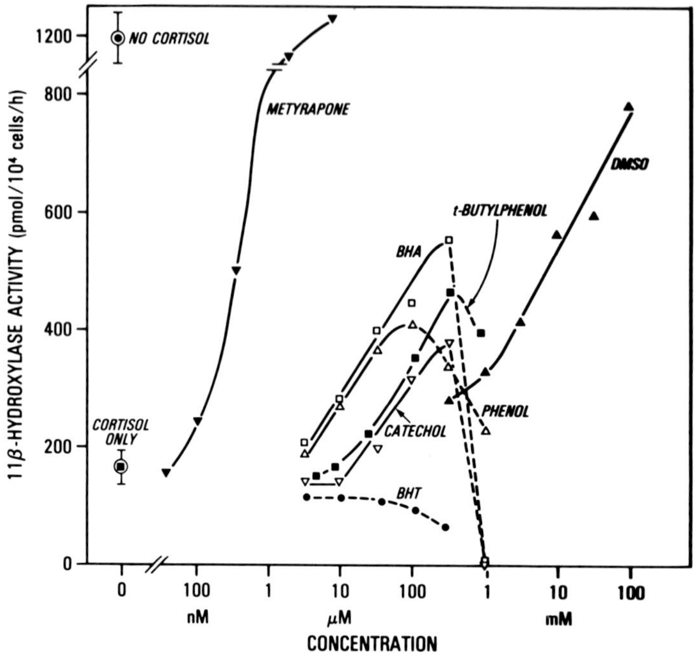

FIGURE 2. Comparison of prevention of loss of 11β-hydroxylase by metyrapone, phenols, and Me_2SO; 50 μM cortisol was added to primary bovine adrenocortical cells cultured in a serum-free defined medium, together with various concentrations of the indicated substances. After 24 hr, compounds were removed and 11β-hydroxylase activity was measured. (From Hornsby, P. J., Aldern, K. A., and Harris, S. E., *Biochem. Pharmacol.*, 34, 865, 1985. With permission from Pergamon Press, Ltd.)

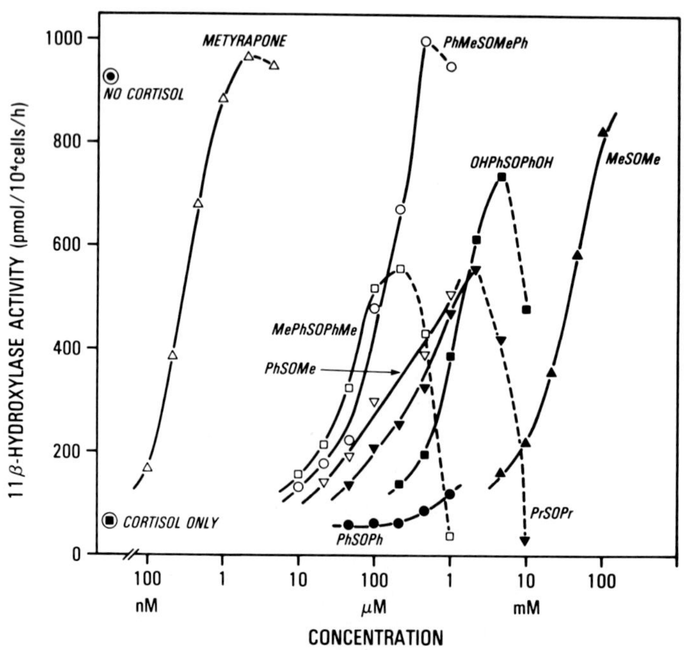

FIGURE 3. Prevention of loss of 11β-hydroxylase activity by Me_2SO, other sulfoxides, and metyrapone. The experiment was performed as described for Figure 2. (From Hornsby, P. J., Harris, S. E., and Aldern, K. A., *Chem. Biol. Interact.*, 51, 335, 1984. With permission)

Mode of Action of Phenols and Sulfoxides

Phenols Do Not Act Solely as Antioxidants

Our results indicated that BHA probably does not act as an antioxidant as such, because the order of potency of the protective phenolic compounds does not correspond to the known order of reactivity of these substances with peroxy radicals.[31] For example, phenol was as effective as BHA and phenol is not a good antioxidant. On the other hand, for prevention of aminooxyacetate toxicity in bovine adrenocortical cells, the order of effectiveness of phenolic compounds was the same as that for chemical antioxidant potency.[33] Thus, BHA is unlikely to protect 11β-hydroxylase by direct inhibition of lipid peroxidation.

Phenols as Peroxidase Donors

An alternative model was proposed in which a bound oxygen-centered radical is reduced by the identified protective compounds, acting as peroxidase donors.[31] In the interaction of hemeproteins with peroxides, various substances that are able to serve as peroxidase donors react with radicals formed, compete with the heme moiety as target for attack by the radicals, and thus protect the hemeprotein from degradation.[15,25,34] Phenols are good peroxidase donors[35,36] and are observed to protect hemeproteins in this way. The protective action of phenols on prostaglandin synthetase has been well studied.[37,38]

Sulfoxides Do Not Act by Scavenging Hydroxyl Radicals

Me_2SO has been extensively investigated as a reactant for HO· radicals produced by cytochrome P450s.[17,39,40] However, although several sulfoxides protected 11β-hydroxylase in the presence of pseudosubstrates, OH·-scavengers other than Me_2SO were ineffective.[32] Although this rules out a role for free hydroxyl radicals, there may be hydroxyl or other radicals bound to the enzyme to which the tested OH· scavengers do not have access. It has been shown that bound oxy radicals may have a reactivity similar to that of OH· and may mimic its activity in several biological systems.[14,41]

Phenols and Sulfoxides Act Partially by Inhibiting 11β-Hydroxylase

Because of the protective effects of an 11β-hydroxylase inhibitor, metyrapone, and because neither phenols nor sulfoxides appeared to protect 11β-hydroxylase by direct radical scavenging, we investigated whether these compounds might act by inhibiting 11β-hydroxylase. Possible direct inhibitory effects of phenols and sulfoxides were investigated and compared to the inhibitory effect of metyrapone. Phenols and sulfoxides were directly inhibitory to 11β-hydroxylase.[31,32] However, BHA and Me_2SO do not significantly inhibit 11β-hydroxylase at concentrations that offer significant protection against loss of activity (100 μ*M* and 10 m*M*, respectively).[8] Both phenols and sulfoxides showed ED_{50}s for inhibition of 11β-hydroxylase that were substantially higher than the ED_{50}s for protection, e.g., for BHA, the ED_{50} for inhibition was 600 μ*M* compared with 300 μ*M* for protection; for phenol, the ED_{50} for inhibition was 3 m*M* compared with 300 μ*M* for protection. Me_2SO showed an ED_{50} for inhibition of 11β-hydroxylase activity of >1 *M*, in contrast to its ED_{50} for protection of 34 m*M*. In contrast, the ED_{50}s for protection and inhibition of 11β-hydroxylase in the case of metyrapone were approximately equal (~250 n*M* for both actions).

Although protective phenols and sulfoxides do not appear to act by scavenging free peroxy radicals, i.e., do not appear to act as antioxidants, an oxygen radical nevertheless appears to be involved because of the synergism of the effective compounds with a lowered concentration of oxygen.[8,31,32] Thus, phenols and sulfoxides, typified by BHA and Me_2SO, respectively, may have a mixed mode of protection, acting partially as inhibitors and partially as reactants for destructive oxygen radicals produced by the cytochrome P450s. Inhibitors may also protect cytochrome P450 against damage by preventing the peroxidase activity of these enzymes and thus preventing the formation of destructive radicals from lipid peroxides.[42] Phenols that protect hemeprotein enzymes apparently must have the properties of a hydrogen donor (peroxidase donor) and also have access to the active site of the enzyme (the site of generation of the radical resulting from peroxidase activity).[15,34] Such compounds may therefore show biphasic effects, protecting the enzyme at low concentrations and inhibiting at higher concentrations.[43] Sulfoxides may also have a mixed mode of action; it is unlikely that sulfoxides act simply as inhibitors, but they may need to bind close to the active site of the enzyme. Dibenzyl sulfoxide has been shown to produce a reverse type I spectrum on incubation with partially purified bovine 11β-hydroxylase.[81] This indicates that such sulfoxides may have access to the substrate site of the enzyme and this may account both for inhibition of the enzyme and ability to react with a radical generated at the substrate site.

Thus, the original suggestion that the protective effect of BHA may result from inhibition of membrane lipid peroxidation and that Me_2SO may act through scavenging of free hydroxyl radicals[8,9,14] requires some modification. For both phenolic and sulfoxide antioxidants, a more probable mechanism for their protective activity involves binding to the active site of the enzyme and local scavenging of an oxygen or lipid radical.[31,32]

Ascorbic Acid

During an investigation of the regulation of 11β-hydroxylase activity in cultured bovine adrenocortical cells, we found that ascorbate, like dimethyl sulfoxide and BHA, could prevent the loss of activity of 11β-hydroxylase in the presence of steroid pseudosubstrates and enhance the induction of 11 β-hydroxylase activity by ACTH, but under the serum-containing conditions used was less effective than the other compounds.[8] We reinvestigated the effect of ascorbate under defined, serum-free conditions, which have been developed for bovine adrenocortical cells.[44,45] Ascorbate specifically protected 11β-hydroxylase activity in the presence of steroid pseudosubstrates, without direct inhibition of the enzyme, and enhanced reinduction of 11β-hydroxylase by ACTH, whereas vitamin E and selenium were without effect in both actions.[46]

Ascorbate Protects Cytochrome $P450_{11\beta}$ in the Presence of Pseudosubstrates

Figure 4 shows that when primary bovine adrenocortical cells were cultured in a serum-free, defined medium in the absence of ACTH, 11β-hydroxylase activity declined with a time course similar to that previously demonstrated in serum-containing medium (halftime of ~ 40 hr).[8] When 50 μ*M* cortisol was added to such cultures, 11β-hydroxylase activity declined with an accelerated time course (halftime of ~ 6 hr), as also previously demonstrated in serum-containing medium.[8] Addition of 5 m*M* ascorbate together with cortisol resulted in a marked diminution of the rate of loss of 11β-hydroxylase activity (halftime of ~ 28 hr). Ascorbate had no effect, however, on the rate of decline of 11β-hydroxylase activity in the absence of cortisol.

Ascorbate Synergizes with Lowered Oxygen

Ascorbate showed a strong synergism with a lowered concentration of oxygen in preventing loss of 11β-hydroxylase activity in the presence of pseudosubstrates.[46] When the effect of ascorbate was examined at three different oxygen concentrations (2, 5, and 19%) ascorbate demonstrated a more marked protection of 11β-hydroxylase at 2% oxygen and much less at 19% oxygen (Figure 5). At 2% oxygen, 5 m*M* ascorbate almost completely prevented loss of 11β-hydroxylase activity in the presence of cortisol.

Ascorbate Does Not Inhibit 11β-Hydroxylase

Unlike phenols and sulfoxides, ascorbate did not behave as an inhibitor of 11β-hydroxylase in protection of 11β-hydroxylase. The observations that phenols and sulfoxides protected

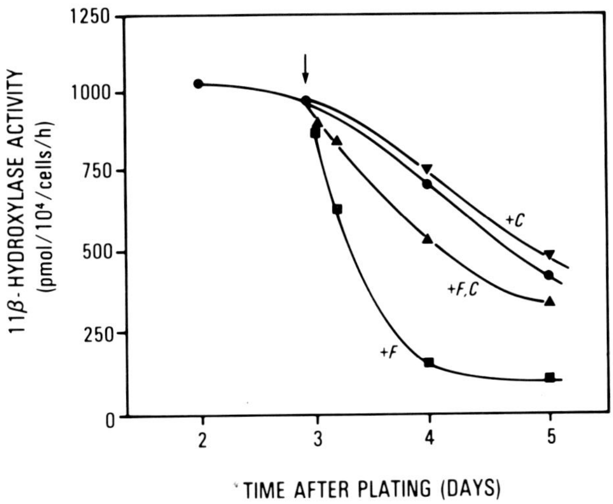

FIGURE 4. Time course of loss of 11β-hydroxylase activity in adrenocortical cells in the presence of cortisol and ascorbate. Primary bovine adrenocortical cell cultures were prepared and were changed to defined, serum-free medium 24 hr after plating. On day 3, some plates received 50 μ*M* cortisol (F; ■), or 5 m*M* ascorbate (C; ▼), or both (▲), or no additions (●). 11β-Hydroxylase activity was measured 1, 4, 24, and 48 hr after the additions. (From Hornsby, P. J., Aldern, K. A., and Harris, S. E., *Endocrinology*, 117, 1264, 1985. With permission.)

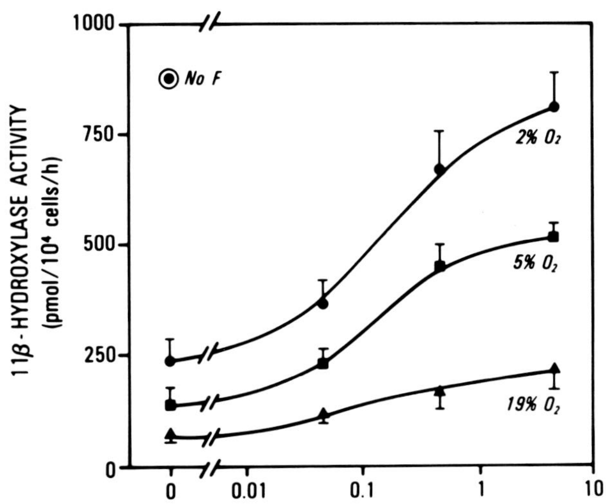

FIGURE 5. Prevention of loss of 11β-hydroxylase activity by ascorbate at different oxygen concentrations; 50μ*M* cortisol (F) was added to primary bovine adrenocortical cells cultured in serum-free, defined medium, together with the concentrations of ascorbate indicated on the abscissa in the indicated concentrations of oxygen. After 24 hr compounds were removed and 11β-hydroxylase activity was measured. (From Hornsby, P. J., Aldern, K. A., and Harris, S. E., *Endocrinology*, 117, 1264, 1985. With permission.)

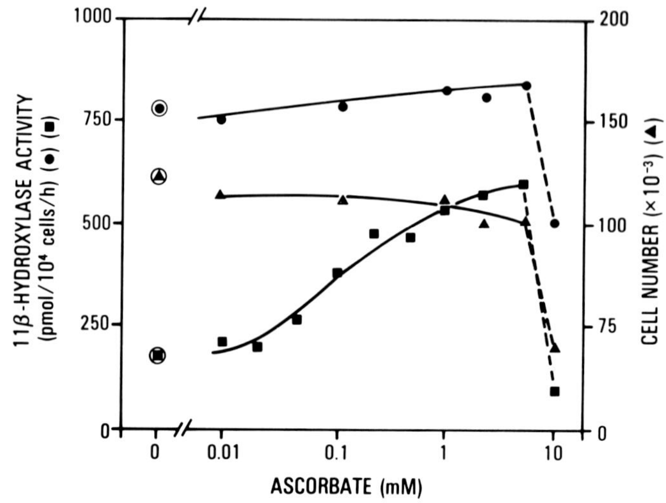

FIGURE 6. Absence of direct inhibition of 11β-hydroxylase by ascorbate over a range of concentrations which protect 11β-hydroxylase in the presence of cortisol. ●: Primary bovine adrenocortical cells were cultured for 3 days in serum-free medium. The cells were then incubated for 1 hr in serum-free medium containing the indicated concentrations of ascorbate together with 10 μ*M* DOC as 11β-hydroxylase substrate. Conversion to corticosterone was then measured. ■: 50 μ*M* cortisol was added to 3-day primary bovine adrenocortical cells in serum-free, defined medium, together with the indicated concentrations of ascorbate. After 24 hr, compounds were removed and 11β-hydroxylase activity was then measured in a 1 hr incubation. ▲: 10^4 primary bovine adrenocortical cells were plated and, after 24 hr, changed to medium with 10% fetal bovine serum and the indicated concentrations of ascorbate, added daily. After 5 days, cell numbers were estimated. (From Hornsby, P. J., Aldern, K. A., and Harris, S. E., *Endocrinology*, 117, 1264, 1985. With permission.)

partially by inhibiting the enzymes necessitated a reevaluation of the possible inhibition of 11β-hydroxylase by ascorbate. Several reports in the literature indicate that ascorbate may inhibit 11β-hydroxylase, but such studies are complicated by pro-oxidant activity of ascorbate under certain in vitro conditions. Ascorbic acid is difficult to use in in vitro experiments, including cell culture, because of its extreme instability to air[47] and because of a pro-oxidant effect that predominates at low ascorbate concentrations in the presence of trace levels of metals of variable valence such as iron and copper.[48] This causes the effects of ascorbic acid to be unpredictable in in vitro systems such as homogenates or preparations containing damaged cells, where such metals are universal contaminants, and may reduce effectiveness of ascorbate in nondefined culture media, which were used in previous experiments. Other antioxidants have also been noted to have a pro-oxidant effect in some in vitro systems where their ability to reduce transition metals predominates.[49]

Primary adrenocortical cells after 3 days in culture were incubated for 1 hr in serum-free medium containing various concentrations of ascorbate together with deoxycorticosterone (DOC) as 11β-hydroxylase substrate (Figure 6). Concentrations of ascorbate from 10 μ*M* to 5 m*M* showed no inhibition of 11β-hydroxylase activity. The same range of ascorbate concentrations added to cells during a 24-hr preincubation with cortisol showed increasing prevention of loss of 11β-hydroxylase activity. However, losses of activity of 11β-hydroxylase were observed when 10 m*M* ascorbate was included in the 1-hr incubation with DOC and when it was included in the 24-hr preincubation with cortisol. This was shown to be due to nonspecific toxicity of ascorbate at this concentration; when cells were grown in the presence of various concentrations of ascorbate for 3 days, there was no inhibition of the growth rate by ascorbate concentrations of 5 m*M* and below, but cells at 10 m*M* showed morphological signs of cytotoxicity and growth was inhibited. Thus, ascorbate provides the only example of a substance that prevents pseudosubstrate-mediated loss of 11β-hydrodxylase activity which does not show any direct inhibitory action on the enzyme, at concentrations that are noncytotoxic.

In tertiary cultures of bovine adrenocortical cells in serum-free, defined medium, ascorbate enhanced ACTH induction of 11β-hydroxylase activity, as previously noted for primary cultures in serum-containing medium.[8] Additionally, as in primary cultures, 50 μ*M* cortisol caused a loss of the induced 11β-hydroxylase activity, which was reduced by the presence of ascorbate.[46]

The Possible Mode of Protection of Cytochrome $P450_{11\beta}$ by Ascorbate

Ascorbate may have the unique property of being able to bind close to the site of formation of destructive radicals without interference with substrate binding and metabolism; it has been observed to be a ligand for liver cytochrome P450s, resulting in a type II spectral change.[50] Ascorbate may be protective for liver cytochrome P450s, since cytochrome P450 is low in ascorbate deficiency in the liver and the action of ascorbate on liver cytochrome P450 is consistent with complexing to P450 and protection by local scavenging of destructive radicals.[14]

Ascorbate is a powerful antioxidant in biological systems, under some circumstances.[51,52] In vivo, ascorbate inhibits lipid peroxidation resulting from iron overload or carbon tetrachloride poisoning.[51,53] In vitro, ascorbate potently inhibits lipid peroxidation of microsomal membranes, synergizing with α-tocopherol, when air oxidation of Fe^{2+} is suppressed,[54] and inhibits lipid oxidation in the presence of hemeprotein catalysts.[21,55] Ascorbate is a two-electron reductant with an intermediate one-electron oxidized form (monodehydroascorbate) produced readily on reaction of ascorbate with peroxy radicals; the remarkable stability of monodehydroascorbate may provide a sink for various forms of oxygen radical production in the cell.[14] Although ascorbate can also act as a pro-oxidant, it is likely that at the high levels of ascorbate present in the adrenocortical cell,[56] in the presence of abundant metal-binding sites, the antioxidant action of ascorbate, rather than the pro-oxidant effect, will predominate.[9] There are cellular mechanisms for the regeneration of ascorbate from its oxidation products, monodehydroascorbate and dehydroascorbate,[14,57] and the adrenocortical cell has specific uptake mechanisms for ascorbate.[9,58]

The Role of Ascorbate in the Adrenal Gland In Vivo

The protective action of ascorbate for cytochrome $P450_{11\beta}$ and the high concentrations of ascorbate in the adrenocortical cell, together with specific mechanisms for its uptake and regeneration, suggest that ascorbate may be a very important intracellular antioxidant in the adrenocortical cell. However, ascorbate likely has multiple roles in the function of this tissue.

Despite the fact that the high adrenal content of ascorbic acid has been known for over 50 years, no consensus on its function in the adrenal cortex has emerged. Szent-Gyorgyi first isolated "hexuronic acid", later identified as vitamin C, from the adrenal glands.[59] Many of the subsequent experiments that have attempted to elucidate the role of this vitamin in the adrenal cortex have indicated that ascorbate exerts an inhibitory effect on steroidogenesis. These results gave rise to the formerly prevalent belief that the role of ascorbate in the adrenal cortex was to act as a restraining influence on steroidogenesis and that it might be involved in the physiological regulation of steroidogenesis by removal of that influence under the action of ACTH.[60,61] This concept appears to have arisen from experiments on adrenal preparations where the pro-oxidant, rather than antioxidant, effect of ascorbate predominates. Under these conditions, ascorbate may initiate lipid peroxidation and result in destruction of cytochrome P450s or interfere with steroidogenesis in other ways.[9]

A protective role for ascorbate must be distinguished from a role as a cofactor for steroidogenesis. The adrenocortical cytochrome P450s do not require ascorbate as a cofactor, although steroidogenesis in isolated guinea pig adrenocortical cells was shown to be enhanced by the presence of ascorbate.[62] In vivo, adrenocortical function continues even in prolonged vitamin C deficiency, both in guinea pigs and in humans, but with lowered activities of 11β-hydroxylase and 3β-hydroxysteroid dehydrogenase.[9] The increased pituitary ACTH stimulation, adrenal hypertrophy, and elevated plasma steroids observed in prolonged scurvy in the guinea pig may result from the combination of a lower level of adrenocortical function (per unit mass of tissue) and the stress of the scorbutic condition itself.[9]

Ascorbate may act as an antioxidant in the adrenal medulla as well as in the cortex. Apart from its role as an antioxidant, the major role of ascorbate in the adrenal gland is probably as cofactor for dopamine-β-hydroxylation in the medulla.[63] This provides an example of the functioning of the adrenal gland — cortex and medulla — as a single unit, with the cortex storing ascorbate for release in high levels to the medulla during the reaction to stress; like cortical cells, medullary cells have specific ascorbate uptake mechanisms.[64]

Vitamin E, Selenium, and Other Biological Antioxidants

We compared the effects of ascorbate with those of the other major biological antioxidants, selenium and α-tocopherol (vitamin E). It was important to investigate the possible effects of these compounds, because of several studies showing antioxidant effects of these substances in this cell type[33,44] and because of the influence of these nutrients on other cytochrome P450 activities.[65-68] Culture conditions deficient in these nutrients were necessary to test for protective effects. Adrenocortical cells may be grown under conditions of deficiency of these nutrients.[44] Previously, possible effects of vitamin E on 11β-hydroxylase were investigated in primary cultures of bovine cells.[8] However, primary and early passage cultures of bovine adrenocortical cells are not deficient in vitamin E and selenium, but gradually become deficient in these compounds over the first few doublings in culture.[33,44,69] The deficiency may be accelerated by ensuring that the medium in which the cells are grown is completely deficient in vitamin E and selenium.[44] Deficiency of both selenium and α-tocopherol is required because these compounds have complementary actions in biological systems, making effects of either nutrient difficult to observe in the presence of the other.[70] Under these conditions, neither α-tocopherol nor selenium prevented loss of 11β-hydroxylase activity in the presence of cortisol or enhanced induction of 11β-hydroxylase activity by ACTH.[46]

Other biological antioxidants, the enzymes catalase and superoxide dismutase (SOD), are also without effect on this process when added externally to cultures. However, all of the antioxidant enzymes (SOD, catalase, glutathione peroxidase, and others) are present at high activity in the adrenal cortex in vivo.[9]

POSSIBLE IN VIVO ROLE OF CYTOCHROME P450/PSEUDOSUBSTRATE EFFECTS

It is speculated that the activity of certain adrenocortical cytochrome P-450s may be regulated in the intact cell by production of destructive bound radicals and that this may participate in the aging of the adrenal cortex and its functional zonation under pathological and perhaps normal conditions.[9,71]

The Adrenal Vascular System and the Regulation of Zonation

All parenchymal cells in the adrenal cortex are of the same cell type. Cell culture experiments provide direct evidence of interconversion of zonal cell types. Zona glomerulosa, zona fasciculata, zona reticularis, fetal zone, and definitive zone cells placed in culture acquire common characteristics, usually resembling those of the fasciculata-reticularis region of the cortex. Presumably this is the result of removal from the in vivo environment, dominated by the adrenal vascular system.[7,10,72-75] The restriction of the zona glomerulosa and aldosterone secretion to the arterial side of the capillary bed implies a central role for the unusual adrenocortical vascular system in the maintenance of structure and function.[9,12,71] It was first proposed by Greep and Deane in 1949[76] that the division of the cortex into the mineralocorticoid- and glucocorticoid-secreting zones may be the result of a requirement for different local cellular environments, as provided by the centripetal capillary system, for the synthesis of the different steroid hormones. This concept is still the central idea for the regulation of functional zonation.

We and others have hypothesized that adrenocortical cells produce a gradient of a substance or substances in the bloodstream which alters adrenocortical cell function and morphology to create the zonation of the adrenal cortex.[7-13,30,71,72,77-80] It is simpler to construct a model for zonation if the gradient substance is assumed to be a steroid or steroids, particularly one secreted by all the zones, such as corticosterone, but this is not critical for the general concept of zonation resulting from the action of a gradient substance.

Cytochrome P450/Pseudosubstrate Interactions as Mediators of Zonation: Modulation by Antioxidants

If the gradient substance does comprise steroids, then the cell culture experiments discussed here suggest that radical production by cytochrome P450s on interaction with pseudosubstrates could be a mediator of steroid action in producing zonation. It is also possible, of course, that steroids act through a protein receptor to regulate zonation.[79] If zonation of CMO is in fact the direct result of cytochrome P450s/pseudosubstrate interactions, then presumably the pseudosubstrate steroids reach a critical concentration at the zona glomerulosa/zona fasciculata boundary, thereby restricting CMO to the glomerulosa, and reach a higher critical concentration in the inner cortex, possibly resulting in lower 11β-hydroxylase in the zona reticularis.[71] Under this hypothesis, the distribution of CMO and 11β-hydroxylase activities in the cortex results from their differential sensitivity to cytochrome P450/pseudosubstrate interactions. If this model is valid, then it may be that zonation in the adrenal cortex is modulated by intracellular levels of antioxidants, principally ascorbate. It is possible that the level of antioxidants in the adrenocortical cell permits loss of CMO activity at a certain steroid concentration, but is high enough to protect 11β-hydroxylase activity. This implies a unique role for radical production by enzymes and radical scavenging by antioxidants in modulating the phenotype of a differentiated cell type.

REFERENCES

1. **Hornsby, P. J., Simonian, M. H., and Gill, G. N.,** Aging of adrenocortical cells in culture, *Int. Rev. Cytol. Suppl.,* 10, 131, 1979.
2. **Hornsby, P. J., Aldern, K. A., and Harris, S. E.,** Adrenocortical cultures as model systems for investigating cellular aging, in *Aging,* Vol. 27, Armstrong, D., Sohal, R. S., Cutler, R. G., and Slater, T. F., Eds., Raven Press, New York, 1984, 203.
3. **Wada, A., Okamoto, M., Nonaka, Y., and Yamano, T.,** Aldosterone biosynthesis by a reconstituted cytochrome $P\text{-}450_{11\beta}$ system, *Biochem. Biophys. Res. Commun.,* 119, 365, 1984.
4. **Yanagibashi, K., Hanui, M., Shively, J. E., Shen, W. H., and Hall, P. F.,** The synthesis of aldosterone by the adrenal cortex. Two zones (fasciculata and glomerulosa) possess one enzyme for 11β-, 18-hydroxylation, and aldehyde synthesis, *J. Biol. Chem.,* 261, 3556, 1986.
5. **Finkelstein, M. and Shaefer, J. M.,** Inborn errors of steroid biosynthesis, *Physiol. Rev.,* 59, 353, 1979.
6. **Kramer, R. E., Gallant, S., and Brownie, A. C.,** The role of cytochrome P-450 in the action of sodium depletion on aldosterone biosynthesis in rats, *J. Biol. Chem.,* 254, 3953, 1979.

7. **Hornsby, P. J. and O'Hare, M. J.,** The roles of potassium and corticosteroids in determining the pattern of metabolism of [^{3}H]deoxycorticosterone by monolayer cultures of rat adrenal zona glomerulosa cells, *Endocrinology,* 101, 997, 1977.
8. **Hornsby, P. J.,** Regulation of cytochrome P-450-supported 11β-hydroxylation of deoxycortisol by steroids, oxygen, and antioxidants in adrenocortical cell cultures, *J. Biol. Chem.,* 255, 4020, 1980.
9. **Hornsby, P. J. and Crivello, J. F.,** The role of lipid peroxidation and biological antioxidants in the function of the adrenal cortex. II, *Mol. Cell. Endocrinol.,* 30, 123, 1983.
10. **Crivello, J. F., Hornsby, P. J., and Gill, G. N.,** Metyrapone and antioxidants are required to maintain aldosterone synthesis by cultured bovine adrenocortical cells, *Endocrinology,* 111, 469, 1982.
11. **Crivello, J. F., Hornsby, P. J., and Gill, G. N.,** Suppression of cultured bovine adrenocortical zona glomerulosa cell aldosterone synthesis by steroids and its prevention by antioxidants, *Endocrinology,* 113, 235, 1983.
12. **Neville, A. M. and O'Hare, M. J.,** *The Human Adrenal Cortex,* Springer-Verlag, Berlin, 1982.
13. **Kahri, A. I., Voutilainen, R., and Salmenpera, M.,** Different biological actions of corticosteroids, corticosterone and cortisol, as a base of zonal function of adrenal cortex, *Acta Endocrinol.,* 91, 329, 1979.
14. **Hornsby, P. J. and Crivello, J. F.,** The role of lipid peroxidation and biological antioxidants in the function of the adrenal cortex. I. A background review, *Mol. Cell. Endocrinol.,* 30, 1, 1983.
15. **O'Brien, P. J. and Rahimtula, A. D.,** The nature of the activated oxygen involved in the hydroxylation reactions catalyzed by cytochrome P450 from a study of its peroxidase properties, in *Microsomes, Drug Oxidations, and Chemical Carcinogenesis,* Vol. 1, Coon, M. J., Conney, A. H., Estabrook, R. W., Gelboin, H. V., Gillette, J. R., and O'Brien, P. J., Eds., Academic Press, New York, 1980, 263.
16. **Griffin, B. W.,** The role of hemeproteins in autoxidation processes, in *Oxygen and Oxy-Radicals in Chemistry and Biology,* Rogers, M. A. J. and Powers, E. L., Eds., Academic Press, New York, 1981, 141.
17. **Mannering, G. J.,** Hepatic cytochrome P-450-linked drug-metabolizing systems, in *Concepts in Drug Metabolism,* Part B, Jenner, P. and Testa, B., Eds., Marcel Dekker, New York, 1982, 53.
18. **Kuhn, H., Gotze, R., Schewe, T., and Rapoport, S. M.,** Quasi-lipoxygenase activity of haemoglobin: a model for lipoxygenases, *Eur. J. Biochem.,* 120, 161, 1981.
19. **Bryant, R. W. and Bailey, J. M.,** Rearrangement of 15-hydroperoxy-eicosatetraeroic acid (15-HPETE) during incubations with hemoglobin: a model for platelet lipoxygenase metabolism, *Prog. Lipid Res.,* 20, 279, 1982.
20. **Svingen, B. A., Buege, J. A., O'Neal, F. O., and Aust, S. D.,** The mechanism of NADPH-dependent lipid peroxidation. The propagation of lipid peroxidation, *J. Biol. Chem.,* 254, 5892, 1979.
21. **Trotta, R. J., Sullivan, S. G., and Stern, A.,** Lipid peroxidation and haemoglobin degradation in red blood cells exposed to *t*-butyl hydroperoxide. The relative roles of haem- and glutathione-dependent decomposition of *t*-butyl hydroperoxide and membrane lipid hydroperoxides in lipid peroxidation and haemolysis, *Biochem. J.,* 212, 759, 1983.
22. **Egan, R. W., Gale, P. H., Baptista, E. M., Kennicott, K. L., Van den Heuvel, W. J. A., Walker, R. W., Fagerness, P. E., and Kuehl, F. A., Jr.,** Oxidation reactions by prostaglandin cyclooxygenase-hydroperoxidase, *J. Biol. Chem.,* 256, 7352, 1981.
23. **Boyd, J. A., Harvan, D. J., and Eling, T. E.,** The oxidation of 2-aminofluorene by prostaglandin endoperoxide synthetase, *J. Biol. Chem.,* 258, 8246, 1983.
24. **Josephy, P. D., Eling, T., and Mason, R. P.,** Co-oxidation of benzidine by prostaglandin synthase and comparison with the action of horseradish peroxidase, *J. Biol. Chem.,* 258, 5561, 1983.
25. **O'Brien, P. J.,** Hydroperoxides and superoxides in microsomal oxidations, *Pharmacol. Ther.,* 2, 517, 1978.
26. **De Matteis, F., Cantoni, L., and Gibbs, A. H.,** Some recent studies on the abnormal pathways of heme degradation not involving production of carbon monoxide, in *Biochemical and Clinical Aspects of Oxygen,* Caughey, W. S. and Caughey, H., Eds., Academic Press, New York, 1979, 395.
27. **White, R. E., Sligar, S. G., and Coon, M. J.,** Evidence for a homolytic mechanism of peroxide oxygen-oxygen bond cleavage during substrate hydroxylation by cytochrome P-450, *J. Biol. Chem.,* 255, 11108, 1980.
28. **Wang, H.-P. and Kimura, T.,** Ferrous ion-mediated cytochrome P-450 degradation and lipid peroxidation in adrenal cortex mitochondria, *Biochim. Biophys. Acta,* 423, 374, 1976.
29. **Wang, H.-P. and Kimura, T.,** Cytochrome P-450 from mitochondria of bovine adrenal cortex. Comparison of cholesterol side-chain cleavage P-450 with steroid 11β-hydroxylation P-450 and immunochemical cross-reactivity between adrenal mitochondrial and liver microsomal cytochromes P-450, *Biochim. Biophys. Acta,* 542, 115, 1978.
30. **Hornsby, P. J.,** Regulation of 21-hydroxylase activity by steroids in cultured bovine adrenocortical cells: possible significance for adrenocortical androgen synthesis, *Endocrinology,* 111, 1092, 1982.

31. **Hornsby, P. J., Aldern, K. A., and Harris, S. E.,** Mode of action of butylated hydroxyanisole (BHA) and other phenols in preventing loss of 11β-hydroxylase activity in cultured bovine adrenocortical cells, *Biochem. Pharmacol.*, 34, 865, 1985.
32. **Hornsby, P. J., Harris, S. E., and Aldern, K. A.,** Mode of action of sulfoxides in preventing loss of activity of 11β-hydroxylase in cultured bovine adrenocortical cells, *Chem. Biol. Interact.*, 51, 335, 1984.
33. **Hornsby, P. J.,** The role of vitamin E in cellular energy metabolism in cultured adrenocortical cells, *J. Cell. Physiol.*, 112, 207, 1982.
34. **Rahimtula, A. D., Hawco, F., and O'Brien, P. J.,** The effects of antioxidants on hemoprotein function, in *Microsomes, Drug Oxidations, and Chemical Carcinogenesis,* Vol. 1, Coon, M. J., Conney, A. H., Estabrook, R. W., Gelboin, H. V., Gillette, J. R., and O'Brien, P. J., Eds., Academic Press, New York, 1980, 415.
35. **Mason, H. S.,** Mechanisms of oxygen metabolism, *Adv. Enzymol.*, 19, 79, 1957.
36. **Job, D. and Dunford, H. B.,** Substituent effects on the oxidation of phenols and aromatic amines by horseradish peroxidase compound I, *Eur. J. Biochem.*, 66, 607, 1976.
37. **Lands, W. E. M.,** The biosynthesis and metabolism of prostaglandins, *Annu. Rev. Physiol.*, 41, 633, 1979.
38. **Warso, M. A. and Lands, W. E. M.,** Lipid peroxidation in relation to prostacyclin and thromboxane physiology and pathophysiology, *Br. Med. Bull.*, 39, 277, 1983.
39. **Ingelman-Sundberg, M. and Ekstrom, G.,** Aniline is hydroxylated by the cytochrome P-450-dependent hydroxyl radical-mediated oxygenation mechanism, *Biochem. Biophys. Res. Commun.*, 106, 631, 1982.
40. **Winston, G. W. and Cederbaum, A. I.,** NADPH-dependent production of oxy radicals by purified components of the rat liver mixed function oxidase system, *J. Biol. Chem.*, 258, 1508, 1983.
41. **Cederbaum, A. I.,** Organic hydroperoxide-dependent oxidation of ethanol by microsomes: lack of a role for free hydroxyl radicals, *Arch. Biochem. Biophys.*, 227, 329, 1983.
42. **Islam, S. I. and Zaher, F. A.,** Inhibition of lipid peroxide decomposition by compounds which bind with cytochrome P-450, *Life Sci.*, 33, 1113, 1983.
43. **Testa, B. and Jenner, P.,** Inhibitors of cytochrome P-450s and their mechanism of action, *Drug Metab. Rev.*, 12, 1, 1981.
44. **Hornsby, P. J., Pearson, D. W., Autor, A. P., Aldern, K. A., and Harris, S. E.,** Selenium deficiency in cultured adrenocortical cells: restoration of glutathione peroxidase and resistance to hydroperoxides on addition of selenium, *J. Cell. Physiol.*, 123, 33, 1985.
45. **Simonian, M. H., White, M. L., and Gill, G. N.,** Growth and function of cultured bovine adrenocortical cells in a serum-free defined medium, *Endocrinology,* 111, 919, 1982.
46. **Hornsby, P. J., Aldern, K. A., and Harris, S. E.,** The function of ascorbic acid in the adrenal cortex: studies in adrenal cells in culture, *Endocrinology,* 117, 1264, 1985.
47. **Feng, J., Melcher, A. H., Brunette, D. M., and Moe, H. K.,** Determination of L-ascorbic acid levels under conditions of organ culture, *In Vitro,* 13, 91, 1977.
48. **Rowley, D. A. and Halliwell, B.,** Superoxide-dependent and ascorbate-dependent formation of hydroxyl radicals in the presence of copper salts: a physiologically significant reaction?, *Arch. Biochem. Biophys.*, 225, 279, 1983.
49. **Gutteridge, J. M. C. and Fu, X.-C.,** Enhancement of bleomycin-iron free radical damage to DNA by antioxidants and their inhibition of lipid peroxidation, *FEBS Lett.*, 123, 71, 1981.
50. **Sato, P. H. and Zannoni, V. G.,** Ascorbic acid and hepatic drug metabolism, *J. Pharmacol. Exp. Ther.*, 198, 295, 1976.
51. **Kunert, K.-J. and Tappel, A. L.,** The effect of vitamin C on in vivo lipid peroxidation in guinea pigs as measured by pentane and ethane production, *Lipids,* 18, 271, 1983.
52. **Niki, E., Saito, T., Kawakami, A., and Kamiya, Y.,** Inhibition of oxidation of methyl linoleate in solution by vitamin E and vitamin C., *J. Biol. Chem.*, 259, 4177, 1984.
53. **Dillard, C. J., Downey, J. E., and Tappel, A. L.,** Effect of antioxidants of lipid peroxidation an iron-loaded rats, *Lipids,* 19, 127, 1984.
54. **Leung, H.-W., Vang, M. J., and Mavis, R. D.,** The cooperative interaction between vitamin E and vitamin C in suppression of peroxidation of membrane phospholipids, *Biochim. Biophys. Acta.*, 664, 266, 1981.
55. **Porter, W. L.,** Recent trends in food applications of antioxidants, in *Autoxidation in Food and Biological Systems,* Simic, M. G. and Karel, M., Eds., Plenum Press, New York, 1980, 295.
56. **Salomon, L. L. and Stubbs, D. W.,** Some aspects of the metabolism of ascorbic acid in rats, *Ann. N.Y. Acad. Sci.*, 92, 128, 1961.
57. **Diliberto, E. J., Jr., Dean, G., Carter, C. and Allen, P. L.,** Tissue, subcellular, and submitochondrial distributions of semidehydroascorbate reductase: possible role of semidehydroascorbate reductase in cofactor regeneration, *J. Neurochem.*, 39, 563, 1982.
58. **Leonard, R. K., Auersperg, N., and Parkes, C. O.,** Ascorbic acid accumulation by cultured rat adrenocortical cells, *In Vitro,* 19, 46, 1983.

59. **Szent-Gyorgyi, A.,** CLXXIII. Observations on the function of peroxidase systems and the chemistry of the adrenal cortex. Description of a new carbohydrate derivative, *Biochem. J.*, 22, 1387, 1928.
60. **Seib, P. A. and Tolbert, B. M.,** *Ascorbic acid: Chemistry, Metabolism and Uses,* American Chemical Society, Washington, D.C., 1982.
61. **Chalopin, H., Mouton, M., and Ratsimamanga, A. R.,** Some interrelationships between ascorbic acid and adreno-cortical function, *World Rev. Nutr. Diet.*, 6, 165, 1966.
62. **Lambert, A. and Robertson, W. R.,** Characteristics of the response of dispersed guinea-pig adrenal cells to low doses (1-50 pg/mℓ) of α1-24 adrenocorticotrophin, *J. Steroid Biochem.*, 17, 603, 1982.
63. **Diliberto, E. J. and Allen, P. L.,** Mechanism of dopamine-β-hydroxylation. Semidehydroascorbate as the enzymic oxidation product of ascorbate, *J. Biol. Chem.*, 256, 3385, 1981.
64. **Westhead, E. W. and Livett, B. G.,** Storage and secretion by adrenal chromaffin cells, *Trends Neurol. Sci.*, 6, 254, 1983.
65. **Carpenter, M. P.,** Vitamin E and microsomal drug hydroxylations, *Ann. N.Y. Acad. Sci.*, 203, 81, 1972.
66. **Pascoe, G. A., Sakai-Wong, J., Soliven, E., and Correia, M. A.,** Regulation of intestinal cytochrome P-450 and heme by dietary nutrients. Critical role of selenium, *Biochem. Pharmacol.*, 32, 3027, 1983.
67. **Engelmann, G. L., Richardson, A. G., and Fierer, J. A.,** Maintenance and induction of cytochrome P-450 in cultured rat hepatocytes, *Arch. Biochem. Biophys.*, 238, 359, 1985.
68. **Murty, H. S., Caasi, P. I., Brooks, S. K. and Nair, P. P.,** Biosynthesis of heme in the vitamin E-deficient rat, *J. Biol. Chem.*, 245, 5498, 1970.
69. **Hornsby, P. J. and Gill, G. N.,** Regulation of glutamine and pyruvate oxidation in cultured adrenocortical cells by cortisol, antioxidants, and oxygen: effects on cell proliferation, *J. Cell. Physiol.*, 109, 111, 1981.
70. **Tappel, A. L.,** Measurement of and protection from in vivo lipid peroxidation, in *Free Radicals in Biology,* Vol. 4, Pryor, W. A., Ed., Academic Press, New York, 1980, 1.
71. **Hornsby, P. J.,** The regulation of adrenocortical function by control of growth and structure, in *Adrenal Cortex,* Anderson, D. C. and Winter, J. S. D., Eds., Butterworths, London, 1985, 1.
72. **Hornsby, P. J., O'Hare, M. J., and Neville, A. M.,** Functional and morphological observations on rat adrenal zona glomerulosa cells in monolayer culture, *Endocrinology,* 95, 1240, 1974.
73. **O'Hare, M. J., Nice, E. C., and Neville, A. M.,** Regulation of androgen secretion and sulfoconjugation in the adult human adrenal cortex: studies with primary monolayer cell cultures, in *Adrenal Androgens,* Genazzani, A. R., Thijssen, J. H. H., and Siiteri, P. K., Eds., Raven Press, New York, 1980, 7.
74. **Simonian, M. H. and Gill, G. N.,** Regulation of the fetal human adrenal cortex: effects of adrenocorticotropin on growth and function of monolayer cultures of fetal and definitive zone cells, *Endocrinology,* 108, 1769, 1981.
75. **Crivello, J. F. and Gill, G. N.,** Induction of cultured bovine adrenocortical zona glomerulosa cell 17-hydroxylase activity by ACTH, *Mol. Cell. Endocrinol.*, 30, 97, 1983.
76. **Greep, R. O. and Deane, H. W.,** Histological, cytochemical and physiological observations on the regeneration of the rat's adrenal gland following enucleation, *Endocrinology,* 45, 42, 1949.
77. **Neville, A. M. and O'Hare, M. J.,** Cell culture and histopatholgy in relation to hyperhydrocortisolism, in *The Endocrine Function of the Human Adrenal Cortex,* James, V. H. T., Seris, M., Giusti, G., and Martiri, L., Eds., Academic Press, London, 1978, 229.
78. **Neville, A. M. and O'Hare, M. J.,** Aspects of structure, function, and pathology, in *The Adrenal Gland,* James, V. H. T., Ed., Raven Press, New York, 1979, 1.
79. **Anderson, D. C.,** Hypothesis: the adrenal androgen-stimulating hormone does not exist, *Lancet,* 2, 454, 1980.
80. **Gill, G. N., Crivello, J. F., Hornsby, P. J., and Simonian, M. H.,** Growth, function, and development of the adrenal cortex: insights from cell culture, *Cold Spring Harbor Conf. Cell Proliferation,* 9, 461, 1982.
81. **Vickery, L. and Hornsby, P. J.,** unpublished observations.

ESSENTIAL FATTY ACIDS: METABOLISM, INTERACTION WITH FREE RADICALS, AND CLINICAL USE

David F. Horrobin and Michel E. Begin

INTRODUCTION

The polyunsaturated fatty acids (PUFAs), fatty acids containing two or more double bonds, are believed to be one of the keys for understanding the damage that can be done to cells by free radicals. Free radical attack has two major consequences: (1) the loss of PUFAs which play important roles in cell membrane structure and as precursors for eicosanoid (prostaglandin and leukotriene) formation and (2) the formation of lipid peroxides and related compounds which can themselves cause damage to other cellular constituents.

Since the PUFAs play such central roles in the interactions of free radicals with cells, it is important to understand how they gain access to the body and how they are metabolized. There are three main groups of PUFAs, characterized as n-6 (or ω-6), n-3, and n-9 by the position of the double bond which is closest to the methyl-terminal carbon.[2-4] The three groups of PUFAs are believed to be metabolized by the same sequence of alternating desaturating and elongating enzymes (Figure 1), although there have recently been suggestions that in some tissues there may be specific enzymes for one or other PUFA series. The simplest form of nomenclature is easy to understand. It may be illustrated by dihomo-γ-malinolenic acid (DGLA), 20:3 n-6, where 20 indicates the number of carbon atoms in the molecule, 3 the number of conjugated double bonds, and n-6 the position of the first double bond from the methyl end of the molecule. Almost all double bonds in PUFAs in the natural state are in the less stable cis configuration. Small amounts of fatty acids containing trans double bonds are formed by bacteria in the gut of ruminant animals. Large amounts of trans acids are found during food processing in attempts to make vegetable oil products more stable.

The n-3 and n-6 series of PUFAs are normally by far the most abundant in the animal body. They are also the most important because animals cannot synthesize n-6 or n-3 PUFAs *de novo*, nor can they convert PUFAs of one series into PUFAs of another series. The n-6 and n-3 PUFAs are vital both as structural components of all membranes and of the permeability barrier of the skin and as precursors of the eicosanoids, the prostaglandins, leukotrienes, and related compounds which play essential regulatory roles in all tissues. Because the n-3 and n-6 PUFAs are vital to the organism and because they cannot be synthesized *de novo*, they are also called essential fatty acids (EFAs), since they must be present in the diet and they play a vitamin-like role.[2-4] Even a single trans bond destroys the ability of an n-6 or n-3 fatty acid to be essential. The n-9 PUFAs have no EFA activity. Thus all EFAs are polyunsaturates, but not all PUFAs are EFAs. The common use of the term "polyunsaturate" rather than "EFA" when, for example, describing a diet is unfortunate because only the EFAs have the desirable effects which the general public (and indeed many dietitians) attribute to any polyunsaturate.

Although the animal organism cannot make EFAs *de novo*, it can perform limited metabolic transformations on the EFAs which are presented to it. The main EFAs provided to the human body in the normal Western diet are *cis*-linoleic acid (18:2 n-6) and *cis*-α-linolenic acid (ALA) (18:3 n-3) which can be desaturated and elongated by the series of reactions shown in Figure 1. In all species, the first step, Δ-6-desaturation, seems to be rate limiting,[5,6] although this is more so in some species than in others.[7,8] The human, rabbit, and guinea pig all exhibit rather slow Δ-6-desaturation, while in rats and mice the process goes much faster. Elongation seems rapid in all species, but Δ-5-desaturation is again very different, being relatively active in rats and mice, but rather slow in humans and guinea pigs.[7,8]

Outline of essential fatty acid metabolism.

	n-6 EFAs		n-3 EFAs	
Linoleic (LA)	18:2n-6		18:3n-3	Alpha-linolenic (ALA)
	↓	delta-6-desaturase	↓	
Gamma-linolenic (GLA)	18:3n-6		18:4n-3	
	↓		↓	
Dihomogammalinolenic (DGLA)	20:3n-6		20:4n-3	
	↓	delta-5-desaturase	↓	
Arachidonic (AA)	20:4n-6		20:5n-3	Eicosapentaenoic (EPA)
	↓		↓	
Adrenic	22:4n-6		22:5n-3	
	↓	delta-4-desaturase	↓	
	22:5n-6		22:6n-3	Docosahexaenoic (DHA)

FIGURE 1. Outline of essential fatty acid metabolism.

Animals can make the n-9 PUFAs *de novo* by desaturating oleic acid. However, the desaturating and elongating systems of the enzymes normally seem to prefer the n-3 and n-6 PUFAs, so that when adequate amounts of 18:2 n-6 or 18:3 n-3 are present in the diet, only trivial amounts of the polyunsaturated n-9 acids are formed. If there are inadequate amounts of n-3 and n-6 EFAs in the diet, then the enzyme sequence can metabolize oleic acid (18:1 n-9) instead, and its product 20:3 n-9 accumulates. The presence of anything other than trace amounts of 20:3 n-9 in the body is an indicator of dietary deficiency or defective absorption of EFAs.[2-4]

Many factors are able to inhibit Δ-6-desaturation in experimental situations.[5,6,9-11] They include aging, diabetes mellitus, high cholesterol intake, high catecholamine release during stress, zinc deficiency, high alcohol intake, and viral infections. Moreover, there is evidence from measurements of human plasma fatty acids that several illnesses are associated with normal or near-normal blood levels of linoleic acid, but reduced levels of its metabolites, including DGLA and arachidonic acid (AA).[3,6,12] These illnesses include atopic eczema,[13,14] premenstrual syndrome,[15] hypertension,[16] Sjögren's syndrome,[17,18] hyperactivity,[19] and diabetes.[20,21] In other individuals, notably those who drink too much alcohol,[22,23] those with cystic fibrosis,[24] and those at risk of myocardial infarction, plasma linoleic acid levels are reduced, but levels of its metabolites are reduced even further. Thus, in these situations, there appear to be inadequate amounts of linoleic acid, but also there appears to be a defect in the desaturation of the linoleic acid which is available.

This restricted ability to desaturate linoleic acid (and by implication ALA, although it is normally present in only very small amounts) has led to interest in the dietary provision of desaturated metabolites which might be able to by-pass the apparently inhibited Δ-6-desaturation step. This interest has been fueled by the observation that human milk, but not milk from cows, is relatively rich in the desaturated metabolites of linoleic acid and ALA, such as γ-linolenic acid (GLA), DGLA, AA, and eicosapentaenoic acid (EPA, 20:5 n-3).[27,28] It has been suggested that, since the human brain is exceptionally rich in long-chain desaturated PUFAs, the presence of these EFAs in human milk may be of particular importance.[29,30] AA is found in meat, eggs, and some seafoods such as prawns, but other n-6 metabolites are found rarely in commonly consumed foods other than organ meats such as liver, pancreas, kidneys, ovaries, and testes. Recently, GLA has been found in various vegetable oils, including evening primrose oil (EPO), blackcurrant, and borage. Interest in EPO in the medical research community has recently been considerable and over 100 reports on studies of Efamol, a form of EPO prepared from seeds of a variety of *Oenothera biennis* specially bred to yield oil of constant quality and composition, have now been published.[31]

Long-chain, polyunsaturated n-3 EFAs are almost confined to seafood, being present in large amounts in oily fish and meat from seals and whales. Thus, people who do not eat seafood must obtain almost all their EPA and docosahexaenoic acid (DHA) by desaturation and elongation of ALA (Figure 1), while those who do eat seafood can readily obtain these EFAs from the diet. Based on the apparently low levels of heart attacks and certain inflammatory diseases in Eskimos who eat large amounts of oily seafood, there has recently been an explosion of interest in the possible beneficial effects of fish oils.[32-34]

THERAPEUTIC ADMINISTRATION OF EFAS

Since it has been hypothesized that free radicals may have roles in many diseases[1,35] (also see other chapters in this volume), notably inflammatory disorders, cancer, alcoholism, diabetes, and cardiovascular diseases, the low levels of linoleic acid, or its metabolites, or both, in humans with these conditions are of considerable interest. These low levels might indicate increased destruction of PUFAs due to elevated levels of free radicals. Harm might result either from the loss of valuable EFAs or from the production of lipid peroxides and related substances or from both effects. Much research on LPs is stimulated by and funded because of the idea that such toxic products of lipid peroxidation may be either the fundamental causes of, or major contributors to, the disease concerned. If this is so, then provision of PUFAs, which would provide further substrates for free radical attack, might be expected to make all these conditions worse.

These ideas have recently been tested by a group of scientists from the Efamol Research Institute and their clinical colleagues. They have set up many double-blind, placebo-controlled trials in which Efamol EPO has been compared with placebos. The placebos have consisted of liquid paraffin or olive oil (rich in the monounsaturate oleic acid but low in PUFAs) made up in capsules which appear identical to Efamol. Identical amounts of vitamin E have usually been added to both Efamol and placebo capsules to ensure that all patients, in both active and placebo groups, have intakes of vitamin E which are more than adequate. Efamol contains 72% of linoleic acid and, of more importance, 9% of GLA which can bypass the rate-limiting Δ-6-desaturase. If free radicals are important in a disease, Efamol might be expected to make the condition worse, by providing higher levels of substrate for lipid peroxidation.

The actual results of these clinical trials have been uniformly against the idea that PUFAs make worse any of these diseases in which free radicals have been implicated. Blood measurements have consistently shown that administration of Efamol raises levels of DGLA and, to a lesser extent, of AA in phospholipids and other lipid fractions. At the same time that there is an elevation in the concentrations of these prime substrates for lipid peroxidation, there is either no change in the disease state or else a significant improvement. In no trial has PUFA administration been shown to make a disease worse. These studies therefore argue strongly against the idea that products of free radical attack on PUFAs are causing harm in these conditions. The results leave open the possibility that loss of EFAs due to free radical attack is important. It is possible that the observed elevations in DGLA and AA following Efamol could have the effect of replacing material lost as a result of free radical damage. The main trial results to date are as follows.

Inflammatory Disorders

Patients with atopic eczema, an inflammatory skin disorder, were significantly improved by treatment with Efamol as compared to placebo.[36] Patients with Sjogren's syndrome, a disease related to rheumatoid arthritis and characterized by lacrimal gland failure, showed improved tear production.[37,38] Patients with Raynaud's syndrome and systemic sclerosis showed a reduced frequency of vasospastic attacks.[39] Most convincing of all, in a major trial of Efamol in patients with rheumatoid arthritis who were taking nonsteroidal anti-inflammatory drugs 30 of 34 patients on active treatment, but only 6 of 18 on placebo, were able to reduce or stop their drug therapy. When, unknown to the patients, they were switched from active treatment to placebo, 26 of the 30 who had improved relapsed within 8 weeks.[40,41]

Alcoholism

Alcohol impairs the conversion of linoleic acid to GLA and also reduces levels of linoleic acid itself in animal studies. Studies in human alcoholics have now confirmed these observations.[42,43] In two centers, placebo-controlled trials have now been carried out on the effects of Efamol during withdrawal from alcohol and the subsequent rehabilitation period.[42,44-46] During the early phase of withdrawal from alcohol, GLA as Efamol reduced the severity of withdrawal symptoms and accelerated the recovery of liver function. After 6 months, the Efamol-treated group showed better cerebral function on a variety of tests as compared to the placebo group. At the same time, blood fatty acid levels were moved towards normal in the Efamol group.

Diabetic Complications

The idea that PUFAs might help to prevent the long-term complications of diabetes was first proposed by Houtsmuller and colleagues at Erasmus University in Rotterdam. In a major 6-year study in newly diagnosed maturity-onset diabetics, they compared the retinal and cardiovascular complications in patients receiving a normal linoleic acid intake and patients receiving about four times the normal amount.[47] Retinopathy was about three times more common in the low linoleic acid group and 6 of 51 patients in the low linoleic acid group experienced myocardial infarctions, whereas 0 of 48 in the high linoleic acid group did. It seems possible that, because the conversion of linoleic acid to GLA is reduced in diabetics, the provision of very large amounts of linoleic acid in the diet may allow adequate amounts of GLA and its metabolites to be formed. Unfortunately diets containing these very large amounts of linoleic acid are unpalatable to many people and it would be desirable if a similar effect could be achieved without the requirement to make a major change in diet. Theoretically this could be done by providing much smaller amounts of GLA directly. Two trials of Efamol in diabetic neuropathy have now shown this to be the case. In rats made diabetic by streptozotocin, impaired nerve conduction develops rapidly. In these animals, nerve conduction can be normalized by the oral administration of Efamol.[48] In a 6-month double blind study of human diabetics with peripheral neuropathy, eight objective neurophysiological parameters showed modest deterioration in the placebo group. In the Efamol group, in contrast, all eight parameters improved and the differences between the effects of active and placebo treatment were highly significant.[49] Thus, in each of these clinical situations in which it has been hypothesized that lipid peroxidation may play a role in causing the damage, double-blind, placebo-controlled studies have shown that replenishment with EFAs is associated with improvement in the clinical situation and not with deterioration. If free radicals are involved in the pathophysiology of these diseases, these results are consistent with the idea that they exert their harmful effects by destroying EFAs. They do not appear to be consistent with the idea that it is the lipid peroxidation products formed by interaction between free radicals and PUFAs which cause the damage. If the latter were the case, one might have expected that further provision of PUFA substrates would make the disease state worse.

CANCER

The idea that lipid peroxidation products are harmful is particularly prevalent in the field of cancer. Ionizing radiation and most known carcinogenic chemicals are capable of generating free radicals which lead to lipid peroxidation. Both the free radicals themselves and the resulting peroxidation products can damage DNA and it therefore seems reasonable to suggest that PUFAs and the peroxidation products generated from them may play a role in cancer.

This view gained strength because of many experiments in which administration of linoleic acid to rats has been shown to enhance the development of mammary tumors.[50,51] Monounsaturated fatty acids do not have this effect and these results have therefore been interpreted as indicating that PUFAs must promote cancer. However, all the human epidemiological evidence, while strongly supportive of the idea that fat in general and saturated fats in particular may promote cancer, provides no indication that high intakes of PUFAs are associated with any cancer.[52]

If linoleic acid does enhance cancer development, and if it does so because of its readily peroxidized double bonds, then PUFAs containing three or more double bonds should be even more effective than linoleic acid. In fact, the evidence is diametrically opposite. Three independent groups have now shown that GLA, as EPO, inhibits the development of mammary cancer in rats.[53-55] Fish oil, containing 20:5 n-3 and 22:6 n-3, fatty acids considerably more unsaturated and highly susceptible to peroxidation, is even more effective at inhibiting mammary tumor growth.[56-60] The interpretation of the linoleic acid experiments which suggests that it is enhancing tumor growth either because it is a PUFA or because it is a prostaglandin precursor must therefore be rejected. GLA is a better prostaglandin precursor than linoleic acid, and GLA, 20:5 n-3, and 22:6 n-3 are all more unsaturated and more liable to peroxidation. A simpler explanation may be correct.[52] This is that in many of the linoleic acid experiments the control groups contained amounts of linoleic acid in the diet which were inadequate to support even normal mammary growth. A deficit of any essential nutrient will lead to inhibition of tumor growth. The apparent growth stimulation caused by the linoleic acid may have been related simply to replacement of a missing essential nutrient. Whether or not this is the correct interpretation, the explanations linking the linoleic acid effect to peroxidation or to prostaglandins are almost certainly incorrect.

The measurement of thiobarbituric acid reactive material (TBAR) is an inadequate indicator of the level of lipid peroxidation in a cell of tissue. Unfortunately, in most situations to date it is the only evidence we have about the degree of lipid peroxidation. If such peroxidation were important in cancer development, one might expect cancer cells to have above normal levels of TBAR. The results of measuring lipid peroxidation in human tumors have, with few exceptions, shown that TBAR levels are below normal rather than above normal.[61-63]

Low levels of lipid peroxidation products might theoretically be related to lack of PUFA substrate, lack of free radicals because of reduced formation or increased quenching, or increased rates of removal of such products. There is some evidence for increased antioxidant activity in some tumors.[64-68] However, in most tumors there appears to be a lack of PUFAs with three or more double bonds and, as will be demonstrated experimentally later, this appears to be the major reason for the observed low levels of TBAR in cancers.[62,68-72] One important reason for this lack of PUFAs with three or more double bonds could be lack of Δ-6-desaturase activity.[73] Such a deficit has been sought and found in a small number of liver and brain tumors and in a larger number of transformed cell lines.[74-78] It may therefore be that loss of desaturation leads to loss of PUFA substrates and hence to low levels of peroxidation products in cancer cells. This sequence, while an attractive hypothesis, remains to be proven.

Surprisingly, in spite of all the work on peroxides and cancer, until very recently no one had added the missing PUFA substrates to the cancer cells in order to see what happened to levels of TBAR and other related products or, indeed, to the cells themselves. The effects of the PUFAs on normal and cancer cell survival have recently been investigated by at least five independent laboratories.[79-85] The results have proved both very consistent and somewhat surprising. At concentrations which have no important adverse effects on human fibroblasts

Table 1
STIMULATED LIPID PEROXIDATION AND LOSS OF CELL VIABILITY IN HUMAN BREAST TUMOR CELLS GROWN IN VITRO IN THE PRESENCE OF EXOGENOUSLY ADDED POLYUNSATURATED FATTY ACIDS

	Content in					
			PUFA (% change)[d]		% Dead cells[e]	
Fatty acid[a]	Superoxide[b]	TBAR[c]	A	B	A	B
AA	0.02 ± 0.003[f]	33.8 ± 21.0[g]	+426	−44 ± 17	99 ± 1	99 ± 1
GLA	0.03 ± 0.004	27.9 ± 6.1	+4660	−51 ± 18	93 ± 3	99 ± 1
EPA	0.02 ± 0.009	19.2 ± 8.6	+1667	−56 ± 16	64 ± 10	97 ± 3
LA	0.01 ± 0.002	ND[h]	ND	ND	35 ± 10	85 ± 12
DHA	ND	10.4 ± 4.1	ND	ND	28 ± 8	45 ± 6
Control	0.00	0.7 ± 0.9	NA[i]	O[j]	15 ± 2	15 ± 5

[a] Ethyl (AA, GLA, LA, DHA) or methyl forms (EPA) were used.
[b] Values obtained at day 3 after supplementation as measured by nitroblue tetrazolium reduction expressed as O.D. at 490 nm.
[c] Average of amounts detected on each consecutive day from day 3 to day 6 after supplementation; expressed as malonaldehyde-equivalent (picomoles) per 1000 cells.
[d] As measured by change (%) in the ratio of added PUFA content relative to 16:0 in phospholipids in supplemented relative to unsupplemented cells at day 3 (A) and in cells at day 7 relative to that at day 3 after supplementation (B).
[e] Measured as trypan-negative cells 6 days (A) and 10 days (B) after supplementation. Cytopathic effects and cell death start to increase above background level by day 4 after supplementation.
[f] All values are means and SDs.
[g] High SD due to the high amount obtained on day 4.
[h] Not done.
[i] Not applicable.
[j] Values for AA, GLA, and EPA as in footnote d.

and a variety of other normal cell lines, apart from modest inhibition of growth in some cases, PUFAs strongly inhibit the growth of or, more usually, kill malignant cells. No malignant cell line has yet been found which cannot be killed by at least one PUFA, and most cell lines are susceptible to the lethal effects of several different PUFAs. One might have predicted that the greater the number of double bonds in a PUFA, the more easily it would be peroxidized and the more lethal that PUFA would be. However, the observed pattern of activity is curious and could not have been predicted by this hypothesis.[83,84,86] The highest level of lethal activity is consistently shown by fatty acids which have three or four double bonds. PUFAs with two or six double bonds are relatively inactive, while EPA, with five, has intermediate activity. If free radical attack on PUFAs were important, one would have expected the five and especially the six double-bonded fatty acids to be the most effective.

In an attempt to understand further what is happening, we have used human breast cancer cells as a model. We have tested the effects of linoleic acid, GLA, AA, EPA, and DHA (22:6n-3), a range of fatty acids with two, three, four, five, and six double bonds. We measured the content of superoxide anion and of TBAR and also the loss during the culture period of PUFAs as indirect indicators of lipid peroxidation. A summary of the results is shown in Table 1.

It is generally accepted that the superoxide radical plays a major role in the initiation of lipid peroxidation.[87] We found that the PUFAs stimulated peroxide radical formation in the cancer cells and that GLA, EPA, and DHA were two to three times more effective than linoleic acid. Superoxide dismutase, with the ability to scavenge superoxide radicals, when added to the tumor cell cultures in various amounts inhibited the cytotoxic effects of the PUFAs in a dose-dependent manner. This suggests a close relation between the cell-killing ability of a PUFA and its ability to generate superoxide radicals. Results consistent with this view were obtained by measuring TBAR. Considerably larger amounts of TBAR were recovered from PUFA-supplemented cancer cells than from control, unsupplemented cells from the same line (Table 1). Moreover, the effectiveness of a given PUFA in killing the cancer cells was well correlated with its efficacy in increasing TBAR concentrations. AA and GLA generated the most TBAR and were the most effective PUFAs in killing the cells. DHA, in spite of its high degree of unsaturation, was relatively ineffective either in raising TBAR levels or in killing the malignant cells.

In spite of the puzzling lack of effect of DHA, other evidence was consistent with the idea that lipid peroxidation was playing a key role in the cell death. Most of the tumor cells die between days 3 and 7 after supplementation. During this time, the PUFA content of the cells falls to about half (Table 1), suggesting extensive PUFA destruction by peroxidation. Iron is known to stimulate lipid peroxidation, whereas antioxidants such as vitamin E are known to inhibit the propagation of lipid peroxidation. If lipid peroxidation is associated with cell killing, then iron should stimulate cell death and vitamin E should inhibit it. Addition of iron greatly increased the extent of cell death when the cells were grown in the presence of suboptimal levels of GLA, but had no harmful effects on the cells in the absence of additional PUFAs. In the presence of PUFA levels which killed almost all cells, the latter died much faster in the presence of added iron. Vitamin E in various forms and synthetic antioxidants such as butylated hydroxyanisole and butylated hydroxytoluene inhibited the effects of GLA in a dose-dependent manner. Indomethacin which inhibits the enzyme cyclooxygenase and therefore the formation of peroxidation products dependent on the enzyme completely failed to reduce either TBAR content or cell death. This suggests that peroxidation products dependent on cyclooxygenase are not involved in the effects of the PUFA. Other cyclooxygenase inhibitors and several inhibitors of lipoxygenase also failed to inhibit the lethal damage done to malignant cells by the PUFAS. Supplemented tumor cells accumulated yellow-brown pigmented material, probably reflecting polymerization products associated with lipid peroxidation. Preliminary experiments have suggested that levels of conjugated dienes rise rapidly in cancer cells when GLA is added to them. No correlation has been demonstrated between the intracellular content of 4-hydroxynonenal and cell death. Malonaldehyde (2 m*M*) failed to kill the tumor cells. Thus, the ultimate toxic agent(s) causing the lethal effects of the PUFAs has not yet been identified, but it is believed almost certainly to be some product of superoxide action on PUFAs, especially those with three, four, or five double bonds.

For comparison, we performed similar experiments on normal kidney cells. Although superoxide radical and TBAR contents increased in the supplemented cells, the concentrations achieved were two to six times less than in the cancer cells. Moreover, the pattern of TBAR accumulation was completely different, being approximately proportional to the numbers of double bonds. Only very small numbers of cells died in the presence of the PUFAs and there was no correlation between PUFA supplementation, accumulation of TBAR, and cell death. There was no production of yellow-brown pigment. Thus, the responses of normal cells and cancer cells to addition were completely different, both qualitatively and quantitatively. The results obtained so far suggest that PUFAs involved in the effective killing of cancer cells can stimulate the initiation of their own peroxidation by increasing the amounts of superoxide radicals. At the same time, TBAR and other related materials are formed. Because normal cells do not seem to be affected in the same way, this opens up the possibility of a "magic bullet" type of nontoxic treatment of cancer by selectively destroying the malignant cells.

CONCLUSIONS

There have been many suggestions that free radicals play important roles in various human diseases, including inflammatory disorders, diabetes, alcoholism, and cancer. Free radicals could cause damage, either by destroying PUFAs which are necessary for cell structure and function and/or by reacting with PUFAs to form toxic peroxidation products and other substances which actually do the damage. The results of studies which have investigated the effects of PUFAs in inflammatory disorders, alcoholism, and diabetes have consistently shown that PUFAs have highly desirable rather than undesirable effects. These observations are consistent with the idea that in these conditions the free radicals are producing much of their damage by destroying necessary PUFAs.

PUFAs and their peroxidation products do indeed seem to be capable of seriously damaging cancer cells. However, even in this situation the balance of actions seems to be highly favorable because normal cells do not seem susceptible to the same lethal actions. Thus, in all the situations in which peroxides have been thought to be important, PUFA supplementation seems to be helpful rather than harmful. This may be due in part to the quenching of free radicals by the presence of the PUFAs and in part to replacement of PUFAs damaged by peroxidation.

REFERENCES

1. **Halliwell, B. and Gutteridge, J. M. C.,** *Free Radicals in Biology and Medicine,* Oxford University Press, Oxford, 1985.
2. **Mead, J. F. and Fulco, A. J.,** *The Unsaturated and Polyunsaturated Fatty Acids in Health and Disease,* Charles C. Thomas, Springfield, Ill., 1976.
3. **Holman, R. T.,** Control of polyunsaturated acids in tissue lipids, *J. Am. Coll. Nutr.,* 5, 183, 1986.
4. **Sinclair, H. M.,** *The Essential Fatty Acids.,* Academic Press, New York, 1958.
5. **Brenner, R. R.,** Nutritional and hormonal factors influencing desaturation of essential fatty acids, *Prog. Lipid Res.,* 20, 41, 1982.
6. **Horrobin, D. F.,** The regulation of prostaglandin biosynthesis by the manipulation of essential fatty acid metabolism, *Rev. Pure Appl. Pharmacol. Sci.,* 4, 339, 1983.
7. **Horrobin, D. F., Huang, Y.-S., Cunnane, S. C., and Manku, M. S.,** Essential fatty acids in plasma, red blood cells and liver phospholipids in common laboratory animals as compared to humans, *Lipids,* 19, 806, 1984.
8. **Stone, K. J., Willis, A. L., and Hart, M.,** The metabolism of dihomogammalinolenic acid in man, *Lipids,* 14, 174, 1979.
9. **Nervi, A. M., Peluffo, R. O., Brenner, R. R. and Leikin, A. I.,** Effect of ethanol administration on fatty acid desaturases, *Lipids* 15, 263, 1980.
10. **Cunnane, S. C., Horrobin, D. F., and Manku, M. S.,** Essential fatty acids in tissue phospholipids and triglycerides of the zinc deficient rat, *Proc. Soc. Exp. Biol. Med.,* 177, 441, 1984.
11. **Huang, Y.-S., Manku, M. S., and Horrobin, D. F.,** The effects of dietary cholesterol on blood and liver polyunsaturated fatty acids and on plasma cholesterol in rats fed various types of fatty acid diet, *Lipids,* 19, 664, 1984.
12. **Horrobin, D. F., and Manku, M. S.,** Essential fatty acids in clinical medicine, *Nutr. Health,* 2, 127, 1983.
13. **Manku, M. S., Horrobin, D. F. and Morse, N. L.,** Reduced levels of prostaglandin precursors in the blood of atopic patients, *Prostaglandins Leukotrienes Med.,* 9, 615, 1982.
14. **Manku, M. S., Horrobin, D. F., Morse, N. L., Wright, S., and Burton, J. L.,** Essential fatty acids in the plasma phospholipids of patients with atopic eczema, *Br. J. Dermatol.,* 110, 643, 1984.
15. **Brush, M. G., Watson, S. J., Horrobin, D. F., and Manku, M. S.,** Abnormal essential fatty acid levels in plasma of women with premenstrual syndrome, *Am. J. Obstet. Gynecol.,* 150, 363, 1984.
16. **Singer, P., Jaeger, W., Voigt, S., and Thiel, H.,** Defective desaturation and elongation of n-6 and n-3 fatty acids in hypertensive patients, *Prostaglandins Leukotrienes Med.,* 15, 159, 1984.

17. **Horrobin, D. F.,** Essential fatty acid metabolism in diseases of connective tissue with special reference to scleroderma and to Sjögren's syndrome, *Med. Hypotheses,* 14, 233, 1984.
18. **Horrobin, D. F.,** Essential fatty acid and prostaglandin metabolism in Sjögren's syndrome, systemic sclerosis and rheumatoid arthritis, *Scand. J. Rheumatol.,* Suppl. 61, 242, 1986.
19. **Mitchell, E. A., Aman, M. G., Turbott, S. H., and Manku, M. S.,** Clinical characteristics and serum essential fatty acid levels in hyperactive children, *Clin. Pediatr.,* 26, 406, 1987.
20. **Mercuri, O., Peluffo, R. O., and Brenner, R. R.,** Depression of microsomal desaturation of linoleic to gamma-linolenic acid in the alloxan diabetic rat, *Biochim. Biophys. Acta,* 116, 407, 2966.
21. **Jones, D. B., Carter, R. D., Haitas, B., and Mann, J. I.,** Low phospholipid arachidonic acid values in diabetic platelets, *Br. Med. J.,* 286, 173, 1983.
22. **Glen, A. I. M., Glen, E., MacDonell, L., and MacKenzie, J.,** Possible pharmacological approaches to the prevention and treatment of alcohol-related CNS impairment: results of a double blind trial of essential fatty acids, in *Pharmacological Treatments for Alcoholism,* Edwards, G. and Littleton, J., Eds., Croom Helm, London, 1984.
23. **Glen, A. I. M.,** Essential fatty acid abnormalities in alcohol dependent patients: effects on erythrocyte mean cell volume, *Br. J. Addiction.,* 80, R6, 1985.
24. **Rivers, J. P. W. and Hassam, A. G.,** Defective essential fatty acid metabolism in cystic fibrosis, *Lancet,* ii, 642, 1975.
25. **Wood, D. A., Butler, E., Riemersma, R. A., Thomsom, M., Oliver, M. F., Fulton, M., Birtwhistle, A., and Elton, R.,** Adipose tissue and platelet fatty acids and coronary heart disease in Scottish men, *Lancet,* ii, 117, 1984.
26. **Salonen, J. T., Salonen, R., Penttila, I., Herranen, J., Jauhiainen, M., Kantola, M., Lappetalainen, R., Maeanpaa, P. H., Alfthan, G., and Puska, P.,** Serum fatty acids, apolipoproteins, selenium and vitamin antioxidants and the risk of death from coronary heart disease, *Am. J. Cardiol.,* 56, 226, 1985.
27. **Gibson, R. A. and Kneebone, G. M.,** Fatty acid composition of human colostrum and mature breast milk, *Am. J. Clin. Nutr.,* 34, 252, 1981.
28. **Harzer, G., Haug, M., Dieterich, I., and Gentner, P. R.,** Changing patterns of human milk lipids in the course of lactation and during the day, *Am. J. Clin. Nutr.,* 37, 612, 1983.
29. **Crawford, M. A., Hassam, A. G., and Stevens, P. A.,** Essential fatty acid requirements in pregnancy and lactation with special reference to brain development, *Prog. Lipid Res.,* 20, 31, 1982.
30. **Clandinin, M. T.,** Long chain polyenoic essential fatty acids in human milk: are they of benefit to the newborn?, in *Composition and Physiological Properties of Human Milk,* Schaub, J., Ed., Elsevier, Amsterdam, 1985, 213.
31. **Toplack, C.,** Evening Primrose Oil Bibliography, Efamol Research Institute, Kentville, Canada, 1985.
32. **Bang, H. O. and Dyerberg, J.,** Fish consumption and mortality from coronary heart disease, *N. Engl. J. Med.,* 313, 822, 1985.
33. **Kromhout, D., Bosschieter, E. B., and Coulander, C. D. L.,** The inverse relation between fish consumption and 20-year-old mortality from coronary heart disease, *N. Engl. J. Med.,* 312, 1205, 1985.
34. **Phillipson, B. E., Rothrock, D. W., Connor, W. E., Harris, and Illingworth, D. R.,** Reduction of plasma lipids, lipoproteins and apoproteins by dietary fish oils in patients with hypertriglyceridemia, *N. Engl. J. Med.,* 312, 1210, 1985.
35. **Dormandy, T. L.,** An approach to free radicals, *Lancet,* ii, 1010, 1983.
36. **Wright, S. and Burton, J. L.,** Oral evening primrose oil improves atopic eczema, *Lancet,* ii, 1120, 1982.
37. **Manthorpe, R., Peterson, S. H., Prause, J. U.,** Primary Sjogren's syndrome treated with Efamol/Efavit: a double blind cross-over investigation, *Rheumatol. Int.,* 4, 165, 1984.
38. **Oxholm, P., Manthorpe, R., Prause, J. U., and Horrobin, D.,** Patients with primary Sjogren's syndrome treated for two months with evening primrose oil, *Scand. J. Rheumatol.,* 15, 103, 1986.
39. **Belch, J. J. F., Shaw, B., O'Dowd, A., Saniabadi, A., Leiberman, P., Sturrock, R. D., Forbes, C. D.,** Evening primrose oil (Efamol) in the treatment of Raynaud's phenomenon: a double blind study, *Thromb.* Haem., 54, 490, 1985.
40. **Ansell, D., Belch, J. J. F., Curan, L., McLaren, M., O'Dowd, A., and Forbes, C. D.,** The effects of Efamol and Efamol Marine on patients with rheumatoid arthritis: a double-blind, placebo-controlled study, in Abstract Book, 6th Int. Congr. Prostaglandins, Florence, June 1986, 24.
41. **Belch, J. J. F., Ansell, D., Madhok, R., O'Dowd, A., and Sturrock, R. D.,** The effects of altering dietary essential fatty acids on requirements for non-steroidal anti-inflammatory drugs in patients with rheumatoid arthritis: a double-blind placebo-controlled study, *Ann. Rheum. Dis.,* 47, 96, 1988.
42. **Horrobin, D. F. and Manku, M. S.,** Essential fatty acids and clinical medicine, *Nutr. Health,* 2, 127, 1983.
43. **Cairns, S. R. and Peters, T. J.,** Biochemical analysis of hepatic lipid in alcoholic, diabetic and control subjects, *Clin. Sci.,* 65, 645, 1983.

44. **Besson, J., Glen, E., Glen, I., MacDonell, L, and Skinner, F.,** Essential fatty acids, mean cell volume and nuclear magnetic resonance of brains of ethanol-dependent human subjects, *Alcohol Alcoholism,* Suppl. 1., 577, 1987.
45. **MacDonell, L., Skinner, F., Glen, I., and Glen, E.,** Effects of gamma-linolenic acid supplementation on neuropsychological function in alcohol-dependent males, presented at *R. College Psychiatr. Ann. Meet.,* Aberdeen, Scotland, April 7, 1987.
46. **Shaw, G. K.,** Effects of Efamol on acute withdrawal symptoms in alcoholics, paper presented at the Conf. Recent Advances in Psychiatric Treatment, San Francisco, September 1986.
47. **Houtsmuller, A. J., Van Hal-Ferwerda, J., Zahn, K. J. and Henkes, H. E.,** Favorable influences of linoleic acid on the progression of diabetic micro- and macroangiopathy in adult onset diabetes mellitus, *Prog. Lipid Res.,* 20, 377, 1982.
48. **Julu, P.,** Effects of Efamol on diabetic neuropathy in streptozotocin diabetic rats, submitted.
49. **Jamal, G. A., Carmichael, H. and Weir, A. I.,** Gamma-linolenic acid in diabetic neuropathy, *Lancet,* i, 1098, 1986.
50. **Carroll, K. K. and Hopkins, G. J.,** Dietary polyunsaturated fat versus saturated fat in relation to mammary carcinogenesis, *Lipids,* 14, 155, 1979.
51. **Abraham, S. and Hillyard, L. A.,** Effect of dietary 18-carbon fatty acids on growth of transplantable mammary adenocarcinomas in mice, *J. Natl. Cancer Inst.,* 71, 601, 1983.
52. **Horrobin, D. F.,** The role of essential fatty acids and prostaglandins in breast cancer, in *Diet, Nutrition and Cancer: A Critical Evaluation,* Vol. 1, Reddy, B. S. and Cohen, L. A., Eds., CRC Press, Boca Raton, Fla., 1986, 101.
53. **Ghayur, T. and Horrobin, D. F.,** Effects of essential fatty acids in the form of evening primrose oil on the growth of the Rat R3230AC transplantable mammary tumour, *IRCS J. Med. Sci.,* 9, 582, 1981.
54. **Karmali, R. A., Marsh, J. and Fuchs, C.,** Effects of dietary enrichment with gamma-linolenic acid upon growth of the R3230AC mammary adenocarcinoma, *J. Nutr. Growth. Cancer,* 2, 41, 1985.
55. **Bunce, O. R., Prasse, K. W., Upchurch, M., El-Ela, S. A., Feldman, E. and Carroll, R. M.,** Effect of evening primrose oil on DMBA-induced mammary tumors, *Proc. Soc. Exp. Biol. Med.,* 181, 472, 1986.
56. **Jurlowski, J. J. and Cave, W. T.,** Dietary effects of menhaden oil on the growth and membrane lipid composition of rat mammary tumors, *J. Natl. Cancer Inst.,* 74, 1145, 1985.
57. **Karmali, R. A., Marsh, J. and Fuchs, C.,** Effect of omega-3 fatty acids on growth of a rat mammary tumor, *J. Natl. Cancer Inst.,* 73, 457, 1984.
58. **Gabor, H. A.,** Fish consumption and mortality from coronary heart disease, *N. Engl. J. Med.,* 313, 823, 1985.
59. **Gabor, H. and Abraham, S.,** Effect of dietary menhaden oil on tumor cell loss and the accumulation of mass of a transplantable mammary adenocarcinoma in BALB/c mice, *J. Natl. Cancer Inst.,* 76, 1223, 1986.
60. **O'Connor, T. P., Rowbuck, B. D., Peterson, F. and Campbell, T. C.,** Effect of dietary intake of fish oil and fish protein on the development of 1-azaserine-induced preneoplastic lesions in the rat pancreas, *J. Natl. Cancer Inst.,* 75, 959, 1985.
61. **Slater, T. F.,** Lipid peroxidation, *Biochem. Soc. Trans.,* 10, 70, 1982.
62. **Ohnishi, T.,** Lipid peroxide formation and phospholipid in normal and tumor tissues, *Gann,* 49, 233, 1958.
63. **Bratoli, G. M. and Galeotti, T.,** Growth related lipid peroxidation in tumors, microsomal membranes and mitochondria, *Biochim. Biophys. Acta,* 574, 537, 1979.
64. **Burlakova, Y. B. and Molochkina, Y. M.,** Changes in the anti-oxidative activity of mouse liver lipids during experimental carcinogenesis, *Biofizika,* 18, 293, 1973.
65. **Duchesne, J.,** A unifying biochemical theory of cancer, senescence and maximal life span, *J. Theor. Biol.,* 66, 137, 1977.
66. **Bartoli, G. M., Bartoli, S., Galeotti, T., and Bertoli, E.,** Superoxide dismutase content and microsomal lipid composition of tumors with different growth rates, *Biochim. Biophys. Acta,* 620, 205, 1980.
67. **Lash, E.,** The antioxidant and pro-oxidant activity in ascites tumors, *Arch. Biochem. Biophys.,* 115, 332, 1966.
68. **Cheeseman, K. H., Burton, G. W., Ingold, K. U., and Slater, T. F.,** Lipid peroxidation and lipid antioxidants in normal and tumor cells, *Toxicol. Pathol.,* 12, 235, 1984.
69. **Thiele, E. H. and Huff, J. W.,** Lipid peroxide production and inhibition by tumor mitochondria, *Arch. Biochem. Biophys.,* 88, 203, 1960.
70. **Utsumi, K., Yamamoto, G., and Inaba, K.,** Failure of iron-induced lipid peroxidation and swelling in the mitochondria isolated from ascites tumor cells, *Biochim. Biophys. Acta.,* 105, 368, 1965.
71. **Liepkalns, V. A., Icard-Liepkalns, C., and Cornwell, D. G.,** Regulation of cell division in a human glioma cell clone by arachidonic acid and alpha-tocopherolquinone, *Cancer Lett,* 15, 173, 1982.

72. **Slater, T. F., Cheeseman, K. H., and Proudfoot, K.,** Free radicals, lipid peroxidation and cancer, in *Free Radicals in Molecular Biology, Ageing and Disease,* Armstrong, D., Ed., Raven Press, New York, 1984, 293.
73. **Horrobin, D. F.,** The reversibility of cancer: the relevance of cyclic AMP, calcium, essential fatty acids and prostaglandin E1, *Med. Hypotheses,* 6, 469, 1980.
74. **Chiappe, L. E., De Tomas, M. E. and Mercuri, O.,** In vitro activity of Δ-6- and -9-desaturase in hepatomas of different growth rates, *Lipids,* 9, 489, 1974.
75. **Reitz, R. C., Thompson, J. A., and Morris, H. P.,** Mitochondrial and microsomal phospholipids of Morris hepatoma 7777, *Cancer Res.,* 37, 561, 1977.
76. **Yavin, E. Yavin, Z., Menkes, J. H.,** Polyunsaturated fatty acid metabolism in neuroblastoma cells in culture, *J. Neurochem.,* 24, 71, 1975.
77. **Robert J., Rebel, G., and Mandel, P.,** Utilization of polyunsaturated fatty acids supplements by cultured neuroblastoma cells, *J. Neurochem.,* 30, 543, 1978.
78. **Dunbar, L. M. and Bailey, J. M.,** Enzyme deletions and essential fatty acid metabolism in cultured cells, *J. Biol. Chem.,* 250, 1152, 1975.
79. **Dippenaar, N., Booyens, J., and Fabri, D.,** The reversibility of cancer: evidence that malignancy in human hepatoma cells is gamma-linolenic acid deficiency dependent, *S. Afr. Med. J.,* 62, 683, 1982.
80. **Dippenaar, N., Booyens, J., and Fabri, D.,** The reversibility of cancer: evidence that malignancy in melanoma cells is gamma-linolenic acid deficiency dependent, *S. Afr. Med. J.,* 62, 505, 1982.
81. **Booyens, J., Dippenaar, N., and Fabri, D.,** The effect of the prostaglandin precursor gamma-linolenic acid on the rate of proliferation of human osteogenic sarcoma and oesophageal carcinoma cells in culture, *S. Afr. Med. J.,* 65, 240, 1984.
82. **Girao, L. A., Ruck, A. C., Cantrill, R. C., and Davidson, B. C.,** The effects of C18 fatty acids on cancer cells in culture, *Anticancer Res.,* 6, 241, 1986.
83. **Begin, M. E., Das, U. N., Ells, G., and Horrobin, D. F.,** Selective killing of human cancer cells by polyunsaturated fatty acids, *Prostaglandins Leukotrienes Med.,* 19, 177, 1985.
84. **Begin, M. E., Ells, G., Das, U. N., and Horrobin, D. F.,** Differential killing of human carcinoma cells by n-3 and n-6 polyunsaturated fatty acids, *J. Natl. Cancer Inst.,* 77, 1053, 1986.
85. **Fujiwara, F., Todo, S., and Imashuku, S.,** Anti-tumor effects of gamma-linolenic acid on human neuroblastoma cell lines in vitro, paper presented at the Annual Meeting of the Japanese Cancer Society, Tokyo, October 1985.
86. **Begin, M. E., Ells, G., and Das U. N.,** Selected fatty acids as possible intermediates for selective cytotoxic effects of anti-cancer agents involving oxygen radicals, *Anticancer Res.,* 6, 291, 1986.
87. **Kappus, H.,** Lipid peroxidation: mechanisms, analysis, enzymology, and biological relevance, in *Oxidative Stress,* Sies, H., Ed., Academic Press, London, 1985, 273.

MECHANISMS OF ANTIOXIDANT ACTION: PREVENTIVE AND CHAIN-BREAKING ANTIOXIDANTS

Graham W. Burton and Keith U. Ingold

MECHANISM OF AUTOXIDATION

The lipids of many of the different types of membranes enclosing cells and organelles frequently contain large amounts of polyunsaturated fat, usually present in the form of phospholipids. The polyunsaturated fatty acid tails of the lipids are characterized by the presence of two or more carbon-carbon double bonds which are separated by a methylene, $-CH_2-$, unit. The methylene unit in this bis-allylic arrangement, $-CH{=}CH-CH_2-CH{=}CH-$, is very susceptible to attack by radicals such as peroxyl, ROO·, and alkoxyl, RO·, undergoing hydrogen abstraction to give a carbon-centered, resonance-stabilized radical, $-\dot{C}H-$ (Reaction 1).

$$ROO\cdot + -CH_2- \rightarrow ROOH + H{-}\overset{|}{\underset{|}{C}}\cdot \qquad (1)$$

This radical reacts extremely rapidly[1] with oxygen to form a peroxyl radical (Reaction 2) which may attack a methylene unit in a bis-allylic group in another polyunsaturated fatty acid tail, thus beginning a chain reaction.

$$H{-}\overset{|}{\underset{|}{C}}\cdot + O_2 \rightarrow H{-}\overset{|}{\underset{|}{C}}{-}OO\cdot \qquad (2)$$

Such more or less spontaneous oxidation processes are called autoxidations[2] or, more commonly in the biological field, lipid peroxidations.

Autoxidation eventually results in some form of degradation of the material affected. For example, foodstuffs such as corn oil, butter, and margarine become rancid. The process is generally represented by four elementary reactions:

Initiation: production of R· (rate = R_i) (3)

Propagation: $R\cdot + O_2 \rightarrow ROO\cdot$ (4)

$ROO\cdot + RH \rightarrow ROOH + R\cdot$ (5)

Termination: $ROO\cdot + ROO\cdot \rightarrow$ molecular products (6)

In this scheme, RH represents a lipid molecule and R· represents the carbon-centered radical derived from it by removal of a hydrogen atom. (Propagation can also occur by addition of the peroxyl radical to a carbon-carbon double bond.)

The initiating radicals may be produced in a variety of ways[3,4] and many lipid molecules may be damaged, i.e., oxidized, during the propagation phase before two chain-carrying peroxyl radicals eventually terminate two chains by reacting with each other to give stable molecular products. At the cellular level, autoxidation is clearly undesirable because it threatens the integrity and the protective, and organizational properties of the membranes which enclose cells and organelles, thus placing the survival of the cell itself in jeopardy.

MECHANISMS OF ANTIOXIDANT ACTION

The fact that living systems do not autoxidize to any significant degree until after death indicates the existence of an active system of defense against peroxidation. The protection is provided by relatively small amounts (compared to lipid) of certain, specific compounds which are termed antioxidants. Antioxidants may be conveniently divided into two classes depending on their mode of action.[5] *Preventive* antioxidants function by reducing the rate of chain initiation (i.e., the rate of formation of "new" radicals), whereas *chain-breaking* antioxidants act by interfering with one or more of the propagation steps (e.g., by trapping peroxyl radicals).[6,7]

Preventive Antioxidants

The molecular precursor for the initiation process is generally H_2O_2 or the lipid hydroperoxide, ROOH. Although hydroperoxides are thermally stable at ambient temperatures, their decomposition is readily catalyzed by UV light and by certain transition metal ions yielding hydroxyl (HO·), alkoxyl, or peroxyl radicals[3] (Reactions 7 to 9).

$$\mathrm{ROOH} \xrightarrow{\text{UV light}} \mathrm{RO\cdot} + \mathrm{HO\cdot} \qquad (7)$$

$$\mathrm{ROOH} + \mathrm{Fe^{2+}} \rightarrow \mathrm{RO\cdot} + \mathrm{HO^-} + \mathrm{Fe^{3+}} \qquad (8)$$

$$\mathrm{ROOH} + \mathrm{Fe^{3+}} \rightarrow \mathrm{ROO\cdot} + \mathrm{H^+} + \mathrm{Fe^{2+}} \qquad (9)$$

Reactions 8 and 9 form a catalytic cycle so that very small traces of iron (or copper, or cobalt, or manganese, etc.) can initiate many chain reactions. This means, in principle, that the existence of even a very small, initial amount of hydroperoxide (or hydrogen peroxide) in the presence of a catalyst may lead to a branching chain reaction whose effects can quickly reach devastating proportions. (The great reactivity of the hydroxyl radical produced in Reaction 7 will result in its attack on its most immediate neighbors. This will generally lead to the formation of peroxyl radicals and hence to hydroperoxides, Reactions 3, 4, and 5.

Preventive antioxidants reduce the rate at which new chain reactions are started. This they do by a variety of means which include converting the hydroperoxide into a more stable product (e.g., alcohol and aldehyde) or by removing or inactivating the catalytic metal ion.

In living organisms, the conversion of hydrogen peroxide and hydroperoxides to stable nonradical products is achieved by the enzymes catalase and the glutathione peroxidases, respectively. Examples of naturally occurring agents which prevent the catalytic decomposition of hydroperoxides (and peroxides) into free radicals by sequestering transition metal ions are the human serum proteins transferrin and albumin, which bind iron and copper, respectively.[8] Ceruloplasmin, another protein found in human serum, also appears to be a form of preventive antioxidant because it is able to oxidize Fe(II) to the catalytically less active Fe(III).[4,8]

Compounds which inhibit photo-induced autoxidations can also be classed as preventive antioxidants. Substances which inhibit photodecomposition of peroxides and hydroperoxides (Reaction 7) would obviously fall into this category. Somewhat different in their mode of action are compounds which inhibit the dye-sensitized formation of singlet oxygen by visible light and its "ene" reaction with unsaturated fatty acids to form hydroperoxides (Reactions 10 and 11). β-Carotene is an example of this type because it is an extremely efficient quencher of singlet oxygen.[9]

$$\mathrm{Dye} \xrightarrow[\text{light}]{\text{visible}} \mathrm{dye^*} \xrightarrow{^3O_2} \mathrm{dye} + {}^1O_2 \qquad (10)$$

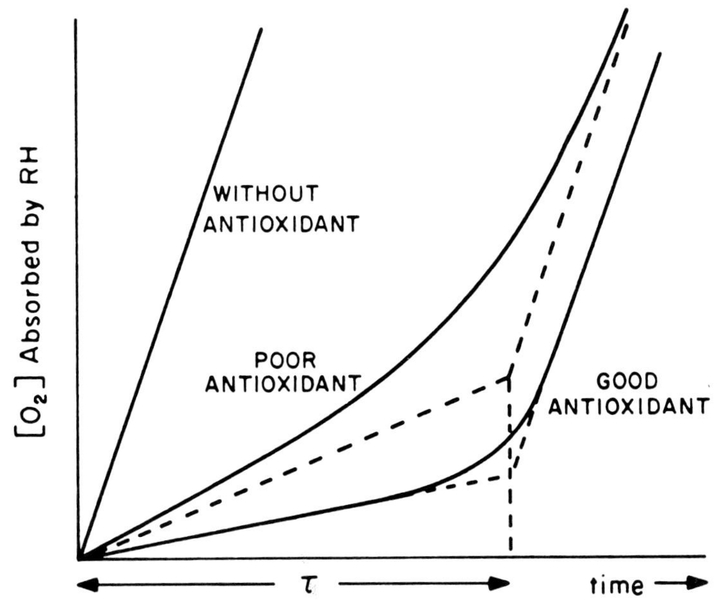

FIGURE 1. The effects of chain-breaking antioxidants on the rate of uptake of oxygen during autoxidation of a hydrocarbon under conditions of a constant and steady rate of production of peroxyl radicals.

$$^1O_2 + RCH{=}CHCH_2R' \rightarrow RCH(OOH)CH{=}CHR' \quad (11)$$

$$^1O_2 + \beta\text{-carotene} \rightarrow {}^3O_2 + \beta\text{-carotene} + \text{heat} \quad (12)$$

Chain-Breaking Antioxidants

Antioxidants in this class interact directly with radicals. For example, many phenols (ArOH) and some aromatic amines can stop a radical chain reaction because they are able to "trap" the chain-carrying peroxyl radicals.[10,11] The initial step involves the transfer of the phenolic hydrogen atom.

$$ROO\cdot + ArOH \rightarrow ROOH + ArO\cdot \quad (13)$$

The chain reaction stops because the unpaired electron in the phenoxyl radical (ArO·) is delocalized over the atoms in the aromatic structure making the radical too unreactive to continue the chain. The phenoxyl radical is eventually removed by reaction with a second peroxyl radical

$$ROO\cdot + ArO\cdot \rightarrow \text{nonradical products} \quad (14)$$

or may, in the presence of ascorbate or other reducing agents, undergo repair to give the starting phenol (see below). Because the chains are broken, the chain length of the autoxidation is reduced and so, of course, is the rate of oxidation.

Protection from autoxidation occurs at the expense of the antioxidant; during the time that the rate of oxidation is reduced (the induction period, τ), the antioxidant is steadily consumed. When it is completely used up, the rate of oxidation returns to its original, uninhibited value (Figure 1). Most phenols react in a clean, stoichiometric manner with peroxyl radicals with each phenol stopping two chains, i.e., each phenol molecule traps two peroxyl radicals (Reactions 13 and 14). This results in a simple and very useful relationship between the rate of chain initiation, R_i, the induction period, and the amount of antioxidant present (Equation 15).

$$R_i = 2[ArOH]/\tau \qquad (15)$$

Figure 1 depicts an idealized situation in which the rate of chain initiation is constant. The rate of uptake of oxygen is therefore constant with time and the induction period measurement can be used, through Equation 15, to obtain either R_i, if [ArOH] is known, or, alternatively, [ArOH] if R_i is already known. In the latter case, the quantity [ArOH] actually corresponds to the total amount of peroxyl-trapping antioxidant present in a sample added to the oxidation system (expressed in terms of the amount of phenol that would give the same τ).

In practice, this situation is conveniently realized by using thermally unstable compounds as initiators, i.e., as the initial source of peroxyl radicals which could start the chain oxidation. These thermally unstable materials, which are usually azo compounds, RN=NR, are useful because in solution (e.g., at 37°C) they decompose very slowly to give carbon-centered radicals at a steady and constant rate (Reaction 16). In the presence of

$$RN{=}NR \rightarrow N_2 + 2R\cdot \qquad (16)$$

oxygen this provides a very convenient means of generating known quantities of peroxyl radicals (Reaction 4). Addition of an oxidizable substrate to the system provides researchers with a very useful, model autoxidation system because free radical chains are now initiated at a known and steady rate, making it very easy to observe and measure the effects of added substances on the autoxidation.

Examples of frequently used initiators are the organic-soluble azobis(-isobutyronitrile [AIBN])and the water-soluble 2,2′-azobis(2-amidinopropane hydrochloride) (ABAP).

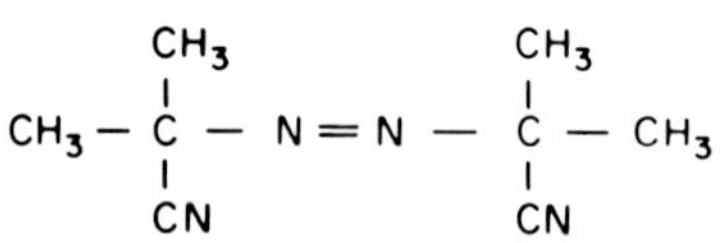

AIBN

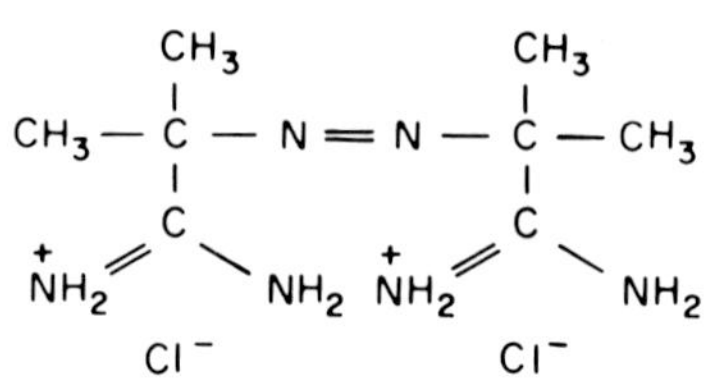

ABAP

One of the most important applications of thermally initiated autoxidations has been in experiments to detect and measure peroxyl radical-trapping antioxidants in biological samples. The peroxyl radicals generated by the initiator are used to ''titrate'' the radical-trapping antioxidants present in the added sample (e.g., a lipid extract from blood plasma). The presence of antioxidant material is clearly indicated by a depressed rate of oxygen uptake (Figure 1) and the total amount of antioxidant (i.e., the phenol equivalent) is calculated from the length of the induction period using Equation 15. Surprisingly, even though this technique has been available since 1955[12] and has been used in other fields,[13] it has seen only very limited application[14] in the biological field until quite recently.

For a phenol to be an effective antioxidant, it must be able to compete successfully for peroxyl radicals against a much greater amount of oxidizable lipid. That is, the rate constant for reaction of a peroxyl radical with the phenol (Reaction 13) must be much larger than the rate constant for reaction with polyunsaturated lipid (Reaction 5). The thermally initiated autoxidation method has proven very useful in this regard also because it has made it possible to carry out in vitro evaluations of the effectiveness of chain-breaking antioxidants. The effectiveness of the antioxidant is readily seen from the extent to which the rate of oxidation is depressed after the antioxidant is added (Figure 1). The rate constant for Reaction 13 can be determined from measurements of the slope of the induction period (Figure 1). Many of the rate constants for phenols have been determined in this way, using AIBN as the initiator and styrene as the (highly) oxidizable substrate.[15-19]

It should be noted that the rate constant for Reaction 13 governs the antioxidant effectiveness of a phenol; the reaction with a second peroxyl radical (Reaction 14) is a radical-radical reaction and is much faster than Reaction 13.[20,21] More detailed information on the thermally initiated autoxidation method is provided elsewhere in the chapter in Volume III by Wayner and Burton, entitled *Measurement of Individual Antioxidants and Radical-Trapping Activity*.

Lipid-Soluble Antioxidants

The thermally initiated autoxidation method has been applied to the determination of the total amount of chain-breaking antioxidant in lipid extracts obtained from human blood plasma and red cells,[22,23] as well as from rat liver and Novikoff hepatoma cells.[24,25] By comparing the results for total antioxidant with the results of HPLC analyses for vitamin E on the same samples, it has been possible to conclude, without exception, that vitamin E (and α-tocopherol, in particular) is the major, if not the only chain-breaking (i.e., radical-trapping), lipid-soluble antioxidant present in these fluids and tissues.

Vitamin E

Although the term vitamin E is understood to include the four tocopherols (α-, β-, γ, δ-), it is commonly equated with α-tocopherol (α-T), the most biologically active[26] and the fully ring-methylated form of these four structurally related phenols.

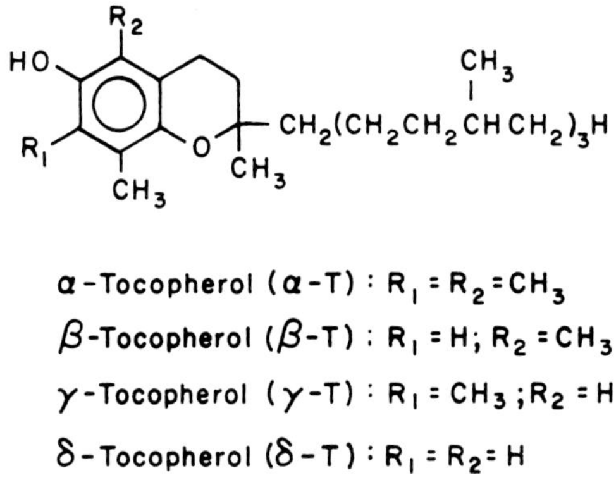

α-Tocopherol (α-T) : $R_1 = R_2 = CH_3$

β-Tocopherol (β-T) : $R_1 = H$; $R_2 = CH_3$

γ-Tocopherol (γ-T) : $R_1 = CH_3$; $R_2 = H$

δ-Tocopherol (δ-T) : $R_1 = R_2 = H$

VITAMIN E

Inspection of the phenolic portion of the α-T structure shows that it comes close to combining all the features that earlier research into simple phenolic antioxidants had indicated are desirable for enhanced antioxidant activity.[15] That is, the earlier work predicted that a para-methoxyphenol that is fully substituted with methyl groups in the ortho and meta positions would have the largest rate constant for reaction with a peroxyl radical (Reaction 13). Several years ago, the phenol substituted in this manner, 4-methoxy-2,3,5,6-tetramethylphenol (4-MTP), was synthesized and the rate constant for its reaction with peroxyl radicals was determined using the thermally initiated autoxidation method. Surprisingly, 4-MTP was found to be only approximately 10% as reactive as α-T.[16,17] (α-T was the most reactive phenol of any of the many phenols tested at that time; in particular, it is more than

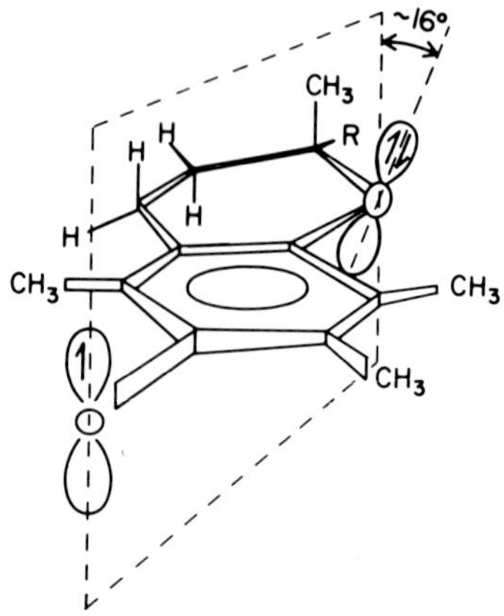

FIGURE 2. A pictorial representation of the operation of the stereoelectronic effect in the α-tocopheroxyl radical (R = phytyl). The diagram illustrates how the chroman ring maintains the lone pair of electrons in the ring oxygen *p*-orbital in an orientation which is almost perpendicular to the plane of the benzene ring. This facilitates interaction with the phenoxyl π-orbital containing the unpaired electron and hence stabilizes the phenoxyl radical. The structure is based on the X-ray structure determined for 2,2,5,7,8-pentamethyl-6-hydroxychroman (R = Me).[16]

200 times as reactive as the major phenolic antioxidant in commercial use, butylated hydroxytoluene, BHT.) The observation that 4-MTP is a relatively poor antioxidant has subsequently played a crucial part in the recognition and understanding of the importance of the heterocyclic ring of the chroman moiety of α-T in optimizing its reactivity toward peroxyl radicals.

As part of the investigation into the origin of the difference between the reactivities of 4-MTP and α-T, X-ray determinations were made of the structures of 4-MTP and a crystalline analog of α-T in which the phytyl tail was replaced by a methyl group. In 4-MTP it was found that the 4-methoxy group is forced by the steric influence of the adjacent methyl groups to adopt a conformation in which the O–CH_3 bond is rotated approximately 90° away from the preferred position (coplanar with the aromatic ring) observed in 3,5-unsubstituted 4-methoxyphenols.[16,17]

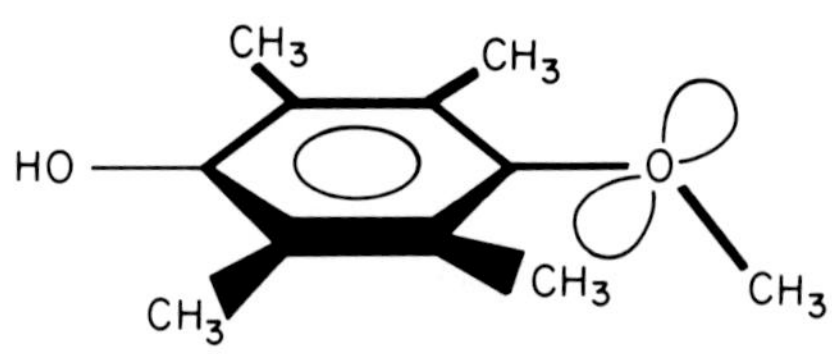

4 - MTP

In contrast to the conformational situation in 4-MTP, it was found that in the α-T analog the chroman heterocyclic ring counteracts the crowding effect of the substituents adjacent to the ring oxygen by holding the ring alkoxy group more or less in the preferred coplanar position. That is, the alkoxy group is oriented so that the *p*-type orbital of oxygen and its lone pair of electrons is nearly perpendicular to the aromatic ring. This conformational arrangement has been found to stabilize the phenoxyl radical intermediate because the *p*-orbital on the para-oxygen atom can interact with the semioccupied orbital that is delocalized from the phenoxyl oxygen through the π-orbitals of the benzene ring. This has the effect of increasing the delocalization of the unpaired electron (Figure 2), stabilizing the phenoxyl radical, and weakening the O–H bond in the parent phenol. A peroxyl radical can therefore more easily abstract the phenolic hydrogen atom (i.e., the rate constant for Reaction 13 will increase). This mechanism of "stereoelectronic enhancement" is not available in 4-MTP because the *p*-orbital on the methoxy oxygen is in the plane of the benzene ring and therefore it cannot stabilize the developing phenoxyl radical by interaction (overlap) with the adjacent, perpendicular *p*-orbitals of the benzene ring.

Combining this chemical evidence for the existence of enhanced antioxidant properties in α-T with the discovery that α-T is the major chain-breaking antioxidant found in the lipids of human plasma and red blood cells,[22,23] and rat liver[24,25] provides convincing proof of the importance of the role of vitamin E as an in vivo antioxidant. It has been demonstrated, in vitro, that α-T does indeed function as an excellent antioxidant when incorporated into the bilayers of model membrane systems (e.g., phospholipid liposomes).[27,28]

A question pertinent to the antioxidant role of vitamin E that could be asked is whether or not the reactivity of a peroxyl radical scavenger (with suitable membrane solubility) correlates with its vitamin E bioactivity. Evidence that it does come from a comparison of the rate constants for the peroxyl radical-trapping reactivities of the four tocopherols. The in vitro reactivities measured by the thermally initiated autoxidation technique are in the order $\alpha > \beta \approx \tau > \delta$ (consistent with the corresponding decrease in the number of aromatic methyl groups)[17] which closely parallels the order of their relative activities in several different bioassays ($\alpha > \beta > \tau > \delta$).[26]

Approaching this question from a different angle, it can be asked: could the peroxyl-radical trapping activity of α-T be improved upon, and if so, would it lead to a compound with greater vitamin E activity than α-T itself? In seeking to optimize antioxidant activity, as measured by the in vitro method, use has been made of the steroelectronic effect, described above, to introduce a subtle alteration into the heterocyclic ring. As indicated in Figure 2, the structure of the heterocyclic ring of the crystalline α-T analog implies that the 2*p*-orbital on the ring oxygen is rotated 16 to 17° away from the plane perpendicular to the aromatic ring. That is, it appears that the stabilization afforded by the steroelectronic effect is not quite optimal in α-T and could be improved upon. For example, if the heterocyclic ring were made flatter, the angle between the oxygen *p*-orbital and the perpendicular would decrease towards zero; this would improve the efficiency of the electronic stabilization and enhance the reactivity of the parent phenol toward peroxyl radicals.

Subsequently, it has been found that the rate of Reaction 13 can actually be increased by decreasing the heterocyclic ring size. Thus, in dihydrobenzofuran (DHBF), there is one less $-CH_2-$ unit in the heterocyclic ring which, as

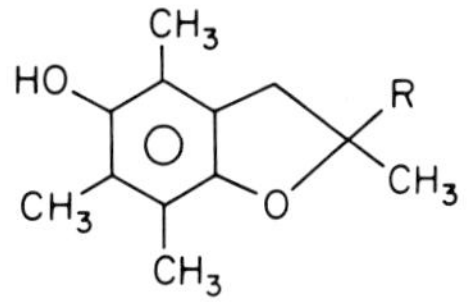

DHBF $[R = H, Me, CH_2(CH_2CH_2CH(CH_3)CH_2)_3H]$

X-ray structural analysis of a model compound has confirmed, causes a significant flattening of the ring, to the extent that the angle between the oxygen *p*-orbital and the plane perpendicular to the aromatic ring is decreased to about 6°.[18] The greater reactivity of the DHBF system toward peroxyl radicals has been confirmed by kinetic measurements on several model compounds, with substantial enhancements of 70 (R = H) to 80% (R = Me) relative to α-T being observed.[18,19]

An animal test of the bioactivity of a DHBF vitamin E analog with an *all-racemic* phytyl tail (R = $C_{16}H_{33}$) has been conducted[29] using the rat curative myopathy bioassay.[30] It was found that the acetate of the DHBF analog was 1.5 to 1.9 times more effective in reducing elevated levels of pyruvate kinase in the plasma of the vitamin E-deficient rats than was the *all-racemic* (i.e., synthetic) form of α-T acetate. This in vivo ratio is similar to the in vitro value of 1.5 obtained for the ratio of the rate constants for the reaction of the unesterified forms with peroxyl radicals (Reaction 13). The enhanced vitamin E activity of this analog together with its superior in vitro peroxyl radical-trapping activity provide additional support for the importance of the link between antioxidant activity (i.e., peroxyl radical-trapping reactivity) and vitamin E activity.

α-T has three chiral carbon atoms giving rise to eight possible stereoisomers. The natural stereoisomer, 2*R*,4′*R*,8′*R*-α-tocopherol (*RRR*-α-T),

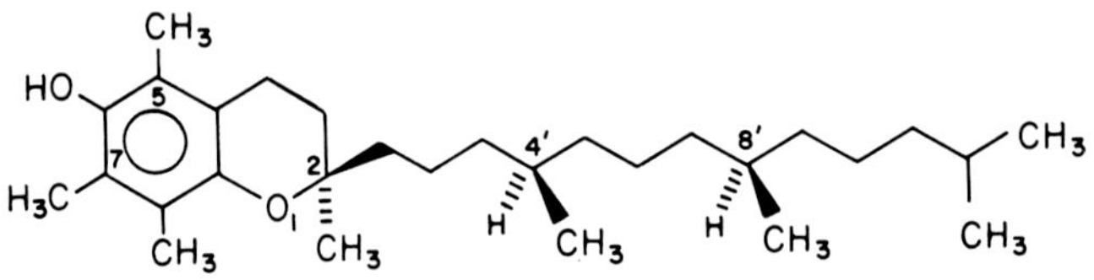

RRR – α –T

is the most active form of vitamin E, having, for example, about three times the bioactivity of the synthetic isomer with inverted stereochemistry at position 2, i.e., *SRR*-α-T.[31]

There cannot be any inherent differences between the stereoisomers in their reactivity toward peroxyl radicals, so the differences in biopotency must result from differences in interactions with their environment. The variation in biopotencies appears too small to be caused by a direct protein-mediated interaction. Instead, it has been proposed that membranes exert a chiral discrimination in favor of the natural isomer, causing more of it to be retained in the membrane.[32]

Support for this idea comes from an experiment conducted with rats fed for several months on a diet in which the sole source of vitamin E was a mixture of equal amounts of *SRR*-α-T and *RRR*-α-T labeled with one and two deuteriomethyl groups, respectively.[32] It was found that the *RRR*/*SRR* ratio in both blood plasma and red cells was always greater than 1 and that the discrimination in favor of the *RRR*-isomer was 1.35 times greater in the red cell than in the plasma lipids. Very recently, it has been shown in exchange experiments between red cells and plasma that the preference of the red cell for the *RRR*-isomer is largely a result of a higher rate of loss of the *SRR*-isomer from the red cell to the plasma lipid.[33]

Chiral discrimination may play an important role in the very high discrimination in favor of the natural isomer shown by rat brain (*RRR*/*SRR*) > 5 after 5 months).[32] It is possible that in traversing the blood-brain barrier α-T may have to diffuse through a series of membranes. This might explain the very slow rate of uptake of α-T into brain and spinal cord tissues.[32]

β-Carotene

There has been considerable recent interest in β-carotene, largely because of epidemiological studies which have indicated a lower incidence of certain types of cancer in groups of individuals with an above average dietary intake of this compound and other carotenoids.[34,35] It was suggested that β-carotene deactivates reactive chemical species, such as singlet oxygen and free radicals, that would otherwise initiate harmful reactions such as lipid peroxidation and, through this process, eventually induce cancer. Although β-carotene has been known for some time to be a very effective quencher of singlet oxygen,[9,36] only recently has it been shown to have a mildly effective, radical-trapping antioxidant capability.[6]

The manner in which β-carotene traps peroxyl radicals places it in a previously unknown class of biological antioxidants. In particular, it is only effective at partial pressures of oxygen which are significantly lower than the 150 torr of normal air. However, low partial pressures of oxygen are found in most tissues under normal physiological conditions (e.g., 15 torr in the capillaries of actively working muscles). The highly unsaturated nature of β-carotene makes it very reactive towards peroxyl radicals; attack probably occurs by addition to one of the unsaturated carbons, rather than by hydrogen abstraction, to give a resonance-stabilized, carbon-centered radical (Structure 6). The stability of this radical, which is much greater than for the

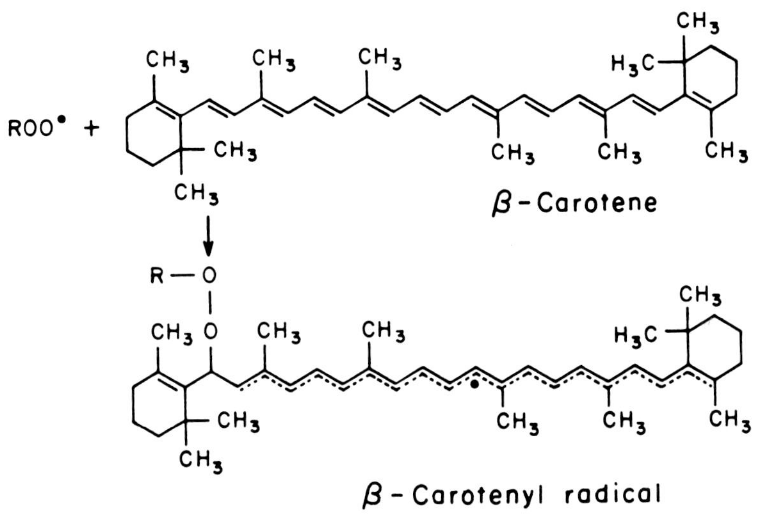

radical intermediate produced by peroxyl attack on polyunsaturated fatty acids, is such that it is able to react reversibly with oxygen at ambient temperatures (Reaction 17).

$$\beta\text{-carotenyl}\cdot + O_2 \rightleftharpoons \beta\text{-caroteneperoxyl radical} \tag{17}$$

At low partial pressures of oxygen, the equilibrium of Reaction 17 shifts in favor of the β-carotenyl radical; this radical acts as an efficient trap for a second peroxyl radical, giving a nonradical (i.e., relatively stable) product (Reaction 18).

$$\beta\text{-carotenyl}\cdot + ROO\cdot \rightarrow \text{molecular products} \tag{18}$$

At higher partial pressures of oxygen, the equilibrium shifts in favor of the β-caroteneperoxyl radical which, because it is a chain-carrying radical, causes β-carotene to behave as a pro-oxidant under these conditions.

The discovery of the radical-trapping antioxidant ability of β-carotene means that there are now, in fact, two known, naturally occurring, lipid-soluble antioxidants with, albeit, very different modes of action. The relative contributions of the two antioxidants in a given tissue will depend upon their concentrations and the partial pressure of oxygen. In human plasma, for example, there is significantly more vitamin E than β-carotene.[37] This implies that vitamin E is the major lipid-soluble, plasma antioxidant.

It should be noted that the mechanism by which β-carotene behaves as a chain-breaking antioxidant is not specific to β-carotene. In general, carotenoids and any other compounds with sufficiently extended systems of conjugated double bonds will also be capable of exhibiting antioxidant activity. As yet, there is little or no in vivo evidence to support this idea.

Water-Soluble Antioxidants

In the aqueous fluids between and within the cells of mammalian tissues, the compounds most frequently associated with antioxidant activity (apart from the preventive antioxidants discussed earlier) have been vitamin C (ascorbate), urate, and compounds with thiol groups (glutathione, in particular). Until

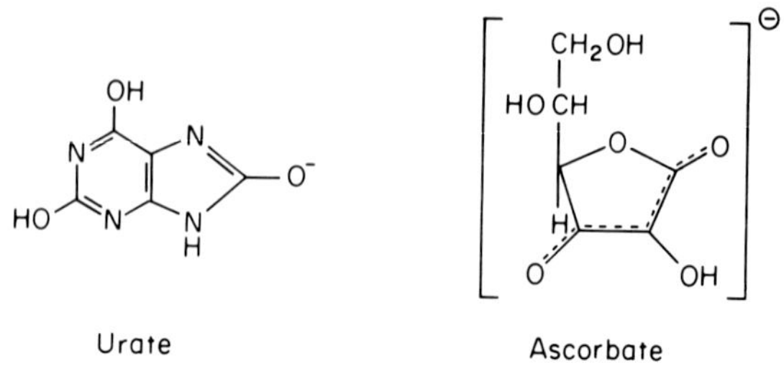

recently, it was far from certain how these compounds exerted their antioxidant effect. The delineation of their modes of action was complicated not only by the question of whether or not the compounds could act independently of one another, but also by the possibility that some of them, if not all of them, could "regenerate" partially oxidized vitamin E. However, the development of the thermally initiated autoxidation method and its extension to aqueous systems (made possible by the use of the water-soluble azo initiator, ABAP) have enabled considerable progress to be made in identifying and quantifying the major contributors to naturally occurring, chain-breaking antioxidant activity, as well as determining the means by which they achieve their action.

The readily water-soluble ABAP has proven to be useful because it generates peroxyl radicals in the aqueous phase at a convenient, known, and fixed rate throughout the duration of the experiment. This makes it possible to tell whether a suspected antioxidant is capable of directly trapping a peroxyl radical and, if so, the stoichiometry of the reaction (i.e., how many peroxyl radicals are trapped by each antioxidant molecule).

Experiments conducted with aqueous, multilamellar, dilinoleoylphosphatidyl choline[28,38] or soybean[39] liposomes using both lipid-soluble and water-soluble initiators to initiate autoxidation have allowed, for the first time, the determination of the abilities of chain-breaking antioxidants, present in either the aqueous or lipid phases, to trap radicals generated in either phase; it has also been possible to begin to examine the question of whether or not these compounds are individually capable of interacting with each other (e.g., the often-suggested synergism between vitamins E and C; see below).

The results obtained from these experiments provided the basis for the first experiments to identify and quantify the chain-breaking antioxidants present in human plasma. The major findings of this latter study are that plasma is very richly endowed with aqueous radical-trapping antioxidants (up to 1 m*M* of peroxyl radical-trapping capacity) and that the predominant contributors to this antioxidant activity are urate and plasma proteins, in that order (35 to 65 and 10 to 50%, respectively).[40,41] The other plasma antioxidants are vitamins C and E, but their contributions are small (0 to 24 and 5 to 10%, respectively) because their plasma concentrations are much smaller. A major portion of the antioxidant activity in the proteins has been attributed to sulfhydryl groups;[41] the origin of the remaining activity has not yet been determined.

The changes in concentrations of the individual antioxidants in plasma during aqueous peroxyl radical attack show that the protein sulfhydryl groups provide the first line of defense, with urate being spared in the initial phase.[41] Significant consumption of vitamin E (α-T) occurs only after most of the urate and protein sulfhydryl have disappeared. (Experimental difficulties prevented the study of the behavior of vitamin C in this experiment.) It is therefore concluded that the proteins and urate play a protective role toward vitamin E. Although the contribution of vitamin E to the overall antioxidant capacity is small in the quantitative sense, its importance lies in the fact that it is probably the last line of defense for the highly vulnerable, polyunsaturated fatty acids present in the plasma lipids.

Vitamin C

Under most physiological conditions, vitamin C (ascorbic acid, $pK_a = 4.25$) exists as its conjugate base, ascorbate (AH−). The autoxidation experiments conducted with multilamellar liposomes[38] have shown that although ascorbate efficiently traps aqueous peroxyl radicals (Reaction 19), it is a poor trap for lipid peroxyls in a phospholipid bilayer (Reaction 20).

$$(ROO\cdot)_{aq} + (AH^-)_{aq} \rightarrow ROOH_{aq} + (A^{\cdot-})_{aq} \tag{19}$$

$$(ROO\cdot)_{lipid} + (AH^-)_{aq} \xrightarrow{\times} \text{no reaction} \tag{20}$$

However, when ascorbate and vitamin E are both present, it is found that the induction periods are significantly longer than those produced by vitamin E alone, irrespective of whether the peroxyl radicals originate in the lipid or the aqueous phases. It is most likely that when the peroxyl radicals are generated in the aqueous phase, the ascorbate simply ''spares'' the vitamin E by directly trapping the radicals before they are able to attack the membrane lipid (i.e., a cooperative effect). However, when the peroxyls are generated within the lipid phase, this is not possible; ascorbate therefore probably regenerates the α-T [α-T(OH)] from the α-tocopheroxyl radical [α-T(O·); Reaction 21] formed by reaction of α-T(OH) with a lipid peroxyl radical (Reaction 13).

$$\alpha\text{-T(O}\cdot)_{lipid} + (AH^-)_{aq} \rightarrow \alpha\text{-T(OH)}_{lipid} + (A^{\cdot-})_{aq} \tag{21}$$

This is described as a synergistic effect and was first discovered to occur between vitamins C and E by Golumbic and Matill.[42]

The delocalized nature of the unpaired electron in the ascorbate radical makes the radical relatively unreactive toward other chemical compounds. The ascorbate radical is destroyed either by self-reaction (i.e., disproportionation; Reaction 22) to give ascorbate and dehydroascorbate (A), or by reaction with other free radicals, thus effectively terminating the incipient chain reaction in which it was formed.[43]

$$2A^{\cdot-} + H^+ \rightarrow AH^- + A \tag{22}$$

Although earlier in vitro experiments[44,45] showed directly that ascorbate does reduce the α-T radical to α-T, yielding the ascorbate radical, these experiments were conducted in homogeneous solution. The results of the experiments conducted with liposomes imply that the reaction between lipid-bound tocopheroxyl radical and aqueous ascorbate does indeed occur. This observation is consistent with the results of a ^{13}C NMR experiment using unilamellar phospholipid vesicles containing ^{13}C-labeled α-T; it was found that the phenolic ''head'' of α-T lies close to the interfacial surface of the membrane.[46]

Studies of the incremental effect of additional ascorbate upon the induction period produced by the antioxidants naturally present in human plasma have shown that as the concentration of ascorbate is increased, its efficiency as an antioxidant decreases.[47] That is, at low ascorbate concentrations, each ascorbate molecule is capable of trapping two peroxyl radicals, whereas at high concentrations the number of radicals trapped per ascorbate tends toward zero. This behavior is explained by the susceptibility of ascorbate to undergo peroxyl radical-catalyzed autoxidative destruction.[47] At higher concentrations of ascorbate, the losses to autoxidation are proportionately higher, leaving a smaller fraction of it available to trap radicals.

Urate

Urate is the conjugate base of uric acid ($pK_a = 3.89$). The concentration of urate in human plasma, which is much higher than its concentration in the plasma of other animals, is close to saturation.[48] It has been proposed for some time, with some experimental support, that urate is an antioxidant[37,48-52] and it has even been suggested that its high concentration in primate plasma, and particularly in human plasma, contributes to the longevity of primates and man.[37]

In experiments with multilamellar phosphatidylcholine liposomes[53] and human plasma samples,[40,41] it has been shown that urate is an excellent trap for peroxyl radicals generated in the aqueous phase with ABAP. Like ascorbate, urate is effectively unable to trap radicals generated in the lipid phase.[53] However, in striking contrast to ascorbate, urate does not produce a synergistic effect with vitamin E in aqueous methyl linoleate emulsions[53] nor in dilinoleoyl phosphatidylcholine liposomes,[41] implying that urate is unable to regenerate α-T from the α-tocopheroxyl radical.

Urate also possesses some preventive antioxidant activity. For example, it has been found to protect ascorbate against oxidation by cupric ion[54] and against iron-induced oxidation.[55] In the latter study, it was found that urate "stabilized" ascorbate in human serum without any apparent concomitant consumption of the urate. This behavior is consistent with metal ion chelation by the urate[56] and rules out the possibility that it regenerates ascorbate from the ascorbyl radical.

Compounds Containing Sulfhydryl Groups

Experiments with human plasma have shown very clearly that sulfhydryl groups on plasma proteins make a very important contribution to the total radical-trapping capacity of plasma.[41] Although model in vitro experiments have indicated that glutathione can regenerate vitamin E from the α-tocopheroxyl radical,[45] it is not at all certain that this result is relevant to plasma which normally contains little glutathione. It also seems likely that the sheer physical size of proteins would tend to inhibit a direct, facile, intermoleular interaction between protein sulfhydryls and lipoprotein-bound α-tocopheroxyl radicals. However, the fact remains that experiments in which glutathione is added to autoxidizing plasma show additional antioxidant protection which does not occur through direct trapping of peroxyl radicals.[41] It is not known at this time what form of synergistic mechanism is responsible for this behavior.

Superoxide Dismutase (SOD)

Under normal physiological conditions, the most common aqueous radical species probably is the superoxide anion, $O_2^{\bar{\cdot}}$.[57] This radical exists in a pH-governed equilibrium with its conjugate acid, the hydroperoxyl radical, $HO_2^{\cdot}$ ($pK_a = 4.9$),[58] the simplest of all peroxyl radicals. Although the concentration of the hydroperoxyl radical is normally at least two orders of magnitude lower than the superoxide anion, it is considerably the more reactive of these two species and it therefore represents a potential threat to the well-being of an organism. The nearly ubiquitous SOD enzymes deal with this threat (and others inherent in the presence of the superoxide anion) by catalyzing the disproportionation of superoxide (Reaction 23), thereby greatly reducing its concentration.[57]

$$2O_2^{\bar{\cdot}} + 2H^+ \rightarrow H_2O_2 + O_2 \tag{23}$$

Thus, by directly interfering with a chain-carrying, free radical species, SOD is classified as another form of chain-breaking antioxidant.

CONCLUSION

Living organisms contain a host of defense systems for handling the ever-present threat of autoxidation that is associated with the presence of free radicals. In unraveling the mechanisms of antioxidant action and interaction, it is important to determine whether the action is one of prevention or one of chain breaking. Preventive antioxidants are sometimes referred to as primary antioxidants because they remove the threat (e.g., a molecular precursor) prior to the initiation of free radical production. Chain-breaking antioxidants are then considered to be a secondary line of defense because they exert their effects at the radical propagation stage.

The aqueous fluids within and between cells contain a variety and abundance of both preventive and chain-breaking antioxidants, suggesting that much of the free radical threat originates in the aqueous phase. Preventive antioxidants include catalase, glutathione peroxidase, albumin, transferrin, ceruloplasmin, and possibly urate. Numbered among the aqueous chain-breaking antioxidants are urate, ascorbate, superoxide dismutase, and proteins and simple peptides with free thiol groups. In stark contrast to this, the lipid compartments of living systems contain only one or two antioxidants. These are the ubiquitous, chain-breaking vitamin E (α-T) and, probably of much less importance, β-carotene.

Progress is also being made in determining the scope and nature of synergistic interactions between the different antioxidants. The vitamin C-vitamin E interaction is now clearly established and there is evidence of an interaction of protein thiols with some, as yet, undetermined agent in human plasma. Future research will more clearly define the quantitative contributions of the individual antioxidants, the scope of synergistic interactions, and the integration of each antioxidant into the overall system of antioxidant defenses of living organisms.

REFERENCES

1. **Maillard, B., Ingold, K. U., and Scaiano, J. C.,** Rate constants for the reactions of free radicals with oxygen in solution, *J. Am. Chem. Soc.*, 105, 5095, 1983.
2. **Howard, J. A.,** Homogeneous liquid-phase autoxidations, in *Free Radicals*, Vol. 2, Kochi, J. K., Ed., John Wiley & Sons, New York, 1973, 3.
3. **Pryor, W. A.,** The role of free radical reactions in biological systems, in *Free Radicals in Biology*, Vol. 1, Pryor, W. A., Ed., Academic Press, New York, 1976, 1.
4. **Halliwell, B. and Gutteridge, J. M. C.,** *Free Radicals in Biology and Medicine*, Clarendon Press, Oxford, 1985.
5. **Ingold, K. U.,** Inhibition of the autoxidation of organic substances in the liquid phase, *Chem. Rev.*, 61, 563, 1961.
6. **Burton, G. W. and Ingold, K. U.,** β-Carotene: an unusual type of lipid antioxidant, *Science*, 224, 569, 1984.
7. **Burton, G. W., Foster, D. O., Perly, B., Slater, T. F., Smith, I. C. P., and Ingold, K. U.,** Biological antioxidants, *Philos. Trans. R. Soc. London Ser. B*, 311, 565, 1985.
8. **Halliwell, B. and Gutteridge, J. M. C.,** Oxygen free radicals and iron in relation to biology and medicine: some problems and concepts, *Arch. Biochem. Biophys.*, 246, 501, 1986.
9. **Foote, C. S. and Denny, R. W.,** Chemistry of singlet oxygen. VII. Quenching by β-carotene, *J. Am. Chem. Soc.*, 90, 6233, 1968.
10. **Ingold, K. U.,** The inhibition of autoxidation by phenols, *Chem. Soc. Spec. Publ.*, 24, 285, 1971.
11. **Mahoney, L. R.,** Antioxidants, *Angew. Chem. Int. Ed. Engl.*, 8, 547, 1969.
12. **Boozer, C. E., Hammond, G. S., Hamilton, C. E., and Sen, J. N.,** Air oxidation of hydrocarbons. II. The stoichiometry and fate of inhibitors in benzene and chlorobenzene, *J. Am. Chem. Soc.*, 77, 3233, 1955.

13. **Mahoney, L. R., Korcek, S., Hoffman, S., and Willermet, P. A.,** Determination of the antioxidant capacity of new and used lubricants; method and applications, *Ind. Eng. Chem. Prod. Res. Dev.,* 17, 250, 1978.
14. **Aristarkhova, S. A., Burlakova, E. B., and Khrapova, N. G.,** Study of the inhibiting activity of tocopherol, *Izv. Akad. Nauk. SSSR Ser. Khim.,* 2714, 1972.
15. **Howard, J. A. and Ingold, K. U.,** The inhibited autoxidation of styrene. I. The deuterium isotope effect for inhibition by 2,6-di-*tert*-butyl-4-methylphenol, *Can. J. Chem.,* 40, 1851, 1962; II. The relative inhibiting efficiencies of meta- and para-substituted phenols, *Can. J. Chem.,* 41, 1744, 1963; III. The relative inhibiting efficiencies of ortho-alkyl phenols, *Can. J. Chem.,* 41, 2800, 1963; IV. Solvent effects, *Can. J. Chem.,* 42, 1044, 1964.
16. **Burton, G. W., Le Page, Y., Gabe, E. J., and Ingold, K. U.,** Antioxidant activity of vitamin E and related phenols. Importance of stereoelectronic factors, *J. Am. Chem. Soc.,* 102, 7791, 1980.
17. **Burton, G. W. and Ingold, K. U.,** Autoxidation of biological molecules. I. The antioxidant activity of vitamin E and related chain-breaking phenolic antioxidants in vitro, *J. Am. Chem. Soc.,* 103, 6472, 1981.
18. **Burton, G. W., Doba, T., Gabe, E. J., Hughes, L., Lee, F. L., Prasad, L., and Ingold, K. U.,** Autoxidation of biological molecules. IV. Maximizing the antioxidant activity of phenols, *J. Am. Chem. Soc.,* 107, 7053, 1985.
19. **Burton, G. W., Hughes, L., and Ingold, K. U.,** Antioxidant activity of phenols related to vitamin E. Are there chain-breaking antioxidants better than α-tocopherol?, *J. Am. Chem. Soc.,* 105, 5950, 1983.
20. **Mahoney, L. R. and DaRooge, M. A.,** The kinetic behavior and thermochemical properties of phenoxyl radicals, *J. Am. Chem. Soc.,* 97, 4722, 1975.
21. **Nikolaev, A. I., Safiullin, R. L., Komissarov, V. D., and Denisov, E. T.,** Kinetics of reaction of alkylperoxy radicals with 2,4,6-tri-tert-butylphenoxyl, *Khim. Fiz.,* 5, 380, 1986.
22. **Burton, G. W., Joyce, A., and Ingold, K. U.,** First proof that vitamin E is a major lipid-soluble, chain-breaking antioxidant in human blood plasma, *Lancet,* ii, 327, 1982.
23. **Burton, G. W., Joyce, A., and Ingold, K. U.,** Is vitamin E the only lipid-soluble, chain-breaking antioxidant in human blood plasma and erthrocyte membranes?, *Arch. Biochem. Biophys.,* 221, 281, 1983.
24. **Cheeseman, K. H., Burton, G. W., Ingold, K. U., and Slater, T. F.,** Lipid peroxidation and lipid antioxidants in normal and tumor cells, *Toxicol. Pathol.,* 12, 235, 1984.
25. **Cheeseman, K. H., Collins, M., Proudfoot, K., Slater, T. F., Burton, G. W., Webb, A. C., and Ingold, K. U.,** Studies on lipid peroxidation in normal and tumour tissues. The Novikoff rat liver tumour, *Biochem. J.,* 235, 507, 1986.
26. **Century, B. and Horwitt, M. K.,** Biological availability of various forms of vitamin E with respect to different indices of deficiency, *Fed. Proc. Fed. Am. Soc. Exp. Biol.,* 24, 906, 1965.
27. **Barclay, L. R. C. and Ingold, K. U.,** Autoxidation of biological molecules. II. The autoxidation of a model membrane. A comparison of the autoxidation of egg lecithin phosphatidylcholine in water and in chlorobenzene, *J. Am. Chem. Soc.,* 103, 6478, 1981.
28. **Barclay, L. R. C., Locke, S. J., MacNeil, J. M., van Kessel, J., Burton, G. W., and Ingold, K. U.,** Autoxidation of micelles and model membranes. Quantitative kinetic measurements can be made by using either water-soluble or lipid-soluble initiators with water-soluble or lipid-soluble chain-breaking antioxidants, *J. Am. Chem. Soc.,* 106, 2479, 1984.
29. **Ingold, K. U., Burton, G. W., Foster, D. O., Zuker, M., Hughes, L., Lacelle, S., Lusztyk, E., and Slaby, M.,** A new vitamin E analogue more active than α-tocopherol in the rat curative myopathy bioassay, *FEBS Lett.,* 205, 117, 1986.
30. **Machlin, L. J., Gabriel, E., and Brin, M.,** Biopotency of α-tocopherols as determined by curative myopathy bioassay in the rat, *J. Nutr.,* 112, 1437, 1982.
31. **Wieser, H. and Vecchi, M.,** Stereoisomers of α-tocopherol acetate. II. Biopotencies of all eight stereo-isomers, individually or in mixtures, as determined by rat resorption-gestation tests, *Int. J. Vitam. Nutr. Res.,* 52, 351, 1982.
32. **Ingold, K. U., Burton, G. W., Foster D. O., Hughes, L., Lindsay, D. A., and Webb, A.,** Biokinetics of and discrimination between dietary *RRR*- and *SRR*-α-tocopherols in the male rat, *Lipids,* 22, 163, 1987.
33. **Cheng, S. C., Burton, G. W., Ingold, K. U., and Foster, D. O.,** Chiral discrimination in the exchange of α-tocopherol stereo-isomers between plasma and red blood cells, *Lipids,* 22, 469, 1987.
34. **Peto, R., Doll, R., Buckley, J. D., and Sporn, M. B.,** Can dietary beta-carotene materially reduce human cancer rates?, *Nature (London),* 290, 201, 1980.
35. **Shekelle, R. B., Liu, S., Raynor, J., Jr. Lepper, M., Maliza, C., Rosoff, A. H., Paul, O., Shryock, A. M., and Stamler, J.,** Dietary vitamin A and risk of cancer in the Western Electric study, *Lancet.* ii, 1185, 1981.
36. **Foote, C. S.,** Photosensitized oxidation and singlet oxidation: consequences in biological systems, in *Free Radicals in Biology,* Vol. 2, Pryor, W. A., Ed., Academic Press, New York, 1976, 85.

37. **Ames, B. N., Cathcart, R., Schwiers, E., and Hochstein, P.,** Uric acid: an antioxidant defense in humans against oxidant- and radical-caused aging and cancer. A hypothesis, *Proc. Natl. Acad. Sci. U.S.A.*, 78, 6858, 1981.
38. **Doba, T., Burton, G. W., and Ingold, K. U.,** Antioxidant and co-antioxidant activity of vitamin C. The effect of vitamin C, either alone of in the presence of vitamin E or a water-soluble vitamin E analogue, upon the peroxidation of aqueous multilamellar phospholipid liposomes, *Biochim. Biophys. Acta.* 835, 298, 1985.
39. **Niki, E., Kawakami, A., Yamamoto, Y., and Kamiya, Y.,** Oxidation of lipids. VIII. Synergistic inhibition of oxidation of phosphatidylcholine liposome in aqueous dispersion by vitamin E and vitamin C, *Bull. Chem. Soc. Jpn.*, 58, 1971, 1985.
40. **Wayner, D. D. M., Burton, G. W., Ingold, K. U., and Locke, S.,** Quantitative measurement of the total, peroxyl radical-trapping antioxidant capability of human blood plasma by controlled peroxidation. The important contribution made by plasma proteins, *FEBS Lett.*, 187, 33, 1985.
41. **Wayner, D. D. M., Burton, G. W., Ingold, K. U., Barclay, L. R. C., and Locke, S. J.,** The relative contributions of vitamin E, urate, ascorbate and proteins to the total radical-trapping antioxidant activity of human blood plasma, *Biochim. Biophys. Acta.*, 924, 408, 1987.
42. **Golumbic, C. and Matill, H. A.,** Antioxidants and the autoxidation of fats. XIII. The antioxygenic action of ascorbic acid in association with tocopherols, hydroquinones and related compounds, *J. Am. Chem. Soc.*, 63, 1279, 1941.
43. **Bielski, B. H.,** Chemistry of ascorbic acid radicals, in *Ascorbic Acid: Chemistry, Metabolism, and Uses,* Seib, P. A. and Tolbert, B. M., Eds., Advances in Chemistry Series 200, American Chemical Society, Washington, D. C., 1982, 81.
44. **Packer, J. E., Slater, T. F., and Willson, R. L.,** Direct observation of a free radical interaction between vitamin E and vitamin C, *Nature London,* 278, 737, 1979.
45. **Niki, E., Tsuchiya, J., Tanimura, R., and Kamiya, Y.,** Regeneration of vitamin E from α-chromanoxyl radical by glutathione and vitamin C, *Chem. Lett.*, p. 789, 1982.
46. **Perly, B., Smith, I. C. P., Hughes, L., Burton, G. W., and Ingold, K. U.,** Estimation of the location of natural α-tocopherol in lipid bilayers by ^{13}C-NMR spectroscopy, *Biochim. Biophys. Acta,* 819, 131, 1985.
47. **Wayner, D. D. M., Burton, G. W., and Ingold, K. U.,** The antioxidant efficiency of vitamin C is concentration-dependent, *Biochim. Biophys. Acta,* 884, 119, 1986.
48. **Cutler, R. G.,** Urate and ascorbate: their possible roles as antioxidants in determining longevity of mammalian species, *Arch. Gerontol. Geriatr.*, 3, 321, 1984.
49. **Howell, R. R. and Wyngaarden, J. B.,** On the mechanism of peroxidation of uric acids by hemoproteins, *J. Biol. Chem.*, 235, 3544, 1960.
50. **Matsushita, S., Ibuki, F., and Aoki, A.,** Chemical reactivity of the nucleic acid bases. I. Antioxidative ability of the nucleic acids and their related substances on the oxidation of unsaturated fatty acids, *Arch. Biochem. Biophys.*, 102, 446, 1963.
51. **Proctor, P.,** Similar functions of uric acid and ascorbate in man?, *Nature (London),* 228, 868, 1970.
52. **Smith, R. C. and Lawing, L.,** Antioxidant activity of uric acid and 3-*N*-ribosyluric acid with unsaturated fatty acids and erythrocyte membranes, *Arch. Biochem. Biophys.*, 223, 166, 1983.
53. **Niki, E., Saito, M., Yoshikawa, Y., Yamamoto, Y., and Kamiya, Y.,** Oxidation of lipids. XII. Inhibition of oxidation of soybean phosphatidylcholine and methyl linoleate in aqueous dispersions by uric acid, *Bull. Chem. Soc. Jpn.*, 59, 471, 1986.
54. **Lam, K.-W., Fong, D., Lee, A., and Liu, K. M. D.,** Inhibition of ascorbate oxidation by urate, *J. Inorg. Biochem.*, 22, 241, 1984.
55. **Sevanian, A., Davies, K. J. A., and Hochstein, P.,** Convservation of vitamin C by uric acid in blood, *J. Free Radical Biol. Med.*, 1, 117, 1985.
56. **Davies, K. J. A., Sevanian, A., Muakkassah-Kelly, S. F., and Hochstein, P.,** Uric acid-iron ion complexes. A new aspect of the antioxidant functions of uric acid, *Biochem. J.*, 235, 747, 1986.
57. **Fridovich, I.,** Oxygen radicals, hydrogen peroxide, and oxygen toxicity, in *Free Radicals in Biology,* Vol. 1, Pryor, W. A., Ed., Academic Press, New York, 1976, 239.
58. **Behar, D., Czapski, G., Rabani, J., Dorfman, L. M., and Schwarz, H. A.,** The acid dissociation constant and decay kinetics of the perhydroxyl radical, *J. Phys. Chem.*, 74, 3209, 1970.

THE ROLE OF COENZYME Q IN FREE RADICAL FORMATION AND ANTIOXIDATION

Robert E. Beyer

INTRODUCTION

Coenzyme Q is an exceedingly versatile molecule functionally involved in a number of distinct, but related, cellular processes. Coenzyme Q, the common name for 2,3-dimethoxy-5-methyl-6-multipenyl-1,4-benzoquinone (Figure 1), has also been called variously Q_{275}, mitoquinone, and ubiquinone. It was first discovered as an obligatory electron transport chain component in the mitochondrial succinoxidase system by Crane et al.[1] in 1957 and was later shown to be similarly involved in the mitochondrial NADH oxidase pathway by Szarkowska.[2] These conclusions were substantiated by Norling et al.[3] in experiments involving a quantitative reincorporation of coenzyme Q-depleted submitochondrial particles. Coenzyme Q thus provides a mobile link allowing reversible interactions between the NADH dehydrogenase, succinate dehydrogenase, and cytochrome b-c_1 portions of the electron transport chain[4-6] (Figure 2). A number of other dehydrogenases donate reducing equivalents to the cytochrome system via coenzyme Q, two examples of which are α-glycerophosphate dehydrogenase and ETF dehydrogenase. In the reduced state, coenzyme Q exerts an influence on the rate of energy liberation from the electron transport chain by regulation of succinate dehydrogenase,[7-10] NADH dehydrogenase,[11-12] and cytochrome b-c_1 activities.[13-15] Coenzyme Q also appears to be involved in the establishment of an inner mitochondrial transmembrane protonmotive gradient, the "protonmotive Q cycle" (Figure 3), resulting in the conservation of energy released by oxidoreduction reactions of the electron transport chain.[16-18] In addition, coenzyme Q has been implicated in both the production of damaging free radicals and as an antioxidant protecting the components of the membrane in which it resides from peroxidative damage. This review will examine the evidence pertaining to the site or sites of free radical production in the mitochondrial electron transport chain and the role of coenzyme Q in this process and will then evaluate the experimental bases in support of, and consistent with, an antioxidant role for coenzyme Q. This subject has been reviewed recently.[19,20]

LIPID PEROXIDATION AND FREE RADICAL FORMATION BY THE MITOCHONDRIAL ELECTRON TRANSFER CHAIN

Free radicals are produced during mitochondrial electron transfer chain activity and the rate of superoxide radical formation is directly proportional to the rate of mitochondrial oxygen utilization.[21] The generation of strong oxidizing agents from $O_2^{\overline{\cdot}}$,[22] which in the presence of iron in several forms promote and propogate lipid peroxidation, results in altered membrane function and thus poses a constant threat to cellular homeostasis. These topics have been the subject of extensive review[23-25] and an opposing view has been published.[26]

Studies of mitochondrial swelling have shown that two types of such volume changes occur — those which are reversible[27] and those which are irreversible. Reversible changes may be initiated by agents such as thyroxine and phosphate[28] and reversed by ATP and the radical scavenger manganese ion.[27] Irreversible swelling may be initiated by pro-oxidants such as glutathione,[29] ascorbate, and iron and is inhibited by the antioxidant α-tocopherol.[30] Pro-oxidant-induced swelling has been shown to be associated with the production of thiobarbituric acid reactive compounds,[30] thus implicating the occurrence of lipid peroxidation. Although an involvement of active oxygen in this process was suspected, no assay for the superoxide radical existed before 1969. The discovery by McCord and Fridovich[31] of the

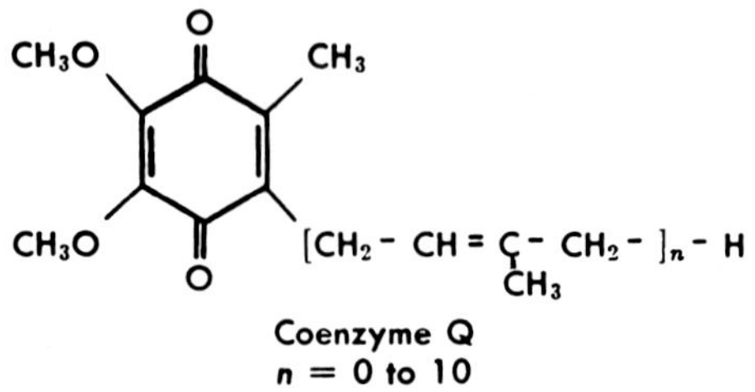

FIGURE 1. The structure of coenzyme Q.

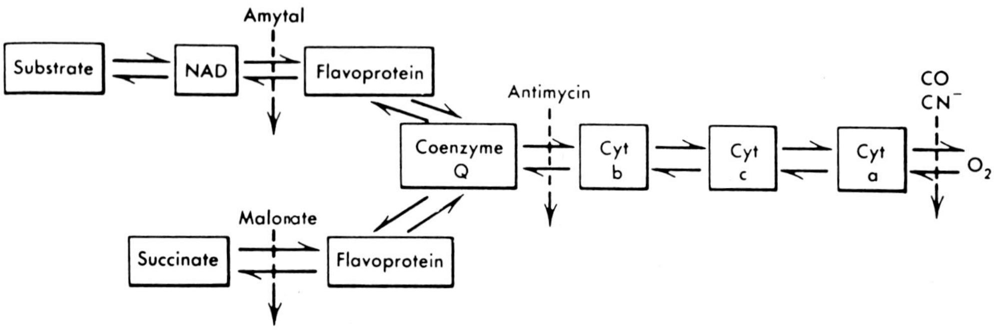

FIGURE 2. Coenzyme Q in the mitochondrial electron transfer chain.

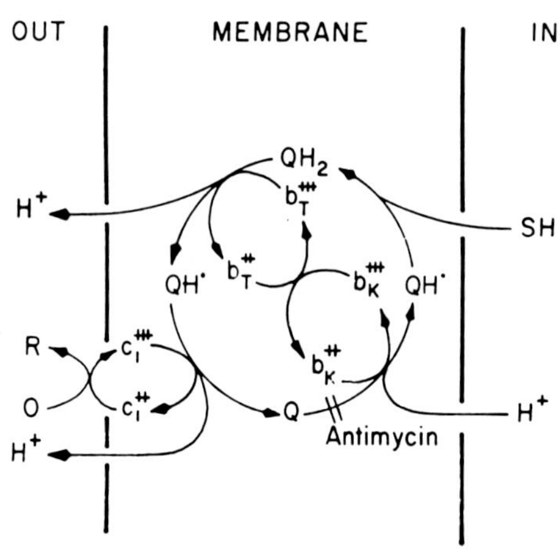

FIGURE 3. Electron transfer and protonmotive reactions of coenzyme Q and the cytochrome *b-c* complex. (From Beyer, R. E. and Starnes, J. W., *Pathobiology of Cardiovascular Injury,* Stone, H. L. and Weglikci, W. B., Eds., Martinus Nijhoff, Boston, 1985, 489. With permission.)

enzyme superoxide dismutase (SOD), catalyzing the removal of superoxide by a dismutase reaction,

$$2O_2^{\overline{\cdot}} + 2H^+ \xrightarrow{\text{SOD}} H_2O_2 + O_2 \qquad (1)$$

allowed a direct connection to be established. Loschen et al.[32] and Boveris and Cadenas[33] were the first to demonstrate the production of $O_2^{\overline{\cdot}}$ by the electron transport chain by comparison of H_2O_2 formation in the presence and absence of SOD. Table 1 shows the rates

Table 1
H_2O_2 AND $O_2^{\bar{\cdot}}$ PRODUCTION CATALYZED BY SUBMITOCHONDRIAL PARTICLES (SMP)[33]

Source of SMPs	Cytochrome *c* reduction (nmol/min/mg protein)	Sensitivity to SOD (%)	$O_2^{\bar{\cdot}}$ (nmol/min/mg protein)	H_2O_2 (nmol/min/mg protein)
Beef heart	6.3	70	4.4	2.5
Rat heart	8.0	62	4.9	3.2
Rat heart	12.8	54	6.9	3.2

of H_2O_2 and superoxide production by submitochondrial particles in the presence of succinate and a concentration of antimycin sufficient to block completely the electron transfer chain. Boveris and Chance[21] had previously observed a high rate of H_2O_2 production by pigeon heart mitochondria respiring with malate + glutamate and succinate + malate as substrates. The rate of H_2O_2 production was elevated by high oxygen pressure or by the addition of antimycin and uncouplers of oxidative phosphorylation.[21] The role of the reduced state of oxidation-reduction components of the electron transfer chain in superoxide formation was shown[34] by the demonstration that mitochondrial H_2O_2 production occurs during state 4 respiration and ceases upon transition to state 3. These results, together with experiments with electron transfer chain inhibitors, narrowed the site(s) of superoxide production to reduced members of the respiratory chain on the substrate side of the antimycin block, i.e., flavoprotein, nonheme iron-sulfur protein, coenzyme Q, and cytochrome *b*. Trumpower and Simmons[35] confirmed the localization of a superoxide-generating site in the span succinate dehydrogenase to cytochrome *b* by showing that the radical could be produced by purified complexes II and III (succinate-cytochrome *c* reductase). They postulated[35] a mechanism in which the semiquinone form of coenzyme Q gave rise to the superoxide radical. Such a role for coenzyme Q in the production of free radicals has been discussed by Boveris and Cadenas[36] and Ernster and Nelson.[37] Boveris and Cadenas[36] and Forman and Boveris[24] have suggested that the semiquinone form of coenzyme Q, i.e., $CoQ^{\bar{\cdot}}$ or CoQH·, is the major mitochondrial autoxidizable component yielding the superoxide radical. The presence of the semiquinone form of coenzyme Q had first been demonstrated[38] by measuring the semiquinone electron spin resonance signal of coenzyme Q-depleted and -replenished submitochondrial particles. These observations have been confirmed.[39]

Coenzyme Q has also been implicated in the etiology of free radical pathology following ischemic episodes[40] and in superoxide anion production during the respiratory burst of human neutrophils by a nonmitochondrial NADH oxidase system.[41] However, it should be noted that no *direct* evidence has been presented in the studies cited above that coenzyme Q, in any form, is responsible for superoxide formation. These studies on mitochondria and submitochondrial particles indicate merely that a site of superoxide production is located between succinate or NADH and cytochrome *b*, but do not discriminate between the several electron-transferring components involved. The report of coenzyme Q involvement in free radical-induced pathology following ischemic episodes[40] is also vague as to evidence that coenzyme Q is responsible. Human neutrophils, which are deficient in cytochromes *a* and *c*, possess a flavin-containing NADPH oxidase, a nonmitochondrial *b* type cytochrome, and coenzyme Q_{10}.[42] Based on the stimulation of both oxygen consumption and superoxide radical production by short isoprenoid chain coenzyme Q analogues and inhibition of both events by quinone antagonists, it has been concluded[42,43] that the endogenous coenzyme Q is necessary for oxygen metabolism and microbicidal events in neutrophils. However, the data,[41-43] although strongly suggestive with respect to a role for coenzyme Q in superoxide radical and H_2O_2 production in neutrophils, do not exclude a role for other components such

Table 2
SUPEROXIDE PRODUCTION BY ACETONE-EXTRACTED AND CONTROL MITOCHONDRIAL MEMBRANES[50]

		Activity (nmol/min/mg protein)		
Assay[a]	SOD	SMPs	CoQ-reconstituted	Percent
$O_2^{\overline{\cdot}}$	–	3.89	0.54	14
	+	0.09	0.06	—
H_2O_2	–	1.86	0.04	—
	+	3.72	0.17	5

[a] Succinate and antimycin present.

as cytochrome *b*, flavin semiquinone, or nonheme iron-sulfur proteins which may also participate in neutrophil electron transfer reactions and superoxide production. This possibility should receive serious consideration as iron-sulfur flavoproteins,[44] flavoproteins,[45] and nonheme iron-sulfur proteins,[46] in addition to quinols,[47,48] have all been shown to have the capacity to generate superoxide radicals. The role of coenzyme Q in the generation of microbicidal radicals by this system has been rendered dubious by the demonstration that coenzyme Q_{10} acts as an effective quencher of singlet oxygen generated by human leukocytes.[49]

More convincing evidence supporting the notion of superoxide production and ensuing hydrogen peroxide formation from mitochondrial coenzyme Q has been presented by Boveris et al.,[50] who report that extraction of submitochondrial particles with acetone, which removes coenzyme Q and a variety of other lipids as well as cytochrome *c*, decreased the rate of superoxide production with succinate as substrate. Reconstitution of extracted membranes with phospholipid and coenzyme Q resulted in a partial restoration of superoxide and hydrogen peroxide production (Table 2), although the degree of restoration was marginal, 14 and 5% for production of $O_2^{\overline{\cdot}}$ and H_2O_2, respectively. These authors also report[50] a linear relationship between reducible coenzyme Q reincorporated into the membranes and the rate of peroxide production in the presence of SOD providing additional evidence for a role of functional coenzyme Q in superoxide production. Boveris et al.[50] suggest that a reduced form of coenzyme Q, either the quinol or semiquinone, functions as a univalent reductant of oxygen, namely,

$$COQH_2 + O_2 \longrightarrow COQH\cdot + H^+ + O_2^{\overline{\cdot}} \quad (2)$$

$$COQH\cdot + O_2 \longrightarrow COQ + H^+ + O_2^{\overline{\cdot}} \quad (3)$$

Despite this strong evidence for an involvement of coenzyme Q in radical production, the reestablishment of electron flow capabilities by the reincorporation of coenzyme Q into the electron transfer chain also involves participation of all of the electron transfer chain components between succinate and the antimycin block. These experiments leave open the question of their participation in oxy radical formation. That Boveris et al. are cognizant of this possibility is evident from their concluding remark:[50] "The present experimental data point to ubisemiquinone and ubiquinol as the main sources of H_2O_2 in mitochondrial inner membrane, but the function of other H_2O_2 generators is not excluded."

Table 3
EFFECT OF RESPIRATORY CHAIN INHIBITORS ON MDA FORMATION BY SMPs[63]

Additions	MDA formation (nmol/min/mg protein) NADH	NADPH
None	2.54	5.10
n-Ethylmaleimide	5.41	5.15
Rotenone	7.11	5.30
Antimycin	0.25	4.54
KCN	0.38	5.30
Rotenone + antimycin	6.12	4.44
Rotenone + KCN	6.86	5.00

Recent evidence suggests that distinct coenzyme Q binding proteins, or apoproteins, in complexes I, II, and III, exhibiting three types of coenzyme Q semiquinone radical signals[51,52] and sensitive to antimycin, thenoyltrifluoroacetone, and rotenone,[52] exist. The existence of such proteins, originally suggested by Green[53] in 1959 and affirmed by Mitchell[54] in 1976, was detected as coenzyme Q semiquinone radicals in purified coenzyme Q-cytochrome *c* reductase[55,56] from beef heart in 1977 and has been identified recently in yeast mitochondrial coenzyme Q-cytochrome *c* reductase.[57] The presence of such coenzyme Q apoproteins appears to stabilize the coenzyme Q semiquinone form[58] and thus renders it less reactive with O_2.

Most convincing evidence for the site of free radical formation in the electron transfer chain has been published by Takeshige and colleagues. They have shown[59] that beef heart submitochondrial particles catalyze a NADH- and NADPH-dependent lipid peroxidation in the presence of ferric ions and ADP and that the site of NADH- and NADPH-dependent production of $O_2^{\cdot-}$ by the respiratory chain is located within the NADH dehydrogenase complex between a mercurial-sensitive and a rotenone-sensitive site,[60] most likely a nonheme iron-sulfur function. The localization of radical formation resulting in lipid peroxidation, measured as malondialdehyde (MDA) formation, in beef heart submitochondrial particles is shown in Table 3. These data indicate that NADH-dependent lipid peroxidation is increased two- to three-fold by *n*-ethylmaleimide and rotenone and inhibited by antimycin and KCN. A direct effect of KCN and antimycin on lipid peroxidation was obviated by their simultaneous addition with rotenone. Turrens and Boveris[61] have confirmed the effect of rotenone on the rate of superoxide production. These data demonstrate that a respiratory chain component between the substrate site and the rotenone-sensitive site of NADH dehydrogenase is required for the reactions leading to lipid peroxidation and that the redox state of the components between the sites inhibited by rotenone and antimycin, i.e., coenzyme Q and cytochrome *b*, strongly influences lipid peroxidation. The absence of inhibition of MDA formation during NADPH oxidation has been explained by the maintenance of the oxidized state of electron transport chain components resulting from the relatively low rate of electron flow from NADPH to NAD^+ via transhydrogenase.[62] Takeshige et al.[63] have shown that during NADPH oxidation catalyzed by submitochondrial particles, 5% of the coenzyme Q is reduced in the presence of antimycin, while during NADH oxidation about 80% of the coenzyme Q is reduced. These observations, together with the demonstration that submitochondrial particles from which coenzyme Q has been extracted catalyze the formation of superoxide radicals at a rate equivalent to that of unextracted particles,[63] render difficult the conclusion that coenzyme Q is involved in superoxide radical formation.

Table 4
INHIBITION OF LIPID PEROXIDATION IN ARACHIDONIC ACID EMULSIONS[66]

Assay	O_2 consumption (nmol/min)	Inhibition (%)
Initial control	58	—
CoQ_6, 1 m*M*	36	38
CoQ_6H_2, 1 m*M*	14	76
α-Tocopherol, 1 m*M*	17	71
Final control	58	—

ANTIOXIDANT PROPERTIES OF COENZYME Q

Evidence consistent with an antioxidant role for coenzyme Q has been derived from experiments on reconstituted membrane systems, submitochondrial particles, and mitochondria to observations on intact animals, as well as in the clinical setting.

In Subcellular Systems

Evidence of a direct nature in support of an antioxidant role for coenzyme Q has come from studies of subcellular systems. The earliest indications that coenzyme Q may function as an antioxidant were by Lea and Kwietny[64] and Mellors and Tappel,[65] who showed that both the quinone and the quinol forms had about 20% of the antioxidant activity of α-tocopherol against lipid peroxidation. However, the methods used to assay antioxidant activity were performed under high oxygen concentrations which favor the conversion of quinols to quinones. Mellors and Tappel[66] then tested coenzyme Q_6 in both the oxidized and reduced forms as antioxidants in the presence of lower concentrations of oxygen more comparable to oxygen availability in the cell. They showed that reduced coenzyme Q_6 is about four times more effective than the oxidized form in reducing the stable free radical diphenyl-*p*-picrylhydrazyl and is almost as reactive in this respect as α-tocopherol. They also showed that CoQ_6H_2 is considerably more potent than the oxidized form as an inhibitor of lipid peroxidation of arachidonic acid emulsions catalyzed by hemoglobin and that the quinol is as effective as α-tocopherol. These data are shown in Table 4.

Mellors and Tappel[66] also tested the antioxidant effects of CoQ_6 on light-catalyzed peroxidation of mitochondrial lipids. These results are presented in Figure 4 which indicates that at 2 hr of light treatment, CoQ_6 inhibited lipid peroxidation by 76%, CoQ_6H_2 by 95%, and α-tocopherol by 100%. As both coenzyme Q and α-tocopherol reside in the hydrophobic milieu of the inner mitochondrial membrane, both would contribute to molecular homeostasis by protecting against oxidative damage. Mellors and Tappel[66] were also pioneers in showing the effect of coenzyme Q, reduced by the electron transfer chain, on lipid peroxidation. Figure 5 shows the effect of β-hydroxybutyrate, in the presence of antimycin, on the ability of exogenous coenzyme Q_6 to inhibit light-induced lipid peroxidation in rat liver mitochondria. Lipid peroxidation was most inhibited when both β-hydroxybutyrate and coenzyme Q_6 were present, suggestive of a strong antioxidant effect of reduced coenzyme Q. Oxidized coenzyme Q was not as effective as the reduced quinone.

Takeshige et al.[63] extended these observations to the submitochondrial particle level. They studied lipid peroxidation in particles from which coenzyme Q had been almost totally extracted and reincorporated to varying, known degrees using methods developed by Ernster and colleagues.[3,67] Figure 6 shows the effect of different contents of coenzyme Q_{10} on lipid peroxidation in submitochondrial particles oxidizing succinate. Reincorporation of coenzyme Q into extracted submitochondrial particles resulted in inhibition of lipid peroxidation and

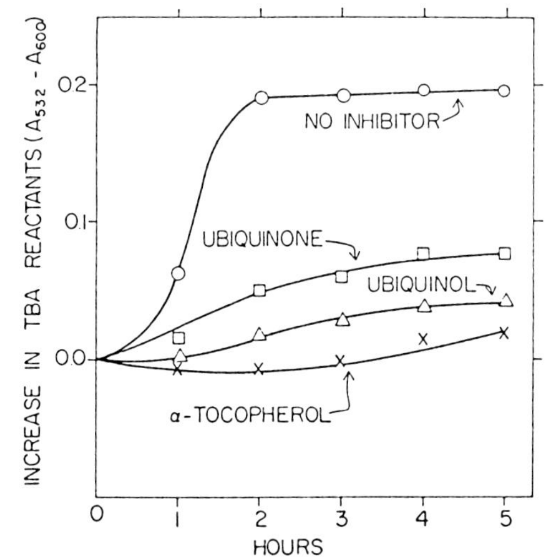

FIGURE 4. Inhibition of mitochondrial photo-oxidation by exogenous coenzyme Q_6 and coenzyme Q_6H_2. Each vial contained 20 mℓ of a suspension of bovine liver mitochondria (2 mg of protein per milliliter of KCl-Tris). α-Tocopherol and oxidized and reduced coenzyme Q_6 were added to 1 m*M;* TBA, thiobarbituric acid. (From Mellors, A. and Tappel, A. L., *J. Biol. Chem.*, 241, 4353, 1966. With permission.)

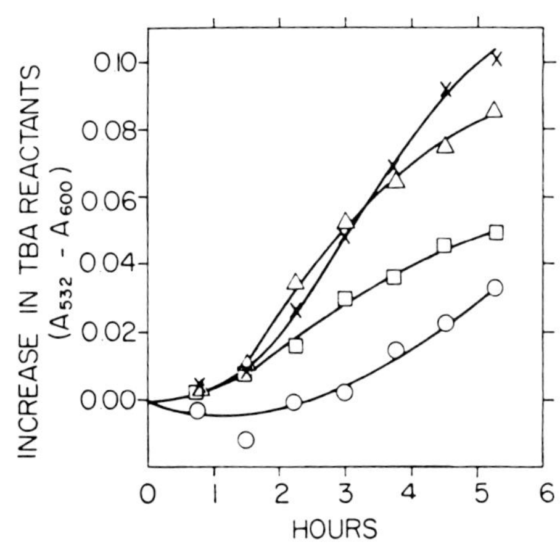

FIGURE 5. Inhibition of photo-oxidation of rat liver mitochondria by coenzynme Q_6H_2 formed by enzymatic reduction of exogenous coenzyme Q_6. The reaction mixture contained 2 mg of coenzyme Q_6 in 0.02 mℓ of ethanol, 25 m*M* sodium β-hydroxy-butyrate, rat liver mitochondria (68 mg of protein), 20 μg of antimycin in 0.02 mℓ of ethanol, and KCl-Tris to make a total volume of 20 mℓ: (○), photo-oxidation of complete mixture; (x), no coenzyme Q_6; (□), no β-hydroxybuturate; (△), no coenzyme Q_6 and no β-hydroxybutyrate; TBA, thiobarbituric acid. (From Mellors, A. and Tappel, A. L., *J. Biol. Chem.*, 241, 4353, 1966. With permission.)

the inhibitory effect of succinate showed a direct relationship with the degree of reincorporation of coenzyme Q. It is also of interest to note the relationship between succinate concentration and inhibition of lipid peroxidation. This relationship, also shown in Figure 6, confirms the more potent antiperoxidant effect of the reduced state of the component involved.

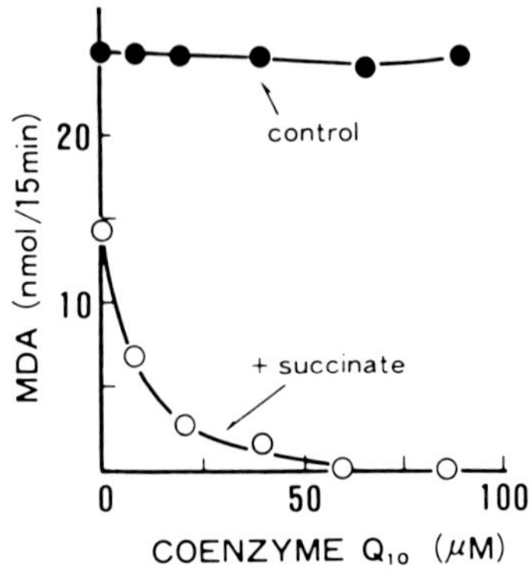

FIGURE 6. Effect of coenzyme Q_{10} extraction and reincorporation on the inhibitory action of succinate on lipid peroxidation. MDA formation was assayed at pH 7.4 using rotenone. The content of enzymatically reducible CoQ_{10} in nmol/mg protein was 0.03 (○), 1.43 (△), 5.12 (□), and 3.6 (●). (From Mellors, A. and Tappel, A. L., *J. Biol. Chem.*, 241, 4353, 1966. With permission.)

Takeshige et al.[63] have interpreted their results on the relationship between the rate of electron transport, the formation of superoxide radical, the redox state of coenzyme Q, and lipid peroxidation as depicted in Figure 7. Under conditions of high coenzyme Q reduction, lipid peroxidation is low. Succinate decreases the rotenone-stimulated rate of lipid peroxidation by reducing coenzyme Q (Figure 7, upper half). At high concentrations of NADPH or low concentrations of NADH (Figure 7, lower half), the rate of electron flow is slow and electron transfer chain components including coenzyme Q tend to be in the oxidized state. Under these conditions, the rate of lipid peroxidation is high and is not affected by respiratory chain inhibitors. Under such conditions, succinate oxidation provides electrons which reduce coenzyme Q and consequently inhibit lipid peroxidation.

Takeshige et al.[63] have provided additional evidence for an antioxidant role for coenzyme Q in its reduced form. They prepared liposomes of mitochondrial lipids into which were incorporated several concentrations of coenzyme Q. The liposomes were incubated with submitochondrial particles, NADH, and rotenone, a condition previously shown to support lipid peroxidation as shown in the upper curve of Figure 8. In the presence of succinate, lipid peroxidation was inhibited in direct proportion to the concentration of coenzyme Q, presumably reduced, in the liposome. This is shown in the lower curve of Figure 8.

Additional support for an antioxidative function for reduced coenzyme Q has been provided by Booth et al.,[68] who report that the reduced form of coenzyme Q is an effective inhibitor of ascorbate-Fe^{2+}-induced peroxidation in egg yolk phosphatidylcholine liposomes. Landi et al.[69,70] also have reported that a number of coenzyme Q homologues from Q_2 to Q_{10}, incorporated into sonic lipid vesicles, protect the lipids against autoxidation over a period of several days. Landi et al.[69,70] report that no difference in the antioxidative potency between oxidized and reduced coenzyme Q was observed in these experiments. This deviation from previous results may be related to the possibility of autoxidation of reduced coenzyme Q during the long period of incubation.

In the course of experiments on the effect of coenzyme Q on lipid peroxidation catalyzed by rat liver microsomes,[71] we have found that reduced coenzyme Q_{10} in solution is readily autoxidized unless special precautions are taken. In these experiments, rat liver microsomes were prepared and lipid peroxidation was induced according to Ernster and Nordenbrand[72] by the addition of NADPH and ADP-Fe^{2+}. Lipid peroxidation was followed by measuring oxygen consumption and MDA formation. The result of an experiment in which CoQ_{10} was added to the oxygen electrode chamber in the fully reduced state is shown in Table 5. As

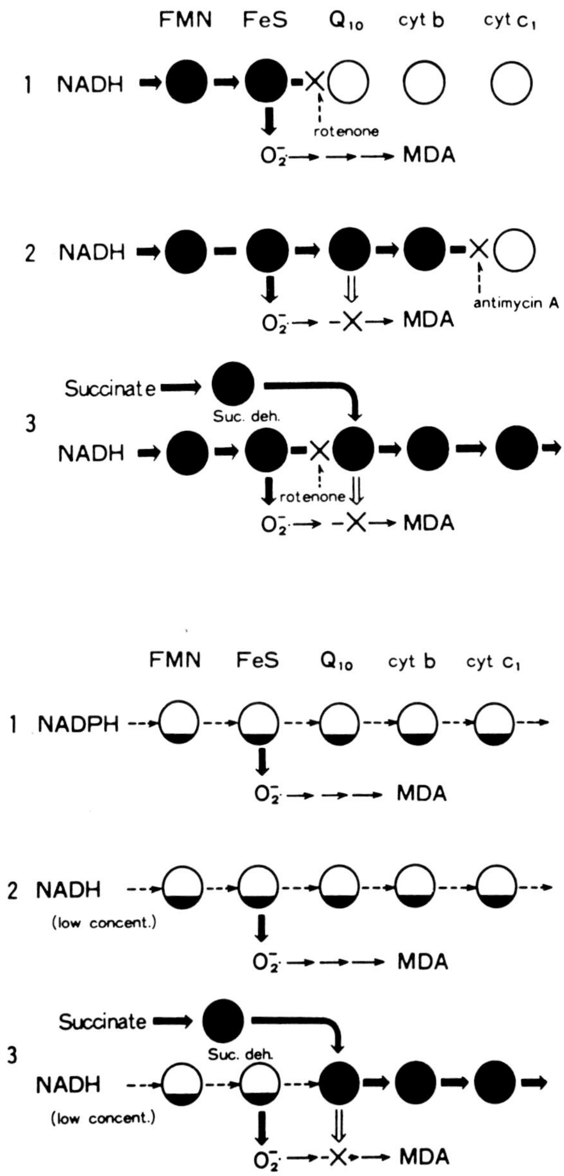

FIGURE 7. The role of the redox state of coenzyme Q on free radical formation. Upper figure: MDA formation at a high rate of electron input from high levels of NADH. Lower figure: MDA formation at low rates of electron input from NADH or NADPH. (From Takeshige, K., Takayanagi, K., and Minakami, S., *Biomedical and Clinical Aspects of Coenzyme Q,* Vol. 2, Yamamura, Y., Folkers, K., and Ito, Y., Eds., Elsevier/North-Holland, Amsterdam, 1980, 15. With permission.)

in liposomes, mitochondria, and submitochondrial particles, reduced coenzyme Q is an effective antioxidant in microsomes. Coenzyme Q in the oxidized form had no antioxidant effect in this particular experiment, although it did have a weak effect occasionally. The presence of oxidized coenzyme Q did not interfere with the antioxidative activity of the reduced form. Inhibition of oxygen consumption and MDA formation by $CoQ_{10}H_2$ over the range 15 to 150 *M* gave an apparent K_i, or I_{50} (concentration giving 50% inhibition), of

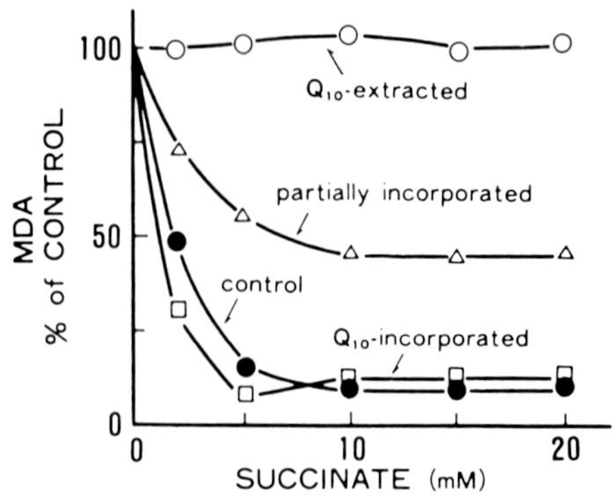

FIGURE 8. Influence of the redox state of submitochondrial electron transfer chain components on lipid peroxidation in liposomes. Experimental conditions as in Figure 6. (●), SMP (0.3 mg) and mitochondrial lipid (0.35 mg) which contains CoQ_{10} in various concentrations; (○), with 15 m*M* succinate. (From Takeshige, K., Takayanagi, K., and Minakami, S., *Biomedical and Clinical Aspects of Coenzyme Q*, Vol. 2, Yamamura, Y., Folkers, K., and Ito, Y., Eds., Elsevier/North-Holland, Amsterdam, 1980, 15. With permission.)

Table 5
EFFECT OF COENZYME Q ON LIPID PEROXIDATION CATALYZED BY RAT LIVER MICROSOMES

Additions	Concentration (μ*M*)	O_2 consumption (nats/min/mg protein)	MDA formed (nmol)
EtOH-cardiolipin	—	224	5.87
CoQ_{10}	150	321	5.71
$CoQ_{10}H_2$	150	59	1.19
$CoQ_{10}H_2$	300	31	0.26
CoQ_{10} + $CoQ_{10}H_2$	150, 75	97	2.01

Note: Reduced coenzyme Q solutions were prepared by reduction in ethanol with $NaBH_4$, extraction with petroleum ether (Boiling Range 40 to 60°C), drying *in vacuo* under N_2, and dissolving the colorless residue in 2% cardiolipin in ethanol. The solution was maintained under N_2 and protected from light. Oxidized coenzyme Q solutions were prepared identically except for the addition of $NaBH_4$.

57 μ*M*. The I_{50} values for two other antioxidant quinones, menadione and vitamin K_5, were also determined and were 20 n*M* for menadione and 37 n*M* for vitamin K_5. The I_{50} obtained for menadione is slightly lower than the 37 n*M* reported by Talcott et al.[73]

In Intact Animals

A number of studies have been published on intact animals which, while not as incisive as experiments at the subcellular biochemical level, are consistent with an antioxidatant function for coenzyme Q. Studies on cold-acclimated rats,[74] rats treated with thyroid hormone,[75] and other conditions also affecting the overall rate of energy metabolism of intact rats,[76] revealed alterations of coenzyme Q concentration in highly aerobic tissues, the direction of change being the same for both metabolic rate and coenzyme Q. Significantly, the increase in tissue coenzyme Q content following thyroid hormone treatment occurred

Table 6
COENZYME Q CONCENTRATION CHANGES WITH ENDURANCE TRAINING[80]

	pmol cytochrome *c*/mg protein		pmol CoQ/mg protein	
Muscle	Sedentary	Trained	Sedentary	Trained
Heart	282	447	1078	1841
Gastrocnemius	64	124	231	457
DVL[a]	64	140	348	715

[a] Deep (red) portion of the vastus lateralis.

Table 7
COENZYME Q CONCENTRATION CHANGES WITH AGE[82]

	Age (months)		
Tissue	2	18	25
Heart	1134	2036	825
Kidney	501	883	688
Gastrocnemius	210	340	233
DVL[a]	365	522	408
Soleus	364	338	240

[a] Deep (red) portion of the vastus lateralis.

after the increase in metabolic rate, suggesting that the increase in coenzyme Q may have been an adaptation to, rather than a cause of, increased oxidative activity.[75] High in vivo oxidative activity such as that which accompanies exhaustive exercise results in a two- to three-fold increase in free radical formation and lipid peroxidation in muscle and liver.[77,78] Adaptation to endurance training reduces the susceptibility of tissue to the damaging effects of free radicals and lipid peroxidation.[79] Beyer et al.[80] have shown that cardiac and other red muscles of endurance-trained rats contain levels of coenzyme Q which are elevated to a greater extent than other components of the mitochondrial electron transport chain (Table 6). Boveris et al.[81] have also reported an example of regulation of coenzyme Q concentrations independent of cytochrome *c* and protein of mitochondria from diabetic animals.

Further supportive evidence consistent with an antioxidant role for coenzyme Q is derived from analyses of coenzyme Q concentrations in tissues from rats over their life span. Beyer et al.[82] have reported that in elderly animals coenzyme Q levels drop drastically in heart, kidney, and highly aerobic (red) muscles (Table 7). This phenomenon is depicted in Figure 9 which shows the severe decline in coenzyme Q of cardiac muscle in the aged rat. These observations on aging animals are consistent with the "free radical theory of ageing"[83] and with a recent report on an inverse correlation between longevity and peroxide-producing potential of mammalian tissues.[84] Further indications of an antioxidant role for coenzyme Q have come from studies in which the concentration of lipoperoxides in mitochondria isolated from hearts and livers of animals treated with coenzyme Q were shown to be decreased when compared to control animals.[85] Administration of coenzyme Q to animals subjected to cardiac ischemia results in improved function of mitochondria isolated from

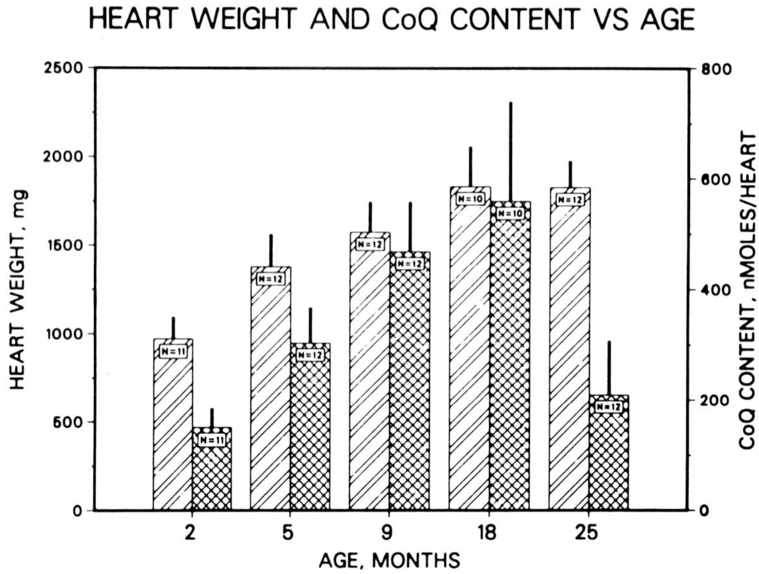

FIGURE 9. Age vs. heart weight and coenzyme Q content in the Sprague-Dawley rat. (From Beyer, R. E., Burnett, B.-A., Cartwright, K. J., Edington, D. W., Falzon, M. J., Kreitman, K. R., Kuhn, T. W., Ramp, B. J., Rhee, S. Y. S., Rosenwasser, J. J., Stein, M., and An, L. C.-I., *Mech. Ageing Dev.*, 32, 267, 1985. With permission.)

the affected area when compared to controls.[86] Beyer and Starnes[87] have reported that coenzyme Q administration partially reverses the severe decline of respiratory metabolism of cardiac tissue seen in animals at advanced ages. This effect was most striking with the oxidation of long-chain fatty acids, the favorite substrate of cardiac muscle.

Further evidence for an antioxidant effect of coenzyme Q in intact animals has been offered by Quinn et al.,[88] who report that coenzyme Q administration protects liver cells against certain changes resulting from exposure to carbon tetrachloride. As it is well established that administration of carbon tetrachloride results in free radical damage,[89] the results of Quinn et al.[88] are consistent with an antioxidant function for coenzyme Q in vivo.

THEORETICAL DISCUSSION: MECHANISMS

In addition to the elimination of preformed oxygen radicals by way of SOD and glutathione peroxidase, there exists a mechanism by which $O_2^{\overline{\cdot}}$ formation can be prevented. It involves the enzyme DTdiaphorase[90] which reduces quinones via a two-electron transfer[91] to produce the relatively stable quinol, thus preventing semiquinone formation and subsequent free radical production.[92] In the cytosol, where the bulk of DTdiaphorase is located, the enzyme most likely serves as a device for the detoxification of exogenous quinones.[92] In microsomes, DTdiaphorase has been shown to serve as a vitamin K reductase in connection with the biosynthesis of prothrombin and related proteins.[93] In mitochondria, which also contain a minor portion of cellular DTdiaphorase,[90] the role of the enzyme is not known, but may well be related to quinone reduction as well. It is conceivable that mitochondrial DTdiaphorase interacts with coenzyme Q and participates in regulation of its redox state, thus influencing its antioxidant function.

The mechanism by which reduced coenzyme Q acts as an antioxidant remains unresolved. Several possibilities exist, however. Reduced coenzyme Q could react with ADP-perferryl ions as suggested by Svingen et al.:[94]

$$ADP\text{–}Fe^{3+}\text{–}O_2^{\overline{\cdot}} + CoQH_2 \rightarrow ADP\text{-}Fe^{2+} + H_2O_2 + CoQ \quad (4)$$

H_2O_2 would be removed by catalase, peroxidase, or glutathione peroxidase. Alternatively, reduced coenzyme Q could react with superoxide directly as a free radical quencher,

$$2O_2^{\overline{\cdot}} + CoQH_2 \rightarrow H_2O_2 + O_2 + CoQ \quad (5)$$

and thus interfere with the initiation of lipid peroxidation. Reduced coenzyme Q might also react with lipid free radicals or lipid peroxide free radicals,

$$2L\cdot + CoQH_2 \rightarrow 2LH + CoQ \quad (6)$$

$$2LOO\cdot + CoQH_2 \rightarrow 2LOOH + CoQ \quad (7)$$

and thus prevent the propagation reaction of lipid peroxidation. As Landi et al.[70] have shown that CoQ_3 and CoQ_9 are equally effective as antioxidants in sonic phospholipid vesicles, the length of the polyisoprenoid chain of CoQ would not appear to be a determining factor in the mechanism of antioxidation.

The question also arises as to whether all of the coenzyme Q in the membrane is available to function as an antioxidant or whether a special coenzyme Q pool, or compartment, exists with a unique antioxidant role. The data of Kröger and Klingenberg[95] suggest that coenzyme Q functions during electron transport as a homogeneous pool. More recent evidence indicates that three distinct coenzyme Q pools exist in pigeon heart mitochondria.[96] One of the three pools is relatively inactive metabolically and consists of both oxidized and reduced coenzyme Q.[96] It is possible that such a pool, in slow equilibrium with the electron transfer chain, may be involved in an antioxidant role. In this regard, it is of interest that the mitochondrial concentration of coenzyme Q may be lowered by 50% after treatment of neuroblastoma cells in culture with mevinolin, an inhibitor of 3-hydroxy-3-methylglutaryl CoA reductase required for the synthesis of the isoprenoid chain of coenzyme Q, without affecting the respiratory function of the mitochondria.[97] As the lowering of the coenzyme Q concentration coincides with inhibition of cell cycling,[97] it is tempting to speculate that the portion of the pool which is reduced in concentration is concerned with protection against oxidative damage and that its loss results in an inability to proliferate. A study of lipid peroxidation catalyzed by such mitochondria and the extent of oxidative damage in cells treated with mevinolin, which is under investigation,[98] should be of interest. It is also possible that a failure to synthesize dolichols required for the glycosylation reaction in the formation of glycoproteins is involved in this loss of proliferative capacity. Carson and Lennarz[99] have shown that compactin, which is also a potent inhibitor of hydroxymethylglutaryl-CoA reductase required for the synthesis of cholesterol and dolichol as well as coenzyme Q, interferes with embryonic development and that the inclusion of exogenous dolichol allows normal development to proceed.

The apoprotein-bound form of coenzyme Q in complexes I, II, and III may be positioned so as to render kinetically favorable the rapid transfer of electrons to the cytochrome system during state 3 respiration. During state 4, apoprotein-bound reduced coenzyme Q may then transfer reducing equivalents to an "antioxidant coenzyme Q pool". As shown recently,[100] oxidized coenzyme Q is restricted positionally in the membrane to a domain where it does not perturb the phase transition of the phospholipids, while its reduction, resulting in an increase in hydrophilicity, allows for greater mobility throughout the hydrocarbon phase of the mitochondrial membrane. Such freedom of movement may be of significance in enhancing the antioxidant role of the reduced form of coenzyme Q and in facilitating the reduction of an "antioxidant coenzyme Q pool" by apoprotein-bound coenzyme Q during state 4 respiration.

The question of the means by which coenzyme Q administration results in the myriad of clinical and experimental benefits reported in several volumes devoted to this subject[101-107] remains essentially unanswered. Whether such administration restores coenzyme Q to levels required to maintain an adequate rate of energy release and conservation or, alternatively, to protect tissue at jeopardy from serious peroxidative damage as an antioxidant, or both, will require further incisive research. Nevertheless, it would appear that enough evidence exists to allow the assignment of an antioxidant role to coenzyme Q in addition to its other well-established cellular functions. While the participation of coenzyme Q in free radical formation in the cell is questionable, the ability of a number of other quinone species to induce oxidative injury or act as antioxidants[108] should be noted. It would appear that coenzyme Q, based on its wide distribution in biological systems and its numerous functions, has been the subject of intense molecular selection during the evolution of most species.[87]

CONCLUSIONS

In this chapter, published experimental data pertaining to the site of free radical formation in the mitochondrial electron transfer chain and the conditions required for its efficient production have been critically reviewed. The evidence suggests that a component of the NADH-coenzyme Q and succinate-coenzyme Q reductases (complexes I and II), most likely a nonheme iron-sulfur protein of each complex, is involved. Studies with electron transfer chain inhibitors appear to eliminate coenzyme Q as a candidate. On the contrary, experiments on subcellular systems, intact animals, and human subjects support the view that coenzyme Q, in its reduced state, may act as a mitochondrial antioxidant.

ACKNOWLEDGMENTS

This review was prepared while the author was Visiting Professor at the Department of Biochemistry, Arrhenius Laboratory, The University of Stockholm, Sweden, and a Senior Fellow of the Fogarty International Center for Advanced Study in the Health Sciences, National Institutes of Health (NIH). The author wishes to express gratitude to the Fogarty International Center of NIH for generous support during his year in Stockholm. Thanks are also due to Ms. Margaret Madouse in Ann Arbor and Ms. Ann Nielsen in Stockholm for their efficient and gracious assistance during preparation of the manuscript and to Peter Löw for a seemingly inexhaustible supply of coenzyme Q_{10} from Dr. Gustav Dallner's cache. I also thank Professor Lars Ernster and Ms. Kerstin Nordenbrand for reading and commenting on early versions of the manuscript.

REFERENCES

1. **Crane, F. L., Hatefi, Y., Lester, R. L., and Widmer, C.,** Isolation of a quinone from beef heart mitochondria, *Biochim. Biophys. Acta,* 25, 220, 1957.
2. **Szarkowska, L.,** The restoration of DPNH oxidase activity by coenzyme Q (ubiquinone), *Arch. Biochem. Biophys.,* 113, 519, 1966.
3. **Norling, B., Glazek, E., Nelson, B. D., and Ernster, L.,** Studies with ubiquinone-depleted submitochondrial particles. Quantitative incorporation of small amounts of ubiquinone and its effects on the NADH and succinate oxidase activities, *Eur. J. Biochem.,* 47, 475, 1974.
4. **Chance, B. and Hollunger, G.,** The interaction of energy and electron transfer reactions in mitochondria. I. General properties and nature of the products of succinate-linked reduction of pyridine nucleotides, *J. Biol. Chem.,* 236, 1534, 1961.
5. **Ernster, L.,** Reaction pathways of succinate-linked acetoacetate reduction in tissue homogenates and isolated mitochondria, *Nature (London),* 193, 1050, 1962.

6. **Löw, H. and Vallin, I.,** Succinate-linked diphosphopyridine nucleotide reduction in submitochondrial particles, *Biochim. Biophys. Acta,* 69, 361, 1963.
7. **Rossi, E., Norling, B., Persson, B., and Ernster, L.,** Studies with ubiquinone-depleted submitochondrial particles. Effects of extraction and reincorporation of ubiquinone on the kinetics of succinate dehydrogenase, *Eur. J. Biochem.,* 16, 508, 1970.
8. **Gutman, M., Kearney, E. B., and Singer, T. P.,** Regulation of succinate dehydrogenase activity by reduced coenzyme Q, *Biochemistry,* 10, 2726, 1971.
9. **Gutman, M., Kearney, E. B., and Singer, T. P.,** Control of succinate dehydrogenase in mitochondria, *Biochemistry,* 10, 4763, 1971.
10. **Gutman, M. and Silman, N.,** The steady state activity of succinate dehydrogenase in the presence of opposing effectors. I. The effect of L-malate and coenzyme QH_2 on the enzymic activity, *Mol. Cell. Biochem.,* 7, 51, 1974.
11. **Gutman, M., Coles, C. J., Singer, T. P., and Casida, J. E.,** On the functional organization of the respiratory chain at the dehydrogenase-coenzyme Q junction, *Biochemistry,* 10, 2036, 1971.
12. **Glazek, E., Norling, B., Nelson, B. D., and Ernster, L.,** Activation of NADH oxidase by succinate in partially ubiquinone-depleted submitochondrial particles, *FEBS Lett.,* 46, 123, 1974.
13. **Ernster, L., Lee, I.-Y., Norling, B., and Persson, B.,** Studies with ubiquinone-depleted submitochondrial particles, essentiality of ubiquinone for the interaction of succinate dehydrogenase, NADH dehydrogenase, and cytochrome *b, Eur. J. Biochem.,* 9, 299, 1969.
14. **Nelson, B. D., Norling, B., Persson, B., and Ernster, L.,** Effect of thenoyltrifluoroacetone on the interaction of succinate dehydrogenase and cytochrome *b* in ubiquinone-depleted submitochondrial particles, *Biochem. Biophys. Res. Commun.,* 44, 1312, 1971.
15. **Nelson, B. D., Norling, B., Persson, B., and Ernster, L.,** Influence of ubiquinone on the rate of antimycin binding to submitochondrial particles, *Biochim. Biophys. Acta,* 267, 205, 1972.
16. **Mitchell, P.,** Protonmotive redox mechanisms of the cytochrome b-c_1 complex in the respiratory chain: protonmotive ubiquinone cycle, *FEBS Lett.,* 56, 1, 1975.
17. **Mitchell, P.,** The protonmotive Q cycle: a general formulation, *FEBS Lett.,* 59, 137, 1975.
18. **Mitchell, P. and Moyle, J.,** Protonmotive mechanisms of quinone function, in *Functions of Quinones in Energy Conserving Systems,* Trumpower, B. L., Ed., Academic Press, New York, 1982, 553.
19. **Ozawa, T.,** Formation of oxygen radicals in the electron transfer chain and antioxidant properties of coenzyme Q, in *Coenzyme Q: Biochemistry, Bioenergetics and Clinical Applications of Ubiquinone,* Lenaz, G., Ed., John Wiley & Sons, Chichester, 1985, 441.
20. **Beyer, R. E., Nordenbrand, K., and Ernster, L.,** The role of coenzyme Q as a mitochondrial antioxidant: a short review, in *Biomedical and Clinical Aspects of Coenzyme Q,* Vol. 5, Folkers, K. and Yamamura, Y., Eds., Elsevier, Amsterdam, 1986, 17.
21. **Boveris, A. and Chance, B.,** The mitochondrial generation of hydrogen peroxide, *Biochem. J.,* 134, 707, 1973.
22. **Gutteridge, J. M. C., Richmond, R., and Halliwell, B.,** Inhibition of the iron-catalyzed formation of hydroxyl radical from superoxide and of lipid peroxidation by desferrioxamine, *Biochem. J.,* 184, 469, 1979.
23. **Chance, B., Sies, H., and Boveris, A.,** Hydroperoxide metabolism in mammalian organs, *Physiol. Rev.,* 59, 527, 1979.
24. **Forman, H. J. and Boveris, A.,** Superoxide radical and hydrogen peroxide in mitochondria, in *Free Radicals in Biology,* Vol. 5, Pryor, W. A., Ed., Academic Press, New York, 1982, 65.
25. **Girotti, A. W.,** Mechanisms of lipid peroxidation, *J. Free Radical Biol. Med.,* 1, 87, 1985.
26. **Fee, J. A.,** On the question of superoxide toxicity and the biological function of superoxide dismutase, in *Proc. 3rd Int. Congr. Oxidases and Related Redox Systems,* King, T. E., Mason, H. S., and Morrison, M., Eds., Pergamon Press, Oxford, 1982, 101.
27. **Beyer, R. E., Ernster, L., Löw, H., and Beyer, T.,** Correlation of optical density and oxidative phosphorylation in reconstructed mitochondrial systems, *Exp. Cell Res.,* 8, 586, 1955.
28. **Lehninger, A. L., Ray, B. L., and Schneider, M.,** The swelling of rat liver mitochondria by thyroxine and its reversal, *J. Biophys. Biochem. Cytol.,* 5, 97, 1959.
29. **Neubert, D. and Lehninger, A. C.,** The effect of thiols and disulfides on water uptake and extrusion by rat liver mitochondria, *J. Biol. Chem.,* 237, 952, 1962.
30. **Hunter, F. E., Jr., Scott, A., Hoffsten, P. E., Gebicki, J. M., Weinstein, J., and Schneider, A.,** Studies on the mechanism of swelling, lysis, and disintegration of isolated liver mitochondria exposed to mixtures of oxidized and reduced glutathione, *J. Biol. Chem.,* 239, 614, 1964.
31. **McCord, J. M. and Fridovich, I.,** Superoxide dismutase: an enzymic function of erythrocuprein (hemocuprein), *J. Biol. Chem.,* 244, 6049, 1969.
32. **Loschen, G., Azzi, A., Richter, C., and Flohe, L.,** Superoxide radicals as precursors of mitochondrial hydrogen peroxide, *FEBS Lett.,* 42, 68, 1974.

33. **Boveris, A. and Cadenas, E.,** Mitochondrial production of superoxide anions and its relationship to the antimycin insensitive respiration, *FEBS Lett.,* 54, 311, 1975.
34. **Loschen, G., Flohe, L., and Chance, B.,** Respiratory chain-linked H_2O_2 production in pigeon heart mitochondria, *FEBS Lett.,* 18, 261, 1971.
35. **Trumpower, B. L. and Simmons, Z.,** Diminished inhibition of mitochondrial electron transfer from succinate to cytochrome *c* by thenoyltrifluoroacetone induced by antimycin, *J. Biol. Chem.,* 254, 4608, 1979.
36. **Boveris, A. and Cadenas, E.,** Production of superoxide radicals and hydrogen peroxide in mitochondria, in *Superoxide Dismutase,* Vol. 2, Oberley, L. W., Ed., CRC Press, Boca Raton, Fla., 1982, 15.
37. **Ernster, L. and Nelson, B. D.,** Function of coenzyme Q, in *Biomedical and Clinical Aspects of Coenzyme Q,* Vol. 3, Folkers, K. and Yamamura, Y., Eds., Elsevier/North-Holland, Amsterdam, 1981, 159.
38. **Bäckström, D., Norling, B., Ehrenberg, A., and Ernster, L.,** Electron spin resonance measurement on ubiquinone-depleted and ubiquinone-replenished submitochondrial particles, *Biochim. Biophys. Acta,* 197, 108, 1970.
39. **Ruzia, F. J., Beinert, H., Schipler, K. L., Dunham, W. R., and Sands, R. H.,** Interaction of ubisemiquinone with a paramagnetic component in heart tissue, *Proc. Natl. Acad. Sci. U.S.A.,* 72, 2886, 1975.
40. **Demopoulos, H. B. and Flamm, E. S.,** Possible pathology induced by CoQ_{10}, in *Biomedical and Clinical Aspects of Coenzyme Q,* Vol. 3, Folkers, K. and Yamamura, Y., Eds., Elsevier/North-Holland, Amsterdam, 1981, 373.
41. **Mollinedo, F. and Schneider, D. L.,** Subcellular localization of cytochrome *b* and ubiquinone in a tertiary granule of resting human neutrophils and evidence for a proton pump ATPase, *J. Biol. Chem.,* 259, 7143, 1984.
42. **Crawford, D. R. and Schneider, D. L.,** Identification of ubiquinone-50 in human neutrophils and its role in microbicidal events, *J. Biol. Chem.,* 257, 6662, 1982.
43. **Crawford, D. R. and Schneider, D. L.,** Evidence that a quinone may be required for the production of superoxide and hydrogen peroxide in neutrophils, *Biochem. Biophys. Res. Commun.,* 99, 1277, 1981.
44. **Fridovich, I.,** Quantitative aspects of the production of superoxide anion radical by milk xanthine oxidase, *J. Biol. Chem.,* 245, 4053, 1970.
45. **Misra, H. P. and Fridovich, I.,** The univalent reduction of oxygen by reduced flavins and quinones, *J. Biol. Chem.,* 247, 188, 1972.
46. **Misra, H. P. and Fridovich, I.,** The generation of superoxide radical during the autoxidation of ferredoxins, *J. Biol. Chem.,* 246, 6886, 1971.
47. **Misra, H. P. and Fridovich, I.,** The role of superoxide anion in the autoxidation of epinephrine and a simple assay for superoxide dismutase, *J. Biol. Chem.,* 247, 3170, 1972.
48. **Oshino, N., Oshino, R., and Chance, B.,** The properties of catalase "peroxidatic" reaction and its relationship to microsomal methanol oxidation, in *Alcohol and Aldehyde Metabolizing Systems,* Thurman, R. G., Yonetani, T., Williamson, J. R., and Chance, B., Eds., Academic Press, New York, 1974, 231.
49. **Littarru, G. P., De Sole, P., Lippa, S., and Oradei, A.,** Study of quenching of singlet oxygen by coenzyme Q_{10} in a system of human leucocytes, in *Biomedical and Clinical Aspects of Coenzyme Q,* Vol. 4, Folkers, K. and Yamamura, Y., Eds., Elsevier, Amsterdam, 1984, 201.
50. **Boveris, A., Cadenas, E., and Stoppani, A. O. M.,** Role of ubiquinone in the mitochondrial generation of hydrogen peroxide, *Biochem. J.,* 156, 435, 1976.
51. **King, T. E.,** Ubiquinone proteins in cardiac mitochondria, in *Function of Quinones in Energy Conserving Systems,* Trumpower, B. L., Ed., Academic Press, New York, 1982, 3.
52. **Suzuki, H. and King, T. E.,** Evidence of an ubisemiquinone radical(s) from the NADH-ubiquinone reductase of the mitochondrial respiratory chain, *J. Biol. Chem.,* 258, 352, 1983.
53. **Green, D. E.,** Structure-function interrelationships in mitochondrial electron transport and oxidative phosphorylation, *Discuss. Faraday Soc.,* 27, 206, 1959.
54. **Mitchell, P.,** Possible molecular mechanisms of the protonmotive function of cytochrome systems, *J. Theor. Biol.,* 62, 327, 1976.
55. **Yu, C. A., Nagoaka, S., Yu, L., and King, T. E.,** Evidence of ubisemiquinone radicals in electron transfer at the cytochromes *b* and c_1 region of the cardiac respiratory chain, *Arch. Biochem. Biophys.,* 204, 59, 1980.
56. **Konstantinov, A. A. and Ruuge, E. K.,** Semiquinone Q in the respiratory chain of electron transport particles, *FEBS Lett.,* 81, 137, 1977.
57. **Yu, L., Yang, F.-D., Yu, C.-A., Tsai, A.-L., and Palmer, G.,** Identification of ubiquinone-binding proteins in yeast mitochondrial ubiquinol-cytochrome *c* reductase using an azido-ubiquinone derivative, *Biochim. Biophys. Acta,* 848, 305, 1986.
58. **Wakabayashi, S., Takao, T., Shimonishi, Y., Kuramitsu, S., Matsubara, H., Wang, T., Zhang, Z., and King, T. E.,** Complete amino acid sequence of the ubiquinone binding protein (QP-C), a protein similar to the 14,000-Dalton subunit of the yeast ubiquinol-cytochrome *c* reductase complex, *J. Biol. Chem.,* 260, 337, 1985.

59. **Takeshige, K. and Minakami, S.,** Reduced nicotinamide adenine dinucleotide phosphate-dependent lipid peroxidation by beef heart submitochondrial particles, *J. Biochem.*, 77, 1067, 1975.
60. **Takeshige, K. and Minakami, S.,** NADH- and NADPH-dependent formation of superoxide anions by bovine heart submitochondrial particles and NADH-ubiquinone reductase preparation, *Biochem. J.*, 180, 129, 1979.
61. **Turrens, J. F. and Boveris, A.,** Generation of superoxide anion by the NADH dehydrogenase of bovine heart mitochondria, *Biochem. J.*, 191, 421, 1980.
62. **Rydström, J., Montelius, J., Bäckström, D., and Ernster, L.,** The mechanism of oxidation of reduced nicotinamide dinucleotide phosphate by submitochondrial particles from beef heart, *Biochim. Biophys. Acta*, 501, 370, 1978.
63. **Takeshige, K., Takayanagi, K., and Minakami, S.,** Reduced coenzyme Q_{10} as an antioxidant of lipid peroxidation in bovine heart mitochondria, in *Biomedical and Clinical Aspects of Coenzyme Q*, Vol. 2, Yamamura, Y., Folkers, K., and Ito, Y., Eds., Elsevier/North Holland, Amsterdam, 1980, 15.
64. **Lea, C. H. and Kwietny, A.,** The antioxidant properties of ubiquinone and related compounds, *Chem. Ind. (London)*, 24, 1245, 1962.
65. **Mellors, A. and Tappel, A. L.,** Quinones and quinols as inhibitors of lipid peroxidation, *Lipids*, 1, 282, 1966.
66. **Mellors, A. and Tappel, A. L.,** The inhibition of mitochondrial peroxidation by ubiquinone and ubiquinol, *J. Biol. Chem.*, 241, 4353, 1966.
67. **Ernster, L., Glaser, E., and Norling, B.,** Extraction and reincorporation of ubiquinone in submitochondrial particles, *Methods Enzymol.*, 53, 573, 1978.
68. **Booth, R. F. G., Galanopoulou, D. G., and Quinn, P. J.,** Protection by ubiquinone and ubiquinol against lipid peroxidation in egg yolk phosphatidylcholine liposomes, *Biochem. Int.*, 5, 151, 1982.
69. **Landi, L., Cabrini, L., Sechi, A. M., and Pasquali, P.,** Antioxidative effect of ubiquinones on mitochondrial membranes, *Biochem. J.*, 222, 463, 1984.
70. **Landi, L., Cabrini, L., Tadolini, B., Sechi, A. M., and Pasquali, P.,** Incorporation of ubiquinone into lipid vesicles and inhibition of lipid peroxidation, *Ital. J. Biochem.*, 34, 356, 1985.
71. **Beyer, R. E., Nordenbrand, K., and Ernster, L.,** unpublished observations, 1986.
72. **Ernster, L. and Nordenbrand, K.,** Microsomal lipid peroxidation, *Methods Enzymol.*, 10, 574, 1967.
73. **Talcott, R. E., Smith, M. T., and Giannini, D. D.,** Inhibition of microsomal lipid peroxidation by napthaquinones: structure-activity relationships and possible mechanisms of action, *Arch. Biochem. Biophys.*, 241, 88, 1985.
74. **Beyer, R. E., Nobel, W. M., and Hirschfield, T. J.,** Coenzyme Q (ubiquinone) levels of tissues of rats during acclimation to cold, *Can. J. Biochem. Physiol.*, 40, 511, 1962.
75. **Pedersen, S., Tata, J. R., and Ernster, L.,** Ubiquinone (coenzyme Q) and the regulation of basal metabolic rate by thyroid hormones, *Biochim. Biophys. Acta*, 69, 407, 1963.
76. **Beyer, R. E., Noble, W. M., and Hirschfield, T. J.,** Alterations of rat-tissue coenzyme Q (ubiquinone) levels by various treatments, *Biochim. Biophys. Acta*, 57, 376, 1962.
77. **Davies, K. J. A., Quintanilha, A. J., Brooks, G. A., and Packer, L.,** Free radicals and tissue damage produced by exercise, *Biochem. Biophys. Res. Commun.*, 107, 1198, 1982.
78. **Salminen, A. and Vihko, V.,** Lipid peroxidation in exercise myopathy, *Exp. Mol. Pathol.*, 38, 380, 1983.
79. **Salminen, A. and Vihko, V.,** Endurance training reduces the susceptibility of mouse skeletal muscle to lipid peroxidation in vitro, *Acta Physiol. Scand.*, 117, 109, 1983.
80. **Beyer, R. E., Morales-Corral, P. G., Ramp, B. J., Kreitman, K. R., Falzon, M. J., Rhee, S. Y. S., Kuhn, T. W., Stein, M., Rosenwasser, M. J., and Cartwright, K. J.,** Elevation of tissue coenzyme Q (ubiquinone) and cytochrome *c* concentrations by endurance exercise in the rat, *Arch. Biochem. Biophys.*, 234, 323, 1984.
81. **Boveris, A. A., Ramos, M. C. P., Stoppani, A. O. M., and Foglia, V. G.,** Phosphorylation, oxidation, and ubiquinone content in diabetic mitochondria, *Proc. Soc. Exp. Biol. Med.*, 132, 170, 1969.
82. **Beyer, R. E., Burnett, B.-A., Cartwright, K. J., Edington, D. W., Falzon, J. J., Kreitman, K. R., Kuhn, T. W., Ramp, B. J., Rhee, S. Y. S., Rosenwasser, J. J., Stein, M., and An, L. C.-I.,** Tissue coenzyme Q (ubiquinone) and protein concentrations over the life span of the laboratory rat, *Mech. Ageing Dev.*, 32, 267, 1985.
83. **Harman, D.,** The free-radical theory of ageing: consequences of mitochondrial ageing, *Age*, 6, 86, 1983.
84. **Cutler, R. G.,** Peroxide-producing potential of tissues: inverse correlation with longevity of mammalian species, *Proc. Natl. Acad. Sci. U.S.A.*, 82, 4798, 1985.
85. **Sugiyama, S., Kitazawa, M., Ozawa, T., Suzuki, K., and Izawa, Y.,** Antioxidative effect of coenzyme Q_{10}, *Experientia*, 36, 1002, 1980.
86. **Sugiyama, S., Ozawa, T., Kato, T., and Suzuki, S.,** Recovery time-course of ventricular vulnerability after coronary reperfusion in relation to mitochondrial function in ischemic myocaridum, *Am. Heart J.*, 100, 829, 1980.

87. **Beyer, R. E. and Starnes, J. W.,** Coenzyme Q and myocardial function in aging and exercise, in *Pathobiology of Cardiovascular Injury,* Stone, H. L. and Weglikci, W. B., Eds., Martinus Nijhoff, Boston, 1985, 489.
88. **Quinn, P. J., Baum, H., Harris, E. J., Franklin, C. S., and Trivedi, P.,** The protective role of coenzyme Q against mercurial and carbon tetrachloride toxicity, in *Biomedical and Clinical Aspects of Coenzyme Q,* Vol. 2, Yamamura, Y., Folkers, K., and Ito, Y., Eds., Elsevier/North-Holland, Amsterdam, 1980, 435.
89. **Brattin, W. J., Glende, E. A., Jr., and Recknagel, R. O.,** Pathological mechanisms in carbon tetrachloride hepatotoxicity, *J. Free Radical Biol. Med.,* 1, 27, 1985.
90. **Ernster, L.,** DT diaphorase, *Methods Enzymol.,* 10, 309, 1967.
91. **Iyanagi, T. and Yamazaki, I.,** One electron reactions in biochemical systems. V. Difference in mechanism of quinone reduction by the NADH dehydrogenase and the NAD(P)H dehydrogenase (DT-diaphorase), *Biochim. Biophys. Acta,* 282, 178, 1970.
92. **Lind, C., Hochstein, P., and Ernster, L.,** DT-diaphorase as a quinone reductase: a cellular control device against semiquinone and superoxide radical formation, *Arch. Biochem. Biophys.,* 216, 178, 1982.
93. **Wallin, R., Gebhard, O., and Prydz, H.,** NAD(P)H dehydrogenase and its role in the vitamin K (2-methyl-3-phytyl-1,4-naphthaquinone)-dependent carboxylation reaction, *Biochem. J.,* 169, 95, 1978.
94. **Svingen, B. A., Buege, J. A., O'Neal, F. O., and Aust, S. D.,** The mechanism of NADPH-dependent lipid peroxidation, *J. Biol. Chem.,* 254, 5892, 1979.
95. **Kröger, A. and Klingenberg, M.,** The kinetics of the redox reactions of ubiquinone related to the electron-transport activity in the respiratory chain, *Eur. J. Biochem.,* 34, 358, 1973.
96. **Jørgensen, M., Rasmussen, H. N., and Rasmussen, U. F.,** Ubiquinone reduction pattern in pigeon heart mitochondria. Identification of three distinct ubiquinone pools, *Biochem. J.,* 229, 621, 1985.
97. **Maltese, W. A. and Aprille, J. R.,** Relation of mevalonate synthesis to mitochondrial ubiquinone content and respiratory function in cultured neuroblastoma cells, *J. Biol. Chem.,* 260, 11524, 1985.
98. **Beyer, R. E. and Hedlund, B.,** research in progress, 1986.
99. **Carson, D. D. and Lennarz, W. J.,** Inhibition of polyisoprenoid and glycoprotein biosynthesis causes abnormal embryonic development, *Proc. Natl. Acad. Sci. U.S.A.,* 76, 5709, 1979.
100. **Aranda, F. J. and Gomez-Fernandez, J. C.,** The interaction of ubiquinone-10 and ubiquinol-10 with phospholipid bilayers. A study using differential scanning calorimetry and turbidity measurements, *Biochim. Biophys. Acta,* 820, 19, 1985.
101. **Folkers, K. and Yamamura, Y., Eds.,** *Biomedical and Clinical Aspects of Coenzyme Q,* Vol. 1, Elsevier/North-Holland, Amsterdam, 1977.
102. **Yamamura, Y., Folkers, K., and Ito, Y., Eds.,** *Biomedical and Clinical Aspects of Coenzyme Q,* Vol. 2, Elsevier/North-Holland, Amsterdam, 1981.
103. **Folkers, K. and Yamamura, Y., Eds.,** *Biomedical and Clinical Aspects of Coenzyme Q,* Vol. 3, Elsevier/North-Holland, Amsterdam, 1980.
104. **Folkers, K. and Yamamura, Y., Eds.,** *Biomedical and Clinical Aspects of Coenzyme Q,* Vol. 4, Elsevier, Amsterdam, 1984.
105. **Folkers, K. and Yamamura, Y., Eds.,** *Biomedical and Clinical Aspects of Coenzyme Q,* Vol. 5, Elsevier, Amsterdam, 1986.
106. **Trumpower, B. L., Ed.,** *Function of Quinones in Energy Conserving Systems,* Academic Press, New York, 1982.
107. **Lenaz, G., Ed.,** *Coenzyme Q: Biochemistry, Bioenergetics and Clinical Applications of Ubiquinone,* John Wiley & Sons, Chichester, 1985.
108. **Smith, M. T., Evans, C. G., Thor, H., and Orrenius, S.,** Quinone-induced oxidative injury to cells and tissues, in *Oxidative Stress,* Sies, H., Ed., Academic Press, London, 1985, 91.

INDUCTION BY SUBSTITUTED PHENOLS OF ENZYMES THAT INACTIVATE ELECTROPHILIC COMPOUNDS

Mary J. De Long

INTRODUCTION

Inhabitants of industrialized countries are continually exposed to two synthetic substituted phenolic antioxidants — BHA, 2(3)-*tert*-butyl-4-hydroxyanisole, and BHT, 3,5-di-*tert*-butyl-4-hydroxytoluene — as additives in food and food packaging, drugs, and cosmetics. These substances are added to extend the shelf life of such products. A National Research Council report estimates consumption in the U.S. of BHA and BHT to be 5.6 and 1.8 mg per person per day, respectively, with Canadian levels even higher for BHA at 7.4 mg per person per day.[1] BHA-containing food-stuffs vary in BHA content from 0.0003% in beverages to 0.1% in chewing gum and candy.[2] The widespread consumption and use of these antioxidants have provoked multiple studies of their potentially beneficial and adverse effects.[3,4]

As antioxidants, BHA and BHT are known to protect against oxygen radical-stimulated lipid peroxidation by serving as chain breakers. In such reactions, BHA and BHT act as electron donors to quench radical species and generate a stable phenoxy radical incapable of further reaction due to the steric bulkiness of tertiary butyl groups ortho to the phenol.[5] In addition to being powerful antioxidants, BHA and BHT are protectors against the acute and/or chronic toxicities of acetaminophen, halogenated compounds, polycyclic aromatics, and other agents which are metabolically activated to electrophilic species.[3,4,6-12]

Initial studies on the mechanism of protection by phenolic antioxidants focused on the possibility that they modulated cytochrome P-450 activities since P-450 inducers such as azo dyes, polycyclic aromatic hydrocarbons, and phenobarbital also protect against a wide variety of noxious agents. Kahl has provided a comprehensive review of the changes in tissue cytochrome P-450 enzyme patterns produced by administration of BHA and BHT.[3] Subsequent studies, however, have failed to demonstrate a strong correlation between P-450 activities or values and protection against toxicity and carcinogenesis, since P-450 levels and activities are either modestly elevated, reduced, or unchanged in response to phenolic antioxidants.[3,13] Nevertheless, it is possible that BHA and/or BHT selectively induce a specific P-450 isozyme which results in heightened protection.

The most compelling explanation of protective action against carcinogens by BHA has been the induction of Phase II* detoxification enzymes.[14-29] Studies in which rodents are fed a diet containing BHA show significant and often striking coordinate increases in glutathione *S*-transferases, quinone reductase, epoxide hydrolase, and UDP-glucuronosyltransferase, as well as elevations of tissue glutathione levels.

The numerous studies of induction of detoxification enzymes by BHA (summary in Tables 1A and 1B) have attempted to provide a link between induction of these enzymes and protection against carcinogenicity, mutagenicity, and acute toxicity of a number of compounds. Five major lines of evidence support enzyme induction by BHA and other substituted phenols as an important mechanism for protection against toxic electrophiles.[4,15,18-22,26-33]

* Enzymes involved in the metabolism of xenobiotics have been grouped into two broad classes:[82] Phase I enzymes (which include the cytochromes P-450) functionalize compounds by oxidation, reduction, or hydrolysis, and Phase II enzymes conjugate functionalized compounds with endogenous ligands such as glucuronic acid, sulfate, and glutathione. The enzymes quinone reductase and epoxide hydrolase are included in the group of detoxification enzymes because they do not introduce new functional groups and rise coordinately with the conjugation enzymes to protect against electrophilic agents. Equating Phase I reactions with activation of xenobiotics to toxic metabolites and Phase II enzymes with detoxification is an acknowledged oversimplification; however, such divisions facilitate discussions of enzyme inductions.

Table 1A
ELEVATIONS OF HEPATIC CYTOSOLIC GLUTATHIONE *S*-TRANSFERASE (GST), CYTOSOLIC QUINONE REDUCTASE (QR), MICROSOMAL EPOXIDE HYDROLASE (EH), AND UDP-GLUCURONOSYLTRANSFERASE (UDPGT) IN MICE AND RATS BY ADMINISTRATION OF 2(3)-*TERT*-BUTYL-4-HYDROXYANISOLE (BHA)

			Enzyme elevations (treated/control)					
			GST					
Animals (species, strain)	Sex	Method of administration	CDNB	DCNB	QR	EH	UDPGT	Ref.
Mice								
1. CD-1	F	Diet, 0.5—0.75%, 7—14 days	7.2—11.1	5.1—10.9	7.8—10	4.3—11.1	1.7—4.7	16-20,30-31, 33,40,74,88
2. CD-1	F	i.p., 100 mg/kg, 7 days	—	—	—	1.6	1.4	16
3. CD-1	F	Gavage, 50—100 μM/ day, 5 days	3.8—4.0	4.4—5.9	4.2—5.8	—	—	21,23,25
4. NMR-1	F	Diet, 0.75%, 14 days	7.6—8.7	—	13.0	2.9	—	41,67,73
	M		2.8	—	—	3.7	—	
5. Swiss-Webster	F	Diet, 0.75—1.0%, 8—10 days	—	—	—	—	1.6	52,72
	M		8.1	—	—	6.5	1.4	
6. A/HeJ	F	Diet, 0.5%, 14 days	5.6	—	—	1.4		24
7. ICR/Ha	F	Diet, 0.5%, 14 days	6.7	—	—	—	—	29
8. CDF_1	F	Diet, 0.5%, 14 days	12.7	—	—	—	1.4	83
Rats								
1. Sprague-Dawley	F	Diet, 0.5—0.75%, 10—14 days	1.5—2.5	2.0—2.1	—	1.6—3.2	2.4—3.1	14,18,19 30,41,53,88
2. Fischer 344	M	Diet, 0.5%, 14 days	1.7	1.5	—	3.2	1.8	13
3. Wistar	M	Diet, 0.8%, 12 days	1.8	—	—	3.8	—	84
4. LEW	M	Diet, 0.45%, 120 days	2.0	1.6	—	—	—	85

Table 1B
ELEVATIONS OF HEPATIC, UDP-GLUCOSE DEHYDROGENASE (UDPGD), GLUCOSE 6-PHOSPHATE DEHYDROGENASE (G6PD), GLUTATHIONE REDUCTASE (GR), AND THIOL LEVELS IN MICE AND RATS BY ADMINISTRATION OF 2(3)-*TERT*-BUTYL-4-HYDROXYANISOLE (BHA)

Animals (species, strain)	Sex	Method of administration	Enzyme and thiol elevations (treated/control)			Thiol levels	Ref.
			UDPGD	G6PD	GR		
Mice							
CD-1	F	Diet, 0.6—0.75%, 7—14 days	2.2—8.3	2.1—3.8	2.6—2.9	1.8	15-17, 88
CD-1	F	i.p., 100 mg/kg, 7 days	1.2	0.71	1.2	—	16
Swiss-Webster	F	Diet, 1.0% 10 days	3.2	—	—	—	52
	M		3.5	—			
CDF_1	F	Diet, 0.5%, 14 days	—	—	2.5	1.4	83
A/HeJ	F	Diet, 0.5%, 14 days	—	—	—	1.3	24
Rat							
Sprague-Dawley	F	Diet, 0.75%, 10 days	1.5	2.3	1.8	1.3	88
Fischer 344	M	Diet, 0.45%, 14 days	—	1.3	1.6	1.1	13
Carworth SPF	F	Gavage, 500 mg/kg		1.4—1.6		—	86,87
	M	Diet, 1% 7 or 28 days	—	0.9—1.1			
LEW	M	Diet, 0.45%, 120 days	—	1.3	1.6	1.3	85

1. Protection against tumor formation occurs in a variety of organs, including forestomach, lung, mammary gland, skin, large bowel, and the hematopoietic system.
2. Induction of detoxification enzymes is a property common to many structurally unrelated chemoprotective compounds, including phenolic antioxidants, suggesting a probable common mechanism of protection among these structurally diverse compounds.[22,30,32]
3. BHA and other phenolic antioxidants can protect against a wide variety of carcinogenic, mutagenic, and toxic agents, suggesting that their action involves a generalized mechanism such as drug metabolism.[3,4]
4. The group of enzymes induced by phenolic antioxidants are each important in the detoxification process of electrophilic metabolites of xenobiotics.
5. Protection is dependent on protein synthesis.[34]

PHASE II DETOXIFICATION ENZYMES

Glutathione *S*-Transferases

The glutathione *S*-transferases are a family of abundant proteins which protect against electrophiles:

1. By catalyzing conjugation of electrophilic substrates with the nucleophile, glutathi-

one,[35,36] resulting in the formation of conjugates which may be excreted in the bile and urine or metabolized further to mercapturic acid conjugates[37]
2. By binding with high-affinity exogenous electrophilic groups of a variety of hydrophobic molecules[38]
3. By undergoing under some circumstances covalent linkage with electrophiles[39]

The glutathione *S*-transferases are widely distributed in both the liver and extrahepatic tissues and in microsomal and cytosolic compartments.[19,36,40,41] They are readily induced by substituted phenols such as BHA and BHT and other antioxidants, including ethoxyquin and disulfiram.[17,20] The glutathione *S*-transferases are also elevated by phenobarbital, 3-methylcholanthrene, 2,3,7,8-tetrachlorodibenzo-*p*-dioxin (TCDD), polychlorinated biphenyls (PCB), and various other pesticides.[36] An excellent correlation between the ability of compounds to elevate glutathione *S*-transferase activity and protection against benzo(*a*)pyrene-induced neoplasia and aflatoxin binding to DNA has been demonstrated.[13,27]

In rat liver, seven dimeric isozymes of glutathione *S*-transferase have been isolated. They are composed of subunits designated Ya, Yb, and Yc. Using cDNA clones complementary to each of the subunits, measurement of translation products, and hybridization studies, it has been demonstrated that increases of enzyme by inducers are due to elevations of mRNA of the subunit proteins and are under control of a multigene family.[42-44] Similar families of multiple isozymes have been found in human liver, rat brain, and mouse liver. Studies in the mouse liver on the major (pI 8.7) and minor (pI 9.3) isozymes show similar increases in mRNA for glutathione *S*-transferase when induced by BHA.[45] BHA also increases thiol levels known to be protective against electrophiles and the enzyme glutathione reductase, which is involved in maintaining reduced glutathione pools.

UDP-Glucuronosyltransferase

The second class of conjugating enzymes comprises the UDP-glucuronosyl-transferases which catalyze the transfer of glucuronic acid from uridine diphosphoglucuronic acid to a variety of endogenous and xenobiotic metabolites. The resulting glucuronides are generally less biologically active and more polar than the aglycone substrates and are readily excreted in bile and urine.[46]

UDP-glucuronosyltransferase activity is found in many species, tissues, and subcellular organelles.[47-49] It is readily induced by BHA and other exogenous compounds such as pesticides and environmental pollutants. A number of isozymes with distinct but overlapping substrate specificities are involved in the conjugation of diverse substrates.[49,50] Hybridization with a cDNA probe complementary to rat liver UDP-glucuronosyltransferase mRNA has demonstrated elevation of a minimum of three distinct mRNAs, two of which may be associated with a common gene or gene family.[51] The induction of UDP-glucuronosyltransferase by BHA has been correlated with increased glucuronidation of noxious substances and inhibition of formation of γ-glutamyl transpeptidase positive foci and hyperplastic nodules induced by diethylnitrosamine and *N*-2-fluorenylacetamide.[52,53] UDP-glucose dehydrogenase which catalyzes the irreversible oxidation of UDP-glucose to UDP-glucuronic acid is also increased to two to eight times control values in the liver by dietary BHA.

Epoxide Hydrolase

Many aromatic and olefinic environmental carcinogens are metabolized by the cytochrome P-450 monoxygenase system to yield arena or alkene oxides.[54] Microsomal epoxide hydrolase plays a central role in the detoxification of arene oxides and epoxides by catalyzing their stereospecific hydration to *trans*-dihydrodiols.[55,56] Epoxide hydrolase also plays a role in the metabolic activation of polycyclic aromatic hydrocarbons to highly toxic diol-epoxides.[54]

The enzyme exists in both microsomal and cytosolic compartments and these forms differ with respect to induction, structure, and substrate specificity.[56-58] Evidence for the existence of isozymes includes differential induction of activities toward various substrates and chromatographic resolution of immunologically distinct proteins with hydrolase activity.[59,60] Despite its role in both activation and deactivation of toxic substances, there is evidence that epoxide hydrolase has a role in reducing carcinogencity, mutagenicity, and other forms of toxicity of a variety of noxious agents.[19,33,61,62]

NAD(P)H:Quinone Reductase

NAD(P)H:quinone reductase has a protective function in the detoxification of quinones and quinoneimines generated from aromatic xenobiotics and endogenous compounds. It has been shown to protect against mutagenicity, reduce covalent binding of carcinogens, and reduce the generation of reactive oxygen species from quinones.[63-67] Quinone reductase is unique among other flavoproteins that reduce quinones in that it promotes obligatory two-electron reductions[68] and thus prevents redox cycling of semiquinones produced by one-electron reductions. Semiquinones are known to react with critical cellular macromolecules or to undergo rapid spontaneous oxidation by molecular oxygen to generate damaging superoxide radicals.[63,64,68]

The production of reactive oxygen species from menadione redox cycling results in light emmision. This chemiluminescence is decreased in hepatic postmitochondrial supernatants of livers of mice fed BHA which increased quinone reductase levels to 13 times control values. Furthermore, this chemiluminescence was enhanced by dicoumarol, a known quinone reductase inhibitor.[66] Such studies support the postulate that elevation of quinone reductase activities by BHA in various tissues contributes to its protective role against toxic electrophiles. Quinone reductase is readily induced by many other classes of chemoprotectors in addition to phenolic antioxidants in both rodent and hepatoma cells.[20,22,31,69,70] Two isofunctional forms of murine quinone reductase have been purified to homogeneity and designated ''hydrophilic'' and ''hydrophobic'' based on order of elution from a phenyl-Sepharose column.[71]

STUDIES OF INDUCTION BY BHA OF DETOXIFICATION ENZYMES IN RODENT TISSUES

Studies in the Liver

In the last 7 years, the induction of detoxification enzymes by BHA has been intensively studied. The liver has been most frequently examined because of its marked response to BHA and its central role in detoxification of xenobiotics. These data are summarized in Tables 1A and 1B. Table 1A includes ratios (of treated to control) of specific activities of glutathione *S*-transferase, quinone reductase, epoxide hydrolase, and UDP-glucuronosyltransferase. Table 1B contains ratios of enzymes involved in maintenance of sulfhydryl and glucuronidation potential: glucose 6-phosphate dehydrogenase, UDP-glucose dehydrogenase, and glutathione reductase and thiol levels. An examination of the studies of induction by BHA in liver presented in Table 1A and 1B results in the following generalizations.

1. Detoxification enzymes are coordinately induced by BHA, but not necessarily to the same degree. Glutathione *S*-transferase, quinone reductase, and epoxide hydrolase show greater elevations than UDP-glucuronosyltransferase.
2. In general, mice show a greater response to induction by BHA than rats.[18,19,41] A 0.75% diet of BHA for 14 days gives an 11-fold induction of glutathione *S*-transferase and quinone reductase in CD-1 mice, but only a 2-fold induction in Sprague-Dawley rats. Similar differences are seen for epoxide hydrolase.
3. Isozyme are induced differentially in BHA-fed mice. UDP-glucuronosyltransferase

exhibits a selective induction by BHA of isozymes specific for group 1 (1-naphthol) substrates.[52] Variations in induction are also seen in enzymes with different subcellular localization, such as cytosolic glutathione *S*-transferase which exhibits a 7.6-fold increase in activity by BHA compared to the microsomal enzyme which increases only 1.7-fold.[41] Also, BHA induces a 12-fold greater increase in the microsomal than the cytosolic form(s) of epoxide hydrolase.[72]

4. The extent of response to induction by BHA depends on the sex of the animal.[73] Glutathione *S*-transferase activity was increased by BHA nine-fold in female and only three-fold in male NMRI mice. A reverse pattern was observed with epoxide hydrolase which is more highly inducible in males than in females.[73]
5. The degree of induction is dependent not only on the dose, but also on the method and duration of administration of BHA. Dietary 0.75% BHA gave a three-fold greater increase in epoxide hydrolase and UDP-glucuronosyltransferase when compared to i.p. injection at 100 mg/kg (an approximately equivalent dose).[16] Significant increases in hepatic glutathione *S*-transferase activity toward CDNB and DCNB and of quinone reductase were observed after 1 day of feeding with 0.75% BHA and maximal increases were obtained at day 9.[74] The messenger RNA levels for glutathione *S*-transferase reach their maximal level after 4 days of BHA feeding.[74] When administered for 5 days by gavage, doses of BHA as low as 10 μmol/day per mouse increased glutathione *S*-transferase and quinone reductase significantly; maximal levels were achieved at 100 μmol/day per mouse (equivalent to 0.75% diet).[21]
6. Increases in levels of the glutathione *S*-transferase detoxification enzymes by BHA are modulated via *de novo* synthesis of the enzyme. The tenfold or greater elevations of hepatic glutathione *S*-transferase activities in the CD-1 mouse in response to BHA administration is reflected in increases of enzyme precipitated by antibodies to the purified major transferase (pI 8.7).[19,45] The quantity of the immunoreactive transferase rises to about 18% of total liver cytosol protein in BHA mice in comparison to 1.5% in animals on control diets. A 20- to 25-fold increase in messenger RNA for glutathione *S*-transferase was also observed after BHA feeding.[45,74] Similar increases by dietary BHA of immunoreactive protein have been demonstrated with antisera directed against the enzymes epoxide hydrolase and quinone reductase.[60,71]

Studies in Extrahepatic Tissues

The literature on the induction of detoxification enzymes by BHA in extrahepatic tissues is modest in comparison to that of liver. For extrahepatic tissues, detailed information is available for CD-1 mice fed a diet of 0.75% BHA for 14 days. In these animals, specific activities of glutathione *S*-transferase, quinone reductase, and epoxide hydrolase have been measured. Review of the results summarized in Figure 1 allows the following generalizations.[19,20]

1. Basal level specific activities of quinone reductase, glutathione *S*-transferases, and epoxide hydrolase vary greatly among the extrahepatic tissues. The urinary bladder has an especially high level of all three enzymes.
2. No tissue examined is totally unresponsive to enzyme induction. BHA produces the greatest enzyme induction in the liver and upper intestinal mucosa. All enzymes are generally coordinately induced. The enzyme inductions vary greatly (e.g., 1,2-and 14.9-fold in the bladder and intestine for glutathione *S*-transferase measured with DCNB).
3. Selected isozymes are preferentially induced by BHA in certain tissues. Isoelectric and immunological studies in small intestine show two prominent peaks of glutathione *S*-transferase with pI 8.7 and pI 9.0, indistinguishable from corresponding liver trans-

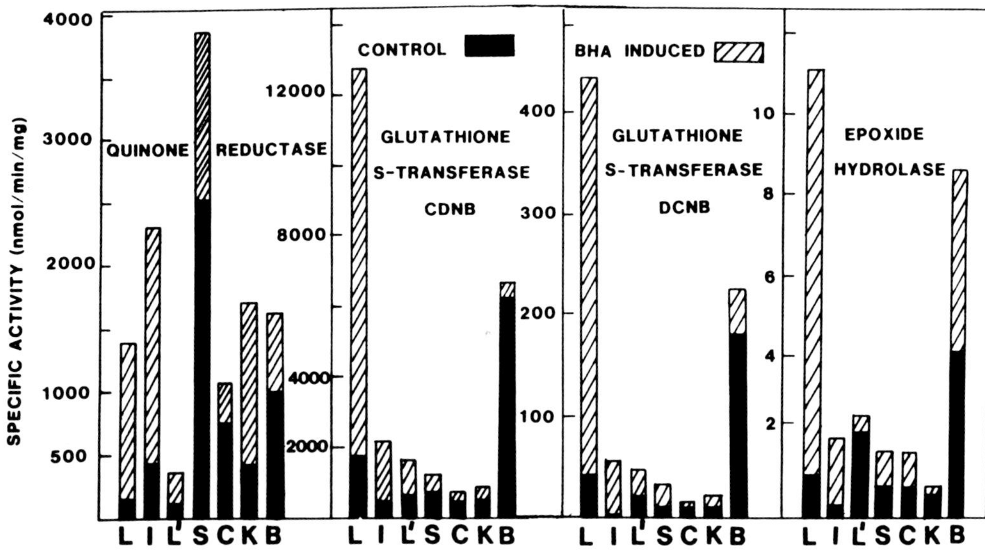

FIGURE 1. Basal and BHA-induced levels of detoxification enzymes in hepatic and extrahepatic tissues of CD-1 mice;[19,20] symbols: L, liver; I, intestine; L′, lung; S, stomach; C, colon; K, kidney; B, urinary bladder.

ferases. Dietary BHA preferentially increases the intestinal transferase activity with pI 8.7.[75] This selective increase in the pI 8.7 peak is supported by the 15-fold increase of glutathione *S*-transferase activity in the upper small intestine with DCNB as compared to the 4.4-fold increase with CDNB. The rate of increase in activity of these enzymes is more rapid in the small intestine than in the liver and may be a consequence of the rapid exfoliation of the intestinal lining.[74] Thus, the protective effect of enzyme induction may reflect both the participation of the liver which plays a major role in the metabolism of xenobiotics as well as the gastrointestinal (GI) tract which serves as a portal of entry for exogenous chemicals and is a possible site of detoxification of ingested carcinogens.

Induction Studies with BHA Analogs

Wattenberg et al. in 1980 reported an extensive study of the structural requirements of 18 phenols, including BHA for inhibition of neoplasia in the forestomach of mice by benzo(*a*)pyrene. Protection against the action of the carcinogen ranged from 80% to no effect, indicating a correlation between phenolic structure and inhibition of neoplasia.[76]

Rahimtula et al.[40] compared induction of glutathione *S*-transferase, epoxide hydrolase, and UDP-glucuronosyltransferase in liver by five phenolic analogs of BHA and found differences in elevation of enzyme activity dependent on structure. In our laboratory, a series of phenolic analogs of BHA were administered by gavage to CD-1 mice for 5 days and enzyme elevations were determined in hepatic and extrahepatic tissues (Figure 2).[38,99] The level of response to the administered phenolic inducer depended on the structure of the compound, the tissue, and the enzyme studied. Despite the complexity of these factors, certain generalizations could be drawn:

1. The liver and upper small intestine were the most responsive tissues to all inducers.
2. The glutathione *S*-transferases and quinone reductase were generally coordinately induced in all tissues, except the kidney and lung, where quinone reductase is more responsive.
3. The most potent inducers in the liver and upper small intestine were 3-BHA (the major

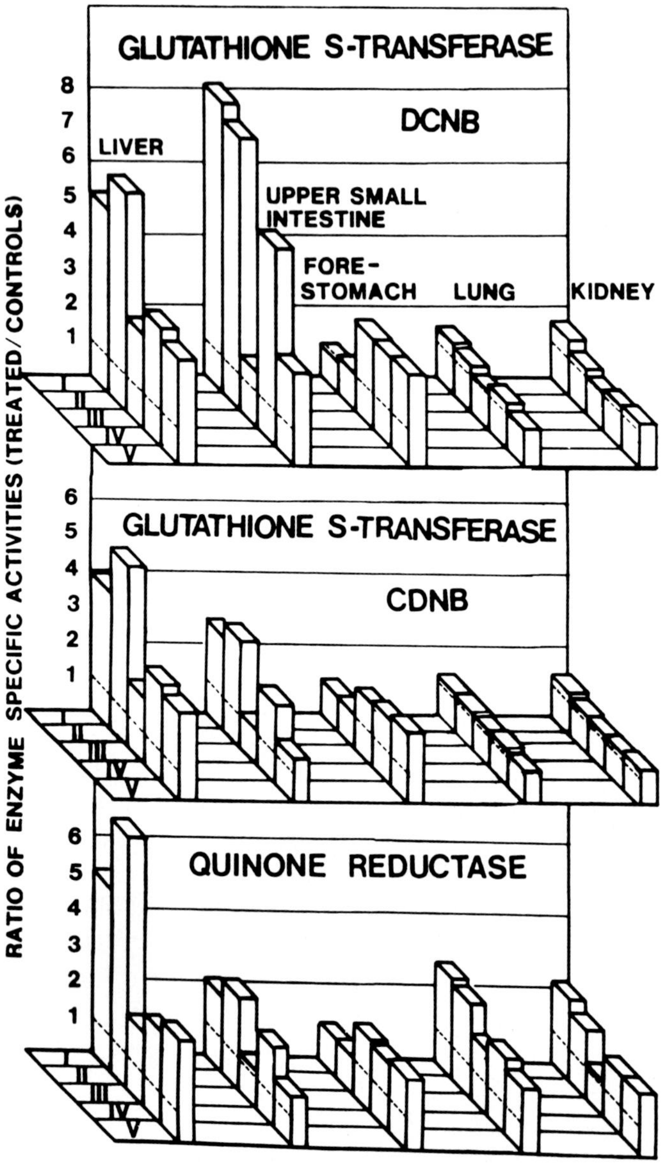

FIGURE 2. Induction patterns of glutathione *S*-transferases measured with CDNB and DCNB and of quinone reductase in cytosols of five tissues of the CD-1 mouse in response to administration of I, 3-BHA; II, methyl-BHA; III, 2-BHA; IV, *tert*-butyl-hydroquinone; and V, 4-hydroxyanisole.

isomer) and methyl-BHA, but these compounds are relatively inactive in the fore-stomach and glandular stomach were 2-BHA (the minor isomer) was the most active compound.

4. *Tert*-butylhydroquinone was promiscuous in its ability to induce in all tissues.

Prochaska et al.[25] tested 12 analogs of 3-BHA, including 2 synthetic analogs of the dimer known to be a BHA metabolite.[77] For all compounds in both the liver and upper small intestine, induction of the glutathione *S*-transferase and quinone reductase differed only by a factor of two, despite the presence of a variety of phenolic and alkoxy substitutions at the

Table 2
INDUCTION OF CYTOSOLIC QUINONE REDUCTASE IN HEPA 1c1c7 CELLS BY PHENOLIC ANTIOXIDANTS[22,26]

Inducing agent (compound)	Conc (μ*M*)	Ratio of specific activity (treated/control)
3-*Tert*-butyl-4-hydroxyanisole (3-BHA)	30	1.32 ± 0.15
	60	1.62 ± 0.10
Tert-butylhydroquinone	10	1.37 ± 0.04 — 2.12 ± 0.13
	30	2.60 ± 0.14 — 4.36 ± 0.22
	50	4.64 ± 0.15
3,5-Di-*tert*-butylhydroxytoluene (BHT)	30	1.01 ± 0.04
	50	2.17 ± 0.21
Pyrogallol	10	2.05 ± 0.17
	60	2.42 ± 0.21
1,2-Diphenol (Catechol)	10	1.39 ± 0.04
	30	1.79 ± 0.10
4-*Tert*-butyl-1,2-diphenol	10	1.26 ± 0.04
	30	1.65 ± 0.05
1,3-Diphenol (Resorcinol)	10	0.86 ± 0.03
	30	0.88 ± 0.02
1,4-Diphenol (Hydroquinone)	10	1.67 ± 0.07
	30	1.92 ± 0.06

1 and 4 position. This clearly demonstrated that the inductive capacity of the 3-BHA analogues tested was not highly dependent on the structure of the compounds administered but possibly was dependent on their rate of conversion to a common metabolite. Considering that *tert*-butylhydroquinone is an inducer in all tissues examined and that it is a known BHA metabolite, the common active enzyme-inducing metabolite of BHA analogs may be related to *tert*-butylhydroquinone.[78]

MECHANISM OF DETOXIFICATION BY ENZYME INDUCTION

To simplify structure-activity studies in animals and to provide a system in which the mechanism of enzyme induction by chemoprotectors could be studied, the hepatoma cells, Hepa 1c1c7, was established as a model system.[79] This cell line possesses competent Ah locus receptors and inducible P-450 monooxygenase and detoxification enzymes.[22,80] Quinone reductase was used as a marker enzyme for induction of detoxification enzymes since it is coordinately induced with other detoxification enzymes and is easily measured. Structure-activity studies with phenolic analogs (Table 2) demonstrated that BHA and BHT were much less efficient enzyme inducers in the cell line than those compounds with unsubstituted phenolic groups, such as *tert*-butylhydroquinone. Again these findings suggested that the ultimate inducers of detoxification enzymes were probably diphenols.[22,26] Testing a similar series of 1,2-, 1,3-, and 1,4-diphenols in the Hepa 1c1c7 model system (Table 2) demonstrated that the *tert*-butyl group on the phenolic analogs was not critical for inductive activity. The most important finding, however, was that all catechols (1,2-diphenols) and hydroquinones (1,4-diphenols) were found to be inducers and all 1,3-diphenols were inactive. Such results strongly indicated that the essential property of phenolic compounds for induction of detoxification enzymes was their ability to undergo facile oxidation-reductions.[22,26] Such a mechanism could involve the signaling of enzyme induction by chemoprotectors undergoing either one- or two-electron valency changes. This in turn may lead to possible perturbations of oxidation-reduction pools. The proposed mechanism provides a common mechanistic pathway to explain induction of detoxification eymzes not only by phenolic antioxidants, but also by the structurally diverse array of chemoprotectors.[22,26] Studies in aryl hydrocarbon hydroxylase mutants of Hepa 1c1c7 deficient in Ah locus function or aryl hydrocarbon hydroxylase expression exhibited no induction of quinone reductase with compounds requiring cytochrome P-450 metabolism for conversion to redox-labile compounds (e.g., azo dyes and polycyclic aromatic hydrocarbons).[81] This again supports the postulate that the inducing signal for induction of detoxification enzymes is mediated via a redox-labile property of the enzyme inducer.

CONCLUSIONS

The phenolic antioxidants are a group of compounds known to protect against electrophilic compounds not only through direct interaction, as with radical chain breaking, but also by induction of the detoxification enzymes such as glutathione *S*-transferase, quinone reductase, epoxide hydrolase, and UDP-glucuronosyltransferase. Induction of these enzymes by BHA and its analogs in rodents is dependent on the structure of the inducer, the species and strain tested, the type of tissue examined, the sex of the animal, and the enzyme induced. Preferential isozyme induction is also seen in certain tissues. The induction of these detoxification enzymes appears to involve *de novo* protein synthesis. Studies of the mechanism of induction by phenolic compounds in the hepatoma, Hepa 1c1c7, a model cell system, suggest that the critical requirement for a phenol (or other compound) to induce these enzymes is that it can undergo facile oxidation-reduction reactions. Further studies are needed to define the exact nature of the inducing signal.

ABBREVIATIONS

BHA, 2,(3)-*tert*-butyl-4-hydroxyanisole
2-BHA, minor isomer
3-BHA, major isomer
BHT, 2,3-di-*tert*-butyl-4-hydroxytoluene
CDNB, 1-chloro-2,4-dinitrobenzene
DCNB, 1,2-dichloro-4-nitrobenzene

ACKNOWLEDGMENTS

I am most grateful to Dr. Paul Talalay for his invaluable assistance in the discussion and preparation of this manuscript. I am also grateful to Darlene Weber for typing this manuscript.

REFERENCES

1. Symp. Food Antioxidants: International Perspectives, Washington, D.C., International Life Sciences Institute Nutrition Foundation, April 21 to 23, 1986.
2. **Anon.,** Chemicals Used in Food Processing, No. 1274, National Academy of Sciences-National Research Council, Washington, D.C., 1965.
3. **Kahl, R.,** Synthetic antioxidants: biochemical actions and interference with radiation, toxic compounds, chemical mutagens and chemical carcinogens, *Toxicology,* 33, 185, 1984.
4. **Wattenberg, L. W.,** Chemoprevention of cancer, *Cancer Res.,* 45, 1, 1985.
5. **Ranney, M. W.,** *Antioxidants, Recent Developments,* Noyes Data Corporation, Parkridge, N.J., 1979, 1.
6. **Ansher, S. S., Dolan, P., and Bueding, E.,** Chemoprotective effects of two dithiolthiones and of butylhydroxyanisole against carbon tetrachloride and acetaminophen toxicity, *Hepatology,* 3, 932, 1983.

7. **Miranda, C. L., Henderson, M. C., Buhler, D. R., and Schmitz, J. A.,** Comparative effects of antioxidants on the toxicity of mixed pyrrolizidine alkaloids from *Seneco jacobaea* in mice, *J. Toxicol. Environ. Health,* 9, 933, 1982.
8. **Miranda, C. L., Henderson, M. C., and Buhler, D. R.,** Dietary butylated hydroxyanisole reduces covalent binding of acetaminophen to mouse tissue proteins *in vivo, Toxicol. Lett.,* 25, 89, 1985.
9. **Moldeus, P., Dock, L., Cha, Y.-N., Berggren, M., and Jernstrom, B.,** Elevation of conjugation capacity in isolated hepatocytes from BHA-treated mice, *Biochem. Pharmacol.,* 31, 1907, 1982.
10. **Wattenberg, L. W.,** Inhibition of carcinogenic and toxic effects of polycyclic hydrocarbons by phenolic antioxidants and ethoxyquin, *J. Natl. Cancer Inst.,* 48, 1425, 1972.
11. **Kim, H. L., Anderson, A. C., Terry, M. K., and Bailey, E. M., Jr.,** Protective effect of butylated hydroxyanisole on acute hymenoxon and bitterweed poisoning, *Res. Commun. Chem. Pathol. Pharmacol.,* 33, 365, 1981.
12. **Kim, H. L. and Jones, J. P.,** Protective effects of butylated hydroxyanisole, ethoxyquin, and disulfiram on acute pyrrolizidine alkaloids poisoning in mice, *Res. Commun. Chem. Pathol. Pharmacol.,* 36, 341, 1982.
13. **Kensler, T. W., Egner, P. A., Trush, M. A., Bueding, E., and Groopman, J. D.,** Modification of aflatoxin B_1 binding to DNA *in vivo* in rats fed phenolic antioxidants, ethoxyquin, and a dithiolthione, *Carcinogenesis,* 6, 759, 1985.
14. **Kahl, R. and Wulff, U.,** Induction of rat hepatic epoxide hydratase by dietary antioxidants, *Toxicol. Appl. Pharmacol.,* 47, 217, 1979.
15. **Cha, Y.-N. and Bueding, E.,** Effect of 2(3)-*tert*-butyl-4-hydroxyanisole administration of the activities of several hepatic microsomal and cytoplasmic enzymes in mice, *Biochem. Pharmacol.,* 28, 1917, 1979.
16. **Cha, Y.-N., Heine, H. S., and Moldeus, P.,** Differential effects of dietary and intraperitoneal administration of antioxidants on the activities of several hepatic enzymes of mice, *Drug Metab. Dispos.,* 10, 434, 1982.
17. **Cha, Y.-N., Heine, H. S., and Ansher, S.,** Comparative effects of dietary administration of antioxidants and inducers on the activities of several hepatic enzymes in mice, *Drug Nutr. Interact.,* 2, 35, 1983.
18. **Cha, Y.-N., Martz, F., and Bueding, E.,** Enhancement of liver microsome epoxide hydratase activity in rodents by treatment with 2(3)-*tert*-butyl-4-hydroxyanisole, *Cancer Res.,* 38, 4496, 1978.
19. **Benson, A. M., Cha, Y.-N., Bueding, E., Heine, H. S., and Talalay, P.,** Elevation of extrahepatic glutathione *S*-transferase and epoxide hydratase activities by 2(3)-*tert*-butyl-4-hydroxyanisole, *Cancer Res.,* 39, 2971, 1979.
20. **Benson, A. M., Hunkeler, M. J., and Talalay, P.,** Increase of NAD(P)H:quinone reductase by dietary antioxidants: possible role in the protection against carcinogenesis and toxicity, *Proc. Natl. Acad. Sci. U.S.A.,* 77, 5216, 1982.
21. **De Long, M. J., Prochaska, H. J., and Talalay, P.,** Tissue-specific induction patterns of cancer-protective enzymes in mice by *tert*-butyl-4-hydroxyanisole and related substituted phenols, *Cancer Res.,* 45, 546, 1985.
22. **De Long, M. J., Prochaska, H. J., and Talalay, P.,** Induction of NAD(P)H:quinone reductase in murine hepatoma cells by phenolic antioxidants, azo dyes and other chemoprotectors: a model system for the study of anticarcinogens, *Proc. Natl. Acad. Sci. U.S.A.,* 83, 787, 1986.
23. **De Long, M. J., Prochaska, H. J., and Talalay, P.,** in *Protective Agents in Cancer,* McBrien, D. C. H. and Slater, T. F., Eds., Academic Press, London, 1983, 175.
24. **Lam, L. K. T., Sparnins, V. L., Hochalter, J. B., and Wattenberg, L. W.,** Effects of 2- and 3-*tert*-butyl-4-hydroxyanisole on glutathione *S*-transferase and epoxide hydrolase activities and sulfhydryl levels in liver and forestomach of mice, *Cancer. Res.,* 41, 3940, 1981.
25. **Prochaska, H. J., Bregman, H. S., De Long, M. J., and Talalay, P.,** Specificity of induction of cancer protective enzymes by analogues of *tert*-butyl-4-hydroxyanisole (BHA), *Biochem. Pharmacol.,* 34, 3909, 1985.
26. **Prochaska, H. J., De Long, M. J., and Talalay, P.,** On the mechanisms of induction of cancer-protective enzymes: a unifying proposal, *Proc. Natl. Acad. Sci. U.S.A.,* 82, 8232, 1985.
27. **Sparnins, V. L. and Wattenberg, L. W.,** Enhancement of glutathione *S*-transferase activity of the mouse forestomach by inhibitors of benzo(a)pyrene-induced neoplasia of the forestomach, *J. Natl. Cancer Inst.,* 66, 769, 1981.
28. **Sparnins, V. L., Chuan, J., and Wattenberg, L. W.,** Enhancement of glutathione *S*-transferase activity of the esophagus by phenols, lactones and benzylisothiocyanate, *Cancer Res.,* 42, 1205, 1982.
29. **Sparnins, V. L., Venegas, P. L., and Wattenberg, L. W.,** Glutathione *S*-transferase activity: enhancement by compounds inhibiting chemical carcinogenesis and by dietary constituents, *J. Natl. Cancer Inst.,* 68, 493, 1982.
30. **Talalay, P., Batzinger, R. P., Benson, A. M., Bueding, E., and Cha, Y.-N.,** Biochemical studies on the mechanisms by which dietary antioxidants suppress mutagenic activity, *Adv. Enzyme Regul.,* 17, 23, 1979.

31. **Talalay, P. and Benson, A. M.,** Elevation of quinone reductase activity by anticarcinogenic antioxidants, *Adv. Enzyme Regul.*, 20, 287, 1982.
32. **Talalay, P., De Long, M. J., and Prochaska, H. J.,** Molecular mechanisms in protection against carcinogenesis, in *Cancer Biology and Therapeutics*, Cory, J. G. and Szentivani, A., Eds., Plenum Press, New York, 1986, 197.
33. **Benson, A. M., Batzinger, R. P., Ou, S.-Y. L., Bueding, E., Cha, Y.-N., and Talalay, P.,** Elevation of hepatic glutathione-*S*-transferase activities and protection against mutagenic metabolites of benzo(a)pyrene by dietary antioxidants, *Cancer Res.*, 38, 4486, 1978.
34. **Huggins, C. and Fukunishi, R.,** Induced protection of adrenal cortex against 7,12-dimethylbenz[a]anthracene, *J. Exp. Med.*, 119, 923, 1964.
35. **Jakoby, W. B.,** The glutathione *S*-transferases: a group of multifunctional detoxification proteins, *Adv. Enzymol. Relat. Areas Mol. Biol.*, 46, 383, 1978.
36. **Jakoby, W. B. and Habig, W. H.,** Glutathione transferases, in *Enzymatic Basis of Detoxification*, Vol. 2, Jakoby, W. B., Eds., Academic Press, New York, 1980, 63.
37. **Chasseaud, L. F.,** The role of glutathione and glutathione *S*-transferases in the metabolism of chemical carcinogens and other electrophilic agents, *Adv. Cancer Res.*, 29, 175, 1979.
38. **Arias, I. M., Fleischner, G., Kirsch, R., Mishkin, S., and Gatmaitan, Z.,** On the structure, regulation and function of ligandin, in *Glutathione: Metabolism and Function*, Arias, I. M. and Jakoby, W. B., Eds., Raven Press, New York, 1976, 175.
39. **Jakoby, W. B. and Keen, J. H.,** A triple-threat in detoxification: the glutathione *S*-transferases, *Trends Biochem. Sci.*, 2, 229, 1977.
40. **Rahimtula, A. D., Jernstrom, B., Dock, L., and Moldeus, P.,** Effects of dietary and *in vitro* 2(3)-*t*-butyl-4-hydroxyanisole and other phenols on hepatic enzyme activities in mice, *Br. J. Cancer*, 45, 935, 1982.
41. **Morgenstern, R. and Dock, L.,** A comparison of induction of microsomal glutathione *S*-transferase activity in the liver of the mouse and rat by dietary 2(3)-*tert*-butyl-4-hydroxyanisole (BHA), *Acta Chem. Scand.*, 36, 255, 1982.
42. **Pickett, C. B., Telakowski-Hopkins, C. A., Ding, G.J.-F., Argenbright, L., and Lu, A. Y. H.,** Rat liver glutathione *S*-transferases: complete nucleotide sequence of a glutathione *S*-transferase mRNA and the regulation of the Ya, Yb, and Yc mRNAs by 3-methylcholanthrene and phenobarbital, *J. Biol. Chem.*, 259, 5182, 1984.
43. **Telakowski-Hopkins, C. A., Rodkey, J. A., Bennett, C. D., Lu, A. Y. H., and Pickett, C. B.,** Rat liver glutathione *S*-transferases: construction of cDNA clone complementary to a Yc mRNA and prediction of the complete amino acid sequence of a Yc subunit, *J. Biol. Chem.*, 260, 5820, 1985.
44. **Ding, G. J.-F., Ding, V. D.-H., Rodkey, J. A., Bennett, C. D., Lu, A. Y. H., and Pickett, C. B.,** Rat liver glutathione *S*-transferases. DNA sequence analysis of Yb_2 cDNA clone and regulation of the Yb_1 and Yb_2 mRNAs by phenobarbital, *J. Biol. Chem.*, 261, 7952, 1986.
45. **Pearson, W. R., Windle, J. J., Morrow, J. F., Benson, A. M., and Talalay, P.,** Increased synthesis of glutathione *S*-transferases in response to anticarcinogenic antioxidants: cloning and measurement of messenger RNA, *J. Biol. Chem.*, 258, 2052, 1983.
46. **Kasper, C. B. and Henton, D.,** Glucuronidation, in *Enzymatic Basis of Detoxification*, Vol. 2, Jakoby, W. B., Ed., Academic Press, New York, 1980, 3.
47. **Antoine, B., Magdalou, J., and Siest, G.,** Functional heterogeneity of UDP-glucuronosytransferases in different membranes of rat liver, *Biochem. Pharmacol.*, 32, 2629, 1983.
48. **Chowdhury, J. R., Novikoff, P. M., Chowdhury, N. R., and Novikoff, A. B.,** Distribution of UDP-glucuronosyltransferase in rat tissue, *Proc. Natl. Acad. Sci. U.S.A.*, 82, 2990, 1985.
49. **Dutton, G. J.,** *Glucuronidation of Drugs and Other Compounds*, CRC Press, Boca Raton, Fla., 1980, 3.
50. **Finley, B. L., Ashley, P. J., Neptune, A. G., and Yost, G. S.,** Substrate-selective induction of rabbit hepatic UDP-glucuronosyltransferases by ethanol and other xenobiotics, *Biochem. Pharmacol.*, 35, 2875, 1986.
51. **Mackenzie, P. I., Gonzalez, F. J., and Owen, I. S.,** Cloning and characterization of DNA complementary to rat liver UDP-glucuronosyltransferase mRNA, *J. Biol. Chem.*, 259, 12153, 1984.
52. **Hazelton, G. A., Hjelle, J. J., and Klaassen, C. D.,** Effects of butylated hydroxyanisole on hepatic glucuronidation capacity in mice, *Toxicol. Appl. Pharmacol.*, 78, 280, 1985.
53. **Sato, K., Kitahara, A., Yin, Z., Waragai, F., Nishimura, K., Hatayama, I., Ebina, T., Yamazaki, T., Tsuda, H., and Ito, N.,** Induction by butylated hydroxyanisole of specific molecular forms of glutathione *S*-transferase and UDP-glucuronyltransferase and inhibition of development of gamma-glutamyl transpeptidase-positive foci in rat liver, *Carcinogenesis*, 5, 473, 1984.
54. **Sims, P. and Grover, P. L.,** Epoxides in polycyclic aromatic hydrocarbons metabolism and carcinogenesis, *Adv. Cancer Res.*, 20, 165, 1974.
55. **Lu, A. Y. H. and Miwa, G. T.,** Molecular properties and biological functions of microsomal epoxide hydrase, *Annu. Rev. Pharmacol. Toxicol.*, 20, 513, 1980.

56. **Seidegard, J. and DePierre, J. W.,** Microsomal epoxide hydrolase. Properties, regulation and function, *Biochim. Biophys. Acta,* 695, 251, 1983.
57. **Hammock, B. D. and Hasagawa, L. S.,** Differential substrate selectivity of murine hepatic cytosol and microsomal epoxide hydrolases, *Biochem. Pharmacol.,* 32, 1155, 1983.
58. **Oesch, F., Timms, C. W., Walker, C. H., Guenthner, T. M., Sparrow, A., Watabe, T., and Wolf, C. R.,** Existence of multiple forms of microsomal epoxide hydrolases with radically different substrate specificities, *Carcinogenesis,* 5, 7, 1984.
59. **Bresnick, E., Mukhtar, H., Stoming, T. A., Dansette, P. M., and Jerina, D. M.,** Effect of phenobarbital and 3-methylcholanthrene administration on epoxide hydrase levels in liver microsomes, *Biochem. Pharmacol.,* 26, 891, 1977.
60. **Guengerich, F. P., Wang, P., Mitchell, M. B., and Mason, P. S.,** Rat and human liver microsomal epoxide hydratase: purification and evidence for the existence of multiple forms, *J. Biol. Chem.,* 12, 248, 1979.
61. **Gozukara, E. M., Belvedere, G., Robinson, R. C., Deutsch, J., Coon, M. J., Guengerich, F. P., and Gelboin, H. V.,** The effect of epoxide hydratase on benzo(a)pyrene diol epoxide hydrolysis and binding to DNA and mixed-function oxidase proteins, *Mol. Pharmacol.,* 19, 153, 1981.
62. **Oesch, F. and Guenthner, T. M.,** Effects of the modulation of epoxide hydrolase activity on the binding of benzo(a)pyrene metabolities to DNA in the intact nuclei, *Carcinogenesis,* 4, 57, 1983.
63. **Chesis, P. L., Levin, D. E., Smith, M. T., Ernster, L., and Ames, B. N.,** Mutagenicity of quinones: pathways of metabolic activiation and detoxification, *Proc. Natl. Acad. Sci. U.S.A.,* 81, 1696, 1984.
64. **Lind, C., Hochstein, P., and Ernster, L.,** DT-Diaphorase as a quinone reductase: a cellular control device against semiquinone and superoxide radical formation, *Arch. Biochem. Biophys.,* 216, 178, 1982.
65. **Smart, R. C. and Zannoni, V. G.,** DT-Diaphorase and peroxidase influence the covalent binding of the metabolities of phenol, the major metabolites of benzene, *Mol. Pharmacol.,* 26, 105, 1984.
66. **Wefers, H., Komai, T., Talalay, P., and Sies, H.,** Protection against reactive oxygen species by NADPH:quinone reductase induced by the dietary antioxidant butylated hydroxyanisole (BHA), *FEBS Lett.,* 169, 63, 1984.
67. **Wefers, H. and Sies, H.,** Hepatic low-level chemiluminescence during redox cycling of menadione and the menadione-glutathione conjugate: relation to glutathione and NAD(P)H:quinone reductase (DT-Diaphorase) activity, *Arch. Biochem. Biophys.,* 224, 568, 1983.
68. **Iyanagi, T. and Yamazaki, I.,** One-electron-transfer reactions in biochemical systems. V. Differences in the mechanism of quinone reduction by the NADH dehydrogenase and the NAD(P)H dehydrogenase (DT-Diaphorase), *Biochem. Biophys. Acta,* 216, 282, 1970.
69. **De Long, M. J., Dolan, P. Santamaria, A. B., and Bueding, E.,** 1,2-Dithiol-3-thione analogs: effects of NAD(P)H:quinone reductase and glutathione levels in murine hepatoma cells, *Carcinogenesis,* 7, 977, 1986.
70. **Benson, A. M., Barretto, P. B., and Stanley, J. S.,** Induction of DT-Diaphorase by anticarcinogenic sulfur compounds in mice, *J. Natl. Cancer Inst.,* 76, 467, 1986.
71. **Prochaska, H. J. and Talalay, P.,** Purification and characterization of two isofunctional forms of NAD(P)H:quinone reductase from mouse liver, *J. Biol. Chem.,* 261, 1372, 1986.
72. **Hammock, B. D. and Ota, K.,** Differential induction of cytosolic epoxide hydrolase, microsomal epoxide hydrolase, and glutathione *S*-transferase activities, *Toxicol. Appl. Pharmacol.,* 71, 254, 1983.
73. **Dock, L., Cha, Y.-N., Jernstrom, B., and Moldeus, P.,** Differential effects of dietary BHA on hepatic enzyme activities and benzo[a]pyrene metabolism in male and female NMRI mice, *Carcinogenesis,* 3, 15, 1982.
74. **Benson, A. M., Hunkeler, M. J., and Morrow, J. F.,** Kinetics of glutathione transferase, glutathione transferase messenger RNA, and reduced nicotinamide adenine dinucleotide (phosphate):quinone reductase induction by 2(3)-*tert*-butyl-4-hydroxyanisole in mice, *Cancer Res.,* 44, 5256, 1984.
75. **Parchment, R. E. and Benson, A. M.,** Glutathione *S*-transferase isozymes of mouse intestine: differential induction by 2(3)-*tert*-butyl-4-hydroxyanisole, *Biochem. Biophys. Res. Commun.,* 119, 1015, 1984.
76. **Wattenberg, L. W., Coccia, J. B., and Lam, L. K. T.,** Inhibitory effect of phenolic compounds on benzo(a)pyrene-induced neoplasia, *Cancer Res.,* 40, 2820, 1980.
77. **Guarna, A., Corte, L. D., Giovanni, M. G., De Sarlo, F., and Sgaragli, G.,** 2,2′-Dihydroxy-3,3′-di-*t*-butyl-5,5′-dimethoxydiphenyl a new metabolite of 2-*t*-butyl-4-methoxyphenol in the rat, *Drug Metab. Dispos.,* 11, 581, 1983.
78. **El-Rashidy, R. and Niazi, S.,** A new metabolite of butylated hydroxyanisole in man, *Biopharm. Drug Dispos.,* 4, 389, 1983.
79. **Darlington, G. J., Bernhard, H. P., Miller, R. A., and Rudde, F. H.,** Expression of liver phenotypes in cultured mouse hepatoma cells, *J. Natl. Cancer Inst.,* 64, 809, 1980.
80. **Hankinson, O.,** Single-step selection of clones of a mouse hepatoma line deficient in aryl hydrocarbon hydroxylase, *Proc. Natl. Acad. Sci. U.S.A.,* 76, 373, 1979.

81. **De Long, M. J., Santamaria, A. B., and Talalay, P.,** Role of cytochrome P-450 in the induction of NAD(P)H: quirone reductase in a murine hepatoma cell line and its mutants, *Carcirogenesis,* 8, 1549, 1987.
82. **Williams, R. T.,** Comparative patterns of drug metabolism, *Fed. Proc.,* 26, 1029, 1967.
83. **Ip, C.,** Comparative effects of antioxidants on enzymes involved in glutathione metabolism, *Life Sci.,* 34, 2501, 1984.
84. **Aspry, K. E. and Bjeldanes, L. F.,** Effects of dietary broccoli and butylated hydroxyanisole on liver-mediated metabolism of benzo(a)pyrene, *Food Chem. Toxicol.,* 21, 133, 1983.
85. **Roebuck, B. D., MacMillan, D. L., Bush, D. M., and Kensler, T. W.,** Modulation of azaserine-induced pancreatic foci by phenolic antioxidants in rats, *J. Natl. Cancer Inst.,* 72, 1405, 1981.
86. **Gaunt, I. F., Feuer, G., Fairweather, F. A., and Gilbert, D.,** Liver response tests. IV. Application to short-term feeding studies with butylated hydroxytoluene (BHT) and butylated hydroxyanisole (BHA), *Food Cosmet. Toxicol.,* 3, 433, 1965.
87. **Feuer, G., Gaunt, I. F., Goldberg, L., and Fairweather, F. A.,** Liver response tests. VI. Application to a comparative study of food antioxidants and hepatotoxic agents, *Food Cosmet. Toxicol.,* 3, 457, 1965.
88. **Cha, Y.-N. and Heine, H. S.,** Comparative effects of dietary administration of 2(3)-*tert*-butyl-4-hydroxy-anisole and 3,5-di-*tert*-butyl-4-hydroxytoluene on several hepatic enzymes in mice and rats, *Cancer Res.,* 42, 2609, 1982.

ANTI-INFLAMMATORY EFFECTS OF NATURAL PHENOLIC COMPOUNDS

Maria José Alcaraz and Angel Villar

INTRODUCTION

Phenolic derivatives are widespread in the plant kingdom, the occurrence of flavane (2-phenylbenzo[*a*]pyrane or 2-phenylchromane) derivatives and related compounds (flavonoids) being especially outstanding. Flavonoids appear mainly in the form of glycosides which can form water-soluble complexes including different polyphenols and ascorbid acid.[1] Their biological actions are known since 1936, when the capacity of *Capsicum annuum* and *Citrus limon* fruits to cure S.C. capillary bleeding was discovered, and subsequently the synergism of these "vitamin P" principles with ascorbic acid in scurvy was demonstrated. Nevertheless, the therapeutic efficacy of "bioflavonoids" was not sufficiently proven to the U.S. Food and Drug Administration and they were removed from the market in that country and considered to be secondary food factors.[2]

The pharmacological interest of flavonoids has been defended by some authors and attacked by others, the very complexity of the subject probably being at the root of such contradictions. It has been claimed that the expected pharmacological activity of isolated products would disappear after purification of plant extracts on account of active principle instability or due to the formation of inactive polymers.[1,3] A relevant point could be the insufficient knowledge of their pharmacokinetic properties. Some flavonoids, such as rutin, show a very poor absorption following oral administration and can undergo ring fission by intestinal micro-organisms.[4,5]

Presently, *Vaccinium myrtillus* anthocyanins are used to treat diabetic retinopathy,[6] and leukocyanidin, diosmin, or rutin derivatives, *O*-(β-hydroxyethyl)rutin and troxerutin, are used to treat conditions with peripheral vascular insufficiency and edema due to impaired venous drainage, such as peripheral arterial occlusive disease, venous insufficiency, and post-thrombotic edema.[7]

ANTI-INFLAMMATORY EFFECTS

There is a great deal of evidence on the anti-inflammatory effects of phenolic derivatives, mainly of flavonoid structure, although most research done in this field is academically oriented and information on clinical trials is scarce yet. Antiedematous activity has been reported for numerous compounds, the rat paw edema induced by carrageenan being the most widely used test. The structures and origin of some of those substances are shown in Figure 1 and Table 1.[6,8-21]

Phenolic derivatives seem to be less potent than the classical nonsteroidal anti-inflammatory (NSAI) drugs in inhibiting edema. Exceptions would include hypolaetin-8-*O*-glucoside[10] (ED_{50} = 57 mg/kg i.p.), gossypin (ED_{50} = 66 mg/kg i.p.),[12,13] or (−)epicatechin (ED_{50} = 74 mg/kg i.p.).[19] There are compounds active in other models of acute inflammation, as reviewed by Gabor.[22]

Less information is available on the action of natural products of phenolic structure in tests reproducing the development of the granulation tissue (e.g., the cotton pellet granuloma in rats) or arthritic conditions (Freund's adjuvant arthritis in rats). Among the flavone derivatives, nepetrin (20 to 40 mg/kg/day i.p.) inhibits both animal models,[9] while luteolin and its 7-*O*-glucoside (20 mg/kg/day i.p.) are active in the cotton pellet granuloma,[23] and hypolaetin-8-*O*-glucoside in the carrageenan-induced abscess (30—90 mg/kg/day i.p.).[10] Recently, the inhibitory activity of baicalein, baicalin, and wogonin in adjuvant arthritis[8]

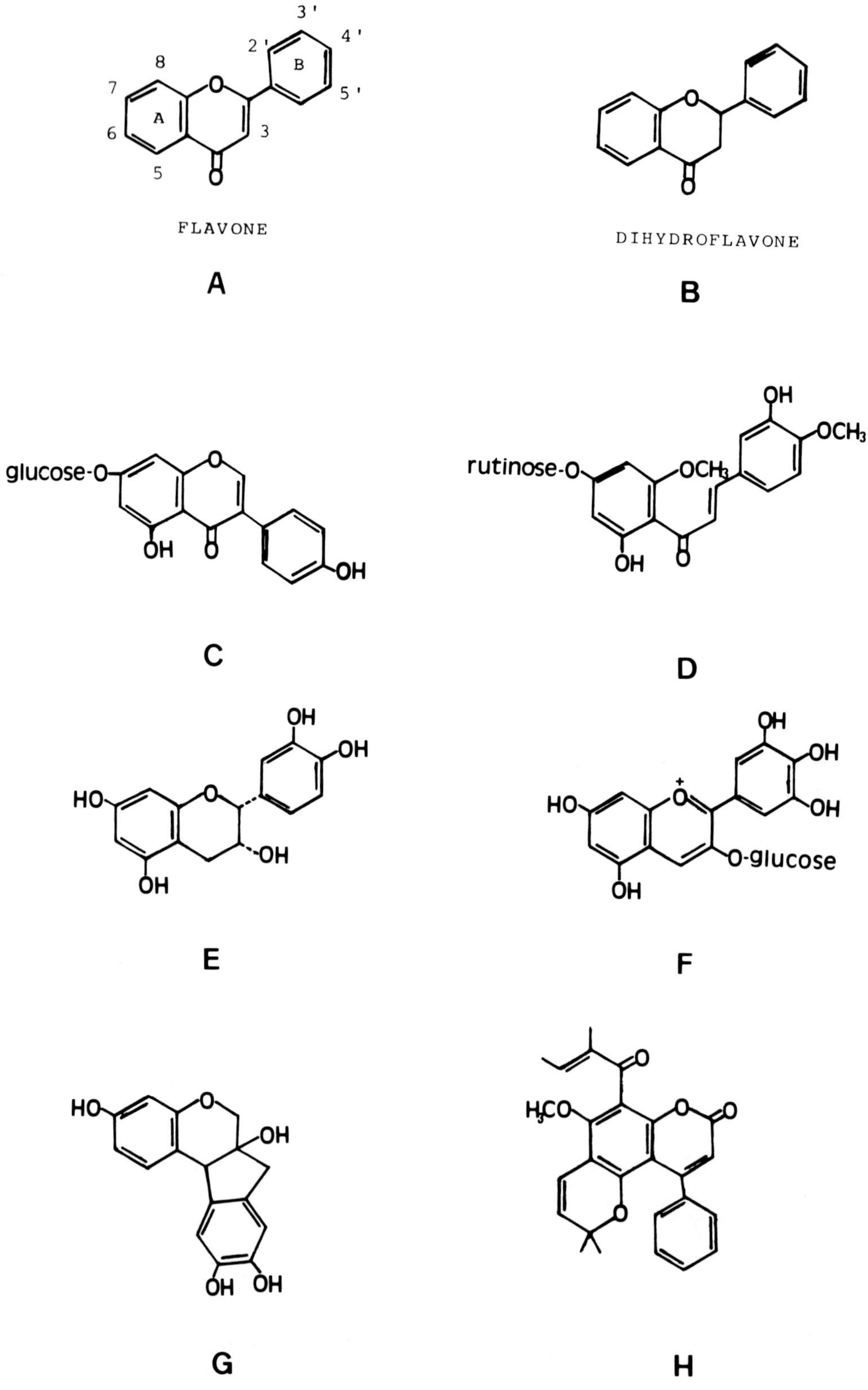

FIGURE 1. Structures of the compounds indicated in Table 1.

Table 1
SOME EXAMPLES OF PHENOLIC COMPOUNDS POSSESSING ANTIEDEMATOUS ACTIVITY

FLAVONES

Basic structure A[a] with substituents at the positions:								
5	**6**	**7**	**8**	**3′**	**4′**		**Plant**	**Ref.**
OH	OH	OH	H	H	H	Baicalein	*Scutellaria baicalensis*	8
OH	OCH_3	*O*-Glucose	H	OH	OH	Nepetrin	*Nepeta hindostana*	9
OH	H	OH	*O*-Glucose	OH	OH	Hypolaetin-8-*O*-glucoside	*Sideritis mugronensis*	10

FLAVONOLS

Basic structure A[a] with substituents at the positions:									
3	**5**	**6**	**7**	**8**	**3′**	**4′**			
OH	OH	H	OH	H	OH	OH	Quercetin	*Chamomilla recutita*	11
O-Rutinose	OH	H	OH	H	OH	OH	Rutin	*C. recutita*	11
OH	OH	H	*O*-Glucose	OH	OH	OH	Gossypin	*Hibiscus vitifolius*	12-14

FLAVANONES

Basic structure B[a] with substituents at the positions:								
5	**6**	**7**	**8**	**3′**	**4′**			
H	Prenyl	OCH_3	H	H	OH	Bavachinin-A	*Psoralea corylifolia*	15

FLAVANONOLS

Basic Structure B[a] with substituents at the positions:							
3	**5**	**7**	**3′**	**4′**			
OH	OH	OH	OH	OH	Taxifolin	Madhuca butyracea	16

Isoflavones	(C)[a]	Genistin	*Genista* spp.	17
Chalcones	(D)[a]	Hesperidin methylchalcone	Semisynthetic	18
Catechins	(E)[a]	(−)-Epicatechin	*Cinnamomum sieboldii*	19
Anthocyanins	(F)[a]	Delphinidin-3-*O*-glucoside	*Vaccinium myrtillus*	6
Arylchromans	(G)[a]	Brazilin	*Caesalpinia sappan*	20
4-Arylcoumarins	(H)[a]	Calophyllolide	*Calophyllum inophyllum*	21

[a] See Figure 1.

has been reported. A few examples of flavonols inhibiting the cotton pellet granuloma could be quercetin and its glycosides, rutin, quercimeritrin, and hyperoside, as well as kaempferol with its 7-*O*-rhamnoside and its 3,7-di-*O*-rhamnoside at a dose level of 20 mg/kg/day i.p. or 50 mg/kg/day p.o.[22-24]

Several flavanones, flavanonols, isoflavones, and catechins can inhibit the proliferation of granulation tissue (e.g., naringenin and its 7-*O*-glucoside[24] taxifolin[16] genistin[17] and (−)-epicatechin with its related polymers isolated from *Cinnamomum sieboldii*.[19,25] Additionally, (−)-epicatechin has shown beneficial effects in adjuvant arthritis.[19]

Phenolic derivatives possessing different structures were found to exert anti-inflammatory effects on the proliferative phase of experimental inflammation. Mention could be made of magnoshinin and magnosalin, neolignans from *Magnolia salicifolia*;[26] the arylchroman brazilin;[20] the 4-arylcoumarin calophyllolide, which is also active in adjuvant arthritis;[21] and the xanthones calophyllin B, jacareubin, 6-deoxyjacareubin, and related principles isolated from *Calophyllum inophyllum* and *Mesua ferrea*.[27] Another xanthone, mangostin, inhibits the primary and secondary responses of adjuvant arthritis.[28]

Antipyretic and analgesic actions are associated with the anti-inflammatory ability of NSAI drugs. In the case of natural phenolic derivatives, the activity seems to be less than that reported for NSAI drugs.[9,10,15]

A shortcoming of the investigations on the anti-inflammatory properties of natural phenolics is the absence of data on toxicity studies. Only in a few cases were there reports of GI side effects of the compounds under screening. Such an aspect is especially interesting in the work on anti-inflammatory drugs, as GI bleeding and ulcer are the most common side effects of NSAI drugs. Some phenolic compounds are devoid of ulcerogenic effects and may even protect the GI tract from damage, as demonstrated for gossypin,[12] (+)-catechin,[12] naringenin,[29] and hypolaetin-8-*O*-glucoside.[10] Joint administration of flavonoids and NSAI drugs has been suggested to reduce the occurrence of digestive lesions.[14]

MECHANISMS OF ACTION

Many reports provide data on the interference of flavonoids with mechanisms involved in the inflammatory process, thereby raising a different hypothesis on how these compounds act. In the following paragraphs, an overview of findings that would explain some aspects of their mechanism of action is presented.

Synthesis and Release of Inflammatory Mediators

Histamine

Some flavonoids inhibit histidine decarboxylase from different tissues or can interact with mechanisms leading to histamine release.[22,29] Quercetin, chrysin, fisetin, taxifolin, phloretin, myricetin, and kaempferol inhibit mast cell histamine secretion induced by receptor-directed agonists.[30,31] Additionally, quercetin is very potent in inhibiting antigen-induced histamine release in human basophils.[32] Flavonoids do not act through increasing cAMP concentrations in those cells, although this point has not been definitely established and such a mechanism would appear to operate in mast cells.[33] The inhibitory activity of flavonoids in human basophils seems to be related to their ability to chelate Cu^{2+}[34]. Several investigations point to the participation of some lipoxygenase (LO) products in the process of histamine release, suggesting that arachidonic acid metabolites may promote Ca^{2+} uptake.[32] Inhibition of 5-LO by quercetin may contribute to its inhibitory effects,[35,36] which are believed to depend on blockade of agonist-induced entry of Ca^{2+}, maybe acting at the receptor-directed calcium channels in the plasma membrane.[31]

Cyclooxygenase Products

According to Vane's hypothesis, inhibition of cyclooxygenase (CO) activity is the major mechanism to explain the anti-inflammatory action of NSAI drugs.[37] As a consequence, the reduction in the synthesis of all arachidonic acid metabolites via CO would control the inflammatory process. Prostaglandins (PGs) potentiate the effects of other mediators (histamine, serotonin, bradykinin, and complement fraction C_{5a}) in inducing vasodilatation, increase in capillary permeability, and pain.[38] Furthermore, PGs largely produced by macrophages can be modulators of chronic inflammation.[39,40]

Phenolic compounds can have either stimulating or inhibiting influences on CO activity. The minimum structural requirement for activation is a phenol possessing a polar side chain in either meta or para position.[41] Corresponding with this hypothesis, rutin and hypolaetin-

8-*O*-glucoside are enzyme activators.[42] On the other hand, it has been postulated that flavonoids possessing the ortho-dihydroxyl group(pyrocatechin) are able to inhibit CO.[43] In fact, quercetin, fisetin, luteolin, taxifolin, morin, (+)-catechin, and (−)-epicatechin were found to be inhibitors in preparations of rat renal medulla. The concentrations used, however, were rather high (1 m*M*).[44] Hypolaetin-8-*O*-glucoside enhances PG formation,[42] a fact that could account for the gastroprotective effects exerted in vivo, as PGI_2 and PGE_2 are formed in the gastric tissues and have antisecretory, antiulcer, and cytoprotective properties.[45]

The mechanism by which phenolic substances can enhance PG formation are thought to depend upon their free radical scavenging activity. It is believed that during the conversion of PGG_2 to PGH_2 an oxygen-dependent free radical is generated, presumably hydroxyl (·OH), which is capable of inactivating CO and other enzymes of the arachidonic acid cascade. Free radical scavengers would trap such oxygen species and therefore they would prevent its destructive action. The result would be the stimulation of PG synthesis.[46-48]

This theory has not been definitely proven in the case of flavonoids. Baumann et al.[44] did not find any correlation between the influence on PG synthesis and the free radical scavenging ability in quite a numerous series of phenolic compounds.

There are a few data concerning the influence of flavonoids on PG inactivation. The inhibition of 15-hydroxyprostaglandin dehydrogenase by baicalein has been related to its inhibitory action on the release of mediators during anaphylaxis.[49]

LO Products

This enzyme generates hydroxy fatty acids (HETEs) and leukotrienes (LTs), which can be chemotactic for leukocytes during inflammation. LTC_4, LTD_4, and LTE_4 cause exudation of plasma, while LTB_4 is a potent chemotactic agent, comparable to the complement-derived peptide C_{5a} and the platelet-activating factor.[50]

Classical NSAI drugs cannot influence LO activity at doses able to inhibit CO and inflammatory reaction. The potential interest of drugs that cause selective inhibition of LO (with therapeutic value in anaphylaxis) as well as agents that would be dual inhibitors of both CO and LO, able to control the inflammatory conditions, with similar properties to the anti-inflammatory corticosteroids, but devoid of the steroid-related toxicity[51] has been noted.

Many flavonoids inhibit LOs from different sources. The presence of the keto group at 4 position and the hydroxyl at 4′ position with absence of substitution at 2′ position has been stated as the minimum structural requirement for inhibition of soybean LO.[43] Studies by Yamamoto et al.[52] showed the presence of the ortho-dihydroxyl group at the ring B as the condition to exert inhibitory effects on rat basophilic leukemia cells 5-LO. The existence of the orthodihydroxyl group at 5,6 has been noted to facilitate the inhibitory activity of flavonoids.[53] Effective inhibitors of LOs are cirsiliol, quercetin, and baicalein, with IC_{50} of 10^{-7}, 3.5×10^{-7}, and 7.1×10^{-7} *M*, respectively, in the rat basophilic leukemia cell system.[52,54] The inhibitory activity of baicalein and related flavonoids would explain the antiallergic and anti-inflammatory uses of *Scutellaria baicalensis* roots in folk medicine.[53]

Antagonism of Inflammatory Mediators

Some flavonoids can inhibit edema, increase in capillary permeability, and isolated organ contractions induced by inflammatory mediators, such as histamine, serotonin, bradykinin, PGE_2, and $PGF_{2\alpha}$, although they do not appear to be potent antagonists.[22,55,56]

Leukocyte Functions

Previously known as the stabilizing effect of some flavonoids on the lysosomes,[57,58] more recently their ability to act on neutrophil functions has been demonstrated. Quercetin inhibits the release of lysosomal enzymes, oxygen consumption, generation of superoxide anion (O_2^-), and chemotaxis, while its glycoside rutin had no significant effect.[31,59,60] On the other hand, there exists the possibility of a direct inhibition of lysosomal hydrolases by flavonoids.[61]

The inhibition exerted by hypolaetin-8-*O*-glucoside on leukocyte migration and release of β-glucuronidase in the carrageenan air pouch in rats[62], as well as the stabilizing effects of (+)-catechin and *O*-(β-hydroxyethyl)rutin on the lysosomes of arthritic rats have been reported,[63] while quercetin and hesperidin have similar effects in animals receiving a hypercholesterolemic diet.[64] It seems that flavonoids could act on different points:

1. By inhibiting phospholipases — It is known that quercetin inhibits activation of neutrophil phospholipase A_2 by zymosan-activated serum. This mechanism might be responsible for the inhibition of enzyme release and for other effects of quercetin on neutrophils.[65] The compounds could interfere with the uptake of Ca^{2+} by the enzyme, as suggested for NSAI drugs.[66]
2. Blocking the stimulus-induced calcium influx through the cell membrane[35] — Flavonoids might inhibit Ca-ATPases associated with exclusion of calcium from the cytosol. Such mechanisms are controversial since, in addition to the lack of correlation between inhibition of secretory functions and inhibition of Ca-ATPase, quercetin and other flavonoids that inhibit neutrophil functions did not significantly alter calcium movements across the plasma membrane.[31]
3. Inhibiting metabolization of arachidonic acid[35,67]—Flavonoids are able to inhibit CO and LO pathways, as indicated in other parts of this chapter.

Catechol-*O*-Methyltransferase

Flavonoids with a catechol structure of the side ring B are potent inhibitors of this enzyme, while glycosides and compounds lacking the Δ2,3 double bond or the keto group at 4 show reduced inhibitory effects.[68,69] Inhibition of catechol-*O*-methyltransferase slows the inactivation of adrenaline and noradrenaline, which in turn could promote antihistaminic and anti-inflammatory effects, improving the tonus of precapillary sphincters and having a role in the reduction of capillary permeability.[61]

Connective Tissue

Increased capillary permeability is a feature of early inflammatory reactions. Flavonoids can decrease capillary permeability and strengthen the blood vessel wall, and thus vascular fragility is also prevented.[1] These compounds have a great affinity for the connective tissue,[3] having at this level a vitamin C-sparing activity which may be due to a nonspecific antioxidant action,[1] to a specific role in reactions of electron transfer between ascorbic acid and a flavonoid molecule,[1] or to their ability to chelate Cu^{2+}, an essential cofactor for oxidation of vitamin C.[70]

Some flavonoids stimulate proline hydroxylation and collagen cross-linking.[1,71] Such action has been demonstrated for (+)-catechin and *O*-(β-hydroxyethyl)rutin in rats made arthritic by adjuvant administration.[72] As an additional mechanism, flavonoids could modify the anionic charges of polysaccharide chains and in this way they would facilitate the approximation of collagen precursors.[1] Furthermore, flavonoids or flavonoid-Cu^{2+} complexes can impede the depolymerizing activity of hyaluronidase, an enzyme that makes the ground substance of connective tissue more permeable.[5,61]

Antioxidant Properties of Phenolic Compounds

The occurrence of phenolic hydroxyls at the 3 or 5 positions and the keto function at 4 position are determinant factors for the antioxidant properties of flavonoids, being increased by the presence of multiple phenolic functions in the molecule and reduced by glycosylation.[73] The scavenger activity for oxygen-derived radicals has been demonstrated in a numerous group of flavonoids and other phenolic compounds.[44,74] Such property would justify, at least in part, their anti-inflammatory effects, as they could prevent the deleterious results of free radical interactions with the tissues during the inflammatory process. Thus, phenolic compounds would protect against lipid peroxidation and destruction of cellular and organelle membranes, release of lysosomal enzymes, chemotactic factors, and altered tissue components that perpetuate the inflammatory condition,[48,66] as well as protect against the depolymerizing action of free radicals on the hyaluronic acid of the synovial fluid.[75] The antioxidant ability of phenolic derivatives would also explain the influence of such compounds on enzymes having a role in inflammation.[53]

REFERENCES

1. **Durandeau, C.,** Flavonoïdes et Tissu Conjunctif, Ph.D. thesis, University of Bordeaux II, Bordeaux, 1978.
2. **McClure, J. W.,** Physiology and functions of flavonoids, in *The Flavonoids* (Part 2), Harborne, J. B., Mabry, T. J., and Mabry, H., Eds., Academic Press, New York, 1975, 970.
3. **Laparra, J., Michaud, J., Lesca, M. F., Blanquet, P., and Masquelier, J.,** Étude autoradiographique de la localisation du tetrahydroxyflavanediol-^{14}C chez la souris, *C.R. Acad. Sci. Paris,* 276, 2847, 1973.
4. **Griffiths, L. A. and Smith, G. E.,** Metabolism of myricetin and related compounds in the rat metabolite formation in vivo and by the intestinal microflora in vitro, *Biochem. J.,* 130, 141, 1972.
5. **Laparra, J., Michaud, J., and Masquelier, J.,** Étude pharmacocinétique des oligomères flavanoliques, *Plant. Med. Phytother.,* XI, 133, 1977.
6. **Pourrat, H.,** Drogues à anthocyanes et maladies vasculaires, *Plant. Med. Phytother.,* XI, 143, 1977.
7. **Gerdin, B. and Svensjö, E.,** Inhibitory effect of the flavonoid *O*-(β-hydroxyethyl)rutoside on increased microvascular permeability induced by various agents in rat skin, *Int. J. Microcirc. Clin. Exp.,* 2, 39, 1983.
8. **Kubo, M., Matsuda, H., Tanaka, M., Kimura, Y., Okuda, H., Higashino, M., Tani, T., Namba, K., and Arichi, S.,** Studies on *Scutellariae radix.* VII. Anti-arthritic and anti-inflammatory actions of methanolic extract and flavonoid components from *Scutellariae radix, Chem. Pharm. Bull.,* 32, 2724, 1984.
9. **Agarwal, O. P.,** The anti-inflammatory action of nepitrin, a flavonoid, *Agents Actions,* 12, 298, 1982.
10. **Villar, A., Gascó, M. A., and Alcaraz, M. J.,** Anti-inflammatory and anti-ulcer properties of hypolaetin-8-glucoside, a novel plant flavonoid, *J. Pharm. Pharmacol.,* 36, 820, 1984.
11. **Della Loggia, R., Tubaro, A., and Zilli, C.,** Topical antiinflammatory activity of some chamomile flavonoids, in *Proc. 32nd Annu. Cong. Medical Plant Research,* Farmaceutish Tijdschrift VoorBelgië, Antwerp, 2, 1984, 250.
12. **Parmar, N. S. and Ghosh, M. N.,** The antiinflammatory and anti-gastric ulcer activities of some flavonoids, *Bull. Jawaharlal Inst. Post-Grad. Medical Educ. Res.,* 1, 6, 1976.
13. **Parmar, N. S. and Ghosh, M. N.,** Effect of gossypin, a bioflavonoid, on the temporal characteristics of paw oedema and increase in vascular permeability induced by carrageenin in rats, *Bull. Jawaharlal Inst. Post-Grad. Medical Educ. Res.,* 2, 42, 1977.
14. **Chateree, G. K. and Pal, S. P.,** Search for antiinflammatory agents from Indian medicinal plants — a review, *Indian Drugs,* 21, 413, 1984.
15. **Anand, K. K., Sharma, M. L., Singh, B., and Ghatak, B. J. R.,** Antiinflammatory, antipyretic and analgesic properties of bavachinin-A, flavanone isolated from seeds of *Psoralea corylifolia* Linn. (Babchi), *Indian J. Exp. Biol.,* 16, 1216, 1978.
16. **Gupta, M. B., Bhalla, T. N., Gupta, G. B., Mitra, C. R., and Bhargava, K. P.,** Anti-inflammatory activity of taxifolin, *Jpn. J. Pharmacol.,* 21, 377, 1971.
17. **Ilarionov, I., Rainova, L., and Nakov, N.,** Antiinflammatory and antiulcer effects of some flavonoids isolated from the genus *Genista, Farmatsiya (Sofia),* 29, 39, 1979.
18. **Tarayre, J. P. and Lauressergues, H.,** Advantages of a combination of proteolytic enzymes, flavonoids and ascorbic acid in comparison with nonsteroidal antiinflammatory agents, *Arzneim. Forsch.,* 27, 1144, 1977.
19. **Swarnalakshmi, T., Gomathi, K., Sulochana, N., Baskar, E. A., and Parmar, N. S.,** Antiinflammatory activity of (−)epicatechin, a bioflavonoid isolated from *Anacardium occidentale* Linn., *Indian J. Pharm. Sci.,* 43, 205, 1981.
20. **Hikino, H., Taguchi, T., Fujimura, H., and Hiramatsu, Y.,** Antiinflammatory principles of *Caesalpinia sappan* wood and of *Haematoxylon campechianum* wood, *Planta Med.,* 31, 214, 1977.

21. **Chaturvedi, A. K., Parmar, S. S., Bhatnagar, S. C., Mistra, G., and Nigam, S. K.**, Anticonvulsant and antiinflammatory activity of natural plant coumarins and triterpenoids, *Res. Commun. Chem. Pathol. Pharmacol.*, 9, 11, 1974.
22. **Gabor, M.**, Anti-inflammatory substances of plant origin, in *Handbook of Experimental Pharmacology: Antiinflammatory Drugs*, Vane, J. R. and Ferreira, S. H., Eds., Springer, New York, 1979, 698.
23. **Kalashnikova, N. A. and Gerashchenko, G. I.**, Antiphlogistic activity of several flavonoids, *Aktual. Vopr. Farm.*, 2, 353, 1974.
24. **Roshchin, Y. V. and Gerashchenko, G. I.**, Antiinflammatory activity of some flavonoids, *Vopr. Farm. Dal'nem Vostoke*, 1, 134, 1973, *Chem. Abstr.*, 83, 37, 683q, 1975.
25. **Otsuka, H., Fujioka, S., Komiya, T., Mizuta, E., and Takamoto, M.**, Studies on antiinflammatory agents. VI. Antiinflammatory constituents of *Cinnamomum sieboldii* Meissn., *Yakugaku Zasshi*, 102, 162, 1982.
26. **Kimura, M., Suzuki, J., Yamada, T., Yoshizaki, M., Kikuchi, T., Kadota, S., and Matsuda, S.**, Anti-inflammatory effect of neolignans newly isolated from the crude drug "shin-i" *(Flos magnoliae)*, *Planta Med.*, 291, 1985.
27. **Gopalakrishnan, C., Shankaranarayanan, D., Nazimudeen, S. K., Viswanathan, S., and Kameswaran, L.**, Antiinflammatory and CNS depressant activities of xanthones from *Calophyllum inophyllum* and *Mesua ferrea*, *Indian J. Pharmacol.*, 12, 181, 1980.
28. **Gopalakrishnan, C., Shankaranarayanan, D., Kameswaran, L., and Nazimudeen, S. K.**, Effect of mangostin, a xanthone from *Garcinia mangostana* Linn. in immunopathological and inflammatory reactions, *Indian J. Exp. Biol.*, 18, 843, 1980.
29. **Parmar, N. S.**, The gastric anti-ulcer activity of naringenin, a specific histidine decarboxylase inhibitor, *Int. J. Tiss. React.*, V, 415, 1983.
30. **Fewtrell, C. M. S. and Gomperts, B. D.**, Effect of flavone inhibitors of transport ATPases on histamine secretion from rat mast cells, *Nature (London)*, 265, 635, 1977.
31. **Bennett, J. P., Gomperts, B. D., and Wollenweber, E.**, Inhibitory effects of natural flavonoids on secretion from mast cells and neutrophils, *Arzneim. Forsch.*, 31, 433, 1981.
32. **Middleton, E., Jr., Drzewiecki, G., and Krishnarao, D.**, Quercetin: an inhibitor of antigen-induced human basophil histamine release, *J. Immunol.*, 127, 546, 1981.
33. **Middleton, E., Jr. and Drzewiecki, G.**, Flavonoid inhibition of human basophil histamine release stimulated by various agents, *Biochem. Pharmacol.*, 33, 3333, 1984.
34. **Middleton, E., Jr. and Drezewiecki, G.**, Effects of flavonoids and transitional metal cations on antigen-induced histamine release from human basophils, *Biochem. Pharmacol.*, 31, 1449, 1982.
35. **Showell, H. J., Naccache, P. H., Walenga, R. W., Dalecki, M., Feinstein, M. B., Sha'afi, R. I., and Becker, E. L.**, The effects of quercetin, 1-tosylamido-2-phenylethyl chloromethylketone, cytochalasin A and nordihydroguaiaretic acid on lysosomal enzyme secretion, arachidonic acid metabolism and Ca^{2+} fluxes in rabbit neutrophils, *J. Reticuloendothel. Soc.*, 30, 167, 1981.
36. **Hope, W. C., Welton, A. F., Fiedler-Nagy, C., Batula-Bernardo, C., and Coffey, J. W.**, In vitro inhibition of the biosynthesis of slow reacting substance of anaphylaxis (SRS-A) and lipoxygenase activity by quercetin, *Biochem. Pharmacol.*, 32, 367, 1983.
37. **Vane, J. R.**, The mode of action of aspirin-like drugs, *Agents Actions*, 8, 430, 1978.
38. **Williams, T. J.**, Interactions between prostaglandins, leukotrienes and other mediators of inflammation, *Br. Med. Bull.*, 39, 239, 1983.
39. **Bonta, I. L. and Parnham, M. J.**, Prostaglandins and chronic inflammation, *Biochem. Pharmacol.*, 27, 1611, 1978.
40. **Lewis, G. P.**, Immunoregulatory activity of metabolites of arachidonic acid and their role in inflammation, *Br. Med. Bull.*, 39, 243, 1983.
41. **Baumann, J. von Bruchhausen, F., and Wurm, G.**, A structure-activity study on the influence of phenolic compounds and bioflavonoids on rat renal prostaglandin synthetase, *Naunyn-Schmiedeberg's Arch. Pharmakol.*, 307, 73, 1979.
42. **Alcaraz, M. J. and Hoult, J. R. S.**, Actions of flavonoids and the novel antiinflammatory flavone, hypolaetin-8-glucoside on prostaglandin biosynthesis and inactivation, *Biochem. Pharmacol.*, 34, 2477, 1985.
43. **Wurm, G., Baumann, J., and Geres, U.**, Beeinflussung des Arachidonsäurestoffwechsels durch Flavonoide, *Dtsch. Apoth. Ztg.*, 122, 2062, 1982.
44. **Baumann, J., Wurm, G., and von Bruchhausen, F.**, Hemmung der Prostaglandin-synthetase durch Flavonoide und Phenolderivate im Vergleich mit deren O_2^- Radikalfängereigenschaften, *Arch. Pharm. (Weinheim)*, 313, 330, 1980.
45. **Waller, S. L.**, Prostaglandins and the gastrointestinal tract, *Gut*, 14, 402, 1973.
46. **Kuehl, F. A., Humes, J. L., Egan, R. W., Ham, E. A., Beveridge, G. C., and van Arman, C. G.**, Role of prostaglandin endoperoxide PGG_2 in inflammatory processes, *Nature (London)*, 265, 170, 1977.

47. **Payne, T. G., Dewald, B., Siegl, H., Gubler, H. U., Ott, H., and Baggiolini, M.,** Radical scavenging and stimulation of prostaglandin synthesis not anti-inflammatory?, *Nature (London)*, 296, 160, 1982.
48. **Hertz, F. and Cloarec, A.,** Pharmacology of free radicals; recent views on their relation to inflammatory mechanisms, *Life Sci.*, 34, 713, 1984.
49. **Iijima, Y., Kawakita, N., and Yamazaki, M.,** Inhibition of 15-hydroxyprostaglandin dehydrogenase by antiallergic agents, *Biochem. Biophys. Res. Commun.*, 93, 912, 1980.
50. **Piper, P. J.,** Pharmacology of leukotrienes, *Br. Med. Bull.*, 39, 255, 1983.
51. **Higgs, G. A. and Vane, J. R.,** Inhibition of cyclo-oxygenase and lipoxygenase, *Br. Med. Bull.* 39, 265, 1983.
52. **Yamamoto, S., Yoshimoto, T., Furukawa, M., Horie, T., and Watanabe-Kohno, S.,** Arachidonate 5-lipoxygenase and its new inhibitors, *J. Allergy Clin. Immunol.*, 74, 349, 1984.
53. **Kimura, Y., Okuda, H., and Arichi, S.,** Studies on *Scutellariae radix*. XIII. Effects of various flavonoids on arachidonate metabolism in leukocytes, *Planta Med.*, 132, 1985.
54. **Masters, D. J., Spruce, K. E., Vickers, V. C., and McMillan, R. M.,** Measurement of arachidonate metabolism using cyanopropyl minocolumns: effects of cyclo-oxygenase and lipoxygenase inhibitors, *Br. J. Pharmacol.*, 84, 45P, 1985.
55. **Chau, T. T. and Haley, T. J.,** Flavonoid antagonism of the spasmogenic effect of angiotensin, bradykinin and eledoisin on guinea pig ileum, *J. Pharm. Sci.*, 58, 621, 1969.
56. **Metzner, J., Bekemeier, H., and Weber, F. G.,** Influence of flavonoids on capillary permeability, carrageenin edema and histamine and PGE_2 spasms, *Agents Actions*, Suppl. 10, 129, 1982.
57. **van Caneghem, P.,** Influence of some hydrosoluble substances with vitamin P activity on the fragility of lysosomes in vitro, *Biochem. Pharmacol.*, 21, 1543, 1972.
58. **Niebes, P. and Ponard, G.,** Stabilization of rat liver lysosomes by (+)-cyanidol-3 in vivo, *Biochem. Pharmacol.*, 24, 905, 1975.
59. **Berton, G., Schneider, C., and Romeo, D.,** Inhibition by quercetin of activation of polymorphonuclear leucocyte functions, *Biochim. Biophys. Acta*, 595, 47, 1980.
60. **Busse, W. W., Kopp, D. E., and Middleton, E., Jr.,** Flavonoid modulation of human neutrophil function, *J. Allergy Clin. Immunol.*, 73, 801, 1984.
61. **Brown, J. P.,** A review of the genetic effects of naturally occurring flavonoids, anthraquinones and related compounds, *Mutat. Res.*, 75, 243, 1980.
62. **Villar, A., Gascó, M. A., and Alcaraz, M. J.,** Some aspects of the inhibitory activity of hypolaetin-*S*-glucoside in acute inflammation, *J. Pharm. Pharmakol.*, 39, 502, 1987.
63. **Rao, C. N., Rao, V. H., Verbruggen, L., and Orloff, S.,** Effect of bioflavonoids on lysosomal acid hydrolases and lysosomal stability in adjuvant-induced arthritis, *Scand. J. Rheumatol.*, 9, 280, 1980.
64. **Rathi, A. B., Nath, N., and Chari, S. N.,** Action of vitamin P-like compounds on lysosomal status in hypercholesterolemic rats, *Acta Vitaminol. Enzymol.*, 5, 255, 1983.
65. **Lee, T. P., Matteliano, M. L., and Middleton, E. Jr.,** Effect of quercetin on human PMNL lysosomal enzyme release and phospholipid metabolism, *Life Sci.*, 31, 2765, 1982.
66. **Lewis, D. A.,** Endogenous anti-inflammatory factors, *Biochem. Pharmacol.*, 33, 1705, 1984.
67. **Ward, P. A., Sulavik, M. C., and Johnson, K. J.,** Activated rat neutrophils. Correlation of arachidonate products with enzyme secretion but not with O_2^- generation, *Am. J. Pathol.*, 120, 112, 1985.
68. **Gugler, R. and Dengler, H. J.,** Inhibition of human liver catechol-*O*-methyl-transferase by flavonoids, *Naunyn Schmiedebergs Arch. Pharmacol.*, 276, 223, 1973.
69. **Borchardt, R. T. and Huber, J. A.,** Catechol-*O*-methyltransferase. v. Structure-activity relationships for inhibition by flavonoids, *J. Med. Chem.*, 18, 120, 1975.
70. **Takamura, K. and Ito, M.,** Effects of metal ions and flavonoids on the oxidation of ascorbic acid, *Chem. Pharm. Bull.*, 25, 3218, 1977.
71. **Ronzière, M. C., Herbage, D., Garrone, R., and Frey, J.,** Influence of some flavonoids on reticulation of collagen fibrils in vitro, *Biochem. Pharmacol.*, 30, 1771, 1981.
72. **Rao, C. N., Rao, V. H., and Steinmann, B.,** Bioflavonoid-mediated stabilization of collagen in adjuvant-induced arthritis, *Scand. J. Rheumatol.*, 12, 39, 1983.
73. **Hudson, B. J. F. and Lewis, J. I.,** Polyhydroxyflavonoid antioxidants for edible oils. Structural critiera for activity, *Food Chem.*, 10, 47, 1983.
74. **Monboisse, J. C., Braquet, P., Randoux, A., and Borel, J. P.,** Non-enzymatic degradation of acid-soluble calf skin collagen by superoxide ion: protective effect of flavonoids, *Biochem. Pharmacol.*, 32, 53, 1983.
75. **McCord, J. M.,** Protection of synovial fluid by superoxide dismutase, *Science*, 185, 529, 1974.

CENTROPHENOXINE AS OH· FREE RADICAL SCAVENGER

I. Zs.-Nagy

INTRODUCTION

Centrophenoxine (CPH) is an ester of *p*-chlorophenoxyacetic acid (PCPA) and dimethylaminoethanol (DMAE).[1,2] Formerly it had been considered in human therapy as a brain metabolic stimulant (and the ill-defined term "neuroenergeticum" was applied to it), since it stimulates glucose uptake, oxygen consumption, and carbon dioxide production of the brain in vivo[3,4] and also in vitro in brain slices.[5] However, the classification of the drug has recently changed, since the concept of nootropica has become widely accepted. The Greek word "noos" means "mind", "intellect", or "sense", and drugs being able to improve these brain functions in a broad sense were designated as "nootropica".[6] Nowadays CPH is one of the most frequently used representative of nootropica in various (mostly European) countries. Nevertheless, in spite of a great number of experimental and clinical studies (the number of available publications on CPH actually exceeds 500), the mechanism of action of CPH on the age-dependent decline of mental performance has never been clearly understood. The search for some more realistic explanation in order to understand the nootropic effects of CPH has been going on for a long time. This chapter will summarize the available data on the effect of CPH, explaining its effect as a membrane-protecting, OH· free radical scavenger agent.

EARLY DATA ON THE EFFECTS OF CPH

In human therapy, a number of beneficial effects of CPH have been observed in cases of cerebral atrophy, brain injury, postapoplectic disorders, chronic alcoholism, barbiturate intoxications, etc.[7-10] It has also been observed that prolonged administration of CPH to healthy, old animals significantly reduced the accumulation of lipofuscin in the brain cells and myocardium.[11-24] Furthermore, the medium life span of the CPH-treated old animals increased significantly and, what is more, the learning ability of the old, treated mice improved as compared to their age-matched controls.[14,25-29] On the basis of the these results, CPH can be considered as a potential antiaging drug.

As regards the mechanism of action of CPH, it was suggested that the DMAE moiety of CPH enters the choline synthesis cycle (and consequently improves the acetylcholine supply of the brain).[27,30] However, this explanation has been contradicted by others.[31-33] It must be clear that an increase of choline availability by transforming a fraction of DMAE from CPH into choline, which has been shown to take place in the liver,[34] cannot be the only explanation of the effect of this drug, since in this case a choline-rich diet alone should have the same effect as CPH has, which is not the case.[32,33] Further data on this problem are discussed by Bertoni-Freddari et al.[34]

The explanation of the pharmacological action mechanism of CPH obviously has to start from the distribution and incorporation of the drug (or its metabolites) in the tissues, cells, or subcellular compartments. Important information was revealed in this respect 12 years ago.[35] It has been shown that after i.v. administration of equimolar doses of ^{14}C-labeled DMAE or CPH, much higher levels of DMAE were encountered in the brain after CPH treatment, as compared to DMAE alone, since in agreement with former assumptions, the esterified form of DMAE with PCPA penetrates the blood-brain barrier much better. CPH was hydrolyzed on both sides of the barrier in its two component parts in vivo. However, the DMAE moiety becomes phosphorylated in the brain to yield phosphoryl-DMAE, which is in turn converted to phosphatidyl-DMAE, seemingly the end metabolite of DMAE in the brain.[35] Phosphatidyl-DMAE is incorporated in the nerve cell membranes and persists in this form for a relatively long time in the place of choline.[35] In other words, DMAE forms a special class of phospholipids in the brain cell membranes. The p-chlorophenoxy acetic acid moiety (PCPA) is excreted in the urine rather quickly.[35] It should be noted that although some fraction of DMAE administered either alone or in the form of CPH was found in acid-soluble and lipid cholines in the brain,[35] evidence is available that the trimethylation of DMAE does not take place in the brain, but in the liver.[35]

Therefore, the presence of DMAE in the nerve cell membrane phospholipids has to be considered as the starting point of any approach to the mechanism of action of CPH. In parallel with the above-mentioned studies on CPH, a new concept called membrane hypothesis of aging has been developed for the explanation of normal cell aging.[36,37] Since this hypothesis attributes a primary role in cell aging to a free radical-induced, age-dependent loss of passive potassium (and probably water) permeability of the cell plasma membrane, causing a series of consequences in the intracellular physiocochemical composition and physiological processes including total and mRNA synthesis,[38,39] specially designed experiments have been carried out in order to reveal whether the effects of DMAE on brain functions can be interpreted in terms of this aging hypothesis, i.e., whether DMAE may exert any local oxy radical-protecting effect on the cell membrane components.

THE OXY RADICAL SCAVENGER PROPERTIES OF CPH

Protein Cross-Linking Model

This type of experiments[40] can be carried out by using relatively simple laboratory equipment. The basic concept of the experimental model is that the OH· free radicals generated in the Fenton reaction attack and cross-link the proteins (in the given case, bovine serum albumin $\sqrt{\text{BSA}}$) added to the reaction mixture, so that the water solubility of the protein (originally 100% for BSA) gradually decreases. Therefore, it is sufficient to perform only protein content measurements in the supernatant in order to follow the effect of the OH· radicals and the influence of radical scavengers like DMAE on this reaction. The molecular weight distribution of the still soluble fraction was also analyzed.[40]

These studies revealed that DMAE is efficiently inhibiting the polymerization of BSA induced by the OH· radicals generated in the Fenton reaction.[40] This result can be interpreted only by assuming that DMAE is an efficient OH· radical scavenger under the given conditions. It should also be stressed that the OH· radical-induced polymerization of BSA could be detected not only as a precipitation of an insoluble, glue-like mass from the solution, but also in the still soluble residue of BSA in the form of di-, tri-, and tetramerized molecules.[40] The covalent character of the newly formed intermolecular bonds has been proven by various chemical tests like treatment with concentrated (6M) urea, etc.[40]

Electron Spin Resonance Spectroscopy (ESR) in Spin Trapping Experiments

This technique,[41] which is becoming very important in biological free radical research, applies specially designed compounds (spin traps), the reactions of which with various types of free radicals are well known. They usually form some relatively long-lived radicals (spin adducts). The concentration of spin adducts can be quantatively measured under controlled conditions by means of their paramagnetic properties in the ESR. Using this technique, one can obtain data regarding the competitions for the given type of radical generated in the reaction mixture and, in some cases, also the rate constants of the radical scavenging process can be estimated.[41] The spin trapping method applied for this purpose[41] has been elaborated for answering special questions of cell biology.[42-44]

The spin trapping experiments demonstrated that the CPH molecule and both components of its display a strong affinity toward the OH· radicals.[41] In the case of DMAE, the reaction rate can be estimated at about $10^9/M$/sec with this type of oxy-radicals which was quite comparable with that of the spin trap (DMPO) uses.[41,44] Since according to Walling[45] most of the bio-organic compounds react with OH· radicals at a rate of 10^7 to $10^{10}/M$/sec, DMAE really may represent an essential protection for numerous cell membrane components against OH· free radical attacks.

The OH· scavenger properties of DMAE have been tested in some further spin trapping models using PBN, 4-POBN, proline, or hydroxyproline as spin traps and all of these controls supported the above statement.[41] Therefore, the conclusion can be drawn that DMAE incorporated into the plasma membrane of nerve cells can really be considered as an OH· free radical scavenger, i.e., the observed effects of CPH should be interpreted on the basis of this point of view.

Although these experiments also revealed an OH· radical scavenger capacity of the PCPA moiety of the CPH molecule, PCPA cannot be responsible for the nootropic effects of CPH, since no binding of this compound is known in the brain. It cannot be excluded, however, that this moiety of CPH may have some physiologically meaningful effect in the blood or in some other tissues (e.g., kidney or liver) during its excretion.

Chemical Methods for Superoxide Radical Reactions

In this group of experiments,[46] well-established biochemical reactions were used for the detection of some typical superoxide radical-induced processes. The influence of some competitions or scavenging could also be detected quantitatively. This group of methods has been applied to study the eventual superoxide radical scavenger properties of CPH and its components, DMAE and PCPA. These experiments revealed that although CPH and its components display some affinity toward the superoxide radicals, their rate constant is only in the order of magnitude of $10^2/M$/sec; therefore, there is no reason to assume that either components or the CPH molecule itself would represent any efficient protection in vivo against superoxide radical-induced damage.[46]

DISCUSSION

Considering that oxygen free radicals do occur in biological systems[47-50] and are implicated as etiological factors in a series of biological phenomena like mutagenesis, aging, and some pathologies,[51-59] and furthermore, that biological compounds (lipids, proteins, amino acids, etc.) proved to be rather sensitive toward this type of radicals in vitro,[60,61] a general role can be attributed to the possibility of OH· radical formation in vivo.

A general summary of the radical forming pathways was given recently.[38] It is evident from the available data that the OH· radical-induced damage is the most serious one. It takes place extremely rapidly and therefore always very near the site of OH· radical formation. It is no wonder that, according to present knowledge, the living systems do not possess any active defense mechanism against OH· radicals, but only the possibility of a continuous replacement of their damaged components.[38] In the light of these data, the presence of a component in the cell membrane possessing an elevated reactivity with OH· radicals as compared to its surroundings can obviously represent a kind of site-specific protection for the macromolecules. Since the DMAE component of CPH forms a class of phospholipids persisting in the nerve cell membranes for long time,[35] one can assume that its OH· radical scavenger ability shown in vitro may really be of significance also in vivo.

Obviously, neither the presence of DMAE in the brain nerve cell membranes nor the OH· radical scavenging ability of this compound as revealed in vitro can unanimously prove the validity of the above assumption in vivo. However, there is a great number of experimental results revealing the effect of CPH treatment on the structure and function of the neuronal membranes in old animals. These data demonstrate that the biological effects of CPH can be interpreted on the basis of the above assumption. The most important experimental observations can be summarized as follows:

1. Synaptosomes prepared from rat brain cortex display an age-dependent decline of membrane lipid fluidity, as revealed by fluorescence anisotropy measurements of DPH-labeled preparations. This parameter returned almost to the adult level, if old rats were treated with CPH for 60 days.[62]
2. In vivo acute Fe^{2+} overload in the cerebrospinal fluid is extremely toxic, or when applying lower doses, it causes an accelerated aging of young rats through an enhanced OH· radical formation. Synaptosomal membranes were protected considerably by CPH pretreatment against the effect of this type of iron overload.[63]
3. The proteins in the synaptosomes of rat brain cortex display an age-dependent shift of their molecular weight distribution toward the higher values, interpreted as a sign of increasing, free radical-induced, intermolecular cross-linking. This shift in distribution was partially reversed in old rats by CPH treatment.[64]
4. Fluorescence recovery after photobleaching (FRAP) experiments[65] performed on hepatocyte membranes of liver smears revealed a very characteristic negative linear correlation of the lateral diffusion constant of membrane proteins with age of the animals.[66] CPH treatment reversed this age-dependent decrease of the lateral mobility of proteins in old rats.[67]
5. Intracellular potassium content increases with age in brain cells, whereas intracellular water content decreases considerably.[68,69] These alterations are due to free radical-induced damage of the cell membrane components, resulting in a loss of passive permeability of the membrane for potassium (and for water).[37-39] CPH treatment of old rats increases the passive potassium permeability of the neuronal cell membrane, i.e., it decreases the intracellular potassium content and rehydrates the cytoplasm considerably.[68,69]
6. Certain synaptic parameters measured by electron microscopic morphometry in the cerebellar glomerulus of rats undergo a characteristic age-dependent change.[34] CPH treatment improves these parameters in old rats.[34]
7. Aging causes a very significant decrease of the rates of total and mRNA synthesis in the brain cortex, which is reversed by CPH treatment of old rats.[39]
8. CPH treatment reverses the age-dependent decrease in the water solubility of proteins and the increase in total protein content in the brain of old rats.[70]
9. CPH treatment is able to prolong the medium life span and to improve the learning ability of several experimental animal species.[39,68]
10. In a recently performed double-blind human clinical trial, CPH treatment of 8 weeks improved the psychometric and behavioral performance of patients with medium level dementia (DSM III, Cat. 1; ICD No. 299); meanwhile a considerable rehydration of the intracellular mass was observed at the expense of the extracellular liquid.[76]

The above results can be interpreted in terms of the membrane hypothesis of aging.[36-39] The central statement of this hypothesis is that aging is caused by the very same effects, i.e., the free radical-induced damages, which are present in young age, too; however, the living system itself changes its physicochemical state during aging so that the radical events become more and more harmful with advancing age. The reason for this phenomenon is to be searched for in the statistical character of the radical reactions, especially the fastest ones. Namely, it is easy to show that the damaging efficiency of an OH· free radical is density dependent, since in the diluted systems one OH· radical can react with the highest probability only with one organic molecule (i.e., with the nearest one, because the next molecule is far away from the site of radical formation). On the other hand, with increasing density we have an increasing probability for bi-, tri-, or multimolecular consequences of one OH· radical event, since the vicinity of the organic molecules shall permit even chain reactions.

Because of the very same role of density in radical processes, due to their relatively compact structure, the membranes of the cells continuously undergo a somewhat faster radical-induced deterioration of their structure and function than the cytosol. Among the membraneous structures of the cells, the plasma membrane has a peculiar position, since apart from the free radical-induced damages, it is also exposed to the residual heat production during the electric discharge accompanying the excitation of the cell membrane.[37] This difference in the rate of damage in the plasma membrane manifests itself in the fact that proteins of the highest turnover rate were found in the cell membrane.[71]

It is also known that the overall protein turnover undergoes a considerable age-dependent decline.[38,72] There must be a regulatory link between the highest rate of damage in the cell membrane and the decrease of protein turnover. In our assumption, this link is that the main intracellular physicochemical parameters such as water content, ionic composition, etc. which exert a regulatory influence for the rates of all enzymatic processes, i.e., also of RNA and protein synthesis, are maintained basically by the cell membrane. However, due to the density differences between cytosol and membranes mentioned above, the cell membrane always becomes deteriorated somewhat faster than the intracellular substances. This means that the cell membrane continuously loses its passive potassium (and probably also water) permeability, as shown by numerous experimental data,[36,37,69,73-75] i.e., there is a continuous increase of intracellular density with advancing age of the cells. This phenomenon represents a part of the maturation process as well; the dry mass content of the tissues and cells obviously increases as compared to the embryonic or young ages, and it seems to be logical to assume that until a certain level of condensation, this process does not seriously inhibit protein synthesis. On the contrary, the system is growing during the maturation period, i.e., its synthetic activity exceeds the decomposition.

On the other hand, the ever ongoing membrane alterations reach certain limits with age when the increase of the intracellular potassium concentration and a consequent condensation of the cellular colloids, i.e., a decrease of the level of intracellular hydration, starts to become rate limiting for the intracellular synthetic processes.[36,37,69] The increasing density of the intracellular space reduces the activity of all existing enzymes including those performing the transcription and translation of the genetic information.[36,37] This process leads to a slowing down of the rates of RNA and protein synthesis and also of the replacement of the components damaged by OH· radical attack. This then causes a further loss of membrane permeability, etc., i.e., a *circulus vitiosus* comes into being which can only be interrupted by increasing the site-specific radical protective efficiency in the cell membrane. This will result in opposite tendencies in all involved processes, i.e., it will improve the physicochemical conditions for the readout[39] and translation of the genetic information again. The reincreased protein synthesis then will be able to eliminate the accumulated end products (lipofuscin) simply by increasing the lysosomal activity. In this frame of thought, practically all effects of CPH can be interpreted without logical or principal contradictions.

CONCLUSIONS

This paper reviews the experimental evidence demonstrating that the nootropic effects of centrophenoxine (CPH) can be interpreted on the basis of the OH· radical scavenger properties of DMAE which is incorporated in the neuronal membranes of the brain in the form of phosphatidyl-DMAE during chronic treatments. The effects of CPH on the membrane physicochemistry correspond to a reversal of the age-dependent deteriorations caused by the OH· radical attacks as explained in terms of the membrane hypothesis of cellular aging, and such a reversal may well be the explanation for the nootropic effects of this drug.

REFERENCES

1. **Pfeiffer, C., Jenney, E. H., Gallagher, W., Blackmore, W., Smith, R. P., Bevan, W., Jr., Killam, K. F., and Killam, E. K.,** Stimulant effect of 2-dimethylaminoethanol: possible precursor of brain acetylcholine, *Science*, 126, 610, 1957.
2. **Thuiller, J., Rumpf, P., and Thuiller, G.,** Derivés des acides regulateurs de croissance des vegetaux. I. Proprietés pharmacologiques de l'ester dimethylaminoethylique de l'acide *p*-chlorophenoxyacetique (235 ANP), *C. R. Séances Soc. Biol. Paris*, 153, 1914, 1959.
3. **Nickel, J., Breyer, U., Claver, B., and Quadbeck, G.,** Zur Wirkung von Aminoethanol-Derivaten auf das Zentralnervensystem, *Arzneim. Forsch.*, 13, 881, 1963.
4. **Oeriu, S., Winterer, D., Dobre, V., and Bruhis, S.,** Glucose transport across blood brain barrier as related to age and meclofenoxate treatment, *J. Pharmacol.*, 4, 497, 1973 (in French).
5. **Herrmann, H.-D. and Dittmann, J.,** Steigerung des Glukose-Stoffwechsels von Kaninchen-Hirnschnitten durch Centrophenoxin, *Arzneim. Forsch.*, 21, 984, 1971.
6. **Giurgea, C. and Sara, S. J.,** Nootropic drugs and memory, in *Practical Aspects of Memory*, Grüenberg, M. M., Morris, P. E., and Sykes, R. N., Eds., Academic Press, London, 1978, 754.
7. **Schmidt, H. and Broicher, H.,** Klinische Erfahrungen bei der Behandlung von Zustanden zerebraler Insuffizienz mit Centrophenoxin (Helfergin), *Med. Welt*, 33, 1432, 1970.
8. **Vojtechovsky, M., Soukupova, B., Safratova, V., and Votava, Z.,** The influence of centrophenoxine (Lucidril) on learning and memory in alcoholics, *Int. J. Psychobiol.*, 1, 49, 1970.
9. **Herrschaft, H., Gleim, F., and Duus, P.,** Die Wirkung von Centrophenoxin auf die regionale Gehirndurchblutung bei Patienten mit zerebrovaskularer Insuffizienz, *Dtsch. Med. Wochenschr.*, 99, 1707, 1974.
10. **Kugler, J.,** *Hirnstoffwechsel und Hirndurchblutung*, Berichstband Centrophenoxin-Arbeitstagung, Schnetztor Verlag, Konstanz, West Germany, 1977.
11. **Nandy, K. and Bourne, G.,** Effects of centrophenoxine on the lipofuscin pigment in the neurons of senile guinea pigs, *Nature (London)*, 210, 313, 1966.
12. **Nandy, K.,** Further studies on the effects of centrophenoxine on the lipofuscin pigment of the neurons in senile guinea pigs, *J. Gerontol.*, 23, 82, 1968.
13. **Nandy, K.,** Properties of neuronal lipofuscin pigment in mice, *Acta Neuropathol.*, 19, 25, 1971.
14. **Nandy, K.,** Centrophenoxine: effects of aging mammalian brain, *J. Am. Geriatr. Soc.*, 26, 74, 1978.
15. **Nandy, K.,** Lipofuscinogenesis in mice early treated with centrophenoxine, *Mech. Ageing Dev.*, 8, 113, 1978.
16. **Chemnitius, K. H., Machnik, G., Löw, O., Arnrich, M., and Urban, J.,** Versuche zur medikamentösen Beeinflußung altersbedingter Veränderungen, *Exp. Pathol.*, 4, 163, 1970.
17. **Meier, C. and Glees, P.,** Der Einfluß des Centrophenoxins auf das Alterspigment in Satellitenzellen und Neuronen der Spinalganglion seniler Ratten, *Acta Neuropathol.*, 17, 310, 1971.
18. **Bourne, G. H.,** Lipofuscin, *Prog. Brain Res.*, 40, 187, 1973.
19. **Hasan, M., Glees, P., and El-Ghazzawi, E.,** Age-associated changes in the hypothalamus of the guinea pig: effect of dimethylaminoethyl-*p*-chloro-phenoxy acetate. An electron microscope and histochemical study, *Exp. Gerontol.*, 9, 153, 1974.
20. **Hasan, M., Glees, P., and Spoerri, P. E.,** Dissolution and removal of neuronal lipofuscin following dimethylaminoethyl-*p*-chloro-phenoxy acetate administration to guinea pigs, *Cell Tissue Res.*, 150, 369, 1974.
21. **Riga, S. and Riga, D.,** Effects of centrophenoxine on the lipofuscin pigments in the nervous system of old rats, *Brain Res.*, 72, 265, 1974.
22. **Spoerri, P. E., Glees, P., and El-Ghazzawi, E.,** Accumulation of lipofuscin in the myocardium of senile guinea pigs. Dissolution and removal of lipofuscin following dimethylaminoethyl-*p*-chlorophenoxyacetate administration. An electron microscope study, *Mech. Ageing Dev.*, 3, 311, 1974.
23. **Spoerri, P. E. and Glees, P.,** The mode of lipofuscin removal from hypothalamic neurons, *Exp. Gerontol.*, 10, 225, 1975.
24. **Glees, P. and Hasan, M.,** Lipofuscin in neuronal aging and diseases, in *Normal and Pathological Anatomy*, Vol. 32, Bargmann, W. and Doerr, W., Eds., Georg Thieme Verlag, Stuttgart, 1976, 44.
25. **Hochschild, R.,** Lysosomes, membranes and aging, *Exp. Gerontol.*, 6, 153, 1971.
26. **Hochschild, R.,** Effect of dimethylaminoethyl-*p*-chlorophenoxyacetate on the life span of male Swiss Webster albino mice, *Exp. Gerontol.*, 8, 177, 1973.
27. **Hochschild, R.,** Effect of dimethylaminoethanol on the life span of senile male A/J mice, *Exp. Gerontol.*, 8, 185, 1973.
28. **Nandy, T.,** Die Wirkung von Centrophenoxin auf Lernfähigkeit, Gadächtnis und Lebensdauer alter Mäuser, in *Hirnstoffwechsel und Hirndurchblutung*, Kugler, J., Ed., Schnetztor Verlag, Konstanz, West Germany, 1977, 17.
29. **Nandy, K. and Lal, H.,** Neuronal lipofuscin and learning deficit in aging mammals, in *Proc. 10th Coll. Int. Neuropsychopharmacol. Congr. (Abstracts)*, Pergamon Press, New York, 1978, 1683.

30. **London, E. D. and Coyle, J. T.,** Pharmacological augmentation of acetylcholine levels in kainate lesioned rat striatum, *Biochem. Pharmacol.*, 27, 2962, 1978.
31. **Haubrich, D. R., Wang, P. F., Clody, D. E., and Wedecking, P. W.,** Increase in rat brain acetycholine induced by choline or Deanol, *Life Sci.*, 17, 975, 1975.
32. **Zaniser, N. R., Chou, D., and Hanin, I.,** Is 2-dimethylaminoethanol (Deanol) indeed a precursor of brain acetylcholine? A gas chromatographic evaluation, *J. Pharmacol. Exp. Ther.*, 200, 545, 1977.
33. **Jope, R. S. and Jenden, D. J.,** Dimethylaminoethanol (Deanol) metabolism in rat brain and its effect on acetylcholine synthesis, *J. Pharmacol. Exp. Ther.*, 211, 472, 1979.
34. **Bertoni-Freddari, C., Giuli, C., and Pieri, C.,** The effect of acute and chronic centrophenoxine treatment on the synaptic plasticity of old rats, *Arch. Gerontol. Geriatr.*, 1, 365, 1982.
35. **Miyazaki, M., Nambu, K., Minaki, A. Y., Hashimoto, M., and Nakamura, K.,** Comparative studies on the metabolism of beta-dimethylaminoethanol in the mouse brain and liver following administration of beta-dimethylaminoethanol and its *p*-chlorophenoxyacetate, Meclofenoxate, *Chem. Pharm. Bull.*, 24, 763, 1976.
36. **Zs.-Nagy, I.,** A membrane hypothesis of aging, *J. Theor. Biol.*, 75, 189, 1978.
37. **Zs.-Nagy, I.,** The role of membrane structure and function in cellular aging: a review, *Mech. Ageing Dev.*, 9, 237, 1979.
38. **Zs.-Nagy, I.,** Aging of the cellular membrane: basic principles and pharmacological interventions, *Geriatrika,* 1(8), 102, 1985.
39. **Zs-Nagy, I. and Semsei, I.,** Centrophenoxine increases the rates of total and mRNA synthesis in the brain cortex of old rats: an explanation of its action in terms of the membrane hypothesis of aging, *Exp. Gerontol.*, 19, 171, 1984.
40. **Zs.-Nagy, I. and Nagy, K.,** On the role of cross-linking of cellular proteins in aging, *Mech. Ageing Dev.*, 14, 245, 1980.
41. **Zs.-Nagy, I. and Floyd, R. A.,** Electron spin resonance spectroscopic demonstration of the hydroxyl free radical scavenger properties of dimethylaminoethanol in spin trapping experiments confirming the molecular basis for the biological effects of centrophenoxine, *Arch. Gerontol. Geriatr.*, 3, 297, 1984.
42. **Floyd, R. A.,** DNA-ferrous iron catalyzed free radical formation from hydrogen peroxide, *Biochem. Biophys. Res. Commun.*, 99, 1209, 1981.
43. **Floyd, R. A.,** Observations on nitroxyl free radicals in arylamine carcinogenesis and on spin-trapping hydroxyl free radicals, *Can. J. Chem.*, 60, 1577, 1982.
44. **Floyd, R. A. and Lewis, C. A.,** Hydroxyl free radical formation from hydrogen peroxide by ferrous iron-nucleotide complexes, *Biochemistry,* 22, 2645, 1983.
45. **Walling, C.,** Fenton's reagent revisited, *Acc. Chem. Res.*, 8, 125, 1975.
46. **Semsei, I. and Zs.-Nagy, I.,** Superoxide radical scavenging ability of centrophenoxine and its salt dependence in vitro, *J. Free Radical Biol. Med.*, 1, 403, 1985.
47. **Fridovich, I.,** The biology of oxygen radicals. The superoxide radical is an agent of oxygen toxicity; superoxide dismutases provide an important defense, *Science,* 201, 875, 1078.
48. **Nohl, H. and Hegner, D.,** Do mitochondria produce oxygen radicals in vivo?, *Eur. J. Biochem.*, 82, 563, 1978.
49. **Nohl, H. and Jordan, W.,** The metabolic fate of mitochondrial hydrogen peroxide, *Eur. J. Biochem.*, 111, 203, 1980.
50. **Nohl, H., Jordan, W., and Hegner, D.,** Identification of free hydroxyl radicals in respiring rat heat mitochondria by spin trapping with the nitrone DMPO, *FEBS Lett.*, 123, 241, 1981.
51. **Harman, D.,** Aging: a theory based on free radical and radiation chemistry, *J. Gerontol.*, 11, 298, 1956.
52. **Harman, D.,** The free radical theory of aging: the effect of age on serum mercaptan levels, *J. Gerontol.*, 15, 38, 1960.
53. **McCord, J. M.,** Free radicals and inflammation: protection of synovial fluid by superoxide dismutase, *Science,* 185, 529, 1974.
54. **McCord, J. M., Wong, K., Stokes, S. H., Petrone, W. F., and English, D.,** Superoxide and inflammation: a mechanism for the anti-inflammatory activity of superoxide dismutase, *Acta Physiol. Scand. Suppl.*, 492, 25, 1980.
55. **Petkau, A.,** Radiation carcinogenesis from a membrane perspective, *Acta Physiol. Scand. Suppl.*, 492, 81, 1980.
56. **Siesjö, B. K., Rehncrona, S., and Smith, D.,** Neuronal cell damage in the brain: possible involvement of oxidative mechanisms, *Acta Physiol. Scand. Suppl.*, 492, 121, 1980.
57. **Tolmasoff, J. M., Ono, T., and Cutler, R. G.,** Superoxide dismutase: correlation with life-span and specific metabolic rate in primate species, *Proc. Natl. Acad. Sci. U.S.A.*, 77, 2777, 1980.
58. **Cutler, R. G.,** Superoxide dismutase, longevity and specific metabolic rate, *Gerontology,* 29, 113, 1983.
59. **Cutler, R. G.,** Peroxide producing potential of tissues: inverse correlation with longevity of mammalian species, *Proc. Natl. Acad. Sci. U.S.A.*, 82, 4798, 1985.

60. **Donato, H. and Sohal, R. S.,** Relationship of lipofuscin accumulation to aging, in *CRC Handbook of Biochemistry in Aging,* Florini, J. R., Adelman, R. C., and Roth, G. S., Eds., CRC Press, Boca Raton, Fla. 1981, 221.
61. **Zs-Nagy, I. and Floyd, R. A.,** Hydroxyl free radical reactions with amino acids and proteins studied by electron spin resonance spectroscopy and spin trapping, *Biochim. Biophys. Acta,* 790, 238, 1984.
62. **Nagy, K., Zs.-Nagy, V., Bertoni-Freddari, C., and Zs.-Nagy, I.,** Alterations of the synaptosomal membrane microviscosity in the brain cortex of rats during aging and centrophenoxine treatment, *Arch. Gerontol. Geriatr.,* 2, 23, 1983.
63. **Nagy, K., Floyd, R. A., Simon, P., and Zs.-Nagy, I.,** Studies on the effect of iron overload on rat cortex synaptosomal membranes, *Biochim. Biophys. Acta,* 820, 216, 1985.
64. **Nagy, K. and Zs.-Nagy, I.,** Alterations in the molecular weight distribution of proteins in rat brain synaptosomes during aging and centrophenoxine treatment, *Mech. Ageing Dev.,* 28, 171, 1984.
65. **Zs.-Nagy, I., Ohta, M., Kitani, K., and Imahori, K.,** An automated method for measuring lateral mobility of proteins in the plasma membrane of cells in compact tissues by means of fluoroescence recovery after photobleaching, *Mikroskopie,* 41, 12, 1984.
66. **Zs.-Nagy, I., Kitani, K., Ohta, M., Zs.-Nagy, V., and Imahori, K.,** Age-dependent decrease of the lateral diffusion constant of proteins in the plasma membrane of hepatocytes as revealed by fluorescence recovery after photobleaching in tissue smears, *Arch. Gerontol. Geriatr.,* 5, 131, 1986.
67. **Zs. Nagy, I., Ohta, M. and Kitani, K.,** Effect of centrophenoxine and BCE-001 treatment on the lateral diffusion constant of proteins in the hepatocyte membrane as revealed by fluorescence recovery after photobleaching in rat liver smears, submitted.
68. **Zs.-Nagy, I., Pieri, C., Giuli, C., and Del Moro, M.,** Effects of centrophenoxine on the monovalent electrolyte contents of the large brain cortical cells of old rats, *Gerontology,* 25, 94, 1979.
69. **Lustyik, Gy. and Zs.-Nagy, I.,** Alterations of the intracellular water and ion concentrations in brain and liver cells during aging as revealed by energy dispersive X-ray microanalysis of bulk specimens, *Scanning Electron Microsc.,* 1, 323, 1985.
70. **Zs.-Nagy, I., Nagy, K., Zs.-Nagy, V., Kalmár, Á., and Nagy, É.,** Alterations in total content and solubility characteristics of proteins in rat brain and liver during aging and centrophenoxine treatment, *Exp. Gerontol.,* 16, 229, 1981.
71. **Tauber, R. and Reutter, W.,** Turnover of plasma membrane proteins and glycoproteins in normal and regenerating liver and Morris hepatoma 7777, *Eur. J. Cell Biol.,* 26, 35, 1981.
72. **Reith, M. E. A., Schotman, P., and Gispen, W. H.,** Measurements of in vivo rates of protein synthesis in brain, spinal cord, heart and liver of young versus old rats, intact versus hypophysectomized rats, *J. Neurochem.,* 30, 587, 1978.
73. **Zs.-Nagy, I., Gyenes, M., Lustyik, Gy., Zs.-Nagy, V., and Jeney, F.,** Age-dependent decrease of cell membrane permeability of rat hepatocytes as revealed by in vivo Rb^+-uptake and release, in *Liver and Aging — 1982,* Kitani, K., Ed., Elsevier, Amsterdam, 1982, 215.
74. **Gyenes, M., Lustyik, Gy., Zs.-Nagy, V., Jeney, F., and Zs.-Nagy, I.,** Age-dependent decrease of the passive Rb^+ and K^+ permeability of the nerve cell membranes in rat brain cortex as revealed by in vivo measurement of the Rb^+ discrimination ratio, *Arch. Gerontol. Geriatr.,* 3, 11, 1984.
75. **Zs.-Nagy, I., Tóth, S., and Lustyik, Gy.,** Verification of the membrane hypothesis of aging on the identified giant neurons of the snail *Lymnaea stagnalis* L. (Gastı opoda, Pulmonata) by a combined application of intracellular electrophysiology and X-Ray microanalysis, *Arch. Gerontol. Geriatr.,* 4, 53, 1985.
76. **Zs.-Nagy, I.,** unpublished results.

INTERACTIONS BETWEEN VITAMINS C AND E AND CYTOCHROME P-450

Emil Ginter

INTRODUCTION

It is known that different antioxidants can influence the biotransformations of drugs, carcinogens, and other xenobiotics in experimental animals and humans. In this survey, recent data on the effect of two redox active vitamins, ascorbic acid and tocopherol, on biotransformation reactions are summarized. As described below, the redox systems of vitamins C and E, localized in different cell compartments, can interact and synergistically influence the detoxication processes.

REDOX SYSTEMS OF VITAMINS C AND E

Vitamin C forms a reversible redox system containing three components: reduced ascorbic acid, which is dissociated at physiological pH and forms an ascorbate anion; (2) ascorbate radical, for which the term ''semidehydro-'' (or ''monodehydro-'') ascorbic acid is frequently used; and (3) dehydroascorbic acid, which exists as hydrated bicyclic monomer in aqueous solutions. Most of the water-soluble vitamin C redox system occurs in the cytosol. By contrast, vitamin C concentration in membrane structures is remarkably lower. Ascorbate is oxidized to dehydroascorbic acid via a one-electron transfer mechanism, yielding ascorbate radical as an intermediate. Ascorbate free radical is much more reactive than reduced ascorbate, but in comparison with other free radicals, it is rather stable. It is assumed that the major portion of free radicals in human blood serum should be ascorbate radical.[1] The vitamin C redox system has, at different redox potentials, both oxidation and reduction effects and is able both to generate and to scavenge oxygen free radicals.

The composition of the vitamin E redox system in the tissues is not exactly known. After administration of labeled vitamin E, almost all of the label is found in tissues as unchanged tocopherol. Some, but not all, authors find in the tissues small amounts of oxidized derivates, such as tocopheryl hydroquinone, and more polar metabolites.[2] It is assumed that α-tocopherol is oxidized to tocopheryl quinone via a one-electron transfer mechanism, yielding chromanoxyl radical as an intermediate. The transformation of tocopherol to radical is reversible, but reduction of tocopheryl quinone to tocopherol has not been proven in tissues. In contrast to ascorbate, vitamin E is transported in the blood plasma in lipoproteins, but there does not appear to exist a specific lipoprotein carrier for tocopherol. It is generally assumed that the chief biological function of vitamin E is to protect the polyunsaturated fatty acids of membrane phospholipids from the disorganizing effect of free radical-mediated peroxidation reactions.[2] Tocopherol is the only lipid-soluble antioxidant present in blood plasma and erythrocyte membranes. The radical scavenging properties of tocopherol reside in the fused chroman ring system, while the phytyl chain provides the vitamin E molecule with an affinity for the hydrophobic environment of the membranes. Both vitamin C and vitamin E are efficient antioxidants and scavengers of free radicals in vitro as well as in vivo and are capable of terminating free radical chain reactions.

In contrast to vitamin E, vitamin C is able to produce hydrogen peroxide in the presence of Fe or Cu ions by reduction of the molecular oxygen. Though the detailed mechanism of this nonenzymatic reaction is not yet clarified, Scarpa et al.[3] have shown that one of the products is superoxide anion ($O_2^{\cdot}$) which reacts with ascorbate or its radical, thus leading to the formation of H_2O_2. Most authors try to explain the favorable influence of vitamin C on different pathologic processes by its ability to scavenge oxygen free radicals. However, the concept of exclusive detrimental effects of oxygen radicals is one-sided as it neglects the important physiological functions of oxygen radicals such as prostaglandin synthesis, membrane transport, and lysosomal enzyme activation. The data summarized in this survey indicate that several of the biochemical roles of vitamin C just arise from its ability to generate oxygen free radicals. Thus, for a normal course of biochemical reactions, an optimum equilibrium between the generation and scavenging of free radicals is desirable.

INTERACTIONS BETWEEN VITAMINS C AND E

Neither molar concentrations nor the turnover of vitamin E in cell membranes are too high. Therefore, the tocopheryl radicals formed during the scavenging process must be reduced back to tocopherol, which permits a single molecule of tocopherol to scavenge a great number of free radicals. According to Tappel,[4] the redox systems of ascorbate and tocopherol interact and participate in the protection of membrane phospholipids against free radical damage. This probably involves a mechanism for the transport of electrons from NADPH, through glutathione, ascorbate, and tocopherol, to the radicals of polyunsaturated fatty acids.

Many studies have been published discussing the possible interactions between vitamins C and E in nonbiological systems or in in vitro reactions involving subcellular factions or tissue homogenates.[5] The data obtained are not conclusive, but most authors confirm the existence of interactions between ascorbate and tocopheryl radicals. Following pulse radiolysis of an aerated solution containg acetone, carbon tetrachloride, isopropanol, α-tocopherol, and ascorbic acid, there is a formation of phenoxy radical, which subsequently disappears with the simultaneous appearance of an ascorbate radical.[6] 8α-Hydroxy-α-tocopherone, formed as a labile intermediate during the oxidation of α-tocopherol to α-tocopheryl quinone by superoxide anion, was reversibly reduced back to α-tocopherol by ascorbic acid.[7] Addition of ascorbic acid to vitamin E radicals generated by oxidizing lipids on silica gel monolayers resulted in the complete quenching of the α-tocopheroxy radical spectrum.[8] Upon exposure of vitamin E radicals to glutathione or ascorbic acid, the electron spin resonance signals of α-chromanoxyl radical disappeared.[9] Vitamin C and vitamin E acted as chain-breaking antioxidants and suppressed the oxidation of methyl lineolate initiated in solutions with azo compounds.[10] When both vitamins were present, the oxidation of linoleate was suppressed quite efficiently; at the initial stage, vitamin E remained almost unchanged and only ascorbic acid was consumed. Tocopherol was consumed only after vitamin C was exhausted.[10] α-Tocopherol and ascorbate act synergistically on the quenching of singlet molecular oxygen; in these experiments, 8α-hydroxy and 8α-ethoxy tocopherones were reduced to tocopherols by ascorbic acid.[11] Synergistic suppression of fatty acid peroxidation by a combination of vitamins C and E was observed in oil-water emulsions and in rat liver microsomes.[12,13] Hydrogen atom exchange was shown to take place between ascorbyl palmitate and vitamin E radical in air-oxidized lipids.[14] In this study, sequential consumption of vitamins C and E, first ascorbyl palmitate then α-tocopherol, was demonstrated by differential pulse polarography. These results suggest that vitamin C may serve as a reservoir of antioxidant potential while vitamin E may function as the specific, highly efficient delivery system.[12]

All these studies were carried out in vitro, mostly in single-phase systems. Apart from a small portion bound to transporting proteins, tocopherol is totally localized in the cell membranes. On the other hand, as noted above, most ascorbate is present in the cytosol and it is not yet determined whether interactions between ascorbate and tocopherol radicals take place in vivo. In guinea pigs administered high amounts of vitamin C, the levels of tocopherol in blood and tissues are quite high.[15-17] However, in the rat, an animal that can synthesize ascorbate, controversial data were obtained.[18] The ascorbate level in the tissues of animals synthesizing vitamin C is only little influenced by exogenous vitamin C. Therefore, for research on the interaction between vitamins C and E, guinea pigs are more suitable, as the ascorbate level in their tissues distinctly depends on the supply of exogenous vitamin C. A significant reduction of α-tocopherol concentration in a number of tissues of guinea pigs, due to chronic vitamin C deficiency, was observed.[16] Tocopherol concetrations in tissue does not depend only on the quantity of available vitamin E, but also on the content of lipid in the cells and on membrane structure. These factors are critical in determining how much vitamin E can be absorbed by the cells. Vitamin C deficiency in guinea pigs is usually connected with liver steatosis, and, in the case of moderate vitamin C deficiency, the tocopherol level in the liver of guinea pigs increases.[19] In very chronic vitamin C deficiency, however, the effect of low ascorbate levels in liver prevails, and thus, in spite of the accumulation of fat, a significant decrease in α-tocopherol level occurs in this organ.[20] It is not probable that higher levels of tocopherol in the tissues of guinea pigs fed a high vitamin C diet would be due to lowered oxidation of vitamin E in the diet. If tocopherol is administered, a gradual corresponding increase of the content of ascorbic acid in the diet results in an increase of the tocopherol levels in the lung.[20] Some, but not all, authors observe an increase in the Vitamin E level in plasma after administration of vitamin C to man.[21-23]

These results indicate that vitamin C is able to reduce vitamin E radicals to tocopherol in vivo. However, it is possible that the vitamin E-sparing effect of vitamin C is indirect, since high doses of ascorbic acid block copper absorption and lower the concentration of active Cu ions in blood and tissues. The report that vitamin C accelerates absorption of peroxide detoxifying selenium in chickens has not been proved conclusively. Chronic vitamin C deficiency can reduce absorption of vitamin E by slowing down the synthesis of bile acids necessary for tocopherol absorption.[24] Although the exact mechanism of the effect of vitamin C on the tocopherol levels remains unknown, the data prove that chronic marginal vitamin C deficiency evokes a relative "endogenous" vitamin E deficiency. Simultaneous drop in the levels of these two efficient antioxidants and scavengers of oxygen free radicals in cytosol (ascorbate) as well as in membranes (tocopherol) can have far-reaching consequences for the liver as well as for the highly oxidized tissues of lung and heart.

EFFECT OF VITAMINS C AND E ON DETOXICATION PROCESSES

Most foreign chemicals are extensively metabolized in the liver or other organs prior to their excretion from the organism. Such biotransformations are carried out by two major enzyme systems: (1) the microsomal hydroxylation system (mixed-function oxidases or monoxygenases) which catalyzes a variety of oxidative reactions in the endoplasmic reticulum (cytochrome P-450 serves in this system as the terminal oxidase.) and (2) a variety of conjugating enzymes localized in the endoplasmic reticulum or cytosol which form various types of conjugates of foreign chemicals or of their cytochrome P-450 oxidized metabolites. Both systems are particularly abundant in the endoplasmic reticulum of liver cells. There is evidence that both vitamin C and vitamin E remarkably influence the activity of the first biotransformation system.

Vitamin C and the Cytochrome P-450 Cycle

It has been known for a long time that vitamin C deficiency decreases the activity of the biotransformation systems. Vitamin C-deficient guinea pigs have prolonged pentobarbital sleeping time, prolonged plasma halflife of several xenobiotics, decreased hydroxylation of cholesterol and of many but not all environmental chemicals, decreased levels of cytochromes P-450 and b_5, and decreased activity of NADPH-cytochrome P-450 reductase in liver microsomes.[25,26] The mechanism of this phenomenon has not been elucidated. In vitamin C-deficient guinea pigs, *de novo* microsomal protein synthesis operates, activities of key enzymes involved in heme synthesis and degradation are not changed, the turnover of cytochrome P-450 heme is normal, and there is no change in the spin state of cytochrome P-450 and no significant change in cytochrome P-450 affinity for drug substrates.[26] In severely scorbutic guinea pigs, cytochrome P-450 concentration in the liver never decreases below 40 to 60% of normal values. In guinea pigs with marginal vitamin C deficiency and normal body weight, the level of microsomal cytochrome P-450 is often unchanged, but simultaneously the activities of some hydroxylation enzymes are significantly inhibited. Hypothetically, mild vitamin C deficiency may selectively affect only a small portion of cytochrome P-450 isoenzymes with concomitant decrease in the activity of some monooxygenases without any pronounced change in total cytochrome P-450 level.

Several mechanisms have been discussed through which marginal vitamin C deficiency might affect the function of the cytochrome P-450 cycle.[24] Cytochrome P-450 is thought to bind substrate (e.g., environmental chemicals, drugs or cholesterol) and oxygen in a sequential manner. In the presence of oxidizing agents (e.g., peroxides), the cycle may run along a shunt by-passing the need for oxygen and electron donors. According to our working hypothesis, vitamin C stimulates the activity of the cytochrome P-450 cycle by producing activated oxygen, hence the cycle proceeds by this shunt.[24] The ability of vitamin C to activate oxygen was demonstrated in nonbiological model systems. According to Visser,[27] ascorbate binds on heme iron with the formation of a ferrous-ascorbate complex which could be a central catalytic unit in the oxygen fixation system. Chrastil and Wilson assume that the ascorbate-cytochrome protein complex is the principal source of oxygen free radicals in the cytochrome system.[28] The relatively consistent quantitative relationship of microsomal ascorbate to cytochrome P-450 during dialysis and during purification of the cytochrome, as well as the fact that ascorbate protects cytochrome P-450 from inhibition by Fe^{2+} chelators, point to a possible existence of ascorbate-cytochrome P-450 complexes involving the heme iron.[26] However, ascorbate concentration in endoplasmic reticulum is the lowest of all organelles in the liver cell. The decrease of monooxygenase activity in the liver of vitamin C-deficient guinea pigs was related to the decrease in cytosolic ascorbate.[29] Hence, it may be assumed that oxygen activation by ascorbate takes place at the division of the smooth endoplasmic reticulum membranes and the cytosol of the liver cell.

If the hypothesis on the vitamin C stimulation of the cytochrome P-450 cycle through oxygen activation is correct, then ascorbate ought to be effective also in vitro. Microsomal hydroxylation of xenobiotics was increased by ascorbate addition in vitro and this increase was dependent on ascorbate concentration.[30] Artificial and microsomal systems need ascorbate addition for activation.[28] According to other authors, ascorbate addition to vitamin C-deficient liver microsomes did not restore the decreased monooxygenase activity.[26] During the preparation of microsomes, destruction of hepatocyte compartmentalization takes place. Compartmentalization of the cell plays a key role in the control of production and scavenging of oxygen radicals. It is possible that the conflicting data from some in vitro experiments can be due to a different degree of destruction of the metabolic compartments of a liver cell.

The concept by which vitamin C stimulates the cytochrome P-450 cycle through the activation of oxygen would help to explain the paradoxical biphasic action of increasing vitamin C doses on the activity of monooxygenases. In vitamin C deficiency, when the levels of ascorbate in liver are extremely low, the activity of microsomal monooxygenases is significantly lowered.[19,25,26,29] At higher intakes of vitamin C, the activity of monooxygenases rises, but at very high vitamin C intake, this activity is inhibited.[19,31] There are a number of other biochemical systems responsive to the supply of vitamin C, parabolically, both in vitro and in vivo, e.g., lipoperoxidation of fatty acid micelles, microsomes, and various brain preparations;[32,33] ligand binding to neurotransmitter receptors;[33] steroidogenesis;[34] activity of cholesterol 7-α-hydroxylase, 3-hydroxy-3-methyl-glutaryl coenzyme A reductase, and cholesterol level in blood serum and liver;[19,35,36] and carnitine biosynthesis.[37] Ascorbic acid exerts both mutagenic and antimutagenic effects.[38] Many of these parabolic responses can be explained in such a way that vitamin C is able both to generate and scavenge oxygen free radicals at different concentrations. In case of deficiency, i.e., at low vitamin C concentrations in the tissues, the generation and scavenging of oxygen free radicals are low and the activity of many systems including the cytochrome P-450 cycle is low. With increasing supply of vitamin C to physiological levels, both generation and scavenging of oxygen radicals rise and the activity of many systems increases. At extremely high vitamin C megadoses, the generation of oxygen radicals is suppressed, scavenging of free radicals is extremly high, and activity of many systems drops back to the values observed at vitamin C deficiency. Not only uncontrolled production of free radicals, but also overloading with free radical scavengers may be harmful. It is possible that the basic biological function of vitamin C resides in maintaining an optimum dynamic equilibrium in the formation and scavenging of free radicals and that both a vitamin C deficiency and also extremely high intake of ascorbic acid present a threat to this optimum equilibrium.

The ability of vitamin C to scavenge free radicals can have favorable effects in detoxication of reactive intermediates formed in the cytochrome P-450 cycle. This cycle is in fact "blind": it detoxicates xenobiotics, but on the other hand, it can activate relatively harmless molecules transforming them into highly reactive intermediates which could be covalently bound to cell macromolecules and induce mutagenesis, carcinogenesis, or necrosis. A typical example is the transformation of benzo(*a*)pyrene to carcinogenic benzo(*a*)pyrendiol epoxide. Vitamin C prevents the binding of these toxic intermediates to nucleic acids or proteins not only in vitro, but also in vivo. Pretreatment of rats with a single high dose of ascorbic acid resulted in a substantial decrease in the in vivo covalent binding of benzo(*a*)pyrene metabolites to nuclear DNA in the liver.[39] Bone marrow ascorbate levels could be modulated tenfold by placing guinea pigs on different dietary intakes of vitamin C. Bone marrow with a high ascorbate concentration displayed a fourfold less covalent binding of phenol equivalents than those with low vitamin C concentrations.[40] It is probable that in these studies ascorbate or its radical reacted with intermediates, thus preventing their toxic effects. Reduction of the frequency of chromosome aberrations in blood lymphocytes in coal tar-exposed workers after long-term administration of ascorbic acid was probably due to a similar mechanism[41] Short-term marginal vitamin C deficiency did not affect antipyrine pharmacokinetics in man, but administration of 500 mg of ascorbic acid per day for 12 months significantly shortened the plasma halflife of antipyrine in diabetics.[42]

Vitamin E and Detoxication Processes

It has been repeatedly proven that vitamin E deficiency causes a decrease in microsomal drug-metabolizing functions in laboratory animals. In liver microsomes of vitamin E-deficient rats, the demethylation of aminopyrine and ethylmorphine and hydroxylation of benzo(*a*)pyrene were slowed down. The liver microsomal epoxide hydrolase activity was increased, while the heme hydroperoxidase activity of liver and lung microsomes was significantly decreased in vitamin E-deficient rats.[43] In vitro addition of α-tocopherol had no effect on the decreased enzyme activity. In some studies, the vitamin E deficiency was linked with a decrease in cytochrome P-450 level in hepatic microsomes.[44] The idea of a vitamin E role in the biosynthesis of the heme part of cytochrome P-450 had to be revised, since in the majority of studies, vitamin E deficiency had little effect on the level of liver cytochromes and it was concluded that there was no direct influence of tocopherol on heme synthesis in the liver.[45]

Combination of tocopherol with essential fatty acids has synergistic effects on biotransformation enzymes. It is probable that vitamin E is required as a structural component of

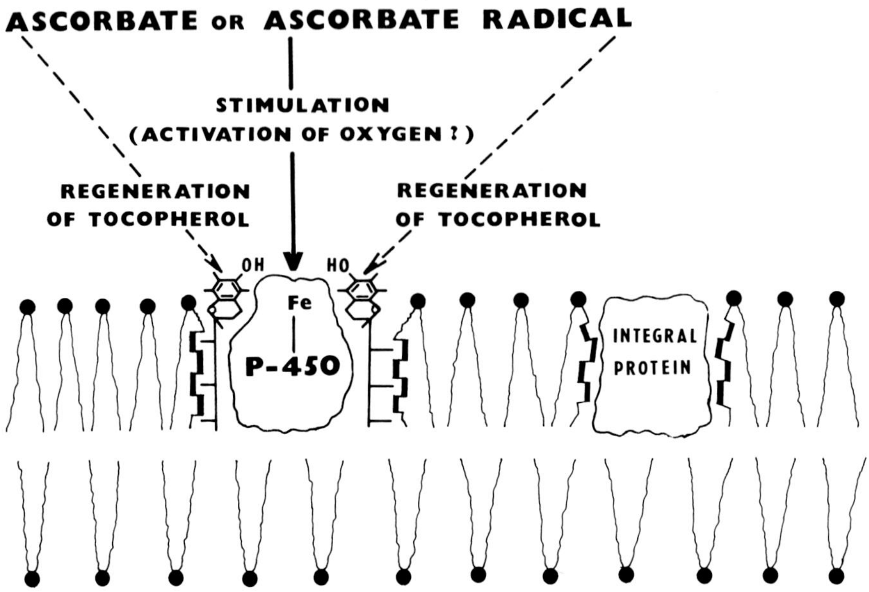

FIGURE 1. Hypothetic scheme of the synergistic effects of vitamins C and E on microsomal, cytochrome P-450-dependent biotransformation systems in the liver cell. Ascorbate or ascorbate radical localized in the cytosol stimulates the cytochrome P-450 cycle by oxygen activation through interaction of vitamin C with cytochrome P-450 iron. Tocopherol, localized near cytochrome P-450 in the phospholipid membranes of the endoplasmic reticulum, scavenges free radicals formed in the cytochrome P-450 cycle during degradation of xenobiotics. Then ascorbate reduces tocopheryl radicals formed in this process back to tocopherol.

the endoplasmic reticulum. The configuration of α-tocopherol is ideal for stabilizing membranes containing polyunsaturated fatty acids. The C-4 and C-8 methyl groups of the tocopherol phytol side chain fit tightly into the pockets provided by the cis-double bonds of polyunsaturated fatty acids (Figure 1). This may influence membrane fluidity and at the same time protect membrane phospholipids against peroxidation. Molenaar et al.[46] submitted an attractive hypothesis on the localization and the role of vitamin E in the membranes of endoplasmic reticulum which helps to explain also the role of tocopherol in biotransformation reactions. According to their idea, the molecules of tocopherol are inserted between the polyunsaturated fatty acids of phospholipids and membrane proteins which produce free radicals (e.g., cytochrome P-450 system). The structural proteins which do not produce free radicals are not surrounded by vitamin E (Figure 1). The chromanol rings anchored in the direct vicinity of monooxygenases scavenge free radicals produced by the cytochrome P-450 cycle, protecting those enzymes against peroxidative self-destruction.

Vitamin E also plays an important role in the protection against air pollutant intoxication. Tocopherol affects the morbidity of laboratory animals exposed to ozone and nitrogen dioxide. Rats with a low vitamin E status are much more susceptible to both NO_2 and O_3 than rats fed a high tocopherol diet. Vitamin E lowered the formation of peroxides in the lungs of ozone-exposed rats. In addition to oxidizing air pollutants, there are a number of other environmental substances able to induce lipid peroxidation, such as carbon tetrachloride, ethanol, alloxan, ascorbate plus Fe or Cu ions, toxic metals (Pb, Hg, Ag, and Cd), cigarette smoke constituents, and many others. Vitamin E, as a chain-breaking antioxidant, is the main protector against in vivo lipoperoxidation induced by these substances.[47] Tocopherol provides protection against the hepatotoxicity of carbon tetrachloride and ethanol. Rats fed high tocopherol diets were totally protected against alloxan- and alloxan-ascorbate-induced peroxidation.[48]

The dose of vitamin E ensuring maximum efficiency of the biotransformation systems in humans is not known. When considering the data found in laboratory animals, it will probably be higher than the officially recommended doses, but at the same time lower than frequently used megadoses of tocopherol. Vitamin E, like vitamin C, inhibits the activity of microsomal monoxygenases in the liver of animals when extremely high doses are administered.[49,50]

Synergism Between Vitamins C and E

Synergistic antioxidation effects of mixtures of vitamins C and E for protection of polyunsaturated fatty acids in oils is well known. The combination of solubilized ascorbyl palmitate with vitamin E shows marked synergism, which allows the amount of antioxidant mixture to be reduced. Cooperative interaction between vitamins C and E was manifested in vitro in suppression of the peroxidation of membrane phospholipids from rat liver microsomes.[12] A distinct synergistic effect of vitamins C and E on stimulation of biotransformation systems in the liver was observed.[51] Guinea pigs fed a cholesterol diet, which may lead to elevated hepatic lipoperoxidation, were given a diet marginally deficient both in ascorbic acid and tocopherol. The other two groups had an increased supply of either vitamin C or vitamin E. The fourth group had identically increased intake of both vitamins. The activity of aniline hydroxylase and *p*-nitroanisole *O*-demethylase in hepatic microsomes was very low in the group marginally deficient in both vitamins. An enhanced supply of vitamin C or vitamin E stimulated the activity of both enzymes. Simultaneous supply of vitamins C and E had a distinct synergistic effect, as a result of which the activity of aniline hydroxylase proved to be up to eight times higher than in the group marginally deficient in both vitamins.[51]

Figure 1 illustrates our hypothetical idea on the mechanism of the synergistic effect of vitamin C and E on the biotransformation system of the liver. The role of vitamin C in the system is both oxidative and antioxidative. Ascorbate or ascorbate radicals, localized mostly in cytoplasm, produce (by reduction of dioxygen) free radicals or activate oxygen by interaction with cytochrome P-450 iron. Activation of oxygen increases the activity of monooxygenases, as the cytochrome P-450 cycle may proceed by a peroxide-dependent shunt bypassing the need for oxygen and electron donors.[24] Vitamin E localized in the membranes of endoplasmic reticulum in close vicinity to cytochrome P-450 scavenges free radicals produced during biotransformation of xenobiotics in the cycle of cytochrome P-450.[46] Tocopheryl radicals generated in this process are regenerated back to tocopherol by interaction with ascorbate, which in this case has an antioxidant effect.

The synergism of vitamins C and E is also manifested in the GI tract by their inhibition of the formation of *N*-nitroso compounds which are a potential source of carcinogenesis in humans. Newmark and Mergens characterize vitamins C and E as a complementary pair of agents blocking very effectively nitrosation reactions by two mechanisms: (1) vitamin C in aqueous phase and vitamin E in lipid or micellar phase reduce the nitrosating agent (nitrous acid anhydride) and (2) vitamins C and E block free radical propagation.[52] The protective effect of both vitamins aimed against formation of *N*-nitroso compounds was confirmed in dozens of studies not only with experimental animals, but also with humans.[52-54]

CONCLUSIONS

The redox system of vitamins C and E function both in vitro and in vivo as efficient antioxidants, scavengers of free radicals, and terminators of free radical chain reactions. The water-soluble ascorbate is localized mostly in the cytosol and the lipid-soluble tocopherol is localized almost exclusively in membranes. In contrast to vitamin E, ascorbate can also have oxidative effects as it forms (by reduction of dioxygen in the presence of Fe or Cu ions) hydrogen peroxide, superoxide anion, and hydroxyl radicals. It is highly probable that, between the redox systems of vitamins C and E, interactions take place (not only in vitro) during which ascorbate reduces tocopheryl radicals regenerating tocopherol. In both vitamins

C and E deficiencies the function of biotransformation systems important for the metabolism of xenobiotics is disturbed. The activity of the microsomal cytochrome P-450 dependent monooxygenases in the liver is inhibited. The exact mechanism of action of vitamins C and E on the cytochrome P-450 system is not known. It is assumed that vitamin C stimulates the cytochrome P-450 cycle by activation of oxygen, because of which the cycle may then proceed through a peroxide-dependent shunt. The main role of vitamin E, localized in close vicinity to the cytochrome P-450 complex, consists in the protection of this system against self-destruction by radicals produced during degradation of xenobiotics. Tocopheryl radicals formed in this process are regenerated back to tocopherol by interaction with ascorbate. This hypothetic scheme explains why the simultaneous administration of vitamins C and E has a synergic effect on the activity of biotransformation enzymes. The synergism of vitamins C and E is also manifested in the prevention of the formation of *N*-nitroso compounds in the GI tract. The optimum doses of vitamin C and E enabling maximum efficiency of detoxication systems in man are not known; they will be probably higher than currently recommended dietary intake. Because of the parabolic effect of increasing doses of vitamins C and E on the activity of monooxygenases (low activity in deficiency of both vitamins, increased activity at increased intake of ascorbic acid or tocopherol, and inhibition of activity at extremely high doses of vitamins C and E), permanent application of megadoses of vitamins C and E is not recommended.

REFERENCES

1. **Sasaki, R., Kurokawa, T., and Tero-Kubota, S.,** Nature of serum ascorbate radical and its quantitative estimation, *Tohoku J. Exp. Med.*, 136, 113, 1982.
2. **Machlin, L. J.,** *Vitamin E. A Comprehensive Treatise,* Marcel Dekker, New York, 1980.
3. **Scarpa, M., Stevanato, R., Viglino, P., and Rego, A.,** Superoxide ion as active intermediate in the autoxidation of ascorbate by molecular oxygen. Effect of superoxide dismutase, *J. Biol. Chem.*, 258, 6695, 1983.
4. **Tappel, A. L.,** Will antioxidant nutrients slow aging processes?, *Geriatrics,* 23, 97, 1968.
5. **McCay, P. B.,** Vitamin E: interactions with free radicals and ascorbate, *Annu. Rev. Nutr.*, 5, 323, 1985.
6. **Packer, J. E., Slater, T. F., and Wilson, R. F.,** Direct observation of free radical interaction between vitamin E and vitamin C, *Nature (London),* 278, 737, 1979.
7. **Nishikimi, K. and Machlin, L. J.,** Oxidation of alpha-tocopherol model compound by superoxide anions, *Arch. Biochem. Biophys.*, 170, 684, 1975.
8. **Bascetta, E., Gunstone, F. D., and Walton, J. C.,** Electron spin resonance study of the role of vitamin E and vitamin C in the inhibition of fatty acid oxidation in a model membrane, *Chem. Phys. Lipids,* 33, 207, 1983.
9. **Niki, E., Tsuchiya, J., Tanimura, J., and Kamyia, Y.,** Regeneration of vitamin E from alpha-chromanoxyl radical by glutathione and vitamin C, *Chem. Lett.*, p. 789, 1982.
10. **Niki, E., Saito, T., Kawakami, A., and Kamiya, Y.,** Inhibition of oxidation of methyl linoleate in solution by vitamin E and vitamin C, *J. Biol. Chem.*, 259, 4177, 1984.
11. **Yamauchi, R., Kato, K., and Ueno, Y.,** Reaction of 8α-hydroperoxy tocopherones with ascorbic acid, *Agric. Biol. Chem.*, 45, 2855, 1981.
12. **Leung, H. W., Vang, M. J., and Mavis, R. D.,** The cooperative interaction between vitamin E and vitamin C in suppression of peroxidation of membrane phospholipids, *Biochim. Biophys. Acta,* 664, 266, 1981.
13. **Barclay, L. R. C., Locke, S. J., and MacNeil, J. M.,** The autoxidation of unsaturated lipids in micells. Synergism of inhibitors vitamins C and E, *Can. J. Chem.*, 61, 1288, 1983.
14. **Lambelet, P., Saucy, F., and Loliger, J.,** Chemical evidence for interactions between vitamins C and E, *Experientia,* 41, 1384, 1985.
15. **Chen, L. H. and Chang, M. L.,** Effect of dietary vitamin E and vitamin C on respiration and swelling of guinea pig liver mitochondria, *J. Nutr.*, 108, 1616, 1978.
16. **Hruba, F., Novakova, V., and Ginter, E.,** The effect of chronic marginal vitamin C deficiency on the α-tocopherol content of the organs and plasma of guinea-pigs, *Experientia,* 38, 1454, 1982.

17. **Bendich, A., D'Apolito, P., Gabriel, E., and Machlin, L. J.,** Interaction of dietary vitamin C and vitamin E on guinea pig immune responses to mitogens, *J. Nutr.*, 114, 1588, 1984.
18. **Chen, L. H.,** An increase in vitamin E requirement induced by high supplementation of vitamin C in rats, *Am. J. Clin. Nutr.*, 34, 1036, 1981.
19. **Ginter, E., Kosinova, A., Hudecova, A., and Mlynarcikova, U.,** Parabolic response of hepatic microsomal hydroxylating system and lipids to graded doses of ascorbic acid in guinea pigs on low and high α-tocopherol intake, *J. Nutr.*, 114, 485, 1984.
20. **Ginter, E., Dvorsky, A., Novakova, V., Kosinova, A., Hudecova, A., and Hruba, F.,** The influence of vitamin C on vitamin E levels in the guinea pig tissues, *Vitaminologia*, 2, 109, 1986.
21. **Fidanza, A., Audisio, M., and Mastroiacovo, P.,** Vitamin C and cholesterol, *Int. J. Vitam. Nutr. Res.*, Suppl. 23, 153, 1982.
22. **Arad, I. D., Dgani, Y., and Eyal, F. G.,** Vitamin E and vitamin C plasma levels in premature infants following supplementation of vitamin C, *Int. J. Vitam. Nutr. Res.*, 55, 395, 1985.
23. **Keith, R. E.,** Dietary vitamin C supplementation and plasma vitamin E levels in humans, *Am. J. Clin. Nutr.*, 33, 2394, 1980.
24. **Ginter, E.,** Cholesterol 7 α-hydroxylase: regulation by ascorbic acid and other nutrients, in *Cholesterol 7α-Hydroxylase (7α-Monooxygenase)*, Fears, R. and Sabine, J. R., Eds., CRC Press, Boca Raton, Fla., 1986, 103.
25. **Degkwitz, E., Walsch, S., Dubberstein, M., and Winter, J.,** Ascorbic acid and cytochromes, *Ann. N.Y. Acad. Sci.*, 258, 201, 1975.
26. **Zannoni, V. G., Holsztynska, E. J., and Lau, S. S.,** Biochemical functions of ascorbic acid in drug metabolism, *Adv. Chem. Ser.*, 200, 349, 1982.
27. **Visser, C. M.,** Role of ascorbate in biological hydroxylations: origin of life considerations and the nature of the oxenoid species in oxygenase reactions, *Bioorg. Chem.*, 9, 261, 1980.
28. **Chrastil, J. and Wilson, J. T.,** Contribution of cytochromes and proteins to the effect of ascorbic acid on artificial and microsomal hydroxylation systems containing oxygen and hydrogen peroxide, *Biochem. J.*, 170, 693, 1978.
29. **Omaye, S. T. and Turnbull, J. D.,** Heme oxygenase activity, drug metabolism, and ascorbic acid distribution in the livers of ascorbic acid-deficient guinea pigs, *Biochem. Pharmacol.*, 28, 1415, 1979.
30. **Wilson, J. T., Chrastil, J., Van Boxtel, C. J., Alvan, G., and Sjoquist, F.,** Effect of ascorbic acid on liver drug metabolism in animals and man, *Fed. Proc.*, 35, 709, 1976.
31. **Peterson, F. J., Holloway, D. E., Duquette, P. H., and Rivers, J. M.,** Dietary ascorbic acid and hepatic mixed function oxidase activity in the guinea pig, *Biochem. Pharmacol.*, 32, 91, 1983.
32. **Yau, T. M. and Mencl, J.,** A study of the peroxidation of fatty acid micelles promoted by ionizing radiation, hydrogen peroxide and ascorbate, *Int. J. Radiat. Biol.*, 40, 47, 1981.
33. **Heikkila, R. E., Cabbat, F. S., and Manzino, L.,** Inhibitory effects of ascorbic acid on the binding of (^{3}H)dopamine antagonist to neostriatal membrane preparations: relationship to lipid peroxidation, *J. Neurochem.*, 38, 1000, 1982.
34. **Pintauro, S. J. and Bergan, J. G.,** Effects of ascorbic acid on in vitro steroidogenesis in guinea pigs, *J. Nutr.*, 112, 584, 1982.
35. **Holloway, D. E. and Rivers, J. M.,** Influence of chronic ascorbic acid deficiency and excessive ascorbic acid intake on bile acid metabolism and bile composition in the guinea pig, *J. Nutr.*, 111, 412, 1981.
36. **Greene, Y. J., Harwood, H. J., and Stackpoole, P. W.,** Ascorbic acid regulation of 3-hydroxy-3-methylglutaryl coenzyme A reductase activity and cholesterol synthesis in guinea pig liver, *Biochim. Biophys. Acta*, 834, 134, 1985.
37. **Hulse, J. D., Ellis, S. R., and Henderson, L. M.,** Carnitine biosynthesis, *J. Biol. Chem.*, 253, 1654, 1978.
38. **Shamberger, R. J.,** Genetic toxicology of ascorbic acid, *Mutat. Res.*, 133, 135, 1984.
39. **Shah, G. M. and Bhattacharya, R. K.,** In vivo effect of L-ascorbic acid on benzo(a)pyrene metabolite-RNA adduct formation in rat liver, *J. Biosci.*, 4, 263, 1982.
40. **Smart, R. C. and Zannoni, V. G.,** Effect of ascorbate on covalent binding of benzene and phenol metabolites to isolated tissue preparations, *Toxicol. Appl. Pharmacol.*, 77, 334, 1985.
41. **Šrám, R., Dobiáš, L., Pastorková, A., Rossner, P., and Janča, L.,** Effect of ascorbic acid prophylaxis on the frequency of chromosome aberrations in the peripheral lymphocytes of coal-tar workers, *Mutat. Res.*, 120, 181, 1983.
42. **Ginter, E. and Vejmolova, J.,** Vitamin C-status and pharmacokinetic profile of antipyrine in man, *Br. J. Clin. Pharmacol.*, 12, 256, 1981.
43. **Reddy, C. C., Thomas, C. E., Scholz, R. W., and Massaro, E. J.,** Effects of inadequate vitamin E and/or selenium nutrition on enzymes associated with xenobiotic metabolism, *Biochem. Biophys. Res. Commun.*, 107, 75, 1982.
44. **Whanger, P. D., Tripp, M. J., and Weswig, P. H.,** Effects of selenium and vitamin E deficiencies in lambs on hepatic microsomal hemoproteins and mitochondrial respiration, *J. Nutr.*, 107, 998, 1977.

45. **Bartlett, R. S., Rousseau, J. E., Frier, H. I., and Hall, R. C.,** Effect of vitamin E on δ-aminolevulinic acid dehydratase activity in weanling rabbits with chronic plumbism, *J. Nutr.*, 104, 1637, 1974.
46. **Molenaar, I., Hulstaert, C. E., and Hardonk, M. J.,** Role in function and ultrastructure of cellular membranes, in *Vitamin E*, Machlin, L. J., Ed., Marcel Dekker, New York, 1980, 372.
47. **Tappel, A. L.,** Vitamin E and selenium protection from in vivo lipid peroxidation, *Ann. N.Y. Acad. Sci.*, 355, 18, 1980.
48. **Dillard, C. J., Kunert, K. J., and Tappel, A. L.,** Effects of vitamin E, ascorbic acid and mannitol on alloxan-induced lipid peroxidation in rats, *Arch. Biochem. Biophys.*, 216, 204, 1982.
49. **Kritchevsky, D., Nitzsche, C., Czarnecki, S. K., and Story, J. A.,** Influence of vitamin E supplementation on cholesterol metabolism in rats, *Nutr. Rep. Int.*, 22, 339, 1980.
50. **Dashman, T. and Kamm, J. J.,** Effects of high doses of vitamin E on dimethylnitrosamine hepatotoxicity and drug metabolism in the rat, *Biochem. Pharmacol.*, 28, 1485, 1979.
51. **Ginter E., Kosinova, A., Hudecova, A., and Madaric, A.,** Synergism between vitamins C and E: effect on microsomal hydroxylation in guinea pig liver, *Int. J. Vitam. Nutr. Res.*, 52, 55, 1982.
52. **Newmark, H. L. and Mergens, W. J.,** α-Tocopherol (vitamin E) and its relationship to tumor induction and development, in *Inhibition of Tumor Induction and Development*, Zedeck, M. S. and Lipkin, M., Eds., Plenum Press, 1981, 127.
53. **Tannenbaum, S. R. and Mergens, W.,** Reaction of nitrite with vitamins C and E, *Ann. N.Y. Acad. Sci.*, 355, 267, 1980.
54. **Wagner, D. A., Shuker, D. E. G., Bilmazes, C., Obiedzinski, M., Baker, I., Young, V. R., and Tannenbaum, S. R.,** Effect of vitamins C and E on endogenous synthesis of N-nitrosamino acids in humans: precursor-product studies with (^{15})nitrate, *Cancer Res.*, 45, 6519, 1985.

TAURINE AND HYPOTAURINE AND MEMBRANE LIPID PEROXIDATION

H. Pasantes-Morales and J. H. Fellman

TAURINE AND HYPOTAURINE: GENERAL ASPECTS

Concentration in Tissues

The sulfur amino acids, taurine and hypotaurine, are normal constituents of many animal tissues.[1] They are present as free amino acids and are not incorporated into proteins nor into any other macromolecules. High concentrations of taurine (2 to 40 m*M*) are found in brain, muscle, heart, endocrine glands, retina, lymphocytes and platelets.[2-5] Substantial amounts of taurine (1 to 5 m*M*) are also present in kidney, spleen, liver, and lung[6] (Table 1). Taurine levels in plasma are much lower, in the range of 0.1 to 0.5 m*M*. Taurine is unequally distributed in brain and heart, with differences of more than tenfold between regions containing the lowest and the highest taurine concentration.[2,3] Also, species differences exist: taurine concentration in brain is 7 to 11 m*M* in rat, chick, and mouse and only 1.8 to 2.0 m*M* in humans and cats.[3] In heart (myocardium), taurine concentration is 1.8 m*M* in frog, 5 to 6 m*M* in beef and dog, 11.6 m*M* in cat, 25.3 m*M* in rat, and 39.4 m*M* in mouse.[2] In some tissues, taurine concentration is high during development and decreases in the mature organ. This is observed typically in brain.[7] The concentration of hypotaurine in animal tissues has not been studied systematically. Low amounts, in the micromolar range, have been recently reported in mammalian tissue,[9] but appreciable amounts have been detected in the reproductive tract of marine invertebrates as well as in mammalian sexual tissues.[9]

Sources of Taurine and Hypotaurine

Biosynthesis, Metabolic Pathways, and Regulatory Mechanisms

The metabolic precursor of taurine and hypotaurine in cells is the sulfur amino acid cysteine. Several pathways have been proposed for the synthesis of taurine from cysteine, but the only one of quantitative significance is that involving cysteine sulfinic acid (CSA) and hypotaurine as intermediates (Figure 1). This pathway has been demonstrated to operate principally in mammalian liver and brain, although its efficiency varies among species.[10,11] The first reaction, i.e., the oxidation of cysteine is carried out by the enzyme cysteine dioxygenase (EC 1.13.11.20) and cysteine sulfinic acid is formed.[12] A decarboxylase (EC 1.4.1.1.1.29) converts cysteine sulfinic acid into hypotaurine.[13] Hypotaurine then is subject to oxidation to form taurine.[14,15] The oxidation of hypotaurine occurs rapidly in most tissues, but the nature of this process is still unclarified. Hypotaurine levels are very low, even in tissues possessing a highly active cysteine sulfinate decarboxylase, and no clear correlation exists between hypotaurine levels and the decarboxylating activity. The enzymatic activity proposed for hypotaurine oxidation in animal tissues is very low and probably does not account for the physiological degradation of hypotaurine. The metabolic conversion of hypotaurine into taurine may have an important role in the physiological action ascribed to hypotaurine as an antioxidant.

The regulation of the metabolic pathway for taurine and hypotaurine biosynthesis may occur through the cysteine sulfinate reaction and at the hypotaurine oxidation step. The first enzyme in the pathway of taurine biosynthesis, cysteine dioxygenase, is rate limiting for the overall metabolic fate of the sulfur amino acids. The product, cysteine sulfinate, is subject to transamination by the ubiquitous aspartate aminotransferase leading to an unstable sulfinopyruvate which spontaneously breaks down to SO_2 and pyruvate. As noted, CSA is also a substrate for cysteine sulfinate decarboxylase, the product of which is hypotaurine. Little is known about the means by which this pathway bifurcation is regulated. Cysteine dioxy-

Table 1
TAURINE CONTENT IN VERTEBRATE TISSUES[1,3,6]

	Taurine (m*M*)			
Tissue	**Rat**	**Cat**	**Rabbit**	**Human**
Brain (cortex)	11.1	2.1	1.6	1.4
Heart (ventricles)	23.4		10.8	5.4
Muscle	2.7	4.5		
Intestine	5.5			
Kidney	4.6	3.1	3.6	1.8
Liver	4.5	8.3	3.0	1.5
Lung	6.6		2.9	3.5
Spleen	4.2			11.5
Adenohypophysis	17.9		15.5	
Retina	34.5	43.0	36.2	24.0
Platelets				24.0
Leukocytes				35.0

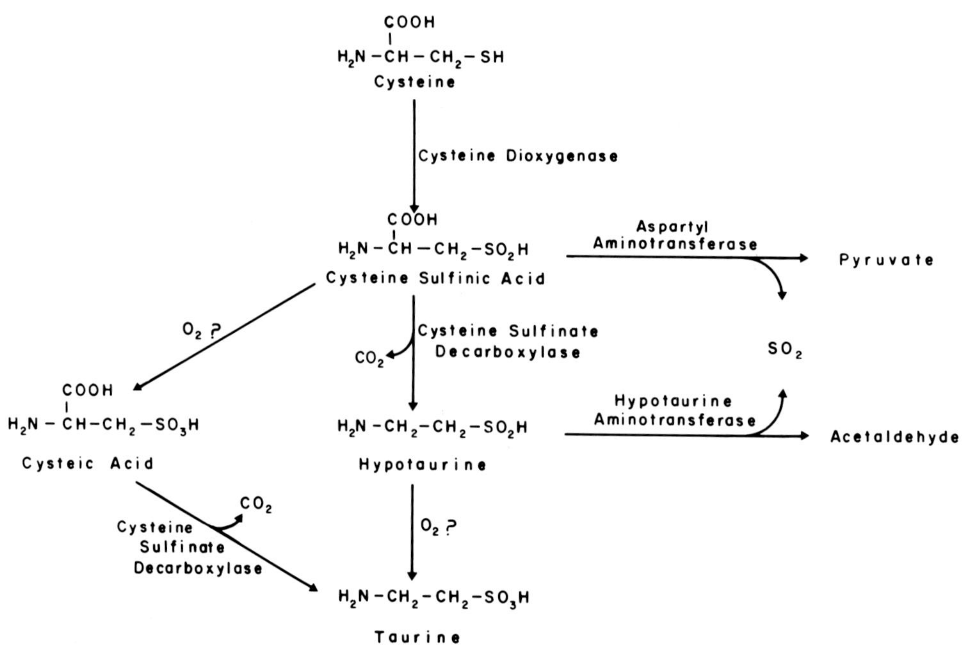

FIGURE 1. Metabolic pathways of taurine and hypotaurine.

genase is an inducible enzyme and animals fed large dietary supplements of cysteine increase both urinary SO_4 and taurine output.[16-18] However, the ratios of the two pathways remain similar and no obvious means of regulating the flow toward either pathway have emerged. Yet when cysteine sulfinic acid is injected into rats, 90% of the material flows through the taurine pathway.[19] The reverse is observed when cysteine is injected; 90% is metabolized via transamination of cysteine sulfinate to sulfate.[19] These observations suggest that the liver is the principal catabolic organ for cysteine metabolism, having been endowed with a regulatory enzyme, cysteine dioxygenase, the rate-limiting enzyme for cysteine oxidation. Cysteine dioxygenase is also detectable in brain tissue and has been reported to be widely distributed in a number of tissues, including spleen (claimed to have the highest activity),

Table 2
DISTRIBUTION OF HYPOTAURINE AMINOTRANSFERASE ACTIVITY IN RAT TISSUES

	Hypotaurine aminotransferase activity (nmol hypotaurine/mg protein/hr)[a]
Liver	13.67 ± 3.49
Kidney	4.70 ± 0.89
Brain	18.53 ± 3.86
Pancreas	0.41 ± 0.25
Heart	0.17 ± 0.17
Testes	7.36 ± 2.64
Seminal vesicles, prostate and epididymis	1.37 ± 0.28

[a] A particulate fraction of tissues was used to determine the enzyme activity. The values are means ±SD from three male rats.

pancreas, kidney, and testis.[20] One of us (JHF), has been unable to confirm this observation; the activity could be found only in liver and brain. Thus, it appears that the major source of hypotaurine and taurine is liver. Cysteine sulfinate and hypotaurine are observed in blood cells and it seems that selective transport may be responsible for the high levels of hypotaurine found in the sex organs[9,21] (see below).

Like cysteine sulfinate, hypotaurine is a substitute for an aminotransferase leading to an unstable intermediate that renders the reaction irreversible. Thus, hypotaurine and an α-ketoacid acceptor such as α-ketoglutarate yield glutamate plus acetaldehyde and SO_2 as the only detectable products.[9,22] The enzyme is widely distributed (Table 2), notably in male sexual tissues which possess large concentrations of hypotaurine and in which aminotransferase may play a regulatory role. The importance of this enzyme in sulfur economy is uncertain. However, at least 10% of the substrate hypotaurine is subject to transamination in mice fed normal or axenic diets.[22,23]

Metabolic Degradation

The degradative reactions described for taurine, mainly its conversion to isethionic acid (which occur only in the large bowel via gut bacterial metabolism),[23] are far too low to account for the metabolic degradation of the amino acid and for its disposal in vivo. It is generally agreed that most taurine is excreted through urine and feces (as taurocholic acid). This mechanism of elimination is sensitive to changes in taurine levels in plasma and/or in tissues, since the immediate response to a decrease in taurine synthesis or availability of dietary supplies is a reduction or even supression of excreted taurine.

Uptake Vs. Synthesis as Sources of Taurine and Hypotaurine

The turnover of taurine and hypotaurine in the whole animal is complex. Each organ appears to have its own distinct set of kinetic parameters for turnover. The large pool of some organs, such as muscle, heart, and brain, turns over very slowly. In contrast, kidney and liver exhibit a very rapid turnover of relatively small taurine pools. The importance of synthesis vs. uptake as the mechanism replenishing endogenous stores of taurine and hypotaurine varies in the different species and in the various tissues. Species such as the cat and primates, including humans, have only a limited ability to form taurine from endogenous precursors[10] and rely almost entirely on the diet supply to maintain their endogenous taurine

Table 3
CONSEQUENCES OF TAURINE DEFICIENCY

Retinal degeneration in cats, rats, monkeys, and humans
Retarded migration of cerebellar granule cells in cats
Abnormalities in migration and organization of neurons from the visual cortex in cats and monkeys
Impaired pregnancy and outcome in cats
Scoliosis and thoracic kyphosis in cats

stores. Others such as the rat and the mouse, have much higher capabilities for the biosynthesis of taurine.[24]

In the same animal, differences also exist among the various organs regarding synthesis. The liver is undoubtedly an active synthetizing organ and the brain also has some biosynthetic capability, whereas the heart and retina depend to a large extent on taurine from extratissue sources, namely, taurine circulating in plasma, proceeding from that formed in other organs or obtained from the diet. These organs possess active mechanisms to concentrate taurine.[32] The relative importance of biosynthesis and uptake in the various organs may be evaluated by the consequences of the administration of a taurine transport blocker, the structural analog of taurine, guanidinoethane sulfonate (GES).[25] Treatment with GES causes taurine depletion in tissues, of a magnitude in accordance to their relative dependence on taurine synthesized elsewhere.[26]

These aspects may have a nutritional interest since, as mentioned before, primates including humans, have a limited ability to form taurine and rely mainly on that obtained from the diet for replenishing their endogenous stores. In cats, which also have such dependence on dietary taurine, and in monkeys, restriction of taurine during pregnancy results in profound alterations in the organization of neuronal layers in the cerebral and cerebellar cortex, with neuroblasts failing to migrate and differentiate properly[27,28] (Table 3).

Interest in this subject is further heightened by the observation that preterm human infants fed a synthetic formula become taurine depleted. Human milk contains large amounts of taurine, 0.4 to 0.8 μmol/mℓ, whereas taurine levels in cow milk or synthetic formulas are only about 0.01 μmol/mℓ. Since a number of reports suggest that dietary taurine affects the taurine pool size of human infants, manufacturers of synthetic human formulas in the U.S., Canada, Japan, and Europe have recently incorporated taurine into the commercial formulas.[28]

PHYSIOLOGICAL AND PHARMACOLOGICAL EFFECTS OF TAURINE AND HYPOTAURINE

In spite of the high amounts of taurine found in animal tissues, its physiological role remains uncertain. A large number of investigations on this subject have been carried out during the last two decades and, although no clear picture has emerged so far, these studies now relate taurine to a number of membrane processes, including ionic fluxes, interaction with receptors, and, more recently, membrane peroxidation. As for hypotaurine, interest has been raised on its participation in oxidative reactions and on its role in sexual functioning.

In the nervous system, taurine has been proposed as an inhibitory neurotransmitter.[3,29] Administered iontophoretically, taurine shows a depressant effect of neuronal activity in practically all the regions examined. Its inhibitory potency is lower than that of glycine or GABA, and it is blocked by antagonists of GABA and glycine receptors. This makes it difficult to differentiate among the effects of taurine mediated by a receptor of its own, yet unidentified, or through an interaction with GABA and glycine receptors. A somewhat specific taurine antagonist recently described[30] may be useful for identifying taurinergic

Table 4
CARDIOVASCULAR ACTIONS OF TAURINE

Positive ionotropy at low calcium concentration
Negative ionotropy at high calcium concentration
Potentiation of digitalis ionotropy
Antiarrhythmia
Protection on calcium paradox
Protection on calcium overload myopathy
Osmotic agent
Hypotensive (central, peripheral, and neural)

pathways in brain. Alternate roles for taurine in the nervous system include a neuromodulatory function, proposed on the basis of the observed release of taurine upon synaptic stimulation and its location in synaptic vesicles.[31-33]

A large number of pharmacological effects of taurine have been described, particularly in the brain and cardiovascular system. In the nervous system, the best known effect of taurine is a generalized anticonvulsant action, observed in practically all models of experimental seizures as well as in some epilepsies.[34] An involvement of taurine in the neurological processes responsible for temperature control is also documented.[35] Interactions of taurine with opioid peptides, suppressing the development of tolerance and increasing the threshold for analgesia, have been recently described.[36]

The cardiovascular effects of taurine (Table 4) include positive ionotropic and antiarrhythmic actions in heart and reduction of blood pressure in several hypertension models in animals as well as in hypertensive patients.[37-39] Beneficial effects of taurine in the treatment of heart disease have been claimed,[39] but those results are still controversial.

The mechanism of these pharmacological effects of taurine has not been yet clarified. For the anticonvulsant action and inotropic effects, a modulatory action of taurine on calcium and/or potassium fluxes is proposed, with basis on a large body of experimental evidence.[37,40-42] Another possible regulatory effect of taurine on the function of excitable tissues involves an interaction with the transduction mechanism of membrane receptors, at the cyclic nucleotide- and phosphoinositide-signaling cascade.[43] The effects of taurine may be exerted directly on the metabolism of the second messengers or indirectly via an action on the calcium pools associated to the mechanisms of receptor transduction.

An insulin-like effect has been recently described for taurine and hypotaurine, consisting of a stimulatory action on the clearance of glucose from plasma.[44] This effect seems to be mediated through a direct interaction with the receptor sites for insulin.[45]

TAURINE AND HYPOTAURINE AS MEMBRANE STABILIZERS

The first suggestion of a stabilizer effect of taurine on membranes derived from the observation of Huxtable and Bressler[46] on the protection afforded by taurine during the isolation of skeletal muscle sarcoplasmic vesicles. Several lines of evidence have been accumulated since this early observation showed stabilizer effects of taurine and hypotaurine on cell membranes of different tissues. The effects of taurine on photoreceptor cells in the retina are the most clear and the best studied. Photoreceptor cells in the vertebrate retina contain very high concentrations of taurine (40 to 60 m*M*),[47] which are maintained to a large extent from that circulating in plasma and concentrated by an active, high-affinity uptake mechanism. This dependence on exogenous taurine makes photoreceptors susceptible to changes in the taurine content of the diet in species such as cat and primates, which have a low capacity to form taurine, or to the effect of antagonists of taurine transport into cells. A reduced availability of taurine to retinal cells results in a depletion of the amino acid,

FIGURE 2. The effect of taurine deficiency on the structure of cat retinal photoreceptors. Young cats (8 to 10 weeks old) were fed a taurine-free diet (Bio-Serv) during 25 weeks. Taurine levels in the retina decreased to 25% of levels in controls. At the end of the treatment, the cats were sacrificed and eyecups were processed for electron microscopy observation. (A) Control. (Magnification × 22,000.) (B) Taurine deficient. (Magnification × 15,000.)

which occurs very slowly, but that eventually leads to decreases of more than 70% of tissue taurine content. The consequence of such reduction in taurine levels in the retina of all the species so far examined is an alteration of the structural and functional integrity of photoreceptors and other retinal cells[48-50] (Figure 2). The electrical response of the retina, the electroretinogram (ERG), is markedly affected with reduced amplitudes in both the b-waves (Figure 3). The structural organization of photoreceptors is also severely disturbed, showing

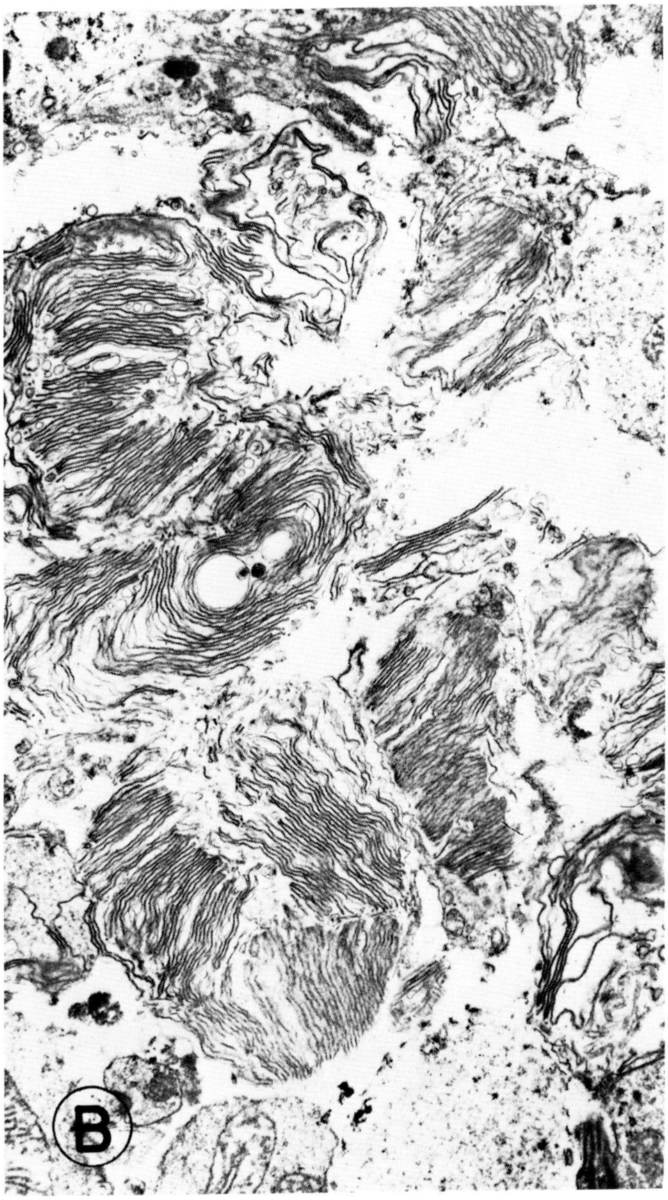

FIGURE 2B.

a marked disarray of the otherwise highly ordered structure of the lamellar disks of the outer segments of photoreceptors.[49-51]

Swelling, disorganization, and vesiculation characterizes the structure in both outer and inner segments in the retina of taurine-deficient animals. Eventually, cell death occurs, resulting in blindness. The disturbance of photoreceptor structure and function associated with taurine deficiency has been observed in cats,[57] monkeys,[52] and rats,[51] regardless of the experimental procedure used to produce the decrease of retinal taurine. These observations strongly support an involvement of taurine in the physiological mechanisms responsible for the integrity of photoreceptor structure and adequate function.

Another experimental approach utilized to investigate the protective effect of taurine and

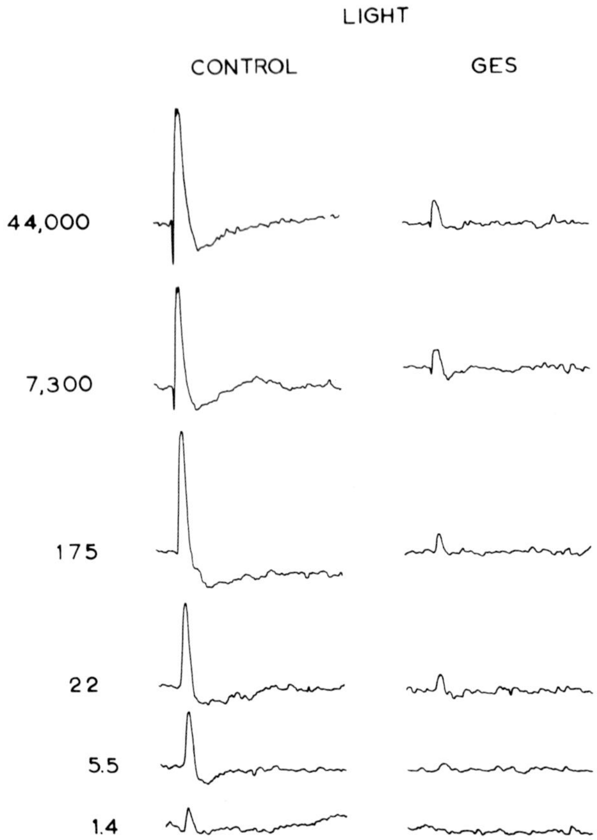

FIGURE 3. The effect of taurine deficiency on the ERG of rats. Taurine deficiency was obtained by treating rats with GES, 1%, in the drinking water, during 6 weeks. Taurine retinal levels in GES-treated rats decreased from 25.3 to 13.4 μmol/g. Recordings were made at the stimulus intensities (lux) indicated in the figure.

hypotaurine on photoreceptors has been the examination of the effect of light on the structure of the isolated outer segment in an in vitro preparation.[53] When isolated rod outer segments are exposed to continuous illumination, a structural disruption occurs, which resembles that of taurine-deficient animals in vivo. The disk membranes, which are barely detected in the intact outer segment, appear protrudent and disoriented. The outer segment is swollen and, after long time exposure to light, the cell membrane is broken and the disks are exposed.[53] The presence of taurine or hypotaurine in the incubation medium provides a complete protection of the outer segment structure.[54] The effect of hypotaurine is somewhat more potent than that of taurine. From other compounds tested by their ability to prevent the disrupting effect of light, the amino acids GABA glycine and cysteine, sucrose, and mercaptoethanol provide partial protection, whereas mannitol, tocopherol, glutathione, 4,4′-Diisothyocyanatostilbone-2,2′-disulfonic acid (DIDS) — an inhibitor of anion transport— and other amino acids are ineffective. The stabilizer effects of taurine and hypotaurine on the outer segment membranes are also observed in other experimental conditions leading to disruption of cell structure. These include exposure to calcium-free media and to oxidative agents. This later situation will be discussed in detail later.

The protective effect of taurine has been also observed in other preparations, such as heart perfused with a calcium-free medium, followed by reperfusion with a physiological medium

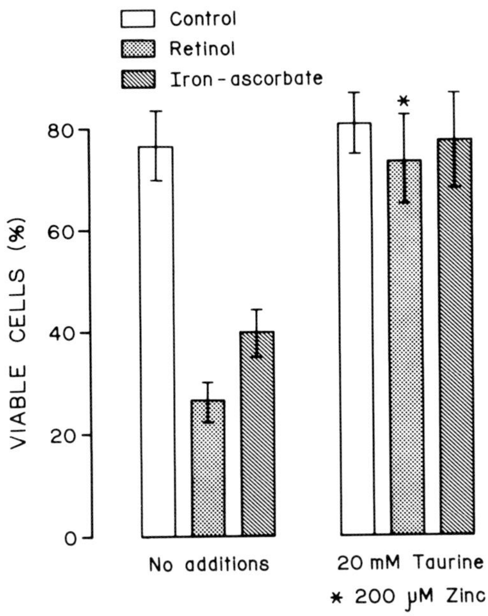

FIGURE 4. Protective effect of taurine on cell viability in human cultured lymphoblastoid cells, exposed to retinol and to iron-ascorbate. Cells were incubated for 30 min in a Krebs-bicarbonate medium in the presence of retinol (10 μ*M*) or 0.2 m*M* ferrous sulfate and 0.4 m*M* ascorbate. Taurine (20 m*M*) or zinc chloride (100 μ*M*) was present when indicated. Cell viability was assessed by Trypan blue exclusion. Results are the means ±SE of six to eight separate experiments.

containing calcium. This induces an extensive cellular and functional damage known as the "calcium paradox",[55] which seems to be a consequence of massive calcium entry during reperfusion. The cell lesion consists of changes in regions of the intercalary disks, the splitting of the glycocalix, and a massive release of intracellular components. The presence of taurine in the perfusion medium largely protects cells from many of the injurious effects of calcium overload. The effect of hypotaurine on this preparation has not been tested.

In human cultured lymphoblastoids, exposure to membrane destabilizers (retinol, retinoic acid, and iron ascorbate) produces cell swelling and a decrease in cell viability. Taurine and hypotaurine provide a protective effect, increasing cell viability and counteracting swelling (Figure 4).[56,57] The possible mechanism of this protective action of the amino acids and its relation to lipid peroxidation are discussed in the next section.

Recently, the effect of taurine in a model of liver injury has been examined. Treatment with CCl_4 induces an increase in lipid peroxides associated to cellular necrosis. Oral administration of taurine decreases membrane lipid peroxidation in hepatocytes and microsomes and protects hepatic cells from the deleterious effect of the solvent.[58] The details of this effect of taurine are given below.

FREE RADICAL GENERATION AND MEMBRANE LIPID PEROXIDATION: EFFECTS OF TAURINE AND HYPOTAURINE

Taurine in Excitable Tissues

The possibility of an antioxidant effect of taurine as the mechanism subserving its protective action on photoreceptor outer segments exposed to light is in agreement with the known effects of light increasing free radical generation and lipid peroxidation in the retina in vivo.[59]

This was confirmed in the in vitro preparation of isolated outer segments exposed to light, where a moderate increase in lipid peroxidation, measured as an enhancement in malonaldehyde formation, is observed associated to the disruptive effect of light.[62] Taurine and hypotaurine are able to prevent this increase in lipid peroxide levels, while affording protection at the same time to the outer segment structure. These results suggest a role for these amino acids as free radical scavengers (see below). However, taurine fails to prevent membrane lipid peroxidation induced by ferrous sulfate.[60] Exposure of photoreceptor outer segments to ferrous sulfate causes a marked disturbance of the cell structure and increases the amount of lipid peroxide more than fivefold. Although taurine provides a partial protection against the structural damage, it does not prevent the increase in lipid peroxides. This observation indicates that the mechanisms responsible for the effects of taurine (protecting membrane and cell structure) are not directly related to the generation and/or propagation of free radicals. The differences observed between the two models of membrane destabilization, i.e., iron and light, may be explained by the involvement of different free radical species in each model. It is known that active species of the various agents inducing lipid peroxidation have specific scavengers which differ for each system. Alternatively, the protective effect of taurine on cell membranes may be exerted at a step preceding the peroxidative process, probably membrane disruptions caused by electrolyte changes and water entry. This membrane alteration may result in the release of agents, such as polyunsaturated fatty acids, directly responsible for the generation of free radicals. The effect of hypotaurine on the iron-induced lipid peroxidation has not been explored yet.

The failure of taurine in preventing iron-induced lipid peroxidation in rod outer segments was confirmed by Anderson and co-workers[61] in a study measuring the regenerability of rhodopsin impaired by iron-induced peroxidation. Also in human lymphoblastoids, the protective effect of taurine on cell viability seems unrelated to the prevention of lipid peroxidation. In cells exposed to iron-ascorbate, a marked increase of malonaldehyde formation is observed, which is not prevented by taurine in spite of its clear effect in reducing cell swelling and maintaining cell viability. In these cells, lipid peroxidation is associated to an increased uptake of calcium of more than fivefold the accumulation in normal cells. This large increase of intracellular calcium is entirely prevented by taurine.[57] A recent proposal by Schanne et al.[62] states that cell death produced as a consequence of lipid peroxidation in different preparations and induced by a variety of toxic agents is due to an increased influx of calcium through the peroxidized membranes and that this intracellular accumulation of calcium which exceeds the buffering capacity of the cell is responsible for cell death. According to this interpretation, the prevention by taurine of enhanced calcium influx occurring during lipid peroxidation may account for its protective effect on cell viability. The validity of this hypothesis awaits confirmation of a generalized increase in calcium influx in other models of membrane lipid peroxidation and of a similar effect of taurine on abnormal calcium uptake. An increased rate of calcium influx is also suspected to be one of the factors mediating cellular necrosis caused by lipid peroxidation in liver following the administration of CCl_4.[58] The effect of taurine in this model of liver injury is dual. Pretreatment with taurine further increases the malonaldehyde formation induced by CCl_4, whereas on the other hand, administration of taurine 24 hr after CCl_4, suppresses the increment in hepatic lipid peroxidation. This effect is also observed in vitro when microsomes are incubated with CCl_4 in the presence of taurine. Treatment with CCl_4 induces an increase in calcium cell content, parallel with the increase in lipid peroxide level. Pretreatment with taurine further enhances the accumulation of calcium and lipid peroxides, whereas administration of taurine after the treatment with CCl_4 has opposite effects, i.e., counteracts the increase in cell calcium and decreases lipid peroxidation. Although these effects of taurine are difficult to interpret, a clear relationship between calcium levels and susceptibility to lipid peroxidation emerges from these studies.

From all the above observations, although a direct role for taurine as an antioxidant cannot be ruled out, the available evidence points to an indirect effect, probably mediated through an action on calcium transport by injured cells. A large number of reports dealing with the effects of taurine on calcium fluxes in various preparations propose taurine as a homeostatic agent which modulates calcium fluxes in such a way that it increases calcium available for cell function and yet exerts a protective action in states of calcium overload.

Hypotaurine in Sexual Tissue

The occurrence of hypotaurine and taurine in mammalian sexual tissues has been well documented. We recently applied a new high-performance liquid chromatography (HPLC) method using a reverse phase column for *o*-phthalaldehyde derivatives of hypotaurine and taurine[63] and confirmed their presence in male rat and human sexual tissue.[21] The remarkably high concentration in these tissues seems to be associated with sexual function. The magnitude of hypotaurine concentration in the uterine washings coincidental to ovulation and the loss of hypotaurine in seminal fluid after castration, with subsequent return to normal concentrations after testosterone replacement,[21,64] is congruent with a role for sulfur amino acids in sexual function. This role was better defined by the observation that sperm motility is maintained in washed sperm cell preparations by addition of hypotaurine and to a lesser degree by taurine.[65] The protection appeared to be related to the vulnerability of the unsaturated fatty acid residues of the membrane to hydroperoxidation. The production of malondialdehyde (as an expression of membrane disruption) was markedly reduced by hypotaurine. Hypotaurine can dramatically decrease the rate of superoxide generation by spermatozoa and prevent the loss of their superoxide dismutase (SOD) activity.[66]

Why sexual tissues specially require high concentration of hypotaurine and taurine is not certain, but one cannot ignore the unique exposure of spermatozoa to oxygen. The passage of sperm from male to female includes a brief but critical exposure to ambient oxygen in the vaginal, cervical, and uterine canals. Therefore, radical species including singlet oxygen, superoxide, hydrogen peroxide, and hydroxyl radical may disrupt cell function. While biologically relevant antioxidants systems are known to quench some of these oxygen species (glutathione-glutathione peroxidase, tocopherol, SOD, and catalase), no specific system for quenching hydroxyl radicals is known and hypotaurine may play this role.

Possible Mechanisms of the Antioxidant Effect

Hypotaurine

Several new lines of evidence support the view that hypotaurine possesses a unique capacity to quench the hydroxyl radical. The structural feature of hypotaurine which may uniquely equip this molecule for the role is the sulfinate moiety. Hydroxyl radicals generate other reactive free radicals when they collide with a number of biologically common functional groups. However, the collision of a hydroxyl radical with the sulfinate moiety of hypotaurine leads, by abstraction of an electron, to the sulfinate radical. The later is stabilized by the resonance distribution of the electrons on the oxygen as well as those other orbitals possessed by the large sulfur atom. Furthermore, the collision of two sulfinate radicals leads to the corresponding stable disulfone. The later, when hydrolysed by the hydroxide ion generated from the hydroxyl radical, leads to 1 mol equivalent of taurine and 1 mol equivalent of hypotaurine (Figure 5).

A variety of biological hydroxyl radical-generating systems have been examined to investigate the role of hypotaurine as a radical trap. For example, xanthine/xanthine oxidase generates superoxide and hydrogen peroxide. The later two generate hydroxyl radical in the presence of iron (Haber-Weiss reaction). The hydroxyl radical can decarboxylate benzoate, thus providing a means to assess the rate of hydroxyl radical formation.[67] Hypotaurine enhances this process.[67] Perhaps of greater biological importance is the involvement of

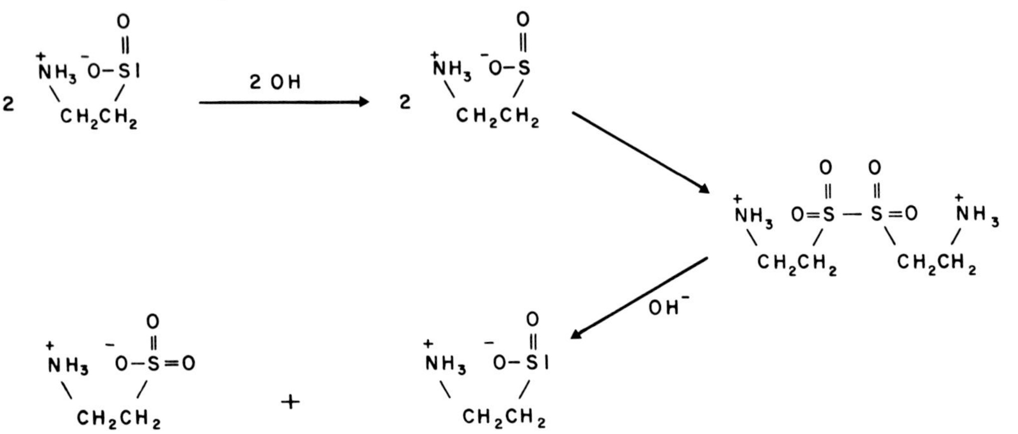

FIGURE 5. Sequence of reactions showing the interaction of hydroxyl radicals with the sulfinate moiety of hypotaurine, the generation of a stable α-disulfone, and the formation of 1 mol of taurine and 1 mol of hypotaurine.

Table 5
HYPOTAURINE QUENCHING OF HYDROXYL RADICALS

Additions	Benzoic acid decarboxylated (nmol)
Xanthine/xanthine oxidase	3.15 ± 0.12
Minus enzyme	0.42 ± 0.01
Plus EDTA/Fe^{2+}	10.22 ± 0.40
Plus EDTA/Fe^{2+}	2.75 ± 0.01
Plus hypotaurine	1.48 ± 0.17

Note: The benzoic acid decarboxylation radical detection system was used for these experiments. The reaction is carried out in 20-mℓ vials, penetrated through the center of their caps with a needle appended to which there is a filter paper disk soaked with 20 μℓ of hyamine hydroxide. The incubation vessels contain 0.01 units of xanthine/xanthine oxidase, 1.0 m*M* xanthine, 0.167 μ*M* EDTA/Fe^{2+}, 5 m*M* hypotaurine, 7-^{14}C-benzoic acid, and 100 μ*M* in 0.1 *M* phosphate to 0.1 m*M* EDTA, pH 7.8. After incubation at 25°C, 20 min, the reaction is terminated by addition of sulfosalicylic through the needle. The released $^{14}CO_2$ was collected on the hyamine-soaked disk for 30 min and the radioactivity was counted.

hypotaurine with liver microsomal NADPH-dependent hydroxyl radical system.[67] Again, hypotaurine can prevent benzoate decarboxylation by this system.[68] The consequence is the observed oxidation of hypotaurine to taurine (Table 5).

Taurine

A major antioxidant property which taurine exhibits is its ability to form a stable chloramine with hypochlorous acid. Neutrophils and eosinophils and several cell lines in culture, which contain the appropriate peroxidase activity, are capable of *N*-chlorotaurine formation.[69] Neutrophils, which contain myeloperoxidase activity, are capable of using the hydrogen peroxide generated by the ''respiratory burst'' to peroxidize the chloride ion to hypochlorous acid, one of the bacteriocidal agents employed by neutrophils. Hypochlorous acid is a powerful oxidizing agent which acts on the thiol groups of certain bacterial proteins to kill the germs.[70] The oxidizing character of the hypochlorous acid is markedly attenuated by its reaction with taurine to form the chloramine derivative.

CONCLUSIONS

Taurine and hypotaurine provide protection to biological membranes in a variety of experimental conditions and pathological situations involving membrane lipid peroxidation. The mechanism underlying this protective action seems to be different for the two amino acids. Taurine effects on membranes altered by lipid peroxidation are probably mediated by a regulating action on membrane permeability, avoiding ion overload and excessive water entry. These effects contribute to maintain the structural and functional integrity of peroxidized membranes, but taurine seems unable to exert a direct effect as an antioxidant. Hypotaurine, in contrast, may act primarily as an antioxidant through a recently demonstrated unique capacity to quench the hydroxyl radical. This reaction, which involves the formation of a disulfone, leads to the conversion of hypotaurine into taurine.

ACKNOWLEDGMENTS

This work was supported in part by Grants No. ER-02540-08 from the National Institutes of Health (NIH) and PCSABNA-030775 from CONACYT.

REFERENCES

1. **Jacobsen, J. G. and Smith, L. H.,** Biochemistry and physiology of taurine and taurine derivatives, *Physiol. Rev.,* 48, 424, 1968.
2. **Koksis, J. J., Kostos, Y. J., and Baskin, S. I.,** Taurine levels in the heart tissues of various species, in *Taurine,* Huxtable, R. and Barbeau, A., Eds., Raven Press, New York, 1976, 145.
3. **Mandel, P. and Pasantes-Morales, H.,** Taurine in the nervous system, *Rev. Neurosci.,* 3, 158, 1978.
4. **Frendo, J. A., Koy, A., and Zglicynski, J. M.,** Taurine in human blood platelets, *Nature (London),* 183, 685, 1959.
5. **Fukuda, K., Hirai, Y., Yoshida, H., Nakajima, T., and Usui, T.,** Free amino acid content of lymphocytes and granulocytes compared, *Clin. Chem.,* 28, 1758, 1982.
6. **Spaeth, D. G. and Schneider, D. L.,** Taurine synthesis, concentration and bile salt conjugation in rat, guinea pig and rabbit, *Proc. Soc. Exp. Biol. Med.,* 147, 855, 1974.
7. **Agrawal, H. C., Davis, J. M., and Himwich, W. A.,** Maturational changes in amino acids in CNS of different mammalian species, *Recent Adv. Biol. Psychiatry,* 10, 258, 1968.
8. **Oja, S. S. and Kontro, P.,** Hypotaurine uptake in mouse brain slices, *Adv. Exp. Med. Biol.,* 139, 115, 1982.
9. **Kochakian, C. D.,** Taurine and sulfur compounds in the reproductive tract of marine animals, in *Taurine,* Huxtable, R. and Barbeau, A., Eds., Raven Press, New York, 1976, 375.
10. **Jacobsen, J. G., Thomas, L. L., and Smith, L. H.,** Properties and distribution of mammalian L-cysteine sulfinate carboxylase, *Biochim. Biophys. Acta,* 85, 103, 1964.
11. **Pasantes-Morales, H., Chatagner, F., and Mandel, P.,** Synthesis of taurine in rat liver and brain in vivo, *Neurochem. Res.,* 5, 91, 1980.
12. **Chapeville, F. and Fromageot, P.,** La formation de l'acide cysteine sulfinique a partir de la cystine chez le rat, *Biochim. Biophys. Acta,* 19, 175, 1955.
13. **Chatagner, F. and Bergeret, B.,** Decarboxylation de l'acide cysteine sulfinique par le foie et le cerveau du rat male, du rat femelle et du rat femelle ovariectomisé, *Bull. Soc. Chim. Biol.,* 38, 1159, 1956.
14. **Kontro, P. and Oja, S. S.,** Hypotaurine oxidation by mouse liver tissue, *Prog. Clin. Biol. Res.,* 179, 83, 1985.
15. **Fellman, J. H. and Roth, E. S.,** Hypotaurine aminotransferase, *Adv. Exp. Med. Biol.,* 139, 99, 1982.
16. **Yamaguchi, K., Sakakibara, S., Koga, K., and Ueda, I.,** Induction and activation of cysteine oxidase of rat liver. I. The effects of cysteine, hydrocortisone and nicotinamide injection on hepatic cysteine oxidase and tyrosine transaminase activities of intact and adrenalectomized rats, *Biochem. Biophys. Acta,* 237, 502, 1971.
17. **Yamaguchi, K., Sakakibara, S., Asamisu, J., and Ueda, I.,** Induction and activation of cysteine oxidase of rat liver. II. The measurement of cysteine metabolism in vivo and the activation of in vivo activity of cysteine oxidase, *Biochem. Biophys. Acta,* 297, 48, 1973.

18. **Kohashi, N., Yamaguchi, K., Hosokawa, Y., Kori, Y., Fujii, O., and Ueda, I.,** Dietary control of cysteine dioxygenase in rat liver, *J. Biochem.*, 84, 159, 1978.
19. **Griffith, O.,** Cysteinesulfinate metabolism, *J. Biol. Chem.*, 258, 1591, 1983.
20. **Misra, C. H.,** In vitro study of cysteine oxidase in rat brain, *Neurochem. Res.*, 8, 1497, 1983.
21. **Kochakian, C. D.,** Hypotaurine: regulation of production in seminal vesicles and prostate of guinea-pig by testosterone, *Nature (London)*, 241, 202, 1973.
22. **Fellman, J. H., Roth, E. S., Avedovech, N. A., and McCarthy, K. D.,** Mammalian hypotaurine aminotransferase: isethionate is not a product, *Life Sci.*, 27, 1999, 1980.
23. **Fellman, J. H., Roth, E. S., Avedovech, M., and McCarthy, K. D.,** The metabolism of taurine to isethionate, *Arch. Biochem. Biophys.*, 204, 560, 1980.
24. **Huxtable, R. J. and Lippincott, S. E.,** Sources and turnover rates of taurine in newborn, weanling and mature rats, *Adv. Exp. Med. Biol.*, 139, 23, 1982.
25. **Quesada, O., Huxtable, R. J., and Pasantes-Morales, H.,** Effect of guanidinoethane sulfonate on taurine uptake by rat retina, *J. Neurosci. Res.*, 11, 179, 1984.
26. **Huxtable, R. J., Laird, H. E., and Lippincott, Y. Y.,** The transport of taurine in the heart and the rapid depletion of tissue content by guanidinoethylsulfonate, *J. Pharmacol. Exp. Ther.*, p. 211, 1979.
27. **Palackal, T., Moretz, R., Wisniewski, H., and Sturman, J.,** Abnormal visual cortex development in the kitten associated with maternal dietary taurine deprivation, *J. Neurosci. Res.*, 15, 223, 1986.
28. **Sturman, J. A.,** Taurine in infant physiology and nutrition, *Pediatr. Nutr. Rev.*, p. 1, 1986.
29. **Oja, S. S. and Kontro, P.,** Taurine, in *Handbook of Neurochemistry*, Vol. 3, Lajtha, A., Ed., Plenum Press, New York, 1983, 501.
30. **Yarbrough, G. G., Singh, D. K., and Taylor, D. A.,** Neuropharmacological characterization of a taurine antagonist, *J. Pharmacol. Exp. Ther.*, 219, 604, 1981.
31. **Kuriyama, K., Muramata, M., Nakagawa, K., and Kakita, K.,** Modulating role of taurine on release of neurotransmitters and calcium transport in excitable tissues, in *Taurine and Neurological Disorders*, Huxtable, R. J. and Barbeau, A., Eds., Raven Press, New York, 1978, 201.
32. **Kuriyama, K., Ida, S., Nishimura, Ch., and Ahkuma, S.,** Distribution and function of taurine in nervous tissues: an overview, *Prog. Clin. Biol. Res.*, 125, 127, 1983.
33. **De Belleroche, J. S. and Bradford, H. F.,** Amino acids in synaptic vesicles from mammalian cerebral cortex: a reappraisal, *J. Neurochem.*, 21, 441, 1973.
34. **Huxtable, R. J.,** Insights on function: metabolism and pharmacology of taurine in the brain, in *The Role of Peptides and Amino Acids as Neurotransmitters*, Lombardini, J. B. and Kenny, A. D., Eds., Raven Press, New York, 1981, 109.
35. **Sgaragli, G. P. and Palmi, M.,** The role and mechanism of action of taurine in mammalian thermoregulation, *Prog. Clin. Biol. Res.*, 179, 343, 1985.
36. **Izumi, K., Fukuda, T., Koja, T., Shimizu, T., Yoshida, M., and Barbeau, A.,** Taurine and tolerance to opioid peptides, *Prog. Clin. Biol. Res.*, 125, 183, 1983.
37. **Welty, J. D. and Read, W. O.,** Studies on some cardiac effects of taurine, *J. Pharmacol. Exp. Ther.*, 144, 110, 1964.
38. **Huxtable, R. J.,** The regulation and function of taurine in the heart and other organs, in *Natural Sulfur Compounds: Novel Biochemical and Structural Aspects*, Cavallini, D., Gaull, G. E., and Zappia, V., Eds., Plenum Press, New York, 1980, 274.
39. **Azuma, J., Hasegawa, N., Awata, A., Sawamura, A., Harada, H., Ogura, K., Ohta, H., Yamaguchi, K., and Kishimoto, S.,** Taurine for treatment of congestive heart failure in humans, *Prog. Clin. Biol. Res.*, 125, 61, 1983.
40. **Azari, J. and Huxtable, R. J.,** The mechanism of the adrenergic stimulation of taurine influx in the heart, *Eur. J. Pharmacol.*, 61, 217, 1980.
41. **Pasantes-Morales, H., Arzate, M. E., and Cruz, C.,** The role of taurine in nervous tissue: its effects on ionic fluxes, in *Taurine in Nutrition and Neurology*, Vol. 139, Huxtable, R. J. and Pasantes-Morales, H., Eds., 1982, 273.
42. **Gruener, R., Markovitz, D., Huxtable, R., and Bressler, R.,** Excitability modulation by taurine. Transmembrane measurements of neuromuscular transmission, *J. Neurol. Sci.*, 24, 351, 1975.
43. **Baba, A., Tatsuno, T., Kumagae, Y., and Iwata, H.,** Inhibitory effect of taurine on the formation of cyclic AMP in guinea pig hippocampus, *Prog. Clin. Biol. Res.*, 125, 161, 1983.
44. **Tojunaga, H., Yoneda, Y., and Kuriyama, K.,** Protective actions of taurine against streptozotocin-induced hyperglycemia, *Biochem. Pharmacol.*, 28, 807, 1979.
45. **Kulakowski, E. C., Maturo, J., and Schaffer, S. W.,** The low affinity taurine-binding protein may be related to the insulin receptor, *Prog. Clin. Biol. Res.*, 179, 127, 1985.
46. **Huxtable, R. J. and Bressler, R.,** Effect of taurine on a muscle intracellular membrane, *Biochim. Biophys. Acta*, 323, 573, 1973.
47. **Orr, H. T., Cohen, A. L., and Lowry, O. H.,** The distribution of taurine in the vertebrate retina, *J. Neurochem.*, 26, 609, 1976.

48. **Lake, N.,** Is taurine an essential amino acid?, *Retina,* 2, 261, 1982.
49. **Hayes, K. C., Carey, R. E., and Smith, S. Y.,** Retinal degeneration associated with taurine deficiency in the cat, *Science,* 188, 949, 1975.
50. **Schmidt, S. Y., Berson, E. L., Watson, G., and Huang, C.,** Retinal degeneration in cats fed casein. III. Taurine deficiency and ERG amplitudes, *Invest. Ophthalmol.,* 16, 673, 1977.
51. **Pasantes-Morales, H., Quesada, Q., Cárabez, A., and Huxtable, R. J.,** Effect of taurine transport antagonists, guanidino ethanesulfonate and β-alanine on the morphology of the rat retina, *J. Neurosci. Res.,* 9, 135, 1983.
52. **Sturman, J. A., Wen, G. Y., Wisniewski, H. M., and Neuringer, M.,** Retinal degeneration in primates raised on a synthetic human infant formula, *Int. J. Dev. Neurosci.,* 2, 121, 1984.
53. **Pasantes-Morales, H., Ademe, R. M., and Quesada, O.,** Protective effect of taurine on the light induced disruption of isolated frog rod outer segments, *J. Neurosci. Res.,* 6, 337, 1981.
54. **Pasantes-Morales, H. and Cruz, C.,** Taurine and hypotaurine inhibit light-induced lipid peroxidation and protect rod outer segment structure, *Brain Res.,* 330, 154, 1985.
55. **Kramer, H. H., Chovan, J. P, and Schaffer, S. W.,** Effect of taurine on calcium paradox and ischemic cardiac failure, *Am. J. Physiol.,* 60, 20, 1981.
56. **Pasantes-Morales, H. H., Wright, C. E., and Gaull, G. E.,** Protective effect of taurine, zinc and tocopherol on retinol-induced damage in human lymphoblastoid cell, *J. Nutr.,* 114, 2256, 1964.
57. **Pasantes-Morales, H., Wright, C. E., and Gaull, G. E.,** Taurine protection of lymphoblastoid cells from iron-ascorbate induced damage, *Biochem. Pharmacol.,* 34, 2205, 1985.
58. **Nakashima, T., Takino, T., and Kuriyama, K.,** Therapeutic and prophylactic effects of taurine administration on experimental liver injury, *Prog. Clin. Biol. Res.,* 125, 449, 1983.
59. **Wiegand, R. D., Giusto, N. M., Rapp, L. M., and Anderson, R. E.,** Evidence of rod outer segment lipid peroxidation following constant illumination of the rat retina, *Invest. Ophthalmol. Visual Sci.,* 10, 1433, 1983.
60. **Pasantes-Morales, H. and Cruz, C.,** Protective effect of taurine and zinc on peroxidation-induced damage in photoreceptor outer segments, *J. Neurosci. Res.,* 11, 303, 1984.
61. **Anderson, R. E., Maude, M., and Nielsen, J. C.,** Effect of lipid peroxidation on rhodopsin regeneration, *Curr. Eye Res.,* 4, 65, 1985.
62. **Schanne, F. A., Kane, A. B., Young, E. E., and Farber, J. L.,** Calcium dependence of toxic cell death: a final common pathway, *Science,* 206, 700, 1979.
63. **Hirschberger, L. L., DeLaRosa, J., and Stipanuk, M. H.,** Determination of cysteinesulfinate, hypotaurine and taurine in physiological samples by reversed-phase high-performance liquid chromatography, *J. Chromatogr.,* 343, 303, 1985.
64. **van der Horst, C. J. G. and Brand, A.,** Occurrence of hypotaurine and inositol in the reproductive tract of the eye and its regulation by pregnenolone and progesterone, *Nature (London),* 223, 67, 1969.
65. **Leibfried, M. L. and Barister, B. D.,** Effects of epinephrine and hypotaurine on *in vitro* fertilization in the golden hamster, *J. Reprod. Fertil.,* 66, 87, 1982.
66. **Alvarez, J. G. and Storey, B. T.,** Taurine, hypotaurine, epinephrine and albumin inhibit lipid peroxidation in rabbit spermatozoa and protect against loss of motility, *Biol. Reprod.,* 29, 548, 1983.
67. **Winston, G. W. and Cederbaum, A. I.,** Oxidative decarboxylation of benzoate to carbon dioxide by rat liver microsomes: a probe for oxygen radical production during microsomal electron transfer, *Biochemistry,* 21, 4265, 1982.
68. **Fellman, J. H. and Roth, E. S.,** The biological oxidation of hypotaurine to taurine: hypotaurine as an antioxidant, *Prog. Clin. Biol. Res.,* 179, 71, 1985.
69. **Wright, C. E., Lin, T. T., Lin, Y. Y., Shurman, J. H., and Gaull, G. E.,** Taurine scavenges oxidized chlorine in biological systems, *Prog. Clin. Biol. Res.,* 179, 137, 1985.
70. **Aldrich, J. M., McCarthy, C. A., and Hurst, J. K.,** Biological reactivity of hypochlorous acid: implications for microbicidal mechanisms of leukocyte myeloperoxidase, *Proc. Natl. Acad. Sci. U.S.A.,* 78, 210, 1981.

THE PHYSIOLOGICAL FUNCTIONS OF GLUTATHIONE

Juan R. Viña, Guillermo T. Sáez, and José Viña

INTRODUCTION

Glutathione is widely distributed in cells. Indeed, it is present in almost all animal cells, at relatively high concentrations.[1] Interest in glutathione has grown steadily in recent years. A survey of available literature shows that the number of scientific papers dealing with glutathione has increased significantly over the last 12 years.[2]

Glutathione is an intracellular reductant and plays major roles in catalysis, metabolism, and transport. It protects cells against free radicals, peroxides, and other toxic compounds and it is important for protection against the deleterious effects of radiations. Many of these physiological functions are derived from its special chemical properties. Indeed, it contains a thiol group and a γ-glutamyl linkage that is resistant to degradation by the usual peptidases.

Several reviews have been written dealing with the biochemical, pharmacological, and clinical aspects of glutathione.[2-4] In this short review, the major physiological functions of glutathione in cells are emphasized.

Glutathione is essential for the normal functioning of many cells — especially in mammals. Indeed, glutathione depletion increases the sensitivity of cells to various aggressions[5] and also has several metabolic effects, e.g., a decrease in the rate of gluconeogenesis[6] or an increase in glycogenolysis.[7] The concept of a glutathione-SH (GSH) threshold for drug detoxication was discussed by Jollow[8] and it may be valid for other functions of glutathione. However, further research is needed to firmly establish this concept. Some of the more relevant functions of GSH in animal cells — mostly in mammals — are discussed below.

PROTECTIVE ROLES OF GLUTATHIONE

Role in Drug Detoxication

Drugs and other foreign lipid-soluble compounds undergo different types of metabolic reactions that lead to the formation of metabolites of less toxic potency which can be excreted by the organism. These reactions have been classified into Phase I and Phase II reactions. The net result of both phases is to decrease the biological reactivity and increase the excretability of the drug.[9] The Phase I reactions include oxidation, reduction, hydroxylation, and hydrolysis, whereas the Phase II reactions include conjugation reactions. Several of the conjugates undergo further transformation before they are excreted. An example of this is the conversion of GSH conjugates to the corresponding mercapturic acid derivatives.[10] An interesting case is the mechanism of detoxication of acetaminophen (paracetamol). This is an analgesic and antipyretic drug that is sold over the counter and that is safe when taken in moderate doses. However, when taken in large amounts, it may cause liver necrosis and death of the patient. Acetaminophen toxicity is mediated by free radicals[11] and is sometimes a means of suicidal attempts. In the U.S., more than 8700 cases of paracetamol overdose were reported between 1976 and 1984.[12] When an overdose of acetaminophen is taken, GSH depletion occurs and binding of a reactive metabolite of paracetamol to cell proteins takes place. This leads to cell death. A rational way to treat acetaminophen overdose is to administer precursors to stimulate GSH synthesis. Several have been tested, like methionine or oral *N*-acetyl cysteine. Cysteine (oral or parenteral) and parenteral *N*-acetyl cysteine may be toxic (see chapter entitled The Role of Free Radicals in the Cytotoxicity of Cysteine, in Volume I). The role of GSH in drug metabolism was reviewed recently.[13]

Protection Against Oxidative Stress

All aerobic cells are subject to oxidative stress. Indeed continuous formation of $O_2^{\bar{}}$ and hydroperoxides has been observed. The antioxidant system has enzymatic and nonenzymatic components. The enzymatic system consists of three basic enzymes: superoxide dismutase (SOD), catalase, and peroxidases. Of the latter group, glutathione peroxidase is the most common form in mammalian cells.[14]

Each of these enzymes is found in several isozymes and in different compartments of the cell. Cytosolic SOD contains copper and zinc and that found in the mitochondria contains manganese. Glutathione peroxidase has two different isozymes: one in the cytosol and the other in the mitochondria. The most abundant form of catalase is found in the peroxisomes.

Superoxide radical ($O_2^{\bar{}}$) produced during metabolism is converted to H_2O_2 by superoxide dismutase. H_2O_2 can be detoxified by the catalase or by glutathione peroxidase. The activity of this enzyme results in oxidation of GSH to (GSSG). The oxidized glutathione, in turn, may be reduced by glutathione reductase at the expense of NADPH. GSSG can also be excreted from cells.[15,16] The biochemistry of oxidative stress was reviewed recently by Sies.[17]

The removal of H_2O_2 and $O_2^{\bar{}}$ from biological systems prevents the iron-catalyzed Haber-Weiss reaction from occurring. There is also an antioxidant defense system that is not enzymatic and consists of small compounds such as vitamin A, β-carotene, retinoic acid, vitamin C, vitamin E, iron chelators, and GSH and other thiols. Furthermore, certain DNA repair enzymes may be considered as part of the antioxidant defense system of the cells because these enzymes repair the DNA strand breaks caused by active oxygen species and thus serve a protective function.

Role in the Prevention of Cataract Formation

GSH metabolism in the ocular lens and in erythrocytes has many similarities, such as the GSH/GSSG ratio, the turnover rate of GSH, or the fact that GSSG is released from erythrocytes and from the lens when the cells are under oxidative stress.[18] The induction of free radical reactions causes a fall in lens GSH that leads to the injury of lens proteins, the crystallines, whose content in the human lens is 35.5%. Precipitation of these proteins and other metabolic alterations leads to the formation of cataracts.[19]

The oxidative damage of lipids, which represents only 2% of the lens mass, plays a role in the formation of cataracts. Recently it has been shown that lens opacification develops along with the accumulation of lipid peroxidation products such as malondialdehyde.[20,21]

The fall of GSH level leads to the formation of S–S bonds between proteins and this is followed by their aggregation, which in turn is followed by lens opacification. Therefore, by maintaining normal GSH levels, it may be possible to avoid further cataract development.[22]

Glutathione and Protection Against Radiation

The involvement of thiols in cell responses to radiation has been known for a long time. Guzman Barrón in 1954[23] already suggested that SH groups, especially those found in proteins, are involved in the mechanisms of radiation-induced cell damage. Indeed, ionizing radiations may interact with water, giving rise to OH·, H^+, and an electron. OH· free radicals are highly reactive and can interact with thiols and give RS· or GS· free radicals. These are much less reactive and can dimerize to give R-SS-R, GSSG, or the mixed disulfide RSSG. If GSH is present, as it is normally at high concentrations, GSSG may be the main product formed, and thus the action of ionizing radiation may be easily reversed in the cell (because GSSG will be reduced by glutathione reductase).

In 1961 Hutchinson was the first to show that a biologically active nucleic extract, when irradiated in dilute solution, was protected by GSH.[24] It has been shown that GSH and other thiol-containing compounds, such as dithiothreitol, cysteamine, and cysteine protect irradiated DNA in a manner dependent on which SH compound is used, the thiol concentration, and the degree of oxygenation.[25,26]

Recently, strains of mutant cells deficient in glutathione have become available. Careful studies are being carried out on the details of the mechanism by which glutathione exerts its radioprotective effects. Meister and colleagues have recently discussed the possibility that selective depletion of glutathione by the use of buthionine sulfoximine may be useful to increase the radiosensitivity of certain tumors.

METABOLIC FUNCTIONS OF GLUTATHIONE

Role in DNA Synthesis

The synthesis of DNA requires a reduction of ribonucleotides to form deoxyribonucleotides. This is a complex reaction in mammals that requires thioredoxin, thioredoxin reductase, other enzymes, and NADPH as a cofactor. The immediate electron donor is the reduced form of thioredoxin which has accepted electrons from NADPH. The reversible oxidation-reduction of thioredoxin is catalyzed by thioredoxin reductase.

Holmgren[27,28] has identified a GSH-dependent hydrogen donor. This is a low molecular weight protein called glutaredoxin and requires GSH to reduce the 2′ carbon in the ribose moiety of the corresponding ribonucleotide. In this reaction, GSH is oxidized to GSSG, which in turn is reduced back to GSH by glutathione reductase.

Strains of *Escherichia coli* have been found that lack thioredoxin activity, but that have the capacity to grow, thus showing that the glutaredoxin pathway is sufficient to maintain an adequate synthesis of deoxyribonucleotides. Both thioredoxin and glutaredoxin function under normal conditions in mammalian cells. However, the contribution of each one remains to be established.

Role in the Regulation of Protein Synthesis

The initiation and elongation of proteins is inhibited when GSH is oxidized. It has been shown in reticulocytes that, when GSH is oxidized, elongation of polypeptide chains is inhibited. When GSSG is reduced, elongation resumes. Therefore, the increase in the concentration of GSSG seems responsible for the inhibition of protein synthesis. It has been shown in ribosomal preparations of brain or liver that the synthesis of polypeptides is inhibited by GSSG. Addition of GSH removes this inhibition.[4]

Role of GSH in the Regulation of Proteolysis

Proteolysis in lysosomes is stimulated by several thiol-containing molecules such as cysteine, GSH, or dithiothreitol (DTT). GSH is effective as intralysosomal proteolytic stimulant in kidney, but not in liver. Recently, Mego studied the mechanism of activation of lysosomal proteolysis by GSH and observed that cysteinyl-glycine, rather than GSH, may be the physiological proteolytic stimulant.[29] This is based on the fact that in kidney lysosomes, GSH is not effective to stimulate proteolysis if inhibitors of γ-glutamyl transpeptidase are present. Thus, the activator of proteolysis may be a product of the interaction of γ-glutamyl transpeptidase and GSH, and the possibility that such an activator is cysteinyl glycine was put forward. This is an interesting observation that links the γ-glutamyl cycle and an important cellular function such as proteolysis, and more work is required to assess its possible physiological implications.

GSH as a Reservoir of Cysteine

The effect of the fasting-refeeding cycle on the steady state concentration and on the turnover of GSH led Tateishi and colleagues to propose that GSH is a cellular reservoir of cysteine. These authors also showed that the cysteinyl moiety of GSH is released when active synthesis is required. This concept involves two GSH pools. One has a fast (2-hr) and another has a slow (30-hr) turnover in the liver.[30]

There are two reasons to explain why the cells develop a reservoir of cysteine instead of maintaining high concentrations of free cysteine. The first one is that GSH performs several physiological functions, as is discussed in this chapter. The second reason is that cysteine, even at moderate concentrations, is highly toxic, as has been shown (see chapter in Volume I mentioned above). Therefore cells accumulate nonprotein thiols as glutathione, rather than as cysteine.

GSH is not toxic to the cells because its rate of autoxidation is various orders of magnitude lower than that of cysteine. When cysteine is required, for instance, for protein synthesis, cysteinyl moieties from glutathione are released. Thus, the idea of GSH as a reservoir of cysteine is supported by the fact that the cell cannot accumulate cysteine because of its toxicity.

Studies on the quantification of amino acid supply to bovine mammary gland for milk protein synthesis have shown that there is a shortage of cysteine to meet the requirements for milk synthesis. Pocius et al. have proposed that blood GSH may be used as a source of amino acid for protein synthesis. Mammary gland is a tissue rich in γ-glutamyl transpeptidase activity, thus capable of degrading GSH in order to use its constituent amino acids for protein synthesis.[31]

The Methionine-Sparing Effect of GSH

The methionine-sparing effect of cysteine was found half a century ago by Rose.[32] Indeed, methionine is converted to cysteine through the cystathionine pathway. Inger and Du Vigneau reported that the injection of GSH restored the growth of rats that were fed on a low-protein diet. This effect has been extensively studied by Tateishi and et al., who have shown that dietary GSH has an L-methionine-sparing effect.[33] These authors have fed animals with (^{35}S)-GSH and have observed that ^{35}S is extensively incorporated into proteins and also into intracellular GSH. The effect of GSH increased when its concentration in the diet increased.

Role in Gluconeogenesis

It has been shown that isolated hepatocytes depleted of GSH exhibit alterations of their carbohydrate metabolism. Depletion of GSH in isolated hepatocytes, with substrates of the GSH-S-transferases (diethylmaleate or phorone), or buthionine sulfoximine, an inhibitor of γ-glutamyl-cysteine synthetase, has been found to result in a decreased rate of glucose synthesis from various precursors not affecting cell integrity. The possible explanation for this inhibition of gluconeogenesis is that some enzymes of the pathway require GSH for normal activity, i.e., phosphoenolpyruvate carboxykinase.[6] It is worthwhile to emphasize that methods of GSH depletion did not cause an oxidation of thiols and that the GSH/GSSG ratio was always maintained.

GSH depletion of isolated rat hepatocytes from 48-hr-starved rats has been shown to induce an inhibition of gluconeogenesis from (2-^{14}C)-propionate plus lactate and an increase in radioactivity in lactate. These results suggest an increase in the pyruvate kinase flux in the GSH-depleted hepatocytes and may explain the observed decrease in gluconeogenesis in GSH-depleted hepatocytes.

Role in the Regulation of the Pentose Phosphate Pathway

The rate-limiting step of the pentose phosphate cycle is the reaction catalyzed by the glucose-6-phosphate dehydrogenase. This enzyme may increase five- to tenfold in response to a high carbohydrate intake. Apart from this type of regulation, a fine control of this enzyme was proposed by Eggleston and Krebs.[34] At physiological concentrations of free $NADP^+$ and free NADPH, the enzyme is completely inhibited. Thus, the regulation of this enzyme is a matter of deinhibition. These authors tested over 100 cell metabolites and only GSSG at low concentrations counteracted the inhibition of NADPH with the help of an unidentified cofactor.

Later on, Rodriguez-Segade and co-workers reported that they had isolated the cofactor from mussel hepatopancreas and from rat liver and found that it was a protein of molecular weight of 10,000 to 15,000 daltons.[35]

Recently the work of Krebs' laboratory has been challenged because GSSG addition has no effect on NADPH inhibition of glucose 6-phosphate dehydrogenase and no evidence was found for the existence of a cofactor that mediates the effect of GSSG on the enzyme. The authors point out that the effect of the oxidized glutathione on the enzyme is an artifact caused by the incorrect application of a control assay.[36] Using lysates of type A spinach chloroplasts, it has not been possible to demonstrate the effect of GSSG on the glucose-6-phosphate dehydrogenase inhibition by NADPH.[37]

Role of GSH in the Aging of Cells

The concentration of GSH in erythrocytes decreases with cell age and mouse age.[38] In the old (31 months) mouse, the GSH content was lower by 30% in the liver, 34% in the kidney, and 20% in the heart than in the mature (17 to 23 months) animals. The GSSG concentrations of tissues did not change with age.[39] Experiments performed in insects showed that the concentration of GSH is decreased by 50% in the old mosquito.[40]

Hazelton and Lang[41] have also found that glutathione peroxidase and glutathione reductase activities decline in the aging mouse. These authors[42] suggested that the cause of the decline in GSH levels is a decreased synthesis rather than increased oxidation because the γ-glutamyl cysteine synthetase activity is decreased with aging. The loss of activity of glutathione reductase may also contribute to the decline in GSH during aging.

Since glutathione serves to protect the cell against the deleterious power of free radicals derived from oxygen, the fact that GSH concentrations in cells from old animals is decreased may be important as regards the free radical theory of aging of Harman[43] and especially as expressed by Miquel and co-workers,[44] who have expanded the theory of Harman in their "oxygen radical-mitochondrial injury" theory of cell aging.

The sulfur-containing amino acids, by acting as precursors in GSH synthesis, are effective in retarding lipofuscin accumulation.[45] It has been reported that GSH supplementation was also effective in retarding lipofuscin storage in human glial cells grown in culture.[46] Therefore the effectiveness of GSH as antioxidant provides evidence supporting the importance of the GSH-utilizing enzymes, GSH peroxidase and GSH-S-transferases, in protecting the cell from oxidative damage.

Role in the Deiodination of Thyroid Hormone

Thyroxine (T_4) is the main hormone secreted by the thyroid gland. However, biologically, the most active product is 3,3′-5-tri-iodothyronine (T_3). Although some T_3 is released from the thyroid gland, the largest proportion of 3,3′-5-tri-iodothyronine circulating in blood derives from deiodination of the circulating T_4 in peripheral tissues. The conversion of T_4 to T_3 involves the removal of the iodine in the 5′ position in the molecule of T_4. This is an enzyme-catalyzed reaction. The liver and kidney are, by far, the most potent tissues able to catalyze such reaction. The enzyme in those tissues has received the name type I iodothyronine 5′-deiodinase (5′D-I). In the pituitary gland, central nervous system (CNS), and brown adipose tissue, another deiodinase has been found. This new type of enzyme has been named type II (5′D-II).[47]

The deiodinase requires a soluble factor for its activity. This soluble factor can be replaced by DTT, thus indicating that a reductive process takes place. GSH is less effective than DTT, but also supports the conversion of T_4 into T_3.

The fact that GSH is involved in T_4 deiodination may have physiological importance. Indeed, Visser suggests that the decreased deiodination of T_4 observed in fasted animals may be due to the fact that GSH levels are decreased during physiological fasting.[48] However, it is possible that GSH level is not the factor that regulates the deiodination, but the changes in the GSH/GSSG ratio. It is clear that other factors regulate the conversion of T_4 into T_3, although the role of GSH in this process seems established.

GSH as a Coenzyme

GSH acts as coenzyme of some enzyme systems, like formaldehyde dehydrogenase or glyoxalase. Glyoxalase I catalyzes the reaction of methylglyoxal and GSH to give *S*-lactoyl glutathione. This is the substrate of glyoxalase II, and D-lactate and GSH are formed. GSH is thus a coenzyme. The glyoxalase system has been known since 1913. The mechanism of action of the glyoxalases has been studied in detail by Mannervik.[49] However, in spite of these efforts, the physiological functions of the glyoxalases are still unknown. In the late 1960s and early 1970s, Szent Gyorgyi suggested a very interesting hypothesis which proposed that methylglyoxal (or other ketoaldehydes) are inhibitors of cell proliferation. These compounds may act by binding to some essential cell components (some containing thiols groups, but also others). Szent Gyorgyi called these ketoaldehydes "retine" due to their function in inhibiting proliferation. Glyoxalase, which promotes methylglyoxal removal, was called "promine" because it relieves the inhibition of proliferation caused by methylglyoxal.[50]

Using isolated hepatocytes from fasted rats, we have found that methylglyoxal may be converted at appreciable rates into L-lactate and glucose.[51] However, if high concentrations of methylglyoxal occur in the cell, most of it will be converted by the glyoxalases to D-lactate. Glyoxalase might be considered as a detoxification mechanism when high concentrations of α-ketoaldehydes are present. More experiments are required to elucidate the physiological function of this ubiquitous and very active enzyme system, which utilizes GSH as a coenzyme, the existence of which has been known for over 70 years, but whose function is still unclear.

Role of GSH in Calcium Homeostasis

The maintenance of cytoplasmic Ca^{2+} is a physiological function of glutathione described a few years ago. Isolated hepatocytes incubated with tert-butyl hydroperoxide show a release of mitochondrial Ca^{2+} that results in a loss of Ca^{2+} both in the intramitochondrial and in the extramitochondrial compartments.

It has been shown that the liberation of mitochondrial Ca^{2+} depends on the oxidation of NADPH, but that the extramitochondrial Ca^{2+} pool depends on the oxidation of thiols. Orrenius and colleagues have suggested that the perturbation of Ca^{2+} homeostasis during hydroperoxide metabolism may be an early event, resulting later in signs of toxicity, such as alterations in the plasma membrane of cells.[52]

Role in Leukotriene Synthesis

Leukotrienes are a group of compounds that were called SRS (slow reacting substance).[53] Some of these substances derive from arachidonic acid. One of those leukotrienes (LTA_4) reacts with GSH to yield LTC_4. Other leukotrienes are derived from polyunsaturated fatty acids, different from arachidonic acid. New leukotrienes may be derived from GSH-containing ones such as LTC_4. γ-Glutamyl transpeptidase reacts with this compound and eliminates the glutamyl moiety, thus resulting in LTD_4 which is identical to LTC_4, except that a cysteinyl glycine residue is now present, instead of a glutathione residue.[54]

ROLE IN THE MAINTENANCE OF MACROMOLECULAR STRUCTURES

Maintenance of Cell Membrane Thiol Status

Glutathione protects membranes against oxidative stress.[55] Cells depleted of glutathione are more prone to membrane damage due to oxidative stress (evidenced, for instance, by lipid peroxidation) than normal cells. Kosower and Kosower, using monobromobimane as a specific reagent, have for the first time analyzed in detail the relationship between GSH and membrane thiol status.[55] Monobromobimane labeling has been successfully used to follow the dynamic changes of the thiol groups in red-cell membranes.

The same authors have established a direct relationship between intracellular GSH status and membrane thiol status. Indeed, GSH is effective for the repair of oxidative damage caused to membrane proteins. GSH can reduce both intrachain and interchain disulfide bridges. This may lead to formation of GSSG, which in turn may be reduced to GSH by glutathione reductase (provided that NADPH is present), and thus the normal glutathione status is recovered. In erythrocytes deficient in glucose 6-phosphate dehydrogenase, membrane damage may, however, occur because the last step in the chain of events leading to disulfide reduction (i.e., the generation of NADPH) is deficient; in this case both intracellular and membrane thiol status will be altered.

Role in the Formation of Disulfide Bonds in a Protein

The shape of a protein molecule is determined by its amino acid sequence.[56] One of the most important factors governing the folding of a protein is the distribution of its polar and nonpolar side chains. The proteins that are secreted outside the cytoplasm or the cell-surface proteins often form additional covalent intrachain bonds (disulfide bonds) between neighboring cysteine-SH groups. This helps to stabilize the three-dimensional structure of extracellular proteins, although they are not essential for the specific folding of the molecule. The S–S bonds are rarely formed in proteins that are still in the cytoplasm. This is due to the high intracellular concentration of GSH that breaks most of such S–S bridges.

Modification of the Structure of Hemoglobin by Glutathione

The covalent binding of glutathione to hemoglobin was reported recently.[57-59] The authors found that when hemoglobin binds to glutathione, its oxygen affinity increases by about 35%. Moreover, in hemoglobin from sickle red cells, the increased oxygen affinity of glutathionyl hemoglobin is associated with an inhibition of sickling of about 70%.

ROLE IN TRANSPORT PROCESSES

Role of GSH on the Secretion of Polypeptide Hormones

Over 20 years ago, it was established that the ratio of NADPH/NADP in endocrine tissues is associated with secretion and/or storage, rather than with the synthesis of intracellular endocrine product.[60,61]

GSH and insulin — Recently, Malaisse has coordinated the publication of an excellent collection of papers on the stimulus-secretion coupling in the pancreatic B-cell.[62] These papers show that nutrient secretagogs such as D-glucose, L-leucine, and 2-ketoisocaproate increase the cytosolic $NADPH/NADP^+$ and/or $NADH/NAD^+$ ratio. Glucose at the concentration of 16.7 m*M* increases the GSH/GSSG ratio from a basal value of 1.0 to 3.3. Other authors found a similar pattern of changes, but the increase was more modest. It was also observed that several nutrients increase the islet content in thiol groups. Thus, D-glucose (16.7 m*M*) increases the thiol content from a basal value of 119 to 170 pmol/μg of protein.[63]

GSH and parathyroid hormone — Using dispersed bovine parathyroid cells, Morrisey[64] found that the total GSH and GSSG contents range from 1.59 to 1.71 μg/mg of cell protein.

An increase in the medium Ca^{2+} concentration from 0.5 to 2.8 caused an increase in GSSG from 0.43 to 1.20 μg/mg of protein, with a concomitant decrease in GSH. The compound 2-cyclohexene-1-one was used to deplete the cells of GSH in a low-calcium medium (0.5 m*M*) to levels found in high-calcium medium (2.0 m*M*). This treatment produced a decrease in PTH secretion in the low-calcium medium, as if the cells were incubated in high-calcium medium.[64]

Role of GSH in Neurotransmitter Release

Addition of the thiol-oxidizing agent diamide to neuromuscular preparations caused a great increase in end-plate potentials. This is probably due to the fact that GSSG may cause a reversible change in the membranes of the cholinergic vesicles at the axon terminal. The effects of diamide on liberation of acetylcholine were not dissociated from possible changes in the availability of intracellular Ca^{2+}. Recently, the role of GSH in Ca^{2+} compartmentation was established and thus it is possible that the stimulation of acetylcholine release due to thiol oxidation is mediated by changes in Ca^{2+} compartmentation in the nerve cell.[4]

Possible Role of the γ-Glutamyl Cycle in Amino Acid Uptake

The γ-glutamyl cycle is responsible for the biosynthesis and degradation of GSH. Therefore, its functions are closely related to the regulation of GSH synthesis and utilization and to a variety of protective, metabolic, and catalytic processes. For details about this metabolic cycle, see References 2 and 3.

γ-Glutamyl amino acids are formed on the cell membrane through interaction of γ-glutamyl transpeptidase, GSH, and extracellular amino acids. That this reaction is involved in amino acid translocation into the cells is strongly supported by the fact that γ-glutamyl amino acids are effectively transported into the cells by a system not shared by free amino acids.[65]

Uptake of L-γ-glutamyl-L-methionine sulfone, a compound poorly metabolized by kidney, was greater than that of L-methionine and uptake was inhibited by γ-glutamyl amino acids, but not by free amino acids. When rat kidney slices were incubated with γ-glutamyl amino acids, the formation of 5-oxoproline was increased. This shows that γ-glutamyl amino acids are translocated into the cells under these conditions. However, it should not be excluded that some γ-glutamyl amino acids might be hydrolyzed at the membrane and that the products of such hydrolysis may be transported.

Animals treated with transpeptidase inhibitors, such as acivicin, and patients with transpeptidase deficiency showed hyperglutathionemia and glutathionuria. It was postulated by some authors[66] that the major function of this enzyme is to hydrolyze GSH. However, this assumption has been challenged by Allison and co-workers, showing that at pH 7.0 in the presence of 50 μmol of GSH and a mixture of amino acids approximating the amino acid composition of plasma, about 50% of the GSH utilized participates in transpeptidation.[67]

Our group has studied the possible role of this cycle on amino acid uptake using the mammary gland of the lactating rat. This tissue is an interesting model in which to study amino acid uptake and its relation to the γ-glutamyl cycle because:

1. The rate of amino acid uptake by the gland is high.
2. γ-Glutamyl transpeptidase activity appears at the onset of lactation and reaches its highest activity at the peak of lactation.
3. Isolated acini can be prepared easily.
4. Amino acid uptake by the gland must be subjected to tight hormonal control.[68]

We have shown that specific inhibition of γ-glutamyl transpeptidase by injection of serine-borate, acivicin, or anthglutin results in a decrease in the arteriovenous differences of amino acids across the mammary glands.[69-71] Furthermore, we have been able to show a correlation between changes in arteriovenous differences of amino acids across the gland and changes in γ-glutamyl transpeptidase activity during lactation. Similar correlations were noted in studies on the effects of milk accumulation.[72] However, other mechanisms, apart from the γ-glutamyl cycle, affect amino acid uptake by the mammary gland. Indeed, we have shown that starvation for 6 to 24 hr produces a decrease in amino acid uptake by the gland. We tried to find out if γ-glutamyl transpeptidase activity in mammary gland was affected by starvation or by ketone bodies, but found that neither affected this activity in the mammary gland, thus showing that the observed effects were independent of the γ-glutamyl cycle.[73]

In other tissues, the possible role of the cycle in the amino acid transport has been questioned. Indeed, inhibition of renal γ-glutamyl transpeptidase does not result in a massive amino aciduria. Also, evidence has been presented showing that the cycle is not involved in amino acid transport in yeast.[74] However, using cerebral endothelial cells in vitro, it has been shown that γ-glutamyl transpeptidase can be related to the transport of amino acids. Experiments that have been carried out in the brain ''in vivo'' support the concept that the cycle is involved in amino acid uptake.

Administration of amino acids to mice led to a decline in the level of kidney GSH. Such effects were not found when mice were also treated with an inhibitor of γ-glutamyl transpeptidase. This experiment indicates a significant relationship between amino acids, export of GSH, and transpeptidation as a major physiological function of γ-glutamyl transpeptidase.[75]

From our data on arteriovenous differences of amino acids across the lactating mammary gland and of the blood flow to the tissue, in controls and in animals treated with inhibitors of γ-glutamyl transpeptidase, the role of net amino acid uptake by the mammary gland mediated by the γ-glutamyl cycle can be estimated at 20.7 μmol/hr/g. Thus, if transpeptidation and amino acid uptake must proceed stoichiometrically, a GSH turnover of 20.7 μmol/hr/g must be postulated. This rate is an order of magnitude higher than GSH turnover in any tissue described so far in any mammalian organ. Thus, transpeptidation and amino acid uptake cannot proceed stoichiometrically, at least in the mammary gland of the lactating rat. Therefore, we have presented an alternative hypothesis. The γ-glutamyl cycle generates a signal which activates amino acid uptake, and we have shown in several experimental models, in which the activity of γ-glutamyl transpeptidase was decreased and thus the amino acid uptake, that the venous injection of γ-glutamyl glutamine produced an increase in the arteriovenous differences of amino acids across the lactating mammary gland. Similar results were obtained when other γ-glutamyl amino acids were injected.[76] This led us to think that intracellular signal which activates amino acid uptake might be a common metabolite of all γ-glutamyl amino acids and that 5-oxoproline might be such a signal. We have performed experiments in vivo on the lactating mammary gland and the placenta that show that 5-oxoproline can play a role in the regulation of amino acid translocation.[77] We are now performing experiments trying to reproduce these data in vitro.

ACKNOWLEDGMENTS

This work was supported by grants by El Fondo de Investigaciones Sanitarias de la Seguridád Social (FISS) 87/1574 to J. R. V. and 87/1415 to J. V. and by La Comisión Asesora y Técnica (CAICYT) 0888/84 to J. R. V., PA 86/0169 to G. T. S., and PR 84/1178 to J. V.

REFERENCES

1. **Jocelyn, P. C.,** *Biochemistry of the SH Group,* Academic Press, London, 1972, 10.
2. **Viña, J., Viña, J. R., and Saez, G. T.,** Glutathione: metabolism and physiological functions, *Life Chem. Rep.,* 4, 1, 1986.
3. **Meister, A. and Anderson, M. E.,** Glutathione, *Annu. Rev. Biochem.,* 52, 711, 1983.
4. **Kosower, N. S. and Kosower, E. M.,** The glutathione status of cells, *Int. Rev. Cytol.,* 54, 109, 1978.
5. **Reed, D. J. and Fariss, M. W.,** Glutathione depletion and susceptibility, *Pharmacol. Rev.,* 36, 25S, 1984.
6. **Saez, G. T., Romero, F. J., and Viña, J.,** Effects of glutathione depletion on gluconeogenesis in isolated hepatocytes, *Arch. Biochem. Biophys.,* 241, 75, 1985.
7. **Viña, J., Catalán, V., Blay, P., Saez, G. T., Perez, C., and Rodriguez, A.,** Metabolic effects of glutathione depletion, *Biochem. Soc. Trans.,* 15, 227, 1986.
8. **Jollow, D. J.,** Glutathione thresholds in reactive metabolite toxicity, *Arch. Toxicol.,* Suppl. 3, 95, 1980.
9. **Hathcock, J. N.,** Metabolic mechanism of drug-nutrient interactions, *Fed. Proc.,* 44, 124, 1985.
10. **Orrenius, S. and Sies, H.,** Compartmentation of detoxication reactions, in *Metabolic compartmentation,* Sies, H., Ed., Academic Press, 1982, 485.
11. **Mason, R. P. and Fischer, V.,** Free radicals of acetaminophen: their subsequent reactions and toxicological significance, *Fed. Proc.,* 45, 2493, 1986.
12. **Black, M.,** Acetaminophen hepatotoxicity, *Annu. Rev. Med.,* 35, 577, 1984.
13. **Orrenius, S. and Moldéus, P.,** The multiple roles of glutathione in drug metabolism, *Trends Pharmacol. Sci.,* 5, 432, 1984.
14. **Oberley, L. W. and Oberley, T. D.,** Free radicals, cancer, and aging, in *Free Radicals, Aging and Degenerative Diseases,* Johnson, J. E., Jr., Walford, R., Harman, D., and Miquel, J., Eds., Alan R. Liss, New York, 1986, 325.
15. **Kondo, T., Dele, G. L., and Beutler, E.,** Glutathione transport by inside-out vesicles from human erythrocytes, *Proc. Natl. Acad. Sci. U.S.A.,* 77, 6359, 1980.
16. **Sies, H. and Summer, K. H.,** Hydroperoxide-metabolizing systems in rat liver, *Eur. J. Biochem.,* 57, 503, 1975.
17. **Sies, H.,** Biochemistry of oxidative stress, *Angew. Chem.,* 25, 1058, 1986.
18. **Beutler, E. and Srivastrava, S. K.,** GSH metabolism in the lens, in *Glutathione,* Flohé, L., Benohr, H. Ch., Sies, H., Waller, H. D., and Wendel, A., Eds., Georg Thieme, Stuttgart, 1974, 201.
19. **Spector, A.,** The search for a solution to senile cataracts (Proctor Lecture), *Invest. Ophthalmol. Visual Sci.,* 25, 125, 1984.
20. **Kolayashi, H., Majima, A., Aritake, T., and Kuwayama, M.,** Experimental studies on the relationship between lipid peroxide and congenital cataract of mice carrying its gene, *Acta Soc. Ophthalmol. Jpn.,* 85(8), 59, 1981.
21. **Chanel, D., Varma, S. D., and Richards, R. D.,** Lens lipid peroxidation conjugated acylpolyene and Schiff bases in human cataracts, *Invest. Ophthalmol. Visual Sci.,* 20(Suppl. 3), 218, 1981.
22. **Vladimirov, Y. A.,** Free radical lipid peroxidation in biomembranes: mechanism, regulation, and biological consequences, in *Free Radicals, Aging and Degenerative Diseases,* Johnson, J. E., Jr., Walford, R., Harman, D., and Miquel, J., Eds., Alan R. Liss, New York, 1986, 141.
23. **Guzman Barrón, E. S.,** The effect of X-rays on systems of biological importance, in *Radiation Biology,* Hollander, A., Ed., McGraw-Hill, New York, 1954, 283.
24. **Hutchinson, F.,** Sulfhydryl groups and the oxygen effect on irradiated dilute solutions of enzymes and nucleic acids, *Radiat. Res.,* 14, 721, 1961.
25. **Held, K. D., Harrop, H. A., and Michael, B. D.,** Effects of oxygen and sulfhydryl-containing compounds on irradiated transforming DNA. II. Glutathione, cysteine and cysteamine, *Int. J. Radiat. Biol.,* 45, 615, 1984.
26. **Held, K. D., Harrop, H. A., and Michael, B. D.,** Effects of oxygen and sulfhydryl-containing compounds on irradiated transforming DNA. III. Reaction rates, *Int. J. Radiat. Biol.,* 45, 627, 1984.
27. **Holmgren, A.,** Hydrogen donor system for *Escherichia coli* ribonucloside diphosphatase reductase dependent upon glutathione, *Proc. Natl. Acad. Sci. U.S.A.,* 73, 2275, 1976.
28. **Holmgren, A.,** Glutathione-dependent synthesis of deoxyribonucleotides. Purifications and characterization of glutaredoxin from *Escherichia coli, J. Biol. Chem.,* 253, 430, 1979.
29. **Mego, J. L.,** Stimulation of intralysosomal proteolysis by cysteinyl glycine, a product of action of gamma glutamyl transpeptidase on glutathione, *Biochim. Biophys. Acta,* 841, 139, 1985.
30. **Tateishi, N., Higashi, T., Naruse, A., Nakashima, K., Shiozaki, H., and Sakamoto, Y.,** Rat liver glutathione: possible role as a reservoir of cysteine, *J. Nutr.,* 107, 51, 1977.
31. **Pocius, P. A., Clark, J. H., and Baumrucker, C. R.,** Glutathione in bovine blood: possible source of amino acids for milk protein synthesis, *J. Dairy Sci.,* 64, 1551, 1981.

32. **Rose, W. C.,** The nutritive significance of the amino acids, *Physiol. Rev.*, 18, 109, 1938.
33. **Tateishi, N., Hirasawa, M., Higashi, T., and Sakamoto, Y.,** The L-methionine-sparing effect of dietary glutathione in rats, *J. Nutr.*, 112, 2217, 1982.
34. **Eggleston, L. V. and Krebs, H. A.,** Regulation of the pentose phosphate cycle, *Biochem. J.*, 138, 425, 1974.
35. **Rodriguez-Segade, S., Freire, M., and Carrión, A.,** Regulation of the oxidative phase of pentose phosphate cycle in mussels, *Biochem. J.*, 170, 577, 1978.
36. **Levy, H. R. and Christoff, M.,** A critical appraisal of the effect of oxidized glutathione on hepatic glucose 6-phosphate dehydrogenase activity, *Biochem. J.*, 214, 959, 1983.
37. **Halliwell, B.,** Effect of oxidized glutathione on the inhibition of glucose-6-phosphate dehydrogenase by NADPH, *Biochem. J.*, 234, 741, 1986.
38. **Abraham, E. C., Taylor, J. F., and Cans, C. A.,** Influence of mouse age and erytrocyte age on glutathione metabolism, *Biochem. J.*, 174, 819, 1978.
39. **Hazelton, G. A. and Lang, C. A.,** Glutathione contents of tissues in the aging mouse, *Biochem. J.*, 188, 25, 1980.
40. **Hazelton, G. A. and Lang, C. A.,** Glutathione levels during the mosquito life span with emphasis on senescence, *Proc. Soc. Exp. Biol. Med.*, 176, 249, 1984.
41. **Hazelton, G. A. and Lang, C. A.,** Glutathione peroxidase and reductase activities in the aging mouse, *Mech. Ageing Dev.*, 29, 71, 1985.
42. **Hazelton, G. A. and Lang, C. A.,** Glutathione biosynthesis in the adult yellow-fever mosquito *(Aedes aegypti), Biochem. J.*, 210, 289, 1983.
43. **Harman, D.,** Aging: a theory based on free radical and radiation chemistry, *J. Gerontol.*, 11, 298, 1956.
44. **Miquel, J. and Fleming, J. E.,** A two-step hypothesis on the mechanism of *in vitro* cell aging: cell differentiation followed by intrinsic mitochondrial mutagenesis, *Exp. Gerontol.*, 19, 31, 1984.
45. **Munkres, K. D. and Rana, R. S.,** Antioxidants prolong life span and inhibit the senescent-dependent accumulation of fluorescent pigment (lipofuscin) in clones of *Podospora anserina, Mech. Ageing Dev.*, 7, 407, 1978.
46. **Thaw, H. H., Collins, V. P., and Brunk, U. T.,** Influence of oxygen tension, pro-oxidants and antioxidants on the formation of lipid peroxidation products (lipofuscin) in individual cultivated human glial cell cells, *Mech. Ageing Dev.*, 24, 211, 1984.
47. **Silva, J. E.,** Brown adipose tissue: an extrathyroidal source of triiodothyronine, *NIPS*, 1, 119, 1986.
48. **Visser, T. J.,** Deiodination of thyroid hormone and the role of glutathione, *Trends Biochem. Sci.*, 5, 222, 1980.
49. **Mannervick, B.,** Glyoxalase I. Kinetic mechanism and molecular properties, in *Glutathione*, Flohé, L., Benohr, H. Ch., Sies, H., Waller, H. D., and Wendel, A., Eds., Georg Thieme, Stuttgart, 1974, 78.
50. **Szent-Gyorgyi, A., Egyud, L. G., and McLaughlin, J. A.,** Ketoaldehydes and cell division. Glyoxal derivatives may be regulators of cell division and open a new approach to cancer, *Science*, 155, 539, 1967.
51. **Saez, G. T., Blay, P., Viña, J. R., and Viña, J.,** Glucose formation from methyl glyoxal in rat hepatocytes, *Biochem Soc. Trans.*, 13, 945, 1985.
52. **Bellomo, G., Jewell, S. A., and Orrenius, S.,** The metabolism of menadione impairs the ability of rat liver mitochondria to take up and retain calcium, *J. Biol. Chem.*, 257, 11,558, 1982.
53. **Samuelsson, B. and Hammarstrom, S.,** Nomenclature for leukotrienes, *Prostaglandins*, 19, 645, 1980.
54. **Hammarstrom, S.,** Leukotrienes: biologically active molecules formed from polyunsatured fatty acids and glutathione, in *Functions of Glutathione: Biochemical, Toxicological and Clinical Aspects*, Larsson, A., Holmgren, A., Orrenius, S., and Mannervick, B., Eds., Raven Press, New York, 1983, 149.
55. **Kosower, N. S. and Kosower, E. M.,** Glutathione and cell membrane thiol status, in *Functions of Glutathione: Biochemical, Toxicological and Clinical Aspects*, Larsson, A., Holmgren, A., Orrenius, S., and Mannervick, B., Eds., Raven Press, New York, 1983, 307.
56. **Anfinsen, C. B.,** Principles that govern the folding of protein chains, *Science*, 181, 223, 1973.
57. **Garel, M.-C. Domenget, Ch., Caburi-Martin, J., Prehu, C., Galacteros, F., and Beuzard, Y.,** Covalent binding of glutathione to hemoglobin. I. Inhibition of hemoglobin S polymerization, *J. Biol. Chem.*, 261, 14,704, 1986.
58. **Craescu, C. T., Poyart, C., Schaeffer, C., Garel, M.-C., Kister, J., and Beuzard, Y.,** Covalent binding of glutathione to hemoglobin. II. Functional consequences and structural changes reflected in NMR spectra, *J. Biol. Chem.*, 261, 14,710, 1986.
59. **Wodak, S. J., De Coen, J.-L., Edelstein, S. J., Demarne, H., and Beuzard, Y.,** Modification of human hemoglobin analyzed by computer modeling. III. Perturbations of hemoglobin conformation analyzed by computer modeling, *J. Biol. Chem.*, 261, 14,717, 1986.
60. **Krass, M. E. and LaBella, F. S.,** Biochemical evidence for a secretory role of the pineal body. Oxidation of (1-^{14}C)-and (6-^{14}C)-glucose in vitro by pineal body, pituitary, and brain from young and adult animals, *J. Neurochem.*, 13, 1157, 1966.

61. **Krass, M. E. and LaBella, F. S.,** Hexosemonophosphate shunt in endocrine tissues. Quantitative estimation of the pathway in bovine pineal body, anterior pituitary, posterior pituitary and brain, *Biochim. Biophys. Acta,* 148, 384, 1967.
62. **Malaisse, W. J.,** Stimulus-secretion coupling in the pancreatic B-cell: introductory remarks, *Experientia,* 40, 1025, 1984.
63. **Anjaneyulu, K., Anjaneyulu, R., Sener, A., and Malaisse, W. J.,** The stimulus secretion coupling of glucose-induced insulin release. Thiol: disulfide balance in pancreatic islets, *Biochimie,* 64, 29, 1982.
64. **Morrisey, J. J.,** Involvement of glutathione oxidation reduction in parathyroid hormone secretion, *Am. J. Physiol.,* 250, (Endocrinol. Metab. 13): E475, 1986.
65. **Griffith, O. W., Bridges, R. J., and Meister, A.,** Transport of γ-glutamyl amino acids: role of glutathione and gamma-glutamyl transpeptidase, *Proc. Natl. Acad. Sci. U.S.A.,* 76, 6319, 1979.
66. **McIntyre, T. M. and Curthoys, N. P.,** Comparison of the hydrolytic and transfer activities of rat renal gamma-glutamyl-transpeptidase, *J. Biol. Chem.,* 254, 6499, 1979.
67. **Allison, R. D. and Meister, A.,** Evidence that transpeptidation is a significant function of gamma-glutamyl-transpeptidase, *J. Biol. Chem.,* 256, 2988, 1981.
68. **Viña, J. and Viña, J. R.,** Role of gamma-glutamyl-transpeptidase in the regulation of amino acid uptake by mammary gland of the lactating rat, in *Functions of Glutathione: Biochemical, Toxicological and Clinical Aspects,* Larsson, A., Holmgren, A., Orrenius, S., and Mannervick, B., Eds., Raven Press, New York, 1983, 25.
69. **Viña, J., Puertes, I. R., Estrela, J. M., Viña, J. R., and Galbis, J. L.,** Involvement of gamma-glutamyl transferase in amino acid uptake by the lactating mammary gland of the rat, *Biochem. J.,* 194, 99, 1981.
70. **Viña, J., Puertes, I. R., Saez, G. T., and Viña, J. R.,** Role of prolactine in amino acid uptake by mammary gland of the rat, *FEBS Lett.,* 126, 250, 1981.
71. **Viña, J., Puertes, I. R., Montoro, J. B., and Viña, J. R.,** Effect of specific inhibition of gamma glutamyl transpeptidase on amino acid uptake by mammary gland of the lactating rat, *FEBS Lett.,* 159, 119, 1983.
72. **Viña, J. R., Puertes, I. R., and Viña, J.,** Effect of premature weaning on amino acid uptake by the mammary gland of the lactating rat, *Biochem. J.,* 200, 705, 1981.
73. **Viña, J. R., Puertes, I. R., Montoro, J. B., and Viña, J.,** Effect of fasting and refeeding on amino acid uptake by mammary gland of the lactating rat. Role of ketone bodies, *Biochem. J.,* 216, 343, 1983.
74. **Payne, G. M. and Payne, J. W.,** The gamma-glutamyltransferase is not involved in the bulk uptake of amino acids, peptides or gamma-glutamyl amino acid in yeast, *Biochem. J.,* 218, 147, 1984.
75. **Meister, A.,** Current status of the gamma-glutamyl cycle, in *Functions of Glutathione in Liver and Kidney,* Sies, H. and Wendel, A., Eds., Springer Verlag, Berlin, 1978, 43.
76. **Viña, J. R., Puertes, I. R., Montoro, J. B., Saez, G. T., and Viña, J.,** The gamma glutamyl amino acids as signals for the hormonal regulation of amino acid uptake by the mammary gland of the lactating rat, *Biol. Neonate,* 48, 250, 1985.
77. **Viña, J. R., Puertes, I. R. Palacín, M., Rodriguez, A., and Viña, J.,** Role of oxoproline in amino acid transfer in placenta and lactating mammary gland, *Biochem. Soc. Trans.,* 14, 1056, 1986.

EFFECTS OF EXERCISE-INDUCED OXIDATIVE STRESS ON LIVER GLUTATHIONE

Alexandre T. Quintanilha

INTRODUCTION

Exercise

Physical fitness is a multibillion dollar industry. People of all ages are encouraged to exercise. Yet the benefits or the deleterious effects of increased levels of physical exercise in humans have not been clearly delineated. Despite the common belief that exercise is always beneficial, a body of literature shows quite clearly that exercise can lead to tissue injury. In this chapter, we will synthesize the evidence that shows how uncommon levels of exercise can lead to damage to a number of different tissues and also how these tissues respond to protect themselves against such damage.

Increased levels of physical activity are possible in animals only at the expense of increased energy consumption. The energy "currency" in most living systems is ATP (adenosine triphosphate); the most efficient cellular mechanism of manufacturing ATP is via oxidative phosphorylation — a process whereby oxygen is consumed (reduced to water) and ADP is simultaneously converted to ATP. Increased demand for ATP is usually accompanied by increased rates of oxygen consumption. It follows therefore that increased levels of physical exercise are closely linked to increased levels of oxygen utilization; in humans, oxygen consumption can increase by one order of magnitude during maximum work loads.[1]

While already at the end of the 18th century, the "discoverers" of oxygen (Lavoisier and Priestley) were well aware of its toxicity, most of the studies on oxygen damage, until quite recently, were done at either high concentrations (>20%) or higher pressures (>1 atm). However, recent work by Sohal[2] has shown that exercise-induced increased rates of oxygen consumption can accelerate the process of aging in insects.

In addition, many researchers have shown that uncommon or strenuous exercise can result in muscle injury. Considering that such exercise can lead to temperature increases in the muscle (as high as 6°C in man), body dehydration, and hormonal alterations, it is not surprising that tissue functions, including those of muscle, are significantly altered by exercise. Vihko and co-workers[3,4] have shown that a single bout of exhaustive exercise causes focal necrosis, inflammation, and increased lysosomal acid hydrolase activity in skeletal muscle. Increased mechanical fragility of rat liver lysosomes and lysosomal enzyme-mediated degradation of proteins has also been demonstrated.[5,6] Exercise-induced increases in serum levels of the intracellular enzyme creatine kinase are well known.[7,8] Structural cellular changes can be observed with both light and electron microscopy;[9,10] decreases in the ability to develop force, that cannot be simply attributed to fatigue, have been reported.[8,11] Yet, delayed muscle soreness, which can last for as long as 5 days after strenuous or uncommon exercise, is probably the most familiar symptom of injury.[12-14] Several causes have been invoked as responsible for exercise-induced injury to skeletal muscles; these fall mainly into two categories (1) increased muscle hydrogen ion and lactate concentrations and (2) mechanical disruption of muscle fibers. These ideas are reviewed in a recent paper by McCully.[15]

The purpose of this chapter is to review the evidence that we and others[16-23] have accumulated in the last few years that supports a different mechanism of exercise-mediated tissue (not only muscle) injury, namely, oxidative injury. In the process, some intriguing questions relating to the role of glutathione as an antioxidant at the systemic level are discussed.

Glutathione

Glutathione is certainly the most ubiquitous and perhaps the most important nonprotein thiol in living organisms; it was isolated and named by Hopkins[24] in 1921; he thought it was a dipeptide of glutamic acid and cysteine. A few years later, it was shown to be a tripeptide.[25] Its structure is γ-L-glutamyl-L-cysteinylglycine; it can exist in a thiol-reduced (GSH) and a disulfide-oxidized (GSSG) form. Since 1980, numerous symposia and reviews have been written about its function and regulation.[26-34] Glutathione is synthesized in two steps catalyzed by γ-glutamylcysteine synthetase (EC 6.3.2.2.) (Equation 1) and glutathione synthetase (EC 6.3.2.3) (Equation 2), respectively.

$$\text{L-Glutamate} + \text{L-cysteine} + \text{ATP} \rightarrow \gamma\text{-L-glutamyl-L-cysteine} + \text{ADP} + \text{Pi} \quad (1)$$

$$\gamma\text{-L-Glutamyl-L-cysteine} + \text{glycine} + \text{ATP} \rightarrow \text{GSH} + \text{ADP} + \text{Pi} \quad (2)$$

Glutathione plays a major role in many cellular and organismal functions: among these it provides cellular protection against oxidative damage; it also plays a critical role in detoxification reactions and amino acid transport, modulates enzymatic activity (of enzymes with critical thiol residues), and plays a role in calcium homeostasis. Conditions that perturb intracellular levels of glutathione have been shown to result in significant alterations in cellular metabolism. GSH is usually present intracellularly in the millimolar range (1 to 10 m*M*), while the concentration of GSSG is two or three orders of magnitude lower.[35] In general, it can safely be said that increased oxidative stress to tissues usually results in a lowering of intracellular GSH and an increase in GSSG. Since high levels of intracellular GSSG can lead to the oxidation of protein thiols, glutathione has been thought to act as a "third messenger" in either enhancing or diminishing the activities of a variety of normal biological processes, such as enzyme catalysis, protein synthesis, and receptor binding.[36] Since GSSG is either exported from cells or reduced back to GSH (in a reaction that consumes NADPH, via glutathione reductase), its intracellular levels usually remain well below the levels of GSH, even under conditions of severe oxidative stress.

METHODS

Only "outbred" animals were used in these studies. Animals were exercised on a Quinton rodent treadmill (Model 42-15) at a grade of 8.5° (15%) and at a speed of 24.1 m/min (0.9 mi/hr); endurance-trained animals were exercised 5 days a week (the training session the first day lasted 20 min, and the duration of each training session was increased 5 min a day so that by the first day of week 5 and thereafter, rats were running for a full 120 min each day until sacrificed). The vitamin E content of the diets is described in Davies et al.[16] and Aikawa et al.[17] For the lipid peroxidation and vitamin E depletion studies, muscle (gastrocnemius, soleus, and plantaris) and liver homogenates (10% w/v) were prepared in 175 m*M* KCl, 15 m*M* Tris, pH 7.4. Free radicals were monitored by electron spin resonance (ESR) in a Varian® E 109E spectrometer in both tissue homogenates. Mitochondrial respiratory control indexes (RCI) were reported[16] as the (rate of uncoupled respiration)/(rate of basal respiration). Basal respiration was measured in a Rank oxygen electrode in the presence of 1 μ*M* dicyclohexylcarbodiimide and either 10 m*M* pyruvate + 2.5 m*M* malate, 10 m*M* succinate + 4 μ*M* rotenone, or 20 m*M* glutamate; uncoupled respiration was achieved with 1 μ*M* carbonyl cyanide *p*-trifluoromethoxyphenylhydrazone. Lipid peroxidation was measured in homogenates by the thiobarbituric acid method. Sarcoplasmic reticulum and endoplasmic reticulum integrity were assessed with latency measurements of alkaline phosphatase activity, using both initial and total (solubilized with Triton® X-100) activities.[16] Vitamin E levels in the homogenates were analyzed by high pressure liquid chromatography.[17]

Table 1
EFFECTS OF EXERCISE TO EXHAUSTION ON LIVER AND SKELETAL MUSCLE TISSUE HOMOGENATES

	Muscle		Liver	
Parameter measured	**Rested**	**Exercised**	**Rested**	**Exercised**
Pyruvate-malate RCI	4.2	3.2	3.5	3.0
Glutamate RCI	3.8	3.0	4.0	3.3
Succinate RCI	1.2	1.0	6.2	3.9
SR or ER % latency	53.5	20.9	52.6	38.4
Lipid peroxidation	27.7	50.1	32.9	77.3
R·concentration	8.0	17.0	8.3	19.5

Note: Respiratory control indexes (RCI) were calculated as the rate of uncoupled respiration/rate of basal respiration. Sarcoplasmic reticulum (SR) and endoplasmic reticulum (ER) percent latencies are expressed as [(total solubilized activity − initial activity)/(total solubilized activity)] × 100. Lipid peroxidation is reported as nanomoles of malondialdehyde per gram of wet tissue. Free radical (R·) concentrations are ESR signal peak heights at g ~ 2.004 in arbitrary units.

For the glutathione studies, tissues were used immediately following decapitation of the animals. Neck blood was collected directly into equal volumes of the following solutions: (A) 17.5 m*M* EDTA, 10 m*M* bis(3-carboxy-4-nitrophenyl) disulfide (DTNB) in 100 m*M* potassium phosphate at pH 7.5 to measure total glutathione (GSH + GSSG); and (B) 17.5 m*M* EDTA, 10 m*M* *N*-ethylmaleimide (NEM) (to minimize the artifactual autoxidation of GSH to GSSG and other mixed disulfides) in 100 m*M* potassium phosphate at pH 6.5 to measure glutathione disulfide (GSSG).[37] No hemolysis was observed under the conditions of our experiments. The blood mixtures were centrifuged at 2000 × g for 6 min and the supranatants were used for the assays.

Representative pieces of each liver lobe were weighed together, minced in a 1:4 w/v dilution of solution (A or B) and then homogenized for 1 min using a polytron tissue homogenizer (Kinematica® Gmbh) at medium setting. The total thigh muscle from one leg was prepared using solution A, while the other was prepared in solution B. Muscle suspensions were homogenized for a total of 2 min. All homogenates were centrifuged at 13,000 × g for 30 min and the supranatants were used for the assays. Fresh aliquots of GSSG were prepared daily and also analyzed for standardization. Tissue concentrations of glutathione were estimated by linear regression from the standard curve.

LIPID PEROXIDATION AND VITAMIN E DEPLETION DURING EXERCISE

Our initial studies[16] showed that extensive exercise results in a lowering of mitochondrial respiratory control, a loss of sarcoplasmic reticulum and endoplasmic reticulum integrity, and an increase in the levels of lipid peroxidation in both the liver and the skeletal muscle of rats (Table 1). A two- to threefold increase in free radical (R·) concentrations was also observed in both tissues following exercise to exhaustion. Subsequent studies have indicated that this radical may arise from mitochondrial semiquinones.

Similar measurements done on vitamin E-deficient rats (Table 2) gave the same general pattern of results; however, the damage was more pronounced for the resting animals, and the endurance capacity of these animals was reduced by about 40% when compared to the (vitamin E-sufficient) control animals. In addition we were able to show[17] that muscle tissue

Table 2
EFFECTS OF EXERCISE TO EXHAUSTION ON LIVER AND SKELETAL MUSCLE TISSUE HOMOGENATES IN VITAMIN E-DEFICIENT ANIMALS

	Muscle		Liver	
Parameter measured	**Rested**	**Exercised**	**Rested**	**Exercised**
Pyruvate-malate RCI	3.5	2.9	3.3	2.7
Glutamate RCI	3.3	2.3	3.7	2.3
Succinate RCI	1.1	1.0	2.8	2.2
SR or ER % latency	37.9	11.8	48.2	14.5
Lipid peroxidation	39.3	45.1	61.0	79.8
R·Concentration	10.6	13.7	11.9	14.7

Note: Parameters are the same as in Table 1.

Table 3
TISSUE LEVELS OF VITAMIN E IN HOMOGENATES OF LIVER AND TOTAL HINDLEG MUSCLE OF RATS FED A VITAMIN E-SUFFICIENT OR A VITAMIN E-DEFICIENT DIET

	Trained (μg/g of wet tissue)	Sedentary (μg/g of wet tissue)
Liver		
Control diet	37.79	34.55
E-deficient diet	0.15	2.96
Muscle		
Control diet	4.91	7.20
E-deficient diet	0.00	1.17

levels of vitamin E were significantly lower in endurance-trained rats than in sedentary animals; this was true whether the animals were fed a vitamin E-deficient or a vitamin E-sufficient diet *ad libitum* (Table 3). Furthermore, in Vitamin E-deficient rats, liver tissue levels of vitamin E were significantly lower in those that were endurance trained compared to the sedentary ones; this was not the case for the animals fed the vitamin E-sufficient diet.

Both the increased levels of lipid peroxidation and depletion of vitamin E in skeletal muscle and in liver that result from various forms of exercise (single bout and/or endurance training) are consistent with increased oxidative stress to these tissues. Whether this stress is a consequence of higher levels of oxygen (or other) free radicals in these tissues, during exercise, remains unresolved. Lipid peroxidation is initiated via hydrogen abstraction; this step does not necessarily involve an oxygen radical. On the other hand, the propagation of lipid peroxidation usually is mediated via peroxy or alkoxy radicals and can be linked to oxygen consumption. Vitamin E is known to stop this chain of events; in the process it is consumed.

GLUTATHIONE STUDIES

Since glutathione is also one of the most important intracellular metabolites involved in protection against increased oxidative stress, we investigated its response to strenuous exercise. Characteristic values for GSSG and total glutathione (GSH + GSSG) in the plasma, liver, and skeletal muscle of sedentary (control) and exercised (90 to 120 min on the treadmill)

Table 4
TISSUE LEVELS OF GLUTATHIONE DISULFIDE (GSSG) AND TOTAL GLUTATHIONE (GSH + GSSG = GS)

Site	Contr.[a]	Exerc.[b]	Change (%)
GSSG			
Plasma[c]	0.13	0.47	+251
Liver[d]	0.007	0.014	+103
Muscle[d]	0.0006	0.0012	+100
GS			
Plasma[c]	3.56	4.60	+29
Liver[d]	3.15	2.58	−18
Muscle[d]	0.33	0.31	−6

[a] Contr., control sedentary values.
[b] Exerc., values after a bout of exercise.
[c] Values expressed in μmol/ℓ.
[d] Values expressed in μmol/g of wet tissue.

Table 5
RATIOS OF GLUTATHIONE DISULFIDE (GSSG) TO TOTAL GLUTATHIONE (GSSG + GSH)

Site	Contr.[a]	Exerc.[b]	Change (%)
Plasma	0.0374	0.1020	+173
Liver	0.0022	0.0054	+145
Muscle	0.0018	0.0039	+117

[a] Contr., control sedentary values.
[b] Exerc., values after a bout of exercise.

rats are shown in Table 4. Our glutathione levels are in good agreement with published results.

As expected from processes that increase oxidative stress to these tissues, higher levels of GSSG were found in the skeletal muscle, liver, and blood of exercised animals compared to controls. Concomitantly we observed a reduction in the GSH levels in both skeletal muscle and liver. In the plasma, the levels of GSH are always low (3 to 6 μ*M* in rats) for resting animals, but increase with the onset of exercise. While the rise in plasma GSSG is not surprising, since most cells will export GSSG when its internal level rises, the increase in plasma GSH with the onset of exercise is perhaps more interesting.

The liver seems to export most of the GSH produced in the hepatic cytosol at rapid rate, mainly into plasma;[33] as such, it may play a crucial role in supplying extrahepatic tissues with GSH and/or its component amino acids. Whereas we cannot exclude the possibility that the rise in plasma GSH is due to its release from damaged cells (e.g., red blood cells are known to be damaged during exercise), we feel that this rise may be the result of an exercise-induced stimulation of GSH export from the liver. Sies and Graf[38] showed that GSH efflux across the sinusoidal plasma membrane in an isolated perfused rat liver can be stimulated upon increasing vasopressin levels in the perfusing fluid (at physiological levels). Physical exercise always results in increases in plasma vasopressin levels. Whether this hormonally controlled mechanism is meant to transport glutathione (as an antioxidant) or its constituent amino acids (as a carbon source for gluconeogenesis in the muscle) to other tissues including skeletal muscle is an important question that we would like to pursue.

As can be seen from Table 5, the ratio of GSSG to (GSH + GSSG) increases in all three tissues that we have studied. This observation could be particularly useful if blood could be used to assess the whole body oxidative stress (especially in humans). As we will show later, however, the plasma changes in glutathione that accompany a bout of physical exercise depend on many factors, namely, the duration and intensity of exercise, age, and level of physical fitness (i.e., whether the animals are endurance trained or not).

Subsequently we studied[22] the time course or the effects of prolonged physical exercise at submaximal levels on total glutathione concentrations in the liver, plasma, and skeletal muscle of rats in two different age groups (45 to 52 and 80 to 87 days old). The results are shown in Figure 1. Skeletal muscle is clearly able to maintain relatively high levels of total glutathione, while the liver is depleted to as much as 20% of control values with prolonged exercise. The initial increase in total glutathione in the plasma, observed at the onset of exercise, is also followed by a progressive decline (to levels that nevertheless remain above 60% of control) with prolonged exercise.

These results are consistent with our proposal that the liver might be the major source of glutathione for the muscle (where it might be needed as an antioxidant during increased and prolonged muscular activity). We should keep in mind, however, that prolonged exercise creates in the body a situation analogous to fasting. In either case, caloric demands are greater than the supply, and glucose requirement increases: amino acids that give rise to pyruvate, malate, and oxaloacetate and therefore become the precursors to phosphoenolpyruvate can then be converted to glucose. Cysteine is one such amino acid; it gives rise to pyruvate. Glutathione is also believed to be an interorgan carrier of cysteine.[33] Fasting is well known to result in the lowering of liver glutathione levels; how the extracellular availability of amino acids may be responsible for this was discussed in a recent review.[34] Glutathione depletion in the liver as a result of prolonged exercise therefore could also be the result of cysteine requirements in other tissues.

Equally important to our understanding of how the liver responds to increased levels of exercise-induced oxidative stress is the question of how fast it recovers from its depletion of glutathione. Figure 2 shows postexercise glutathione repletion in livers of rats submitted to two different periods of running (50 to 99 min and 150 to 200 min). As for Figure 1 each point represents one animal: for an "outbred" strain of animals such as ours, we should not be too surprised at finding levels of glutathione as high as 2.5 μmol/g of wet liver tissue after 150 min of running. Each experimental point in Figure 2 has been matched with a control value (sedentary animal) sacrificed at the same time as the experimental one (exercised animal) to make sure that we were not simply dealing with circadian effects.[39] Our results clearly show that the depletion in liver glutathione progresses further after exercise has ceased and that it may take as long as 3 to 4 hr until glutathione levels have returned to normal. It follows that the liver may be particularly debilitated in its ability to handle any additional oxidant stress (e.g., drug induced) for several hours postexercise.

Whatever the reasons for this phenomenon, and there may be many, the fact is that prolonged or unaccustomed exercise can lead to a severe depletion in liver glutathione that can last for several hours after the end of exercise. If indeed the liver is a major source of glutathione for a variety of tissues and if, in addition to its many functions (described above), glutathione is also involved in the biosynthesis of prostaglandins and leukotrienes,[40] then our results could be important for understanding the effects of physical exercise on a large number of cellular and organismal functions.

EFFECTS OF ENDURANCE TRAINING ON GLUTATHIONE RESPONSE TO OXIDANT STRESS

In an attempt to understand both the harmful and beneficial effects of physical exercise, we studied how endurance training might modify cellular glutathione responses to exercise-

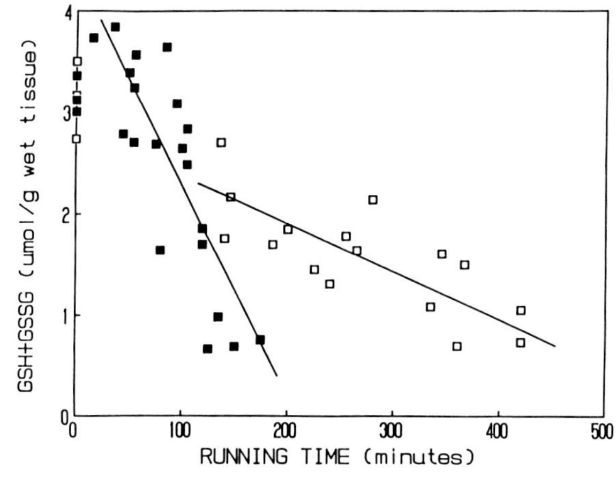

A

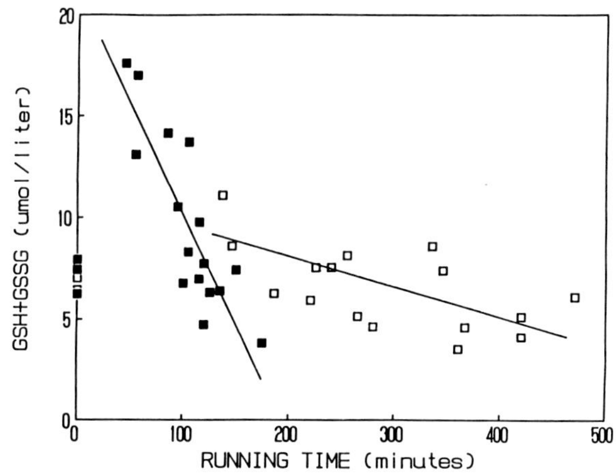

B

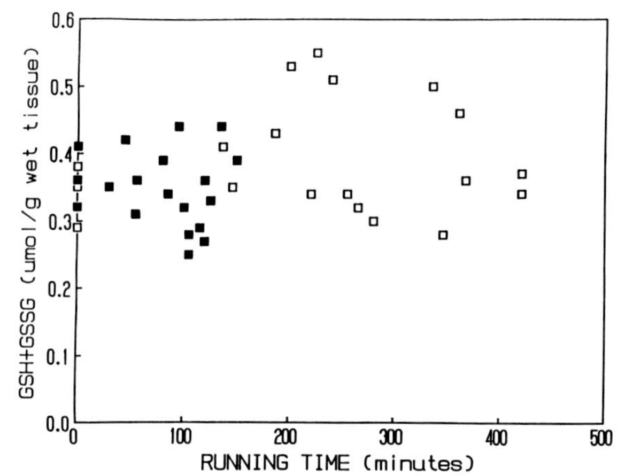

C

FIGURE 1. Total glutathione in (A) liver (μmol/g of wet tissue), (B) plasma (μ*M*), and (C) skeletal muscle (μmol/g of wet tissue) of exercised animals as a function of their running time; values for sedentary animals are indicated at zero time. Filled squares (■) represent 80- to 87-day-old animals (200 to 225-g weight); empty squares (□) represent 45- to 52-day-old animals (130- to 190-g weight). Correlation coefficients: liver, r = 0.82 for heavy animals and r = 0.79 for lighter animals; plasma, r = 0.86 for heavy animals and r = 0.65 for lighter animals; muscle, r = 0.04 for heavy animals and r = 0.27 for lighter animals. (From Lew, H., Pike, S., and Quintanilha, A., in *Bioelectrochemistry and Bioenergetics,* Vol.18, Elsevier, Amsterdam, 1988. With permission.)

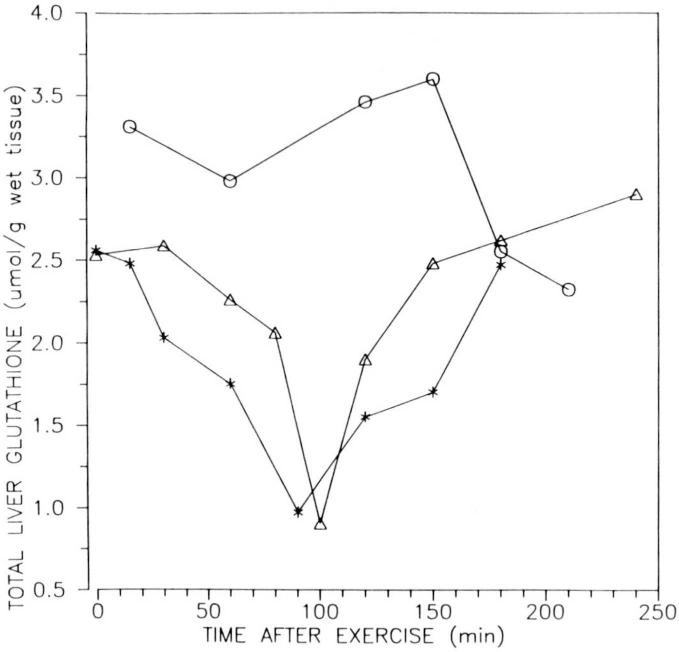

FIGURE 2. Time course of postexercise liver total glutathione (μmol/g of wet tissue) in rats weighing 200 to 225 g that had been exercised for 50 to 99 min (△) or 150 to 200 min (*). Sedentary animals sacrificed at the same time of day as the exercised animals are represented by "○". Each point represents one animal. (From Lew, H., Pike, S., and Quintanilha, A., in *Bioelectrochemistry and Bioenergetics*, Vol.18, Elsevier, Amsterdam, 1988. With permission.)

induced oxidant stress. Data in Table 6 show that following a short running bout of 42 min, endurance-trained rats show no significant increases in tissue levels of GSSG; this is true not only for the tissues studied earlier (namely, skeletal muscle, liver, and plasma), but also for the heart; in fact in the plasma, liver, and heart, the levels of GSSG are lower following this short exercise bout. The other interesting finding is that, in contrast to our earlier observations for untrained animals (confirmed here), the levels of total glutathione in the plasma do not increase at the onset of exercise for the trained animals.

We speculate that a possible reason might be that endurance training has resulted in an increased ability for (1) rapid glutathione cellular turnover and (2) uptake by those tissues where it plays a vital role during periods of increased physical activity. Alternatively, many of these tissues might have developed alternative mechanisms for effective defense against exercise (or other) induced oxidant stress. It seems therefore that, at least during the initial stages of physical exercise, endurance-trained animals show none of the evidence (as judged from the glutathione status of tissues) of increased oxidative stress that we have so clearly shown for untrained animals undergoing comparable workloads. Could this be one of the benefits of training? It is clear that more work is needed to answer this question, but we feel that our results thus far are encouraging.

CONCLUSION

Our initial attempts were directed toward finding evidence of oxidative damage in muscle as a result of increased rates of oxygen consumption during exercise. Not only have we, and others, found such evidence in muscle, but, somewhat to our surprise, we also found such evidence in the liver. Our subsequent work showing severe exercise-induced depletion

Table 6
EFFECT OF A 42-MIN RUNNING BOUT ON THE LEVELS OF TOTAL GLUTATHIONE (GSH + GSSG) AND GLUTATHIONE DISULFIDE (GSSG) OF PLASMA, LIVER, HEART, AND SKELETAL MUSCLE IN ENDURANCE-TRAINED AND UNTRAINED RATS

	Glutathione levels			
	(GSH + GSSG)		(GSSG)	
Tissue, training state	**Rested**	**42-min run**	**Rested**	**42-min run**
Plasma[a]				
Untrained	8.80	20.52	0.54	3.84
Trained	6.24	5.76	0.66	0.36
Liver[b]				
Untrained	3.04	3.14	0.0027	0.0079
Trained	2.80	2.61	0.0036	0.0026
Heart[b]				
Untrained	0.65	0.58	0.0033	0.0041
Trained	0.71	0.66	0.0019	0.0010
Skeletal muscle[b]				
Untrained	0.39	0.29	0.0036	0.0054
Trained	0.41	0.35	0.0023	0.0031

Note: Number of animals used: 7 untrained resting controls; 7 untrained runners; 5 trained resting controls; 6 trained runners.

[a] Values expressed in μmol/ℓ.
[b] Values expressed in μmol/g of wet tissue.

of glutathione levels in the liver gave us a mechanism that enables us to understand this puzzle. Beginning with this finding, we shifted our attention from muscle to the liver.

Whether during prolonged or unaccustomed exercise glutathione is being exported from the liver to other tissues as a source of antioxidant protection or as a source of cysteine or both still needs to be clarified. Whatever the case, as the liver is being increasingly depleted of glutathione, there is little doubt that many metabolic pathways, both at the cellular and organismal level, will be perturbed. How some of these perturbations could help to explain some of the important physiological changes that are associated with exhaustion remains a fascinating and provocative field of inquiry. Finally, the question of how endurance training may help tissues other than the muscle to cope with increased levels of oxidative stress (exercise or nonexercise induced) will still require extensive research efforts from biochemists, physiologists, and nutritionists before any clear advice can be given to the many that seem to be preoccupied with "physical fitness".

REFERENCES

1. **Brooks, G. A. and Fahey, T. D.,** *Exercise Physiology: Human Bioenergetics and Its Applications,* John Wiley & Sons, New York, 1984.
2. **Sohal, R. S.,** *Age Pigments,* Elsevier, Amsterdam, 1981.
3. **Vihko, V., Rantamäki, J., and Salminen, A.,** Exhaustive physical exercise and acid hydrolase activity in mouse skeletal muscle, *Histochemistry,* 57, 237, 1978.
4. **Vihko, V., Salminen, A., and Rantamäki, J.,** Exhaustive exercise, endurance training, and acid hydrolase activity in skeletal muscle, *J. Appl. Physiol. Respir. Environ. Exercise Physiol.,* 47, 43, 1979.

5. **Kasperek, G. J., Dohm, G. L., Tapscott, E. B., and Powell, T.,** Effect of exercise on liver protein loss and lysosomal enzyme levels in fed and fasted rats, *Proc. Soc. Exp. Biol. Med.*, 164, 430, 1980.
6. **Kasperek, G. J., Dohm, G. L., Barakat, H. A., Strausbauch, P. H., Barnes, D. W., and Snider, R. D.,** The role of lysosomes in exercise-induced hepatic protein loss, *Biochem. J.*, 202, 281, 1982.
7. **Armstrong, R. B., Laughlin, M. H., Rowe, L., and Taylor, C. R.,** Metabolism of rats running up and down an incline, *J. Appl. Physiol.*, 55, 518, 1983.
8. **Newman, D. J., Jones, D. A., and Edwards, R. H. T.,** Large delayed plasma creatine kinase changes after stepping exercise, *Muscle Nerve*, 6, 380, 1983.
9. **Fridén, J., Sjöström, M., and Ekblom, B.,** Myofibrillar damage following intense eccentric exercise in man, *Int. J. Sports Med.*, 4, 170, 1983.
10. **Highman, B. and Altland, P. D.,** Effects of exercise and training on serum enzyme and tissue changes in rats, *Am. J. Physiol.*, 205, 162, 1963.
11. **Davies, C. T. M. and White, M. J.,** Muscle weakness following eccentric work in man, *Pfluegers Arch.*, 392, 168, 1981.
12. **Asmussen, E.,** Cardiac output in rest and work at low and high oxygen pressures, *Acta Physiol. Scand.*, 35, 73, 1955.
13. **Hough, T.,** Ergographic studies in muscle soreness, *Am. J. Physiol.*, 7, 76, 1902.
14. **Newman, D. J., Mills, K. R., Quigley, B. M., and Edwards, R. H. T.,** Pain and fatigue after concentric and eccentric muscle contractions, *Clin. Sci.*, 64, 55, 1983.
15. **McCully, K. K.,** Exercise-induced injury to skeletal muscle, *Fed. Proc.*, 45, 2933, 1986.
16. **Davies, K. J. A., Quintanilha, A. T., Brooks, G. A., and Packer, L.,** Free radicals and tissue damage produced by exercise, *Biochem. Biophys. Res. Commun.*, 107, 1198, 1982.
17. **Aikawa, K. M., Quintanilha, A. T., de Lumen, B. O., Brooks, G. A., and Packer, L.,** Exercise endurance training alters vitamin E tissue levels and red blood cell hemolysis in rodents, *Biosci. Rep.*, 4, 253, 1984.
18. **Quintanilha, A. T.,** Effects of physical exercise and/or vitamin E on tissue oxidative metabolism, *Biochem. Soc. Trans.*, 12, 403, 1984.
19. **Quintanilha, A. T.,** Oxidative effects of physical exercise, in *Reactive Oxygen Species in Chemistry, Biology and Medicine*, Quintanilha, A., Ed., Plenum Press, New York, 1988.
20. **Lew, H., Pyke, S., and Quintanilha, A. T.,** Changes in the glutathione status of plasma, liver and muscle following exhaustive exercise in rats, *FEBS Lett.*, 185, 262, 1985.
21. **Lew, H., Pyke, S., and Quintanilha, A. T.,** The effect of physical exercise on the antioxidative capacity of the liver, *J. Bioelectrochem. Bioenergetics*, 18, 231, 1987.
22. **Pyke, S., Lew, H., and Quintanilha, A. T.,** Severe depletion in liver glutathione during physical exercise, *Biochem. Biophys. Res. Commun.*, 139, 926, 1986.
23. **Jackson, M. and Edwards, R. H. T.,** Free radicals, muscle damage and muscular dystrophy, in *Reactive Oxygen Species in Chemistry, Biology and Medicine*, Quintanilha, A., Ed., Plenum Press, 1988, 197.
24. **Hopkins, F. G.,** On an autoxidisable constituent of the cell, *Biochem. J.*, 15, 286, 1921.
25. **Hunter, G. and Eagles, B. A.,** Glutathione. A critical study, *J. Biol. Chem.*, 72, 147, 1927.
26. **McIntyre, T. M. and Curthoys, N. P.,** The interorgan metabolism of glutathione, *Int. J. Biochem.*, 12, 545, 1980.
27. **Kaplowitz, N.,** The importance and regulation of hepatic glutathione, *Yale J. Biol. med.*, 54, 497, 1981.
28. **Meister, A. and Anderson, M. E.,** Glutathione, *Annu. Rev. Biochem.*, 52, 711, 1983.
29. **Meister, A.,** Selective modification of glutathione metabolism, *Science*, 220, 473, 1983.
30. **Orrenius, S., Ormstad, K., Thor, H., and Jewell, S. A.,** Turnover and functions of glutathione studied with isolated hepatic and renal cells, *Fed. Proc.*, 42, 3177, 1983.
31. **Meister, A.,** New developments in glutathione metabolism and their potential application in therapy, *Hepatology*, 4, 739, 1984.
32. **Larsson, A., Orrenius, S., Holmgren, A., and Mannervik, B., Eds.,** *Functions of Glutathione: Biochemical, Physiological, Toxicological and Clinical Aspects*, Raven Press, New York, 1983.
33. **Kaplowitz, N., Aw, T. Y., and Ookhtens, M.,** The regulation of hepatic glutathione, *Annu. Rev. Pharmacol. Toxicol.*, 25, 715, 1985.
34. **Bannai, S. and Tateishi, N.,** Role of membrane transport in metabolism on function of glutathione in mammals, *J. Membr. Biol.*, 89, 1, 1986.
35. **Kosower, N. S. and Kosower, E. M.,** The glutathione status of cells, *Int. Rev. Cytol.*, 54, 109, 1978.
36. **Gilbert, H. F.,** Biological disulfides: the third messenger, *J. Biol. Chem.*, 257, 12,086, 1982.
37. **Adams, J. D., Lauterburg, B. H., and Mitchell, J. R.,** Plasma glutathione and glutathione disulfide in the rat: regulation and response to oxidative stress, *J. Pharmacol. Exp. Ther.*, 227, 749, 1983.
38. **Sies, H. and Graf, P.,** Hepatic thiol and glutathione efflux under the influence of vasopressin, phenylephrine and adrenaline, *Biochem. J.*, 226, 545, 1985.
39. **Isaacs, J. and Binkley, F.,** Glutathione dependent control of protein disulfide-sulfhydryl content by subcellular fractions of hepatic tissue, *Biochim. Biophys. Acta*, 497, 192, 1977.

40. **Grundfest, C. C., Chang, J., and Newcombe, D.,** Acrolein: a potent modulator of lung macrophage arachidonic acid metabolism, *Biochim. Biophys. Acta,* 713, 149, 1982.

VITAMIN E AND RED BLOOD CELL GLUTATHIONE

Ciro Costagliola and Ernesto Rinaldi

INTRODUCTION

The importance of the antioxidant effect of vitamin E (α-tocopherol) in protecting against free radical-induced membrane damage has been well documented.[1] Vitamin E is localized in the membrane components of the cell, and one of its principal roles is to terminate free radical-generated chain reactions, particularly in membranes that are rich in polyunsaturated fatty acids (PUFA).[2] This is borne out by the finding that the cell membrane is the primary site of the pathology observed in vitamin E deficiency.[3-5] However, the exact mechanism of action of vitamin E *in situ* has not yet been completely clarified. The antioxidant action of vitamin E has been demonstrated in in vitro studies, where both enzymic and nonenzymic lipid peroxidation is inhibited by this compound.[6] The feeding of α-tocopherol to animals may produce results different from the in vitro addition of the compound because metabolic incorporation of tocopherol into some unique relationship with cellular membrane lipids may occur.[7] In this light, vitamin E could act as a redox controller of selenium (Se) in membrane proteins.[7]

Altered activity in a number of enzyme systems in vitamin E-deficient animals suggested that vitamin E plays a necessary regulatory role in protein synthesis.[8] Just how the vitamin participates in this sequence of events is not known.

About 35 years ago, Rose and György[9] observed that in vitro vitamin E can act to protect red blood cells (RBC) against hemolysis induced by oxidative agents. It is now well recognized that vitamin E deficiency, both in humans and in animals, is associated with a reduction in RBC survival.[10] In fact, vitamin E-deficient erythrocytes lose their filterability after incubation in vitro[10] and lyse when incubated with 100 μM Cu_2SO_4.[11] Moreover, it is well known that vitamin E deficiency is associated with an enhanced RBC sensitivity to oxidative agent-induced damage.[12-15] In addition, a poor correlation between plasma α-tocopherol concentration and degree of RBC hemolysis in vitro has been observed both in normal subjects and in patients with malabsorption.[7] A partial explanation for this lack of correlation may lie in the observation by Bieri and Evarts[16] that the distribution of vitamin E between RBCs and plasma is influenced by the plasma lipid concentration. Vitamin E exchanges rapidly between RBCs and plasma[17] and apparently the distribution is shifted in favor of the plasma with an increased lipid content of this compartment.

Kitabachi and Wimalasena[18] have suggested that RBC have specific binding sites for α-tocopherol. These sites (one with high affinity and low capacity, the other with low affinity and high capacity) have properties compatible with a biologically significant receptor. It is conceivable that vitamin E binding in RBCs may serve to perform a protective action against oxidative stresses.[19] There is a large body of evidence in support of this effect. Pritchard and Chan,[20] in a study on vitamin E and sulfolipid synthesis, suggested that besides exerting an antioxidant effect on the membrane, vitamin E also produces a direct intracellular modulating effect on the Golgi membranes. More recently, the group of Auricchio,[21] in a study on the in vitro effect of α-tocopherol on lysophosphatidylcholine-induced lens damage, concluded that there are two mechanisms whereby vitamin E could exert its protecting effect against free radical-induced membrane damage: (1) chemically, as an antioxidant, and (2) via a physicochemical interaction between the molecule and the membrane structure.

The alterations of cell membrane enzymes found in vitamin E deficiency and the effect of vitamin E on some enzyme systems are in agreement with the concept of a double mechanism of action of this compound. Diplock hypothesized 10 years ago, and has recently confirmed,[22,23] that vitamin E has two functions: first as an antioxidant and second as an active element in biological membranes. The latter is the most important function, and it is complementary to the former.

SYNERGISM BETWEEN VITAMIN E AND GLUTATIONE

Reduced glutathione (GSH) is the most abundant nonprotein thiol described so far. Its concentration in cells is generally at least one order of magnitude greater than that of the other known thiols. It is a tripeptide made up of three amino acids: glutamic acid, cysteine, and glycine. GSH is synthesized in mammalian erythrocytes, and isotopic-labeling studies have revealed that it has a rapid turnover.[24] The enzymes responsible for GSH synthesis are found in RBC and are very active when isolated, but they are allosterically inhibited in the cell by ADP, GSH, and $NADPH_2$.[24-26] At 37°C RBC synthesize their entire content of GSH in about 10 min.[25]

GSH is easily oxidized to disulfide GSSG.[27] Although the precise role of GSH in many tissues is still obscure, it forms in the RBC, together with glutathione reductase (GSH-rx), an essential component of the enzyme system that protects hemoglobin from oxidation.[25] High levels of GSH are maintained by the reduction of GGSG to GSH that is catalyzed by the NADPH-dependent enzyme, GSH-rx. Glucose also helps to prevent the oxidation of GSH by continually generating $NADPH_2$, which is necessary for the enzymic reduction of GSH.[26] GSH maintains free and protein -SH groups in a reduced state. Its physiological role has not yet been completely clarified, although some of its functions, such as preventing protein thiol groups from oxidation, removing excess H_2O_2, transport of amino acids, and detoxification, have been described.[28] GSH concentration is not reduced by washing RBC or by long anaerobic incubation in isotonic media.[26] Oxidative stress induces an efflux of GSSG from erythrocytes[29,30] and from liver[31] and lens cells.[32]

We have recently demonstrated that there is a difference between arterial and venous plasma GSSG content.[33] The higher levels of GSSG observed in venous plasma cannot be accounted for by RBC alone. Venous plasma GSSG has been shown to reflect the efflux of oxidized glutathione from other tissues,[32,34] and thus it is an indirect measure of their oxidative state. In any case, the main function of GSH is the detoxification of endogenous metabolic peroxides through the glutathione-peroxidase (GSH-px) pathway and of exogenous substances, such as drugs, through the glutathione-*s*-transferase (GSH-tr) pathway.[28] Two GSH-px activities have been described so far: a Se-dependent activity that utilizes H_2O_2 and organic hydroperoxides as substrata and a Se-independent activity that is similar to GSH-tr.[35] It is well known that GSH-px reduces lipid hydroperoxides to less reactive lipid alcohols. One of the roles of GSH could be to act as a line of defense against oxidants.[36,37] GSH-px catalyzes the reaction that leads to the oxidation of GSH, but, in the presence of GSH-rx, GSSG is maintained in the reduced state by $NADPH_2$.[26] $NADPH_2$ plays a major role in the "glutathione system" in the maintenance of GSH in the reduced form.[38] The rate of GSH synthesis in RBC is controlled by a GSH-induced feedback inhibition that acts on the enzymes that catalyze the synthesis of γ-glutamylcysteine.[25] Moreover, the intracellular content of GSH controls pentose-phosphate activity.[39,40] In GSH deficiency, there is an insufficient supply of the NADP coenzyme, which is produced by GSH-px, to the enzymes of the pentose-phosphate pathway.[26] RBCs deficient in either glucose-6-phosphate dehydrogenase (G-6-PDH) or glutathione synthetase (GSH-sy) activities share a common defect of increased sensitivity to oxidants.[41,42] RBC with a reduction or, more rarely, with a congenital absence of GSH and/or of some of its related enzymes (GSH-rx, GSH-px, GSH-S-tr, GSH-sy, and G-6-PDH) show a shortened survival time in vivo and in vitro and an increased susceptibility to hemolysis. Also, hypoxia causes a severe perturbation of GSH metabolism with a decrease of GSH-px activity.[43,44]

As mentioned above, both vitamin E and GSH deficiencies have been reported to be responsible for a reduction in the survival time of RBC, for a modified shape and deformability of RBCs, and for an increased susceptibility to oxidative damage.[10] It has been suggested that GSH protects the microsomes by inhibiting the formation of free radicals and that vitamin E might be involved in the formation of free radicals.[45] There is experimental evidence of a positive relationship between GSH and vitamin E. Levander and Welsh[10] have demonstrated that RBC from vitamin E-deficient rats lose their filterability. This phenomenon is accentuated by incubation with pro-oxidants, it is correlated with lipid peroxidation, and it can be prevented by feeding synthetic antioxidants.[10] It has also been demonstrated that vitamin E-deficient animals have lower erythrocyte GSH levels than do controls (the GSH concentration both in vitamin E-deficient and in control animals was age related), and it was suggested that RBC filterability depends to some extent on intracellular GSH levels.[10]

The foregoing data show that both vitamin E and GSH are involved in the maintenance of the vital functions of RBC, but what are the reciprocal biochemical effects of these two compounds within RBCs?

Our group in Naples has demonstrated that high doses of vitamin E raise RBC glutathione levels both in rabbit[46,47] and in humans.[48] This positive effect of vitamin E on GSH is also found in the aqueous humor of humans[49] and in ocular tissues and fluids of rabbit and rat.[50]

Indirect evidence has pointed to a close relationship between vitamin E and GSH. Increased levels of GSH are generally held to be a secondary consequence of the biochemical changes induced by vitamin E. Our hypothesis is that, in addition to purely antioxidant effects, vitamin E enhances the levels of GSH both in RBCs and in ocular tissues and fluids. Vitamin E probably exerts an active effect on biological membranes, most likely by modulating the activities of some enzymic pathways.[22,23] Exactly where vitamin E interacts with the glutathione system is not known. However, we have postulated that the enhanced RBC glutathione level could be the result of either increased GSH synthesis brought about by GSH-sy stimulation[2,37] or of a reduction of GSH, which serves to detoxify the free radicals produced by cell membrane lipid peroxidation.[2,39] Both mechanisms counteract the GSH feedback on the synthesizing enzymes.[47]

As shown in Figure 1, RBCs contain the hexose monophosphate shunt pathway that, through G-6-PDH, supplies large amounts of NADPH, which is a key element in the glutathione system.[38] In fact, via NADPH, GSH-rx continuously reduces GSSG,[26] thus restoring the levels of GSH. In G-6-PDH deficiency, there is a defect in this system because NADPH is either absent or its concentration is too low to restore the GSH levels.[24,26] In such cases, exposure of RBC to oxidants results in two progressive events. (1) an enhanced release of GSSG from RBC and (2) hemolysis.

In vivo, RBC GSSG is produced both by direct autoxidation of GSH and by the indirect action of oxidizable substances. The latter mechanism is mediated through the GSH-px and GSH-S-tr enzymes.[26,47,48,51] The former is involved, together with catalase, in dealing with naturally produced hydroperoxides.[51] GSH-S-tr binds GSH to exogenous oxidants and is responsible for GSH consumption; it also yields products containing more stable thio complexes.[52]

If one of the six steps shown in Figure 1 is not functioning, the RBCs become more sensitive to lysis. There are two reasons for this. First, high molecular weight aggregates are formed in RBC membranes by intermolecular disulfide bonds, and cell deformability is diminished in cells containing such aggregates.[53] Second, GSH deficiency causes the following changes in hemoglobin properties.

1. Increased oxygen affinity with decreased interaction between the subunits (absence of the sigmoid shape)
2. Reduction of the Bohr effect

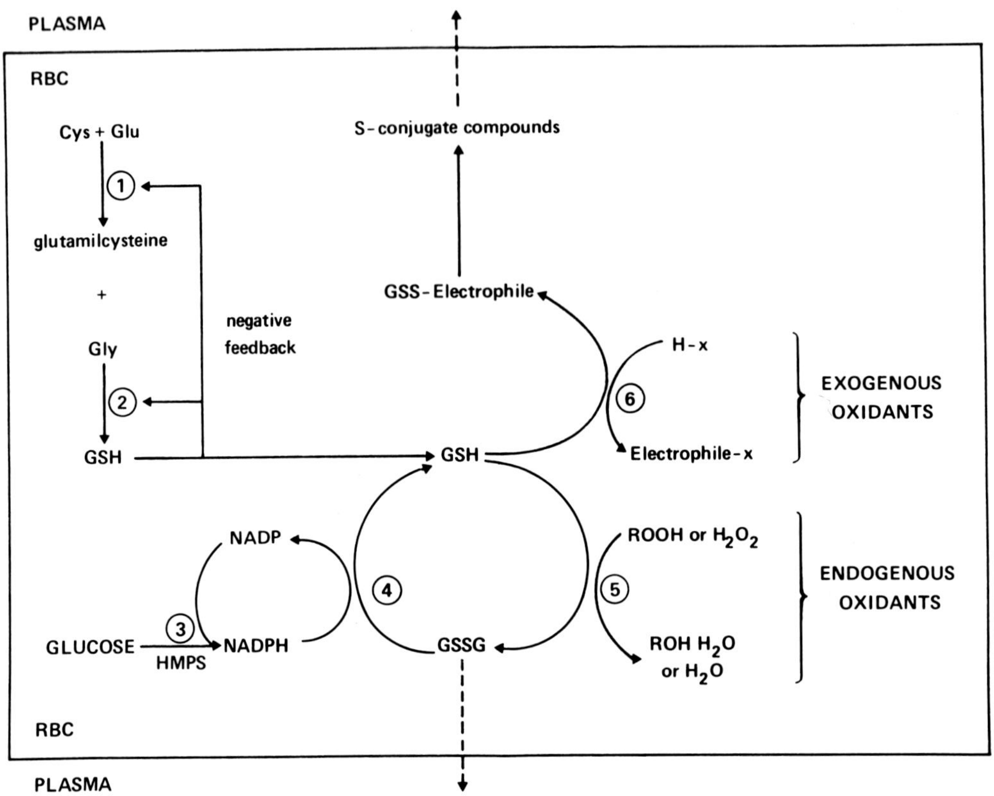

FIGURE 1. Glutathione system. Cys, cysteine; Glu, glutamic acid; Gly, glycine; GSA, reduced glutathione; GSSG, oxidized glutathione; 1, γ-glutamilcysteine-synthetase; 2, GSH-sy; 3, GGPDH; 4, GSH-rx; 5, GSH-px; 6, GSH-S-tr; and HMPS, hexose monophosphate shunt.

3. Easy dissociation of the heme group from globin
4. Increased tendency of hemoglobin to precipitate[25,26]

CONCLUSIONS: THERAPEUTIC IMPLICATIONS

The administration of vitamin E appears to be beneficial in patients with erythrocyte disorders.[54] The role of vitamin E in the treatment of these disorders is matter of speculation. Our investigations on the range of action of vitamin E (Figure 2) suggest how vitamin E could be effective in these disorders. Based on Diplock's findings,[22,23] we think that the glutathione system, in the presence of vitamin E, provides a second system of defense against oxidative damage. In fact, vitamin E acts as an antioxidant, an effect that is exerted outside the RBC, where GSH levels are very low.[24,33] Thus, vitamin E exerts a main protective role by functioning as a free radical scavenger[55] through a nonenzymatic mechanism. In vitamin E deficiency, the free radical scavenger function is flawed, and thus high amounts of reactive species may initiate the peroxidation of membrane PUFA, which results in an overburdened glutathione system. Peroxy radicals that escape destruction by this means are then destroyed by GSH-px, so that the functional and structural integrity of the system is preserved.[22,23,56,57] Nevertheless, tissue degeneration may occur, since a failure in the effectiveness of the glutathione system, an inadequate supply of substrates or bioenergetic compounds, or a severe oxidative insult lead to cell lysis.[10,53]

Evidence is now emerging that vitamin E may incorporate into biological membranes

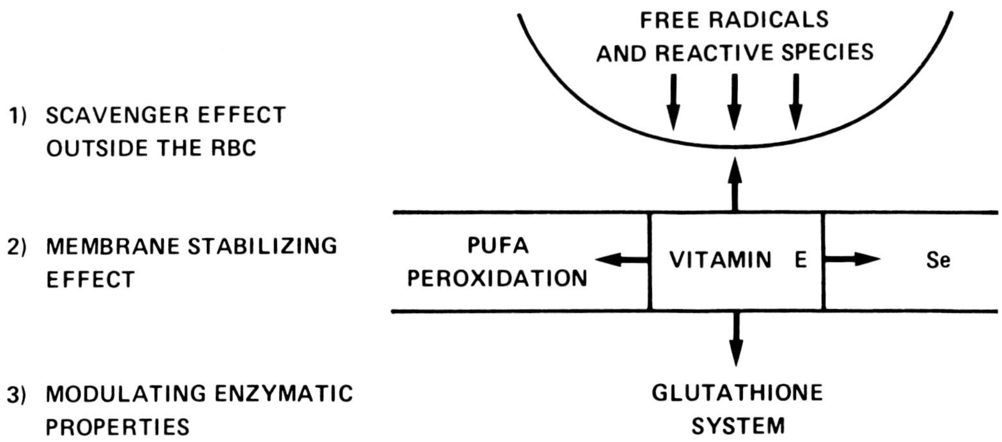

FIGURE 2. Mechanisms of action of vitamin E. See text for details.

through a physicochemical association between the phytil side chain of α-tocopherol and an arachidonyl chain of the cell membrane.[18,19,56] Therefore, vitamin E could directly modulate the activity of membrane-associated enzymes[23,57,58] and, indirectly, influence intracellular enzymes.[44,54,59] The structure of α-tocopherol is such that it can perform these two tasks simultaneously. In fact, while the phenolic hydroxyl group of the chromanol head piece is available at the hydrophilic region of the biological membranes, where it performs protective antioxidant functions, the phytyl portion of the molecule is located within the stereospecific spaces of the intracellular membrane in which the phytyl side chain of tocopherol can interact with the arachidonyl chains of the membrane phospholipids that surround it.[22,23,60] At this level, within the cell membrane, vitamin E could exert a bulk effect on the fluidity of membrane lipids (due to its lipid solubility in RBC and in other tissues) by a direct interaction with PUFA in which peroxidation is inhibited.[10,61,62] This effect is accompanied by changes in membrane-associated enzyme activities.[63]

It is important to note that patients with hereditary disorders of the glutathione system (G-6-PDH and GSH-sy deficiencies) show parallel low levels of serum and RBC α-tocopherol[54,65] and that there are reduced serum vitamin E concentrations in sickle cell anemia and thalassemia.[66,67] This reduction could be ascribed to an increased utilization of vitamin E by the RBC membrane with a consequent reduction of vitamin E serum levels. Reduced serum α-tocopherol would be expected to be associated with reduced RBC tocopherol, because RBC become exposed to increased oxidant damage.[54] From these studies, it is clear that, outside the RBC, vitamin E acts as an antioxidant, and within RBC membranes, it helps to preserve a system of defense against oxidative stresses.

It has also been postulated that vitamin E could act as a redox controller of Se in membrane proteins.[7,22] The importance of Se as a trace element in animal nutrition was first appreciated in the early 1950s.[68] Se is also found in a prosthetic group of GSH-px.[69] Se deficiency results in the depletion of GSH-px in RBCs and in other tissues, which in turn causes a decreased ability to metabolize endogenous oxidants,[27] with a consequent increased cell sensitivity to oxidants.[45-47] This is another possible example of a biochemical interaction between vitamin E and the glutathione system. The primary response of the cell to a deficiency of both vitamin E and Se is a significant increase in GSH levels and synthesis activity.[70] However, this compensatory mechanism for vitamin E deficiency does not provide protection from oxidants because in such cases the membrane barrier shows increased permeability, which leads to disturbances of metabolism within the cell, thus rendering the cell more susceptible to the effects of the oxidative species.[59]

Hitherto we have offered many data in support of the view that vitamin E and GSH are closely related and that the two compounds exert mutually protecting properties: vitamin E prevents the oxidation of GSH and of the protein-SH groups during the redox reactions,[48,63] whereas GSH (probably together with ascorbic acid) plays an important role in the regeneration of oxidized vitamin E.[26,64] At present it is not possible to establish whether the interaction between these compounds is due to a direct modulating effect of vitamin E on some GSH-related enzymes or whether it consists in a first line of cell defense that preserves the glutathione system filters. Studies are underway in our laboratory and in other laboratories to clarify these mechanisms, so it is hoped that we shall soon have an answer to this problem.

ACKNOWLEDGMENTS

We are indebted to Jean Gilder for editing the manuscript.

REFERENCES

1. **Shapiro, S. S. and Mott, D. J.,** Alterations of enzymes in the red blood cell membrane in vitamin E deficiency, *Ann. N.Y. Acad. Sci.*, 393, 263, 1982.
2. **McCay, P. B. and King, M. M.,** Vitamin E: its role as a biological free radical scavenger and its relationship to the microsomal mixed function oxidase systems, in *Vitamin E,* Machlin, L. J., Ed., Marcel Dekker, New York, 1980, 289.
3. **Gross, S. J. and Laudaw, S. A.,** The effect of vitamin E on red blood cell hemolysis and bilirubinemia, *Ann. N.Y. Acad. Sci.*, 393, 315, 1982.
4. **Ritchie, J. H., Fish, M. B., McMasters, V., and Grossman, M.,** Edema and hemolytic anemia in premature infants, *N. Engl. J. Med.*, 279, 1185, 1968.
5. **Farrell, P. M., Bieri, J. G., Fratantoni, J. F., Wood, R. E., and di Sant'Agnese, P. A.,** The occurrence and effects of human vitamin E deficiency: a study in patients with cystic fibrosis, *J. Clin. Invest.*, 60, 233, 1977.
6. **Molenaar, I., Vos, J., and Hommes, F. A.,** Effect of vitamin E deficiency on cellular membranes, *Vitam. Horm. (N.Y.)*, 30, 45, 1972.
7. **Bieri, J. G., and Farrel, P. M.,** Vitamin E, *Vitam. Horm. (N.Y.),* 34, 31, 1976.
8. **Passeri, M. and Provvedini, D.,** La vitamina E nella fisiopatologia geriatrica, *Acta Vitaminol. Enzymol.*, 5, 53, 1982.
9. **Rose, C. S. and György, P.,** Hemolysis with alloxan and alloxan-like compounds and the protective action of tocopherol, *Blood,* 5, 1062, 1950.
10. **Levander, O. A. and Welsh, S. O.,** Cross-linking of membrane proteins in red blood cells from vitamin E-deficient lead-poisoned rats, *Life Sci.*, 28, 147, 1981.
11. **Sayare, M., Fikiet, M., and Paulus, J.,** Effect of Vitamin E on the binding of hemoglobin to the red cell membrane, *Ann. N.Y. Acad. Sci.*, 393, 251, 1982.
12. **Brownlee, N. R., Huttner, J. J., Panganamala, R. V., and Cornewell, D. G.,** Role of vitamin E in glutathione induced oxidant stress: methemoglobin, lipid peroxidation and hemolysis, *J. Lipid Res.*, 18, 635, 1977.
13. **Dallman, P. R.,** The nutritional anemia, in *Hematology of Infancy and Childhood,* Nathan, D. G. and Oski, F. A., Eds., W. B. Saunders, Philadelphia, 1974, 97.
14. **Jacob, H. S. and Lux, S. E.,** Degradation of membrane phospholipids and thiols in peroxide hemolysis: studies in vitamin E deficiency, *Blood,* 32, 549, 1968.
15. **Leonard, P. J. and Losowsky, M. S.,** Effects of tocopherol administration on red cell survival in vitamin E deficient human subjects, *Blood,* 35, 388, 1971.
16. **Bieri, J. G. and Evarts, R. P.,** Tocopherol and polyunsaturated fatty acid in human tissues, *Am. J. Clin. Nutr.*, 28, 117, 1975.
17. **Bieri, J. G., Tolliver, T. T., and Catignani, G. L.,** Simultaneous determination of tocopherol and retinol in plasma or red cells by high pressure liquid chromatography, *Am. J. Clin. Nutr.*, 32, 2143, 1979.
18. **Kitabachi, A. E. and Wimalasena, J.,** Demonstration of specific binding sites for ^{3}H-RRR-α-tocopherol on human erythrocytes, *Ann. N.Y. Acad. Sci.*, 393, 300, 1982.

19. **Kitabachi, A. E., Nathans, A., Barker, J., Kitchell, L., and Watson, B.,** *Tocopherol, Oxygen and Biomembranes,* de Duve, C. and Hayaishi, A., Eds., Elsevier/North-Holland, Amsterdam, 1978, 201.
20. **Pritchard, E. T. and Chan, S. C.,** Effect of dietary vitamin E on sulfolipid synthesis in rat mandibular salivary gland, *Lipids,* 18, 146, 1983.
21. **Libondi, T., Menzione, M., and Auricchio, G.,** *In vitro* effect of α-tocopherol on lysophosphatidylcholine-induced lens damage, *Exp. Eye Res.,* 40, 661, 1985.
22. **Diplock, A. T.,** Vitamin E, selenium and the membrane-associated drug-metabolizing enzyme system of rat liver, *Vitam. Horm. (N.Y.),* 32, 445, 1974.
23. **Diplock, A. T.,** The modulating influence of vitamin E in biological membrane unsaturated phospholipid metabolism, *Acta Vitaminol. Enzymol.,* 4, 303, 1982.
24. **Strivastava, S. K. and Beutler, E.,** Permeability of normal and glucose-6-phospholipid dehydrogenase deficient erythrocytes to glutathione, *Biochem. Biophys. Res. Commun.,* 28, 659, 1967.
25. **Jackson, R. C.,** Studies in enzymology of glutathione metabolism in human erythrocytes, *Biochem. J.,* 111, 309, 1969.
26. **Jocelyn, P. C.,** *Biochemistry of the SH Group,* Academic Press, London, 1972, 247.
27. **Bhattacharya, S. K., Robson, J. S., and Stewart, C. P.,** The determination of glutathione in blood and tissues, *Biochem. J.,* 60, 696, 1955.
28. **Hazelton, G. A. and Lang, A. C.,** Glutathione contents in the tissues in the aging mouse, *Biochem. J.,* 188, 25, 1980.
29. **Srivastava, S. K. and Beutler, E.,** The transport of oxidized glutathione from human erythrocytes, *J. Biol. Chem.,* 244, 9, 1969.
30. **Smith, J. E.,** A relationship of in vivo erythrocyte glutathione flux to the oxidized glutathione transport system, *J. Lab. Clin. Med.,* 83, 444, 1974.
31. **Oshino, N. and Chance, B.,** Properties of glutathione release observed during reduction of organic hydroperoxide, demethylation of aminopyrine and oxidation of some substances in perfused rat liver, and their implication for the physiological function of catalase, *Biochemistry,* 162, 509, 1977.
32. **Srivastava, S. K. and Beutler, E.,** Cataract produced by tyrosinase and tyrosine systems in rabbit lens in vitro, *Biochem. J.,* 112, 421, 1969.
33. **Costagliola, C., Menzione, M., and Romano, L.,** Oxidative state of glutathione in arterial and venous red cells and plasma of rabbit, *Eur. J. Haematol,* 38, 327, 1987.
34. **Bartoli, G. M. and Sies, H.,** Reduced and oxidized glutathione efflux from liver, *FEBS Lett.,* 86, 89, 1978.
35. **Bast, A. and Haenen, G. R. M. M.,** Cytochrome P-450 and glutathione: what is the significance of their interrelationship in lipid peroxidation?, *Trends Biochem. Sci.,* 9, 510, 1984.
36. **Srivastava, S. K., Lal, A. K., and Ausari, N. H.,** Defense system of the lens against oxidative damage: effect of oxidative challenge on cataract formation in glutathione-peroxidase deficient acatalasemic mice, *Exp. Eye Res.,* 31, 1, 1980.
37. **Goosey, J. D., Zigler, J. S., and Kinoshita, J. H.,** Cross-linking of lens crystallins in photodynamic systems: a process mediated by singlet oxygen, *Science,* 208, 1278, 1980.
38. **Costagliola, C., Iuliano, G., Menzione, M., Apponi Battini, G., and Auricchio, G.,** Effects of topical glucocorticoid administration on the protein and non protein sulfhydrylic groups of the rabbit lens, *Ophthalmic Res.,* 19, 351, 1987.
39. **Hosada, S. and Makamura, W.,** Role of glutathione in regulation of hexose-monophosphate pathways in Ehrlich ascite tumor cells, *Biochem. Biophys. Acta,* 222, 53, 1970.
40. **Enesco, H. E. and Verdone-Smith, J.,** Tocopherol increases lifespan in the rotifer, *Exp. Gerontol.,* 15, 335, 1980.
41. **Luzzato, L. and Testa, U.,** Human erythrocyte glucose-6-phosphate dehydrogenase: structure and function in normal and mutant subjects, in *Current Topics in Hematology,* Piomelli, S. and Yachsin, S., Eds., Alan R. Liss, New York, 1978, 1.
42. **Mohler, D. N., Majerus, P. W., Minnick, V., Hess, C. E., and Garrick, M. D.,** Glutathione synthetase deficiency as a cause of hereditary hemolytic disease, *N. Engl. J. Med.,* 283, 1253, 1970.
43. **Farrari, R., Visioli, D., Guarnieri, C., and Caldarera, M.,** Vitamin E and the heart: possible role as antioxidant, *Acta Vitaminol. Enzymol.,* 5, 11, 1983.
44. **Guarnieri, C., Flamingui, F., and Caldarera, M.,** Role of oxygen in the cellular damage induced by reoxygenation of hypoxic heart, *J. Mol. Cell. Cardiol.,* 10, 893, 1978.
45. **Haenen, G. R. M. M. and Bast, A.,** Protection against lipid peroxidation by a microsomal glutathione-dependent labile factor, *FEBS Lett.,* 159, 24, 1983.
46. **Costagliola, C., Libondi, T., Menzione, M., Simonelli, F., and Romano, L.,** Variazioni della concentrazione del glutathione ematico dopo trattamento con alte dosi di vitamina E, in *Proc. Natl. Congr. Soc. Ital. Biol. Sper.,* Idelson, Naples, 1983, 147.
47. **Costagliola, C., Libondi, T., and Menzione, M.,** Livelli ematici di glutathione in conigli trattati con alte dosi di vitamine E, *Boll. Soc. Ital. Biol. Sper.,* 61, 279, 1985.

48. **Costagliola, C., Libondi, T., Menzione, M., Rinaldi, E., and Auricchio, G.,** Vitamin E and red blood cell glutathione, *Metabolism,* 34, 712, 1985.
49. **Costagliola, C., Vito, P., Menzione, M., Rinaldi, E., and Auricchio, G.,** Effect of vitamin E on the glutathione levels on the human aqueous humor, *Proc. Int. Soc. Eye Res.,* 3, 118, 1984.
50. **Costagliola, C., Iuliano, G., Menzione, M., Rinaldi, E., Vito, P., and Auricchio, G.,** Effect of vitamin E on glutathione content in red blood cells, aqueous humor and lens of humans and other species, *Exp. Eye Res.,* 43, 905, 1986.
51. **Srivastava, S. K., Ansari, N. H., and Awasthi, Y. C.,** Lens glutathione depletion by 1-chloro-2,4-dinitrobenzene and oxidative stress, *Curr. Eye Res.,* 3, 117, 1984.
52. **Jacoby, W. B.,** The glutathione-S-transferases: a group of multifunctional detoxification proteins, *Adv. Enzymol.,* 46, 381, 1977.
53. **Johnson, G., Allen, D. W., Cadman, S., Fairbanks, V. F., White, J. G., Lampkin, B. C., and Kaplan, M. E.,** Red cell-membrane polypeptide aggregates in glucose-6-phosphate dehydrogenase mutants with chronic hemolytic disease, *N. Engl. J. Med.,* 301, 522, 1974.
54. **Corash, L. M., Shettz, M., Breri, J. B., Bartsocas, C., Moses, S., Bashan, N., and Schulman, J. D.,** Chronic hemolytic anemia due to glucose-6-phosphate dehydrogenase deficiency or glutathione-synthetase deficiency: the role of vitamin E in its treatment, *Ann. N.Y. Acad. Sci.,* 393, 348, 1982.
55. **Fernsworth, C. C., Stone, W. L., and Aratz, E. A.,** Effects of vitamin E and selenium deficiency on the fatty acid composition of rat retinal tissues, *Biochim. Biophys. Acta,* 552, 281, 1979.
56. **Diplock, A. T. and Lucy, J. A.,** The biochemical modes of action of vitamin E and selenium: a hypothesis, *FEBS Lett.,* 29, 205, 1973.
57. **Galli, C. and Socini, A.,** Biological actions and possible uses of vitamine E, *Acta Vitaminol. Enzymol.,* 4, 245, 1982.
58. **Fukuzawa, K., Takade, S., and Tsukatani, H.,** The effect of concentration on the antioxidant effectiveness of α-tocopherol in lipid peroxidation induced by superoxide free radicals, *Arch. Biochem. Biophys.,* 240, 117, 1985.
59. **Chow, C. K.,** Dietary vitamin E and cellular susceptibility to cigarette smoking, *Ann. N.Y. Acad. Sci.,* 393, 426, 1982.
60. **Lucy, J. A.,** Functional and structural aspects of biological membranes: a suggested structural role for vitamin E in the control of membrane permeability and stability, *Ann. N.Y. Acad. Sci.,* 203, 4, 1972.
61. **Stewart-DeHaan, P. J., Creighton, M. D., Sanwal, M., Ross, W. M., and Trevithick, J. R.,** Effects of vitamin E on cortical cataractogenesis induced by elevated temperature in intact rat lenses in medium 199, *Exp. Eye Res.,* 32, 51, 1981.
62. **Ross, W. M., Creighton, M. D., Inch, W. R., and Trevithick, J. R.,** Radiation cataract formation is diminished by vitamin E in rat lenses in vitro, *Exp. Eye Res.,* 36, 645, 1983.
63. **Arias, J. M. and Jacoby, W. B.,** *Glutathione: Metabolism and Function,* Raven Press, New York, 1966, 15.
64. **Leung, H. W., Vang, M. J., and Mavis, R. D.,** The cooperative interaction between vitamin E and vitamin C in suppression of peroxidation of membrane phospholipids, *Biochim. Biophys. Acta,* 664, 266, 1981.
65. **Stockman, J. A., Landan, S., and Oski, F. A.,** Primaquine induced hemolysis in glucose-6-phosphate dehydrogenase deficiency: effect of vitamin E, *Pediatr. Res.,* 13, 442, 1979.
66. **Natta, C. L. and Machlin, L. J.,** Plasma levels of tocopherol in sickle cell anemia subjects, *Am. J. Clin. Nutr.,* 32, 2359, 1979.
67. **Chin, D. and Lubin, B.,** Abnormal vitamin E and glutathione peroxidase levels in sickle cell anemia: evidence for increased susceptibility to lipid peroxidation in vivo, *J. Lab. Clin. Med.,* 94, 542, 1979.
68. **Cohen, H. J., Chovanicc, M. E., Misketta, D., and Baker,** S. S., Selenium repletion and gluthione peroxidase differential effects on plasma and red blood cells enzyme activity, *Am. J. Clin. Nutr.,* 41, 735, 1985.
69. **Burk, R. F., Nichiki, K., Lawrence, R. A., and Chance, B.,** Peroxide removal by selenium-dependent and selenium-independent glutathione peroxidase in hemoglobin-free perfused rat liver, *J. Biol. Chem.,* 253, 43, 1978.
70. **Yoshida, M., Fukunaga, Y., Iwami, K., and Yasomoto, K.,** Variation of glutathione level and synthesis activity in chick liver due to selenium and vitamin E deficiencies, *J. Biochem.,* 96, 1391, 1984.

INTERACTION BETWEEN ANTIOXIDANT VITAMINS C AND E AND THEIR EFFECT ON IMMUNE RESPONSES

Adrianne Bendich

INTRODUCTION

Vitamins C and E are important antioxidants involved in the protection against free radical damage. These two vitamins can act synergistically to maintain their antioxidant potential. Vitamins C and E, like almost all the micronutrients, are required for the development of optimal immune responses. Although this review cannot discuss in depth the intricacies of immune functions, the following overview is provided to assist in understanding the significance of the interaction of vitamins C (ascorbic acid) and E (tocopherol) in the development of nonspecific as well as specific immune responses.

The immune system has the dual role of protecting against infection by outside pathogens or other substances which would be recognized as foreign (antigenic) and protecting against the development of cancerous cells. Defense against pathogens is initiated by the sequential responses of phagocytic cells (neutrophils and macrophages), complement proteins, opsonins, and other blood-borne substances and would be essentially the same regardless of the pathogen in question. These initial responses are collectively termed nonspecific immune responses.

Following the initial encounter with the pathogen, the immune system recognizes that the substance is foreign (antigenic) and develops a specific immune response which includes the development of a memory response. In many instances, a specific antibody is produced which binds the antigenic pathogen and facilitates the destruction of the pathogen by the immune system. The cells involved in the development of the specific immune response include T- and B-lymphocytes and mononuclear cells (macrophages). T-lymphocytes are associated with the cell-mediated responses that are performed by helper, cytotoxic, memory, and suppressor subpopulations. B-lymphocytes are associated with antibody production, often termed humoral immunity, and the memory response.

The lymphocytes involved in the specific immune response are also involved in the internal surveillance of the cells of the body. It is hypothesized that malignant cells are recognized as ''foreign'' and are destroyed either by cytotoxic T-cells, macrophages, or natural killer cells. These cytotoxic cells are also involved in the rejection of a graft or organ from another individual.

Even though vitamin C is water soluble and vitamin E is fat soluble, numerous experiments using cell-free systems have indicated that vitamin C interacts with vitamin E to maintain the antioxidant potential of vitamin E. Both antioxidant vitamins have been shown to enhance nonspecific and specific immune responses. Vitamin C differs from vitamin E in that most mammals can synthesize vitamin C. The guinea pig is the only common small laboratory animal that, like humans and primates, cannot synthesize vitamin C.

The purpose of this chapter is to review the data indicating that vitamins C and E can act synergistically and that this interaction can result in the protection of the antioxidant function of vitamin E. Data on the ability of these two vitamins to protect phagocytic cells from self-generated free radical damage and protect lymphocytes and macrophages involved in specific immune responses from immunodepressing effects of lipid peroxidation and other free radical damage will be discussed. These findings will be reviewed for possible application for enhancement of human health.

Table 1
EVIDENCE OF THE INTERACTION OF VITAMINS C AND E

	Model systems	Ref.
a.	Reduction of α-tocopheroxyl radicals by ascorbate using pulse radiolysis generated in water/isopropanol/acetone solution containing CCl_4	2
b.	Additive effects of vitamin C and vitamin E to decrease lipid autoxidation reaction: vitamin C regenerates vitamin E	3, 4
c.	Synergistic protection of autoxidation of unsaturated lipids in micelles by vitamins E and C	5
d.	ESR confirmation of the capability of ascorbate to reduce α-tocopheroxyl radical generated by oxidizing lipid in a silica gel	6
e.	Peroxidation of phospholipid liposomes initiated by ferric iron: vitamin C can inhibit free radical generation in lipid matrix and regenerate vitamin E free radical when present	7, 8
f.	Protection against lipid peroxidation of chiken fat by vitamin C regenerating vitamin E	9
g.	Protection against peroxidation in aqueous as well as lipid phases of multilamellar phospholipid liposomes by vitamins E and C	10

EVIDENCE FOR THE INTERACTION OF VITAMINS C AND E

In 1968, Tappel proposed that vitamin C could be involved in the regeneration of reduced tocopherol from the tocopheroxyl free radical.[1] Since this hypothesis was proposed, numerous cell-free model systems have been developed to demonstrate the proposed interaction (Table 1).[2-10] The consensus of data indicated that vitamin C was involved in the maintenance of the reduced form of vitamin E. Recently, McCay[11] has questioned whether vitamin C acted directly or indirectly in the regeneration process. He argued that vitamin C could decrease the level of free radicals by quenching them directly; the result would be a lower level of free radicals available to be quenched by vitamin E. From the data summarized in Table 1, it would seem that vitamin C could be acting both indirectly as well as directly as an antioxidant to maintain the reduced form of vitamin E.

In addition to the finding that vitamin C protected vitamin E in cell-free systems, Bendich et al.[12] found that dietary intake of a high level of vitamin C (10 g/kg diet) increased the level of vitamin E in the lung and plasma of guinea pigs compared to the levels found when the diet contained a low but antiscorbutic level of vitamin C (0.2 g/kg diet). Moreover, the plasma vitamin C level was increased significantly as the level of dietary vitamin E was increased from 0 to 200 mg/kg diet. These data indicated that each nutrient spared the other under certain circumstances. There have been several other reports demonstrating that low vitamin C status resulted in a depression in the vitamin E content of lung, liver, and plasma[13] as well as the adrenal gland[14] of guinea pigs. When guinea pigs were exposed to polychlorinated biphenyls (PCB), feeding of high levels of vitamin C (1 g/kg diet) and vitamin E (2000 mg/kg diet) protected the guinea pigs from growth retardation and lipid peroxidation more effectively than supplementation with either vitamin C or vitamin E.[15]

VITAMINS C AND E AND NONSPECIFIC IMMUNE RESPONSES

The major leukocyte involved in the nonspecific immune response is the neutrophil. Neutrophils respond to chemical signals from pathogenic organisms such as bacteria and move to the site of entry (chemotaxis). The neutrophils are activated by this process. Through the uptake of oxygen, neutrophils initiate the respiratory burst with the consequent generation of several highly reactive free radicals and engulf and then kill the pathogen.[16]

In the process of neutrophil activity, free radicals are released into the surrounding area and cause damage to other cells as well as to the cell that released the free radicals. The

Table 2
VITAMIN C AND VITAMIN E CONTENT OF BLOOD COMPONENTS

	Vitamin C[a] (μg/10⁹ cells)	Vitamin E[b] (μg/10⁹ cells)
Leukocytes	148.0	
Neutrophils		4.5
Lymphocytes		3.9
Erythrocytes	0.55	0.2
Platelets	2.91	0.5
Plasma (μg/mℓ)	6.9	8.82

[a] Adapted from Barkhan, P. and Howard, A. N., *Biochem. J.*, 70, 163, 1958.
[b] Adapted from Hatam, L. J. and Kayden, H. J., *J. Lipid Res.*, 20, 639, 1979.

free radicals can also react with metal ions, hydrogen peroxide, the myeloperoxidase enzyme system, and chloride ions to generate other reactive molecules capable of killing bacteria as well as injuring surrounding cells.[17] Injury to these cells may be due to peroxidation of lipids in the cellular membrane or to the generation of cytotoxic lipid peroxides which are released into the surrounding area or carried to other sites in the body.[18] Lipoproteins in the plasma can also be oxidized and these molecules have recently been shown to be cytotoxic.[19]

To protect against the harmful effects of free radicals or free radical-generated, active metabolites, the neutrophil contains enzymes capable of inactivating some of these reactive molecules. These enzymes include superoxide dismutase, catalase, and glutathione peroxidase. Neutrophils also contain high concentrations of the antioxidant vitamins C and E. The concentration of vitamin E in neutrophils and lymphocytes is approximately 35 times as great as that found in red blood cells and approximately 10 times as great as that in platelets (Table 2).[20] The concentration of vitamin C in leukocytes (all white blood cells) is over 200 times the concentration in red blood cells and almost 50 times as great as in platelets.[21] Both nutrients are at higher concentrations in the while blood cells than in the surrounding plasma. The high concentration of both of these antioxidants may be of importance in the functioning of these cells.

Both vitamins, in separate experiments, have been shown to be of importance in protecting neutrophils from damage by self-generated free radicals. Anderson[22] has reported that vitamin C protected cell membranes from oxidation and protected neutrophils from autoxidative damage. Vitamin C was rapidly taken into neutrophils upon stimulation with a chemotactic agent prior to the respiratory burst[23] as well as during the respiratory burst.[24] Extracellular ascorbate was shown to quench superoxide free radicals generated by activated neutrophils.[24] In addition to self-protection, vitamin C has been shown to stimulate random motility and chemotaxis in neutrophils from normal humans as well as from individuals suffering from impaired neutrophil function.[22]

Neutrophils from vitamin C-deficient guinea pigs demonstrated a significant decline in phagocytic functions, including ingestion and killing of Gram-negative bacteria.[25] In a preliminary study, incubation of neutrophils from premature infants with vitamin C resulted in an increase in chemotaxis.[26] Premature infants have low plasma vitamin C concentrations.[27]

Supplementation of mouse diets with 180 mg/kg of vitamin E (greater than three times the normal dietary level) provided a threefold increase in survival from infection with *Diplococcus pneumoniae* which was attributed to increased clearance of the pathogen.[28]

Chavance et al.[29] in an epidemiological survey of 100 French adults over 60 years of age, demonstrated a statistically significant association of high vitamin E status with low number of infections. Ingestion of vitamin E (1600 mg/day for 7 days) by adult volunteers resulted in an enhanced rate of in vitro neutrophil phagocytosis, but decreased bactericidal activity.[30] These vitamin E-related changes may be the result of decreased influx of hydrogen peroxide into the neutrophil and maintenance of microtubule integrity. These effects were considered to be mediated by an overall decrease in free radical-generated damage to plasma membranes and structural proteins.[31] The practical application of this vitamin E effect was demonstrated in an infant with a congenital deficiency of glutathione synthetase. The resultant loss in glutathione led to an increased accumulation of hydrogen peroxide and oxidative damage to neutrophils. Administration of 400 mg/day of vitamin E decreased the incidence of bacterial infections and in vitro tests indicated that neutrophil functions had been normalized.[32]

As with vitamin C deficiency, vitamin E deficiency resulted in a significant decrease in chemotactic ability of neutrophils. In addition, neutrophils from vitamin E-deficient rats ingested significantly lower numbers of particles than neutrophils from rats fed vitamin E. Repletion with vitamin E caused a normalization of neutrophil function.[33] Premature infants have low plasma vitamin E levels as well as low plasma vitamin C levels.[27] Supplementation with vitamin E resulted in an enhancement of neutrophil chemotaxis, phagocytosis, and bactericidal activity in premature infants.[34]

There are no studies in the literature which examine the effects of both vitamins C and E on neutrophil function. It is predicted, based upon evidence presented in cell-free systems as well as the work to be described in the section on specific immune responses, that there would be a synergistic enhancement of neutrophil antioxidant protection when both nutrients are provided in the diet at optimal concentrations. Since nonspecific immune responses are the first line of defense against pathogens, enhancement of neutrophil function could have important consequences, especially in the premature infant.

VITAMINS C AND E AND SPECIFIC IMMUNE RESPONSES

As discussed in the Introduction, the generation of a specific immune response involves the interaction of macrophages, T- and B-lymphocytes, lymphokines, and antibodies. Assessment of immune status can include both in vivo and in vitro tests. The delayed hypersensitivity skin test is used to measure the in vivo memory of previous exposure to the antigen as well as the interaction of macrophages and T- and B-lymphocytes. Measurement of antibody titers following exposure to antigens such as foreign proteins or bacteria has been used as an indicator of the level of response to pathogens and immune cell interactions.

In vitro exposure of the T- and B-cells to mitogenic stimulation is considered to be an indirect measure of the in vivo response to antigenic challenge. Similarly, in vitro exposure and consequent antibody production has been used to simulate in vivo responses to infection. The in vitro cytotoxic capability of T-lymphocytes has also been used as an indicator of in vivo function.[35]

The immune system has been shown to respond adversely to severe vitamin deficiencies.[36-38] Even when half of the recommended levels of all vitamins was provided to mice, natural killer cell cytotoxicity, lymphokine production, and cytotoxic T-cell activity were significantly diminished within 2 months of the initiation of the diets.[39] Reviews of the literature on the effects of either vitamin C or vitamin E have shown that deficiency of either nutrient adversely affects both in vivo and in vitro immune responses. There have been experiments with either nutrient which have indicated that supplementation of diets with vitamin C or vitamin E enhanced immune responses beyond the level found with the standard diet.[22,40-47] Bendich et al. had previously demonstrated that vitamin E was a potent immunostimulator using an animal model capable of synthesizing vitamin C (rat).[48-50] The level of dietary vitamin E which was necessary to maintain optimal T- and B-lymphocyte responses was greater than the level required to prevent weight loss, red blood cell hemolysis, and testes and muscle degeneration (standard indicators of vitamin E status).[51] The mechanism of immunostimulation by vitamin E was hypothesized to involve: (1) the lipid antioxidant capacity for the maintenance of membrane fluidity and integrity, (2) alteration of prostaglandin production, (3) enhancement of membrane receptor functions, and (4) protection of nuclear material from free radical damage.[47]

The possible synergistic action of both vitamins C and E on specific immune responses of T- and B-lymphocytes to mitogenic stimulation was recently evaluated.[12] Guinea pigs, one of the few mammals incapable of synthesizing vitamin C, were deprived of vitamin E and fed a low but antiscorbutic level of vitamin C. Responses of T- and B-lymphocytes were severely depressed. In contrast, when the diet contained high levels of these nutrients, both T- and B-lymphocyte responses were significantly enhanced compared to the deficient groups.

The mechanisms for the immunoenhancement seen in this experiment may have involved the maintenance of mitochondrial as well as cell membrane integrity. Mitochondrial membranes have a high concentration of polyunsaturated lipids and were therefore potential targets for free radical-mediated peroxidation. Peroxidative damage resulted in abnormal swelling of mitochondria and damage to the endoplasmic reticulum.[52] Lymphocyte membranes from vitamin E-deficient rats show marked ultrastructural changes including mitochondrial swelling.[53] Mitochondrial swelling was lessened and lipid peroxidation was significantly decreased when guinea pig diets were supplemented with vitamins C and E rather than being deficient in either nutrient.[54]

The immune responses of guinea pigs fed a vitamin E-deficient and low vitamin C diet were further depressed when the guinea pigs were exposed to a highly oxidizing environment. In a pilot study, when guinea pigs were placed in an environmental chamber and exposed to 100% oxygen, the immune responses were severely depressed and there was a loss of weight following exposure. If the diets were supplemented with either vitamin, the immune responses were enhanced significantly. However, when both vitamins were supplemented at the highest levels examined (10 g/kg diet, vitamin C; 200 mg/kg diet, vitamin E), not only were the immune responses enhanced significantly, but the weight loss was reversed.[55]

CONCLUSIONS

The immune system has been shown to be more sensitive than other systems to dietary-induced vitamin deficiencies. The immune system has also been shown to be adversely affected by products of lipid peroxidation. Vitamins C and E, like all essential nutrients, are important for the maintenance of immune function. These vitamins have in common the ability to quench free radicals. Data were presented which indicated that vitamin C protected the antioxidant potential of vitamin E by the regeneration of the reduced form of tocopherol using cell-free systems.

Both nonspecific immune responses of neutrophils and specific immune functions were adversely affected when either antioxidant vitamin was not present in the diet. Supplementation of the diet with either vitamin resulted in the enhancement of several important neutrophil and lymphocyte functions.

There are at present, no experimental data on the interaction of vitamin C and E on neutrophil function, however, it is postulated that both are required for optimal neutrophil activity. Since guinea pigs, like man and other primates, are unable to synthesize vitamin C, experiments were undertaken to determine whether vitamins C and E acted synergistically to enhance immune responses. The experiments shwed that guinea pig immune responses were best when the diet contained higher than standard leels of both nutrients. These data indicated that vitamins C and E enhanced immune responses synergistically and were especially important in reversing the immunodepressing effect of a highly oxidizing environment. Moreover, clear evidence for the vitamin E sparing effect of vitamin C in lung and plasma were documented.[12,55] Because of the importance of both arms of the immune response in the maintenance of health, further research is warranted to determine the intake of the antioxidant vitamins C and E required for the generation of optimal immune responses in man.

REFERENCES

1. **Tappel, A. L.,** Will antioxidant nutrients slow aging processes?, *Geriatrics,* 23, 97, 1968.
2. **Packer, J. E., Slater, T. F., and Willson, R. L.,** Direct observation of a free radical interaction between vitamin E and vitamin C, *Nature (London),* 278, 737, 1979.
3. **Niki, E., Tsuchiya, J., Tanimura, R., and Kamiya, Y.,** Regeneration of vitamin E from alpha-chromanoxyl radical by glutathione and vitamin C, *Chem. Lett.,* p. 789, 1982.
4. **Niki, E., Saito, T., and Kamiya, Y.,** The role of vitamin C as an antioxidant, *Chem. Lett.,* p. 631, 1983.
5. **Barclay, L. R. C., Locke, S. J., and MacNeil, J. M.,** The autoxidation of unsaturated lipids in micelles. Synergism of inhibitors vitamin C and E, *Can. J. Chem.,* 61, 1288, 1983.
6. **Bascetta, E., Gunstone, F. D., and Walton, J. C.,** Electron spin resonance study of the role of vitamin E and vitamin C in the inhibition of fatty acid oxidation in a model membrane, *Chem. Phys. Lipids,* 33, 207, 1983.
7. **Leung, H. W., Vang, M. J., and Mavis, R. D.,** The cooperative interaction between vitamin E and vitamin C in supression of peroxidation of membrane phospholipids, *Biochim. Biophys. Acta,* 664, 266, 1981.
8. **Scarpa, M., Rigo, A., Maiorino, M., Ursini, F., and Gregolin, C.,** Formation of alpha-tocopherol radical and recycling of alpha-tocopherol by ascorbate during peroxidation of phosphatidylcholine liposomes. An electron paramagnetic resonance study, *Biochim. Biophys. Acta,* 801, 215, 1984.
9. **Lambelet, P., Saucy, F., and Loliger, J.,** Chemical evidence for interactions between vitamins E and C, *Experientia,* 41, 1384, 1985.
10. **Doba, T., Burton, G. W., and Ingold, K. U.,** Antioxidant and co-antioxidant activity of vitamin C. The effect of vitamin C, either alone or in the presence of vitamin E or a water-soluble vitamin E analogue, upon the peroxidation of aqueous multilamellar phospholipid liposomes, *Biochim. Biophys. Acta,* 835, 298, 1985.
11. **McCay, P. B.,** Vitamin E: interactions with free radical and ascorbate, in *Annual Review of Nutrition,* Vol. 5, Olson, R. E., Ed., Annual Reviews, Calif., 1985, 323.
12. **Bendich, A., D'Apolito, P., Gabriel, E., and Machlin, L. J.,** Interaction of dietary vitamin C and vitamin E on guinea pig immune responses to mitogens, *J. Nutr.,* 114, 1588, 1984.
13. **Hruba, F., Novakova, V., and Ginter, E.,** The effect of chronic marginal vitamin C deficiency on the α-tocopherol content of the organs and plasma of guinea-pigs, *Experientia,* 38, 1454, 1982.
14. **Kanazawa, K., Takeuchi, S., Hasegawa, R., Okada, M., Makiyama, I., Hirose, N., Toh, T., Cho, S. H., and Kobayashi, M.,** Influence of ascorbic acid deficiency on the level of non-protein SH compounds and vitamin E in the blood and tissues of guinea pigs, *Nihon Univ. J. Med.,* 23, 257, 1981.
15. **Kawai-Kobayashi, K. and Yoshida, A.,** Effect of dietary ascorbic acid and vitamin E on metabolic changes in rats and guinea pigs exposed to PCB, *J. Nutr.,* 116, 98, 1986.
16. **Gallin, J. I.,** Abnormal phagocyte chemotaxis: pathophysiology, clinical manifestations, and management of patients, *Rev. Infect. Dis.,* 3(6), 1196, 1981.
17. **Fantone, J. C. and Ward, P. A.,** Polymorphonuclear leukocyte-mediated cell and tissue injury, *Hum. Pathol.,* 16(10), 973, 1985.
18. **Till, G. O., Hatherill, J. R., Tourtellotte, W. W., Lutz, M. J., and Ward, P. A.,** Lipid peroxidation and acute lung injury after thermal trauma to skin, *Am. J. Pathol.,* 119, 376, 1985.
19. **Cathcart, M. K., Morel, D. W., and Chisolm, G. M., III,** Monocytes and neutrophils oxidize low density lipoprotein making it cytotoxic, *J. Leukocyte Biol.,* 38, 341, 1985.
20. **Hatam, L. J. and Kayden, H. J.,** A high performance liquid chromatographic method for the determination of tocopherol in plasma and cellular elements of the blood, *J. Lipid Res.,* 20, 639, 1979.
21. **Barkhan, P. and Howard, A. N.,** Distribution of ascorbic acid in normal and leukaemic human blood, *Biochem. J.,* 70, 163, 1958.

22. **Anderson, R.**, Ascorbic acid and immune functions: mechanism of immunostimulation, in *Vitamin C Ascorbic Acid,* Counsell, J. N. and Hornig, D. H., Eds., Applied Science, London, 1981, 249.
23. **Moser, U. and Weber, F.**, Uptake of ascorbic acid by human granulocytes, *Int. J. Vitam. Nutr. Res.*, 54, 47, 1983.
24. **Hemila, H., Roberts, P., and Wikstrom, M.**, Activated polymorphonuclear leucocytes consume vitamin C, *FEBS Lett.*, 178, 25, 1985.
25. **Shilotri, P. G.**, Glycolytic, hexose monophosphate shunt and bactericidal activities of leukocytes in ascorbic acid deficient guinea pigs, *J. Nutr.*, 107, 1507, 1977.
26. **Vohra, K., Khan, A. J., Rosenfeld, W., Telang, V., and Evans, H. E.**, Correction of defective chemotaxis of neonatal neutrophils with ascorbic acid, *Pediatr. Res.*, 17, 340, 1983.
27. **Arad, I. D., Dgani, Y., and Eyal, F. G.**, Vitamin E and Vitamin C plasma levels in premature infants following supplementation of vitamin C, *Int. J. Vitam. Nutr. Res.*, 55, 395, 1985.
28. **Heinzerling, R. H., Tengerdy, R. P., Wick, L. L., and Lueke, D. C.**, Vitamin E protects mice against *Diplococcus pneumoniae* type I infection, *Infect. Immun.*, 10, 1292, 1974.
29. **Chavance, M., Brubacher, G., Herbeth, B., Vernhes, G., Mikstacki, T., Dete, F., Fournier, C., and Janot, C.**, Immunological and nutritional status among the elderly, *Topics Aging Res. Eur.*, 1(3), 231, 1984.
30. **Baehner, R. L., Boxer, L. A., Allen, J. M., and Davis, J.**, Autooxidation as a basis for altered function by polymorphonuclear leukocytes, *Blood,* 50(2), 327, 1977.
31. **Oliver, J. M., Albertini, D. F., and Berlin, R. D.**, Effects of glutathione oxidizing agents on microtubule assembly and microtubule dependent surface properties of human neutrophils, *J. Cell Biol.*, p. 921, 1976.
32. **Boxer, L. A., Oliver, J. M., Spielberg, S. P., Allen, J. M., and Schulman, J. D.**, Protection of granulocytes by vitamin E in glutathione synthetase deficiency, *N. Engl. J. Med.*, 301, 901, 1979.
33. **Harris, R. E., Boxer, L. A., and Baehner, R. L.**, Consequences of vitamin E deficiency on the phagocytic and oxidative functions of the rat polymorphonuclear leukocyte, *Blood,* 55(2), 338, 1980.
34. **Chirico, G., Marconi, M., Colombo, A., Chiara, A., Rondini, G., and Ugazio, A. G.**, Deficiency of neutrophil phagocytosis in premature infants: effect of vitamin E supplementation, *Acta Paediatr. Scand.*, 72, 521, 1983.
35. **Rose, N. R., Friedman, H., and Fahey, J. L., Eds.**, *Manual of Clinical Laboratory Immunology,* 3rd ed., American Society for Microbiology, Washington, D.C., 1986.
36. **McMurray, D. N.**, Cell-mediated immunity in nutritional deficiency, *Prog. Food Nutr. Sci.*, 8, 193, 1984.
37. **Panush, R. S. and Delafuente, J. C.**, Vitamins and immunocompetence, *World Rev. Nutr. Diet,* 45, 97, 1985.
38. **Beisel, W. R.**, Single nutritents and immunity, *Am. J. Clin. Nutr.*, 35, 417, 1982.
39. **Saxena, Q. B., Saxena, R. K., and Adler, W.**, Effect of feeding a diet with half of the recommended levels of all vitamins on the natural and inducible levels of cytotoxic activity in mouse spleen cells, *Immunology,* 52, 41, 1984.
40. **Chandra, R. K., Ed.**, *Nutrition, Immunity and Illness in the Elderly,* Pergamon Press, New York, 1985.
41. **Watson, R. R., Ed.**, *Nutrition, Disease Resistance, and Immune Function,* Marcel Dekker, New York, 1984.
42. **Gershwin, M. E., Beach, R. S., and Hurley, L. S., Eds.**, *Nutrition and Immunity,* Academic Press, Orlando, Fla., 1985.
43. **Tengerdy, R. P.**, Disease resistance: immune response, in *Vitamin E: A Comprehensive Treatise,* Machlin, L. J., Ed., Marcel Dekker, New York, 1980, 429.
44. **Gross, R. L. and Newberne, P. M.**, Role of nutrition in immunologic function, *Phys. Rev.*, 60, 188, 1980.
45. **Keusch, G. T., Wilson, C. S., and Waksal, S. D.**, Nutrition, host defenses, and the lymphoid system, in *Advances in Host Defense Mechanisms,* Vol. 2, Gallin, J. I. and Fauci, A. S., Eds., Raven Press, New York, 1983, 275.
46. **Levy, J. A.**, Nutrition and the immune system, in *Basic and Clinical Immunology,* Stites, D. P., Stobo, J. D., Fudenberg, H. H., and Wells, J. V., Ed., Medical Publishing, Appleton & Lang, Los Altos, Calif., 1982, 275.
47. **Bendich, A.**, Antioxidant vitamins and immune responses, in *Nutrition and Immunology,* Chandra, R. K., Ed., Alan R. Liss, New York, 1988, 125.
48. **Bendich, A., Gabriel, E., and Machlin, L. J.**, Effect of dietary level of vitamin E on the immune system of the spontaneously hypertensive (SHR) and normotensive Wistar Kyoto (WKY) rat, *J. Nutr.*, 113, 1920, 1983.
49. **Bendich, A., Gabriel, E., and Machlin, L. J.**, Depression of rat and guinea pig lymphocyte blastogenic responses by vitamin E deficiency: a new model for reproducible immune modulation, *Ann. N.Y. Acad. Sci.*, 435, 382, 1984.

50. **Bendich, A., Gabriel, E., and Machlin, L. J.,** Effects of vitamin E on rat lymphocyte mitogen responses, in *Nutrition, Immunity and Illness in the Elderly,* Chandra, R. K., Ed., Pergamon Press, New York, 1985, 173.
51. **Bendich, A., Gabriel, E., and Machlin, L. J.,** Dietary vitamin E requirement for optimum immune responses in the rat, *J. Nutr.,* 116, 675, 1986.
52. **Packer, L., Deamer, D. W., and Heath, R. L.,** Regulation and deterioration of structure in membranes, in *Advances in Gerontological Research,* Vol. 2, Strehler, B. L., Ed., Academic Press, New York, 1967, 77.
53. **Lehmann, J. and McGill, M.,** Biochemical and ultrastructural alterations in platelets, reticulocytes, and lymphocytes from rats fed vitamin E deficient diets, *J. Lipid Res.,* 23, 299, 1982.
54. **Chen, L. H. and Chang, M. L.,** Effect of dietary vitamin E and vitamin C on respiration and swelling of guinea pig liver mitochondria, *J. Nutr.,* 108, 1616, 1978.
55. **Bendich, A., D'Apolito, P., Gabriel, E., and Machlin, L. J.,** Modulation of immune system function of guinea pigs by dietary vitamins E and C following exposure to 100% oxygen, *Fed. Proc.,* 42, 923, 1983.

EFFECT OF VITAMIN E STATUS ON HEPATIC CHOLESTEROL 7α-HYDROXYLASE

Nongnuch (Chupukcharoen) Settasatian, Prapon Wilairat, and Prayad Komaratat

INTRODUCTION

Studies on the relationship of fat-soluble vitamins to cholesterol metabolism indicate that vitamin E status can have a profound effect on cholesterol level in tissues and organs.[1] Several abnormalities in lipid metabolism have been observed in vitamin E-deficient animals, such as elevation of plasma and skeletal muscle cholesterol and increase in muscle gangliosides, sulfatides, neutral glycosylceramides, sphingomyelin, and phosphatidylcholine.[2-4] Elevated muscle cholesterol level in vitamin E deficiency together with the appearance of muscular dystrophy has been observed in rabbit,[4,5] rat,[6] chick,[7] calf,[8] guinea pig,[9] and man.[10]

Every animal tissue seems capable of *de novo* synthesis of cholesterol, with the possible exception of the mature nerve cell.[11] However, liver is considered the dominant site of cholesterol synthesis. This occurs mainly in the microsomal membranes plus components in the soluble fraction.[12] There are approximately 26 separate enzymatic reactions involved in the biochemical conversion of acetate to cholesterol. 3-Hydroxy-3-methylglutaryl-coenzyme-A reductase (HMG-CoA reductase, EC 1.1.1.34) catalyzes the conversion of HMG-CoA to mevalonic acid and is the major regulatory enzyme in the biosynthetic pathway, being subject to feedback inhibition by the end product, cholesterol.[13]

In addition to the endogenous route, liver is able to obtain its supply of cholesterol by uptake from plasma.[14] Cholesterol is carried in the circulation by lipoprotein fractions, namely, very low-density (VLDL), low-density (LDL), and high-density (HDL) lipoproteins. On the surface of liver cells are receptors for LDL which bind and then internalize these molecules in an endocytic process. The endocytosed vesicles then fuse with lysosomes where the lipoproteins are catabolized, releasing cholesterol into the cytosol. The number of LDL receptors on the cell surface is regulated by the intracellular level of cholesterol and thereby modulates cholesterol homeostasis.

In the liver, cholesterol is catabolized to bile acids.[15] This conversion occurs mainly on microsomal membranes of most mammalian species. The first step in the pathway is catalyzed by cholesterol 7α-hydroxylase (EC 1.14.13.17) which introduces a hydroxyl group into the 7α position of cholesterol.[16] This enzyme belongs to a class of enzymes known as mixed-function oxidases whose activity requires NADPH and molecular oxygen, as well as cytochrome P-450 and NADPH-cytochrome P-450 reductase.

Because of the inhibitory effect of carbon monoxide on the activity of cholesterol 7α-hydroxylase, cytochrome P-450 has been implicated as the terminal oxidase for the hydroxylation reaction which binds both substrate and oxygen.[17] A specific form of cytochrome P-450 isolated from liver microsomes of rabbit treated with cholestyramine (a bile salt sequestering agent) is active in the hydroxylation of cholesterol with an activity of 37.5 pmol/min/nmol cytochrome P-450 in a reconstituted enzyme system.[18]

Activity of cholesterol 7α-hydroxylase is enhanced by cholesterol,[19] thyroid hormone,[16] biliary drainage,[20] cholstyramine treatment,[17,19] and glucocorticoid hormones.[21] Factors that inhibit activity include 7α-hydroxycholesterol,[17] bile acids and salts,[17] ethanol feeding,[22] and gallstone formation.[23]

VITAMIN E AND CHOLESTEROL CATABOLISM

Cholesterol 7α-hydroxylase displays a circadian rhythm primarily regulated via the hypothalamo-hypophysis-adrenal axis, and at the cellular level, the most likely agents are the glucocorticoids.[24] Variations in hepatic cholesterol 7α-hydroxylase activity parallel fluctuations in the plasma corticosterone concentration levels. The establishment of the circadian rhythm also requires protein and RNA synthesis since it is blocked by actinomycin D, probably owing to an alternating cycle of repression and derepression of enzyme synthesis.[21]

Cholesterol 7α-hydroxylase activity in microsomes has also been shown to be increased in vitro by ATP, magnesium ions, and a cytosolic protein fraction.[25,26] Loss of enzyme activity is seen in the presence of *Escherichia coli* alkaline phosphatase and is proportional to the amount of phosphatase. A proposed regulation mechanism of this short-term rapid change in the activity of cholesterol 7α-hydroxylase involves a shuttle between a phosphorylated and dephosphorylated state. The enzyme is activated (phosphorylated form) by protein kinase and ATP and is inactivated (dephosphorylated form) by protein phosphatase.[27] This has recently been confirmed in a reconstituted enzyme system from rat.[28]

Experiments carried out on vitamin E-deficient rabbits showed an increase in plasma cholesterol level, in the LDL and VLDL but not HDL fractions,[29] whereas rabbits supplemented with up to 2100 IU of vitamin E per week for a period of 9 weeks became hypocholesterolemic.[30] Elevation in plasma cholesterol level was associated with reduction in cholesterol 7α-hydroxylase activity and, conversely, reduction in cholesterol level was associated with elevation in the hydroxylase activity.

The role of vitamin E as an in vivo antioxidant is well documented (see Reference 31 for a recent review). It serves to protect membrane lipids and proteins from the deleterious effects of such oxidizing agents as superoxide anion, hydrogen peroxide, and hydroxyl radical. One possible sensitive protein target is the liver microsomal cytochrome P-450 system. This is reduced when rabbits are depleted of vitamin E and is increased upon supplementation with the vitamin.[29,30] Other liver microsomal enzymes dependent upon cytochrome P-450 are also affected by lack of vitamin E, e.g., drug hydroxylating activities of rat liver microsomes.[32] Reduction in cytochrome P-450 content in liver microsomes of vitamin E-deficient animals may be due to degradation of the hemoprotein by lipid peroxidation as has been observed when liver microsomes are incubated in lipid peroxidation generating system.[33] However, the effect of vitamin E on microsomal hemoproteins may be a selective one since the level of another microsomal hemoprotein, cytochrome b5, was not similarly affected.[29,30]

Nevertheless, the reduction in cytochrome P-450 level in vitamin E deficiency cannot be the sole cause of reduction in cholesterol 7α-hydroxylase activity since the reduction in cytochrome P-450 was relatively less than that of the hydroxylase activity. There exists the possibility that the enzyme has also been inactivated by a protein phosphatase released from lysosomes known to be leaky in vitamin E-deficient animals.[34]

In addition to its role as an antioxidant, vitamin E may act to stabilize membranes by the specific interactions between the phytyl side chains of tocopherol and fatty acyl chains of arachidonic acid, and this mechanism has been invoked to explain the changes in microsomal membrane-bound enzymes and drug-metabolizing enzyme systems following vitamin E depletion.[35] However, it is difficult to predict *a priori* whether conformational changes to a biological membrane lead to inhibition or activation of membrane-bound enzymes. Changes to the Arrhenius profiles (plot of log enzyme activity vs. reciprocal of absolute temperature) of a number of membrane-bound enzymes in liver and muscle of vitamin E-deficient rabbits have previously been documented.[36] These phenomena have been interpreted as reflecting changes to the microenvironment of membrane-bound enzymes. However, further studies on sarcoplasmic reticulum Ca-ATPase (EC 3.6.1.3) have recently indicated that the protein component is damaged as a result of a peroxidative process occurring in the vitamin E-deficient animal.[37,38]

CONCLUSIONS

What relevance do the animal studies have to the situation in man? There is an increasing evidence linking high cholesterol levels with enhanced risk of atherosclerosis and coronary heart disease.[39] Observations on the redistribution of cholesterol among the lipoprotein fractions with elevation in (beneficial) HDL-cholesterol level as a result of vitamin E supplementation have not been consistent.[40-44] This situation could be attributed to differences in such parameters as age, body weight, and HDL-cholesterol status among the test groups. On the other hand, administration of tocopheryl nicotinate, used widely in Japan for improving blood circulation, to patients with hypercholesterolemia has resulted in a significant lowering of serum cholesterol level, with improvement of subjective symptoms and no reports of side effects.[45]

In summary, activity of cholesterol 7α-hydroxylase, the regulatory enzyme in the pathway of cholesterol catabolism to bile acids, is related to vitamin E status. Depletion of vitamin E results in reduction of cholesterol 7α-hydroxylase activity and supplementation with vitamin E is able to enhance enzyme activity. This relationship is probably linked to the level of cytochrome P-450, a necessary component of the hydroxylation reaction and which is subjected to peroxidative damage in the absence of biological antioxidant. Other contributing factors include dephosphorylation of the enzyme and perturbations resulting from membrane conformational destabilization. Changes in the activity of cholesterol 7α-hydroxylase can affect liver cholesterol level, which in turn regulates both its anabolic pathway and uptake from plasma, and hence the plasma cholesterol level. This chain of events provides a testable mechanism for the hypocholesterolemic effect of vitamin E.

REFERENCES

1. **Alfin-Slater, R. B. and Morris, R. S.,** Vitamin E and lipid metabolism, *Adv. Lipid Res.*, 1, 183, 1963.
2. **Eskelson, C. D., Jacobi, H. P., and Fitch, D. M.,** Some effects of vitamin E deficiency on in vitro cholesterol metabolism, *Physiol. Chem. Phys.*, 5, 319, 1973.
3. **Westrope, K. L., Miller, R. A., and Wilson, R. B.,** Vitamin E in a rabbit model of endogenous hypercholesterolemia and atherosclerosis, *Nutr. Rep. Int.*, 25, 83, 1982.
4. **Albarracin, I., Lassaga, F. E., and Caputto, R.,** Changes of gangliosides and other lipids in skeletal muscle from rabbits with experimental dystrophy, *J. Lipid Res.*, 15, 89, 1974.
5. **Liewsaree, P., Yuthavong, Y., Wilairat, P., and Komaratat, P.,** Protein and lipid composition of sarcoplasmic reticulum from dystrophic muscle of vitamin E-deficient rabbits, *Nutr. Rep. Int.*, 22, 853, 1980.
6. **Heinrich, M. R. and Mattill, H. A.,** Lipids of muscle and brain in rats deprived of tocopherol, *Proc. Soc. Exp. Biol. Med.*, 52, 344, 1943.
7. **Dam, H., Prange, I., and Sondergaard, E.,** Muscular degeneration (white striation of muscles) in chick reared on vitamin E-deficient low-fat diets, *Acta Pathol. Microbiol. Scand.*, 31, 172, 1952.
8. **Blaxter, K. L. and Wood, W. A.,** The nutrition of the young Ayrshire calf. V. The nutritive value of cow whole milk, *Br. J. Nutr.*, 6, 1, 1952.
9. **Deuel, H. J., Jr., Cox, J. R., Alfin-Slater, R. B., and Ershoff, B. H.,** The relationship between vitamin E and cholesterol metabolism, *Biochem. J.*, 61, xiv, 1955.
10. **Bloor, W. R.,** Effect of activity on the phospholipid and cholesterol content of muscle, *J. Biol. Chem.*, 119, 451, 1937.
11. **Dietschy, J. M. and Wilson, J. D.,** Regulation of cholesterol metabolism, *N. Engl. J. Med.*, 282, 1128, 1970.
12. **Gaylor, J. L.,** Microsomal enzymes of sterol biosynthesis, *Adv. Lipid Res.*, 10, 89, 1972.
13. **Redwell, V. W., Nordstrom, J. L., and Mitschelen, J. J.,** Regulation of HMG-CoA reductase, *Adv. Lipid Res.*, 14, 1, 1976.
14. **Brown, M. S. and Goldstein, J. L.,** A receptor-mediated pathway for cholesterol homeostasis, *Science*, 232, 34, 1986.

15. **Danielson, H. and Sjovall, J.,** Bile acid metabolism, *Annu. Rev. Biochem.*, 44, 233, 1975.
16. **Myant, N. B. and Mitropoulos, K. A.,** Cholesterol 7α-hydroxylase, *J. Lipid Res.*, 18, 135, 1977.
17. **Boyd, G. S., Alexander, M. G., and Lawson, M. E.,** Studies on rat liver microsomal cholesterol 7α-hydroxylase, *Eur. J. Biochem.*, 37, 334, 1973.
18. **Chiang, J. Y. L., Malmer, M., and Hutterer, F.,** A form of rabbit liver cytochrome P-450 that catalyzes the 7α-hydroxylation of cholesterol, *Biochim. Biophys. Acta*, 750, 291, 1983.
19. **Hulcher, F. H. and Margolis, R. D.,** Rate limiting, diurnal activity of hepatic microsomal cholesterol 7α-hydroxylase in pigeons with high serum cholesterol, *Biochim. Biophys. Acta*, 712, 242, 1982.
20. **Danielsson, H., Einarsson, K., and Johansson, G.,** Effect of biliary drainage on individual reactions in the conversion of cholesterol to taurocholic acid, *Eur. J. Biochem.*, 2, 44, 1967.
21. **Mitropoulos, K. A. and Balasubramanian, S.,** The role of glucocorticoids in the regulation of the diurnal rhythm of hepatic β-hydroxy-β-methylglutaryl-coenzyme A reductase and cholesterol 7α-hydroxylase, *Biochem. J.*, 160, 49, 1976.
22. **Lakshmanan, M. R. and Veech, R. L.,** Short- and long-term effects of ethanol administration *in vivo* on rat liver HMG-CoA reductase and cholesterol 7α-hydroxylase activities, *J. Lipid Res.*, 18, 325, 1977.
23. **Nicolau, G., Shefer, S., Salen, G., and Mosbach, E. H.,** Determination of hepatic 7α-hydroxylase activity in man, *J. Lipid Res.*, 15, 146, 1974.
24. **Gielen, J., Van Cantfort, J., Robaye, B., and Renson, J.,** Rat liver cholesterol 7α-hydroxylase. III. New results about its circadian rhythm, *Eur. J. Biochem.*, 55, 41, 1975.
25. **Sanghvi, A., Grassi, E., Warty, V., Diven, W., Wight, C., and Lester, R.,** Reversible activation-inactivation of cholesterol 7α-hydroxylase possibly due to phosphorylation-dephosphorylation, *Biochem. Biophys. Res. Commun.*, 103, 886, 1981.
26. **Goodwin, C. D., Cooper, B. W., and Margolis, S.,** Rat liver cholesterol 7α-hydroxylase: modulation of enzyme activity by changes in phosphorylation state, *J. Biol. Chem.*, 257, 64469, 1982.
27. **Scallen, T. J. and Sanghvi, A.,** Regulation of three key enzymes in cholesterol metabolism by phosphorylation/dephosphorylation, *Proc. Natl. Acad. Sci. U.S.A.*, 80, 2477, 1983.
28. **Tang, P. M. and Chiang, J. Y. L.,** Modulation of reconstituted cholesterol 7α-hydroxylase by phosphatase and protein kinase, *Biochem. Biophys. Res. Commun.*, 134, 797, 1986.
29. **Chupukcharoen, N., Komaratat, P., and Wilairat, P.,** Effects of vitamin E deficiency on the distribution of cholesterol in plasma lipoproteins and the activity of cholesterol 7α-hydroxylase in rabbit liver, *J. Nutr.*, 115, 468, 1985.
30. **Komaratat, P., Chupukcharoen, N., and Wilairat, P.,** Effect of vitamin E on cholesterol plasma lipoprotein distribution and metabolism in rabbit, *Int. J. Vitam. Nutr. Res.*, 55, 167, 1985.
31. **Burton, G. W., Cheeseman, K. H., Doba, T., Ingold, K. U., and Slater, T. F.,** Vitamin E as an antioxidant *in vitro* and *in vivo*, in *Biology of Vitamin E*, Ciba Foundation Symp. 101, Porter, R. and Whelan, J., Eds., Pitman, London, 1983, 4.
32. **Carpenter, M. P. and Howard, C. N.,** Vitamin E, steroids, and liver microsomal hydroxylations, *Am. J. Clin. Nutr.*, 27, 966, 1974.
33. **Levin, W., Lu, A. Y. H., Jacobson, M., Kuntzman, R., Poyer, J. L., and McCay, P. B.,** Lipid peroxidation and the degradation of cytochrome P-450 heme, *Arch. Biochem. Biophys.*, 158, 842, 1973.
34. **Zalkin, H., Tappel, A. L., Caldwell, K. A., Shibko, S., Desai, I. D., and Holliday, T. A.,** Increased lysosomal enzymes in muscular dystrophy of vitamin E-deficient rabbits, *J. Biol. Chem.*, 237, 2678, 1962.
35. **Diplock, A. T.,** Possible stabilizing effect of vitamin E on microsomal, membrane-bound, selenide-containing proteins and drug-metabolizing enzymes systems, *Am. J. Clin. Nutr.*, 27, 995, 1974.
36. **Vajanamarhutue, C., Wilairat, P., and Komaratat, P.,** Effects of vitamin E deficiency on the activities of lipid-requiring enzymes in rabbit liver and muscle, *J. Nutr.*, 109, 848, 1979.
37. **Patipaporn, K., Wilairat, P., and Komaratat, P.,** Altered property of sarcoplasmic Ca-ATPase from vitamin E-deficient dystrophic rabbit is associated with the protein and not the lipid component, *Biochem. Int.*, 6, 335, 1983.
38. **Promkhatkaew, D., Komaratat, P., and Wilairat, P.,** Ascorbic acid-Fe^{2+} treatment mimics effect of vitamin E deficiency on sarcoplasmic Ca-ATPase of rabbit muscle, *Biochem. Int.*, 10, 937, 1985.
39. **Keys, A.,** *Seven Countries: A Multivariate Analysis of Death and Coronary Heart Disease*, Harvard University Press, Cambridge, Mass., 1980, 1.
40. **Herman, W. J., Ward, K., and Faucett, J.,** The effect of tocopherol on high density lipoprotein cholesterol, *Am. J. Clin. Pathol.*, 72, 848, 1979.
41. **Hatam, L. J. and Kayden, H. J.,** The failure of α-tocopherol supplementation to alter the distribution of lipoprotein cholesterol in normal and hyperlipoproteinemic persons, *Am. J. Clin. Pathol.*, 76, 122, 1981.
42. **Schwartz, P. L. and Rutherford, I. M.,** The effect of tocopherol on high-density lipoprotein cholesterol, *Am. J. Clin. Pathol.*, 76, 843, 1981.
43. **Kesaniemi, Y. A. and Grundy, S. M.,** Lack of effect of tocopherol on plasma lipids and lipoproteins in man, *Am. J. Clin. Nutr.*, 36, 224, 1982.

44. **Stampfer, M. J., Willett, W., Castelli, W. P., Taylor, J. O., Fine, J., and Hennekens, C. H.,** Effect of vitamin E on lipids, *Am. J. Clin. Pathol.,* 79, 714, 1983.
45. **Wakasa, K., Ohsaka, A., Shibue, K., and Imanura, M.,** Clinical and experimental studies on the effects of α-tocopheryl nicotinate on cholesterol metabolism, *Vitamins (Jpn),* 34, 1, 1966.

FREE RADICAL LIPID PEROXIDATION AND VITAMIN E IN LIVER INJURY

Toshikazu Yoshikawa and Motoharu Kondo

INTRODUCTION

The gastrointestinal (GI) tract is designed to digest and absorb almost anything and everything. There is a large blood flow leading to and away from the absorptive surfaces of the intestines, which before being distributed to the body as a whole is directed into the liver. The liver itself contains an active detoxification system. This system can convert a wide variety of foreign substances, drugs, and poisons to derivatives more readily excreted than the original parent compound. Unfortunately, in some instances this detoxification system works in the inverse manner; it may convert a relatively harmless xenobiotic into a toxic substance.

Free radical mechanisms of liver injury have been proposed for a number of xenobiotic compounds, i.e., synthetic chemicals foreign to biological system, particularly halogenated organic compounds. These proposed hepatotoxic mechanisms generally involve biotransformation of the xenobiotic to a reactive free radical intermediate capable of injurious interaction with cell macromolecules. Activation of xenobiotics to free radical intermediates in a biological system generally requires an enzymatic "molecule-assisted homolysis" process involving some bond formation between the xenobiotic substrate and the responsible enzyme component which probably acts as an electron donor.[1]

Biological systems are also equipped with endogenous defense system to protect cell components against attack by reactive species. The liver cell has a relatively rich supply of both water-soluble antioxidants, chiefly reduced glutathione, and lipid-soluble antioxidants, chiefly vitamin E and ubiquinone. In vitro studies have demonstrated that vitamin E plays a major role in protecting endoplasmic reticulum components from injury by free radical species generated by the endogenous NADPH lipid peroxidation system.[2]

FREE RADICAL DAMAGE IN LIVER

Butler's initial proposal[3] in 1961 that the hepatotoxicity of CCl_4 was related to its biotransformation to a free radical was prompted by his demonstration that CCl_4 was reduced in vivo and in vitro to $CHCl_3$. He proposed that the free radical intermediate of CCl_4 could cause serious biochemical consequences by reacting with tissue components. These secondary mechanisms could involve the generation of toxic products arising directly from CCl_4 metabolism or from peroxidative degeneration of membrane lipids. These are trichloromethyl ($\cdot CCl_3$), trichloromethylperoxy ($\cdot OOCCl_3$), and chlorine ($\cdot Cl$) free radicals, as well as phosgene and aldehydic products of lipid peroxidation.

Halothane anesthesia induces liver damage. Reaction products consistent with addition and abstraction reactions of a halothane free radical with lipids have been detected following halothane administration in vivo. Generation of halothane free radicals within the membranes of the endoplasmic reticulum could account for early biochemical, functional, and ultrastructural aberrations of the organelle associated with injury-producing exposure to halothane.[4]

Therapy with a number of clinically effective hydrazine derivatives has been associated with occasional and sometimes fatal liver injury. It was reported that the highly effective tuberculostatic agent isoniazid and the antidepressant iproniazid form toxic free radical intermediates.[5,6]

Ethanol is an example of a chemical hepatotoxin whose primary site of injury apparently involves mitochondria. MacDonald[7] reported increased lipid peroxidation products in both mitochondrial and microsomal lipids of animals given ethanol.

VITAMIN E AND LIVER INJURY

Vitamin E was discovered in rats first as an antisterility vitamin factor. Ever since extensive work has been done to clarify its biological action, and it has been introduced into clinical use, too. Many reports have been published concerning the relationship between liver disease and vitamin E;[8-10] it has been reported that blood levels of vitamin E are depressed in certain types of liver disease and that vitamin E produces a therapeutic effect on liver disease. Much work has been done in animal experiments with vitamin E.

In vitamin E deficiency, various kinds of liver diseases become aggravated, and vitamin E produces a protective effect on the liver. Spontaneous necrosis sometimes associated with hemorrhage develops in rats raised on diets low in vitamin E and sulfur-containing amino acids.[11-13]

EFFECTS OF VITAMIN E ON CARBON TETRACHLORIDE-INDUCED HEPATOTOXICITY

Liver cell injury induced by carbon tetrachloride (CCl_4) involves initially the metabolism of CCl_4 to trichloromethyl free radical by the mixed-function oxidase system of the endoplasmic reticulum.[14,15] It is postulated that secondary mechanisms link CCl_4 metabolism to the widespread disturbances in hepatocyte function. These secondary mechanisms could involve the generation of toxic products arising directly from CCl_4 metabolism or from peroxidative degeneration of membrane lipids.[16] The liver damage and the increase in thiobarbituric acid (TBA) reactants induced by CCl_4 were remarkable in vitamin E-deficient rats and only slight in those supplemented with vitamin E (Figures 1 and 2).[17,18] These findings seems to bear out the postulation that vitamin E diminishes lysosomal membrane destruction by CCl_4 and protects the liver through its inhibitory action on lipid peroxidation.

EFFECTS OF VITAMIN E ON D-GALACTOSAMINE HEPATITIS

Liver injury produced by galactosamine resembles that seen in viral hepatitis.[15] The liver cell injury appears to be due to a rapid and profound fall in intracellular content of uridine triphosphate and uridine diphosphate glucose (UDP glucose). Adenosine triphosphate (ATP) levels are unchanged.[15] The free radical reaction is not primarily the cause of liver impairment in D-galactosamine-induced hepatitis. Even in this type of hepatitis, once the liver is impaired by a toxic substance such as D-galactosamine, lipid peroxidation can occur, giving rise to secondary and tertiary damage to the liver. The effect of vitamin E on D-galactosamine-induced liver impairment was comparable to the results obtained in CCl_4-induced liver disorder; liver disorders were remarkable in vitamin E-deficient rats and slight in those suplemented with vitamin E (Figure 3).

CONCLUSIONS

Glatskin et al.[8] proposed that as the vitamin E requirement of tissue was increased in liver disease, patients with liver cirrhosis showed better vitamin E absorption than healthy persons. They further stated in their reports that the decrease in serum level of vitamin E in liver disease was due to increased destruction of vitamin E in the small intestine. We reported that serum levels of vitamin E were depressed in various liver diseases (e.g., viral hepatitis, alcoholic hepatitis, and fulminant hepatitis) (Figure 4).[10] There was a significant correlation between serum levels of vitamin E and β-lipoprotein (Figure 5), cholesterol (Figure 6), and phospholipid (Figure 7). McCormic et al.[19] reported that vitamin E was bound to lipoprotein in the blood and that the blood level of β-lipoprotein played an important role in the transport of vitamin E. Little vitamin E is present in sera from abetalipoproteinemia patients,[20] and serum levels of vitamin E are slightly higher than in normal in patients with β-lipoproteinemia.[21] These findings suggest that the diminished serum vitamin E in patients with liver disease is ascribable to the depression in blood levels of β-lipoprotein that results from liver disorders, because the liver is the major supply source of endogeneous β-lipoprotein. Furthermore, serum levels of vitamin E are influenced by such factors as intake, absorption, and catabolism of vitamin E.

It is obvious that vitamin E is closely related to liver diseases. In the hepatitis patients with low serum levels of vitamin E, free radical lipid peroxidation might easily damage the hepatic cells. For protection against free radical damage in these types of hepatitis, vitamin E might be necessary to recover the loss.

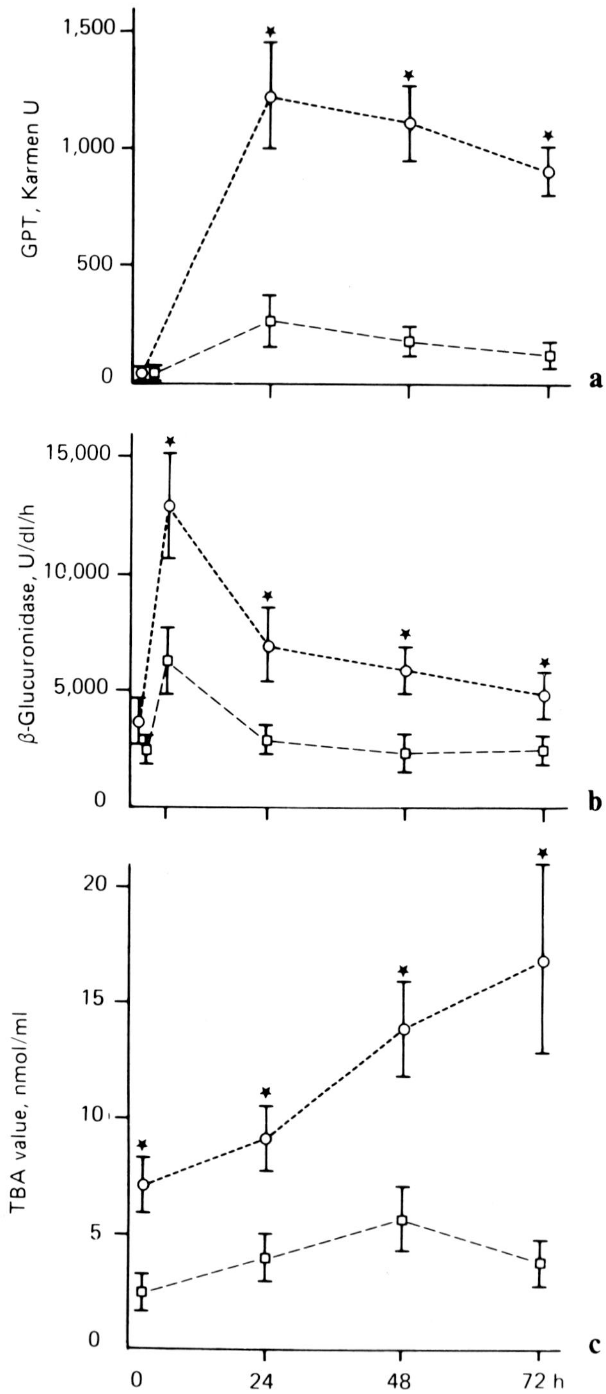

FIGURE 1. Serum levels of GPT (a), β-glucuronidase (b), or TBA-reactive substances (c) before and after injection of 0.5 mℓ/kg of CCl_4. In vitamin E-deficient rats, these levels were remarkably elevated after the injection: ○, vitamin E-deficient rats; □, vitamin E-supplemented rats. Mean values ± SD are shown for ten rats; *, p <0.001 different from the values of vitamin E-supplemented group by Sudent's t test. (From Yoshikawa, T. et al., *Digestion*, 25, 222, 1982. With permission.)

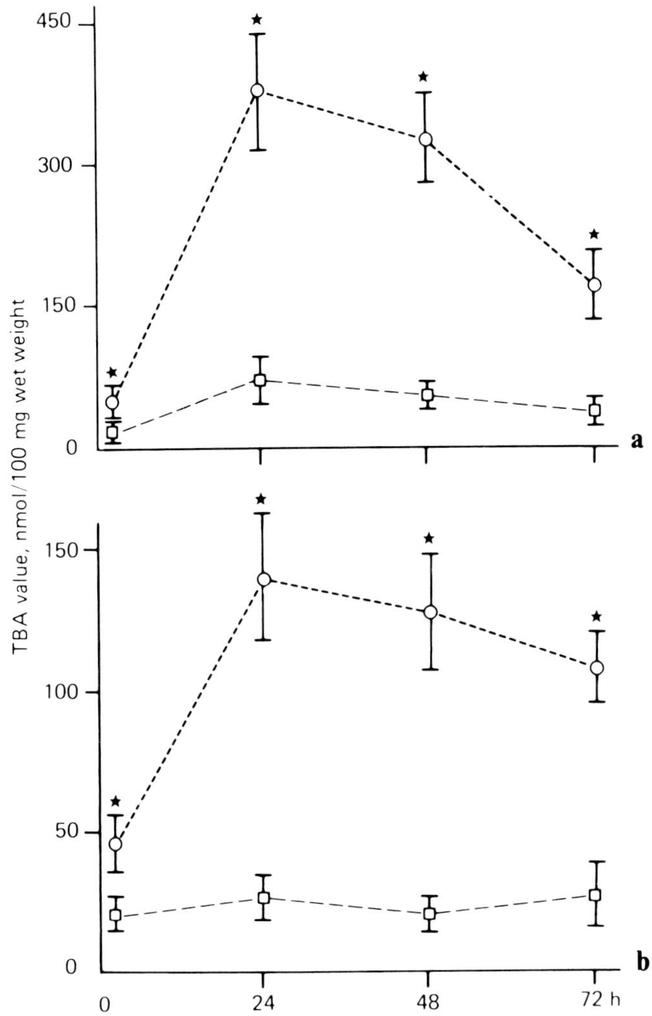

FIGURE 2. Liver tissue levels of TBA-reactive substances before and after injection of 0.5 mℓ/kg of CCl_4 (a) or 750 mg/kg of D-GalN (b). The TBA-reactive substances were elevated markedly after injection of CCl_4 or D-GalN, while vitamin E-supplemented rats only showed a slight change: ○, vitamin E-deficient rats; □, vitamin E-supplemented rats. Mean values ±SD are shown for ten rats; *, $p < 0.001$ different from the values of vitamin E-supplemented group by student's *t* test. (From Yoshikawa, T. et al., *Digestion*, 25, 222, 1982. With permission.)

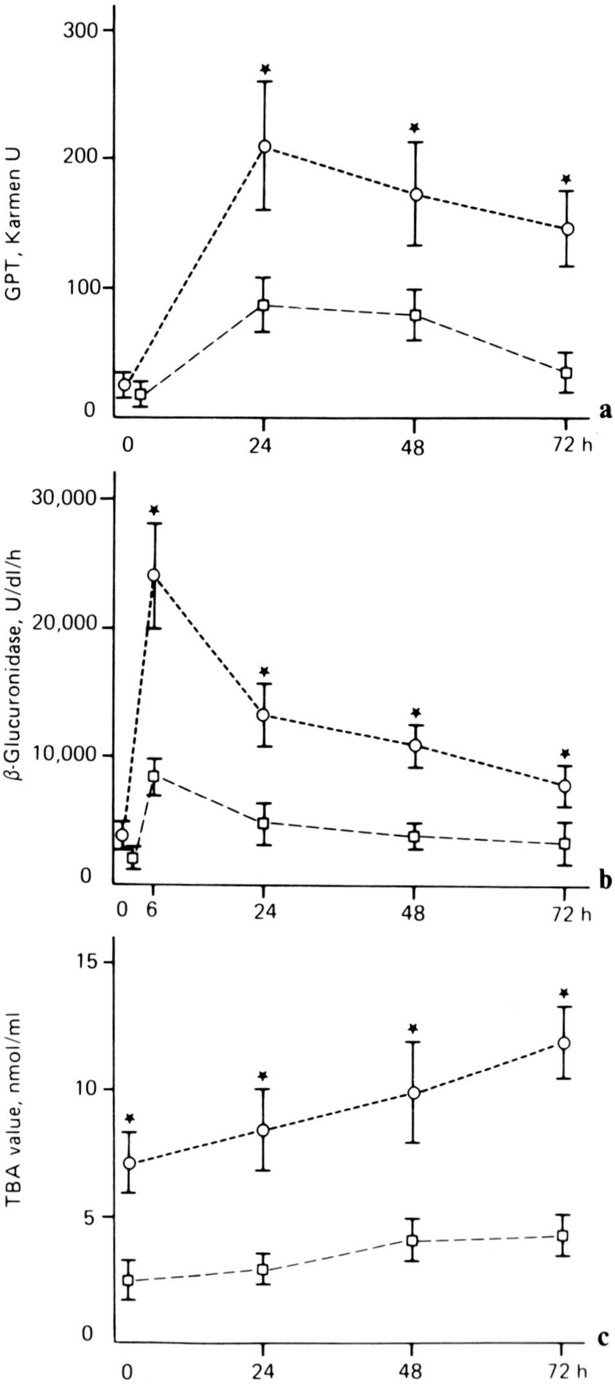

FIGURE 3. Serum levels of GPT (a), β-glucuronidase (b), or TBA-reactive substances (c) before and after injection of 750 mg/kg of D-GalN. In vitamin E-deficient rats, these levels were remarkably elevated after the injection: ○, vitamin E-deficient rats; □, vitamin E-supplemented rats. Mean values ±SD are shown for ten rats; *, $p < 0.001$ different from the values of vitamin E-supplemented group by student's *t* test. (From Yoshikawa, T. et al., *Digestion*, 25, 222, 1982. With permission.)

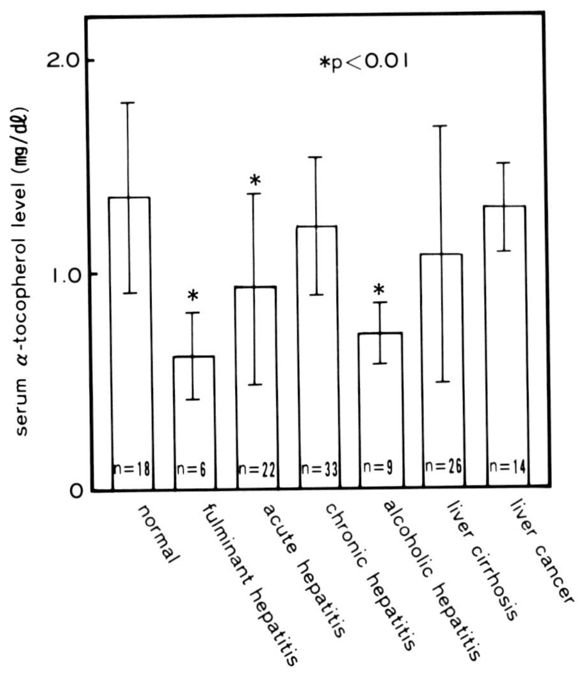

FIGURE 4. Serum α-tocopherol level in liver diseases.

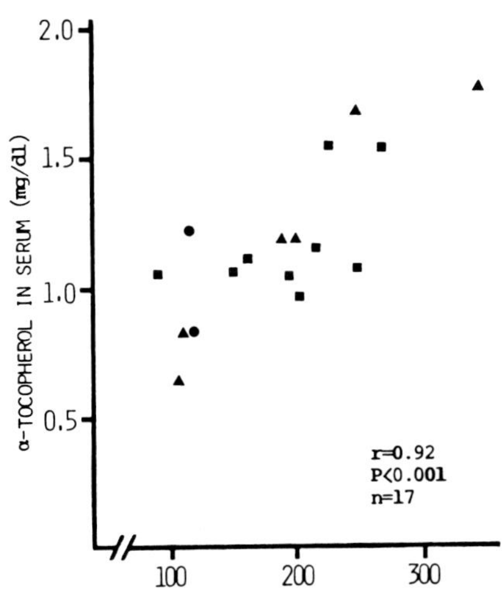

FIGURE 5. Relationship between α-tocopherol and β-lipoprotein: ●, acute hepatitis; ■, chronic hepatitis; and ▲, liver cirrhosis. (From Yoshikawa, T. et al., *Acta Vitaminol. Enzymol.*, 4, 311, 1982. With permission.)

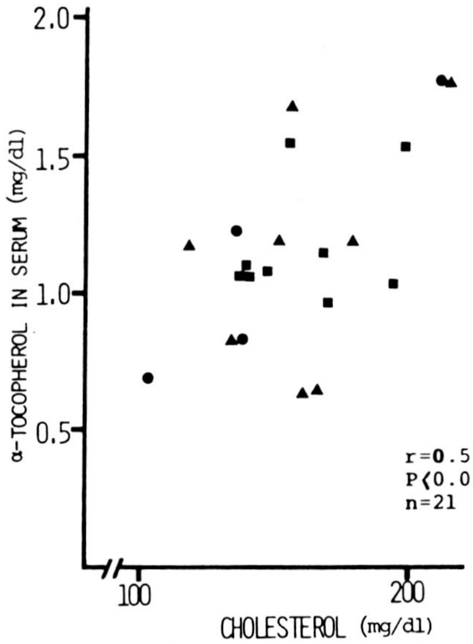

FIGURE 6. Relationship between α-tocopherol and cholesterol: ●, acute hepatitis; ■, chronic hepatitis; and ▲, liver cirrhosis. (From Yoshikawa, T. et al., *Acta Vitaminol. Enzymol.*, 4, 311, 1982. With permission.)

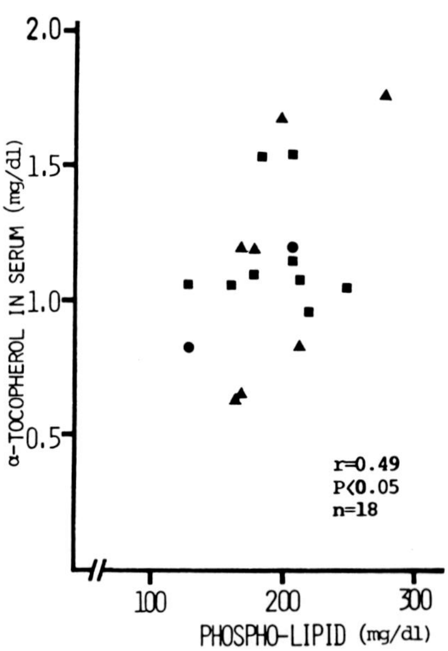

FIGURE 7. Relationship between α-tocopherol and phospholipid: ●, acute hepatitis; ■, chronic hepatitis; and ▲, liver cirrhosis. (From Yoshikawa, T. et al., *Acta Vitaminol. Enzymol.*, 4, 311, 1982. With permission.)

REFERENCES

1. **Reynolds, E. S. and Moslen, M. T.,** Free radical damage in liver, in *Free Radicals in Biology,* Pryor, W. A., Eds., Academic Press, 1980, 49.
2. **McCay, P. B., Pfeifer, P. M., and Stipe, W. H.,** Vitamin E protection of membrane lipids during electron transport functions, *Ann. N.Y. Acad. Sci.,* 203, 62, 1972.
3. **Butler, T. C.,** Reduction of carbon tetrachloride in vivo and reduction of carbon tetrachloride and chloroform in vitro by tissues and tissue constituents, *J. Pharmacol. Exp. Ther.,* 134, 311, 1961.
4. **Reynolds, E. S. and Moslen, M. T.,** Halothane hepatotoxicity: enhancement by polychlorinated biphenyl pretreatment, *Anesthesiology,* 47, 19, 1977.
5. **Mitchell, J. R., Zimmerman, H. J., Ishak, K. G., Thorgeirsson, U. P., Timbrell, J. A., Snodgrass, E. R., and Nelson, S. D.,** Isoniazid liver injury: clinical spectrum, pathology, and probably pathogenesis, *Ann. Intern. Med.,* 84, 181, 1976.
6. **Nelson, S. D., Mitchell, J. R., Timbrell, J. A., Snodgrass, W. R., and Corcoran, III,** Isoniazid and iproniazid: activation of metabolites to toxic intermediates in man and rat, *Science,* 193, 901, 1976.
7. **MacDonald, C. M.,** The effects of ethanol on hepatic lipid peroxidation and on the activities of glutathione reductase and peroxidase, *FEBS Lett.,* 35, 227, 1973.
8. **Klatskin, G. and Krehl, W. A.,** The significance of the plasma tocopherol concentration and of tocopherol tolerance tests in liver disease, *J. Clin. Invest.,* 29, 1528, 1950.
9. **Latner, A. L.,** Regime for treatment of severe and acute liver diseases, *Br. Med. J.,* 2, 748, 1950.
10. **Yoshikawa, T., Takemura, S., and Kondo, M.,** Alpha-tocopherol level in liver diseases, *Acta Vitaminol. Enzymol.,* 4, 311, 1982.
11. **Gyorgy, P. and Goldblatt, H.,** Further observations on production and prevention of dietary hepatic injury in rats, *J. Exp. Med.,* 89, 245, 1949.
12. **Hove, E. L., Copeland, D. H., and Salmon, W. D.,** Fatal vitamin E deficiency disease in rats characterized by massive lung hemorrhage and liver necrosis, *J. Nutr.,* 39, 397, 1949.
13. **Schwarz, K.,** Dietetic hepatic injuries and mode of action of tocopherol, *Ann. N.Y. Acad. Sci.,* 52, 225, 1949.
14. **Rechnagel, R. O.,** Carbon tetrachloride hepatotoxicity, *Pharmacol. Rev.,* 19, 145, 1967.
15. **Recknagel, R. O., Glende, E. A., Jr., and Hruszkewycz, A. M.,** Chemical mechanisms in carbon tetrachloride toxicity, in *Free Radicals in Biology,* Vol. 3, Pryor, W. A., Ed., Academic Press, New York, 1977, 97.
16. **Brattin, W. J., Glende, E. A., Jr., and Recknagel, R. O.,** Pathological mechanisms in carbon tetrachloride hepatotoxicity, *J. Free Radicals Biol. Med.,* 1, 17, 1985.
17. **Yoshikawa, T., Furukawa, Y., Murakami, M., Takemura, S., and Kondo, M.,** Effects of vitamin E on D-galactosamine-induced or carbon tetrachloride-induced hepatotoxicity, *Digestion,* 25, 222, 1982.
18. **Yoshikawa, T.,** Lipid peroxidation damage and its protection by vitamin E, *J. Kyoto Pref. Univ. Med.,* 92, 315, 1983.
19. **McCormic, E. C., Cornwell, D. G., and Brown, J. B.,** Studies on the distribution of tocopherol in human serum lipoproteins, *J. Lipid Res.,* 1, 221, 1960.
20. **Kayden, H. J., Silber, R., and Kossmann, C. E.,** Role of vitamin E deficiency in abnormal autohemolysis of acanthocytosis, *Trans. Assoc. Am. Physicians,* 78, 334, 1965.
21. **Fredrickson, D. S., Levy, R. L., and Lees, S.,** Fat transport in lipoproteins, *N. Engl. J. Med.,* 276, 34, 1967.

CYTOTOXIC ACTIONS OF ISOPROTERENOL IN CARDIAC CELLS: PROTECTIVE EFFECTS OF ANTIOXIDANTS

Kenneth Ramos and Daniel Acosta

OVERVIEW

The cellular mechanisms by which isoproterenol (ISO) causes alterations in calcium homeostasis that result in myocardial cell injury are not fully understood. Traditionally, excessive cardiac stimulation associated with channel-mediated calcium transport has been recognized as the primary mechanism of cardiotoxicity. More recently, however, a number of studies have suggested that oxidative by-products and/or activated oxygen species formed during the oxidation of ISO may play an important role as mediators of myocardial toxicity. Oxidative by-products may cause damage to critical cellular constituents and result in non-specific translocation of calcium across the sarcolemma. This concept is supported by studies which show that antioxidants prevent the oxidation of ISO and reduce the accumulation of calcium and subsequent manifestation of toxicity.

PROPOSED MECHANISMS OF ISO-INDUCED CARDIOTOXICITY

ISO is a synthetic catecholamine (i.e., substituted 3,4-dihydroxyphenylethylamine) with prominent cardiac effects. Its positive inotropic and chronotropic actions are due to activation of cardiac β-1-adrenergic receptors. Activation of β-receptors is associated with stimulation of the adenylate cyclase-cyclic adenosine monophosphate system which in turn results in an increase of the slow calcium inward current.[1,2] High concentrations of ISO or related catecholamines cause alterations in calcium homeostasis that result in extensive myocardial necrosis.[3,4] An increased myocardial calcium content is observed in vivo and in vitro after exposure to 30 mg/kg or 10^{-5} *M* ISO, respectively.[5] The abnormal calcium accumulation within myocardial cells may be due to excessive cardiac stimulation.[1,3] Pharmacological interventions which favor calcium influx potentiate the toxicity of catecholamines.[6] Although calcium plays an important regulatory role in various myocardial processes, the excessive translocation of calcium across the plasma membrane may alter the integrity and function of several membrane systems. Excessive calcium levels inside myocardial cells are associated with sequestration of calcium by various subcellular compartments. In the case of mitochondria, accumulation of large amounts of calcium results in deficient energy production.[7] This leads to failure of calcium regulatory mechanisms, induction of contracture, increased oxygen requirement, excessive ATP breakdown, and activation of degradative proteinases and phospholipases.[8]

ISO is a more potent cardiotoxin than other catecholamines.[9] This difference has been attributed to its effects on the coronary microcirculation. ISO, but not norepinephrine, causes vasodilation that may result in relative myocardial ischemia due to insufficient blood supply to sustain the enhanced cardiac stimulation.[10,11] This concept is supported by previous studies in which the infarct-like lesions induced by ISO were compared to the focal myocarditis induced by norepinephrine.[12]

In contrast to the concepts presented above, Dhalla and co-workers have proposed that oxidative metabolites of catecholamines, and not the parent compound, may be responsible for the production of cardiotoxicity.[13-15] This proposal is supported by studies which show that necrosis and hypercontractility occur upon perfusion of the isolated rat heart with spontaneously oxidized solutions of ISO.[13] In these studies, cardiotoxicity could not be demonstrated if the isolated hearts were perfused with fresh (i.e., unoxidized) ISO. Oxidative metabolites of catecholamines produce cellular damage and contractile failure in the isolated perfused heart.[14] Agents known to protect the heart against catacholamine cardiotoxicity can also be effective against the cardiac damage induced by adrenochrome, the product formed upon oxidative cyclization of adrenaline.[15] Rona[2] has emphasized that major differences exist between adrenochrome- and ISO-induced myocardial toxicity. However, comparisons between preformed oxidative metabolites and oxidizing solutions of catecholamines must be made with caution since multiple oxidation products are present in these solutions.

The major metabolic pathways for the degradation of catecholamines involve two enzyme systems, catechol-*O*-methyl transferase and monoamine oxidase. Catechol-*O*-methyl transferase catalyzes the formation of *O*-methyl derivatives. Monoamine oxidase activity results in the formation of products in which the ethanolamine side chain of the molecule undergoes oxidative deamination. Singal and co-workers[16] have proposed that high levels of catecholamines may exceed the capacity of these enzyme systems and lead to the formation of oxidative metabolites.

Other mechanisms may contribute to the production of cardiotoxicity. These include myocardial hypoxia,[17,18] mobilization of free fatty acids,[19] intracellular acidity,[20] platelet aggregation,[21] and changes in itermediary metabolism.[22,23] In this chapter the role of oxidative by-products and/or activated oxygen species in catecholamine-induced cardiotoxicity is examined. Specifically, the effects of agents which prevent the oxidation of the parent compound upon the manifestation of toxicity are discussed.

OXIDATION OF CATECHOLAMINES

Oxidative metabolites of catecholamines result from the spontaneous or enzymatically catalyzed oxidation of the parent compound. In direct oxidations, such as those catalyzed by metals, an aminochrome (indole-5,6-quinone) results from a one-step reaction involving the withdrawal of two electrons followed by immediate cyclization of the molecule.[24] Oxidation can also occur via the leukoadrenochrome pathway, a two-step process involving an internal redox reaction.[25] The first stage of this reaction is the oxidation of the parent catecholamine to the semiquinone radical which is readily converted to the open chain quinone. The quinone is an electron-deficient species and therefore is highly reactive with respect to nucleophile.[26] The most readily available nucleophile is the side chain amine group of the quinone itself. Hence, an intracyclization can take place at neutral pH resulting in the formation of leukoaminochrome. The leukoaminochrome is much more easily oxidized than the parent catechol and proceeds further to the aminochrome (Figure 1). Aminochromes are very reactive and thus may undergo rearrangement, reduction, oxidation, substitution, condensation product formation, addition product formation, and complex formation.[27] Strong nucleophiles can compete with the side chain moiety to prevent the intracyclization reaction.[28] Open chain quinones can also be activated to free radical forms by flavoproteins such as NADPH cytochrome P-450 reductase and NADH cytochrome c reductase.[29] These free radicals can then transfer a single electron to molecular oxygen to form superoxide and the regenerated quinone. This process is known as redox cycling.

Aqueous solutions of catecholamines autoxidize readily upon exposure to air. Autoxidation of catecholamines increases as a function of pH. Neutral solutions of catecholamines may undergo autoxidation in the absence of traces of heavy metals.[30] ISO can undergo oxidation by means of a self-propagating mechanism that generates highly reactive free radicals.[30-33] At pH 6 to 8, the rate of the reaction and the nature of the final products are dependent upon the catalysts and buffers present in the system.[27] After the reaction is initiated, the aminochrome itself serves as a catalyst. Superoxide anions present in the system can also act as initiators or propagators of the oxidation process.[34] The superoxide anion can abstract a proton from the catechol moeity to form catechol anions that can react with molecular

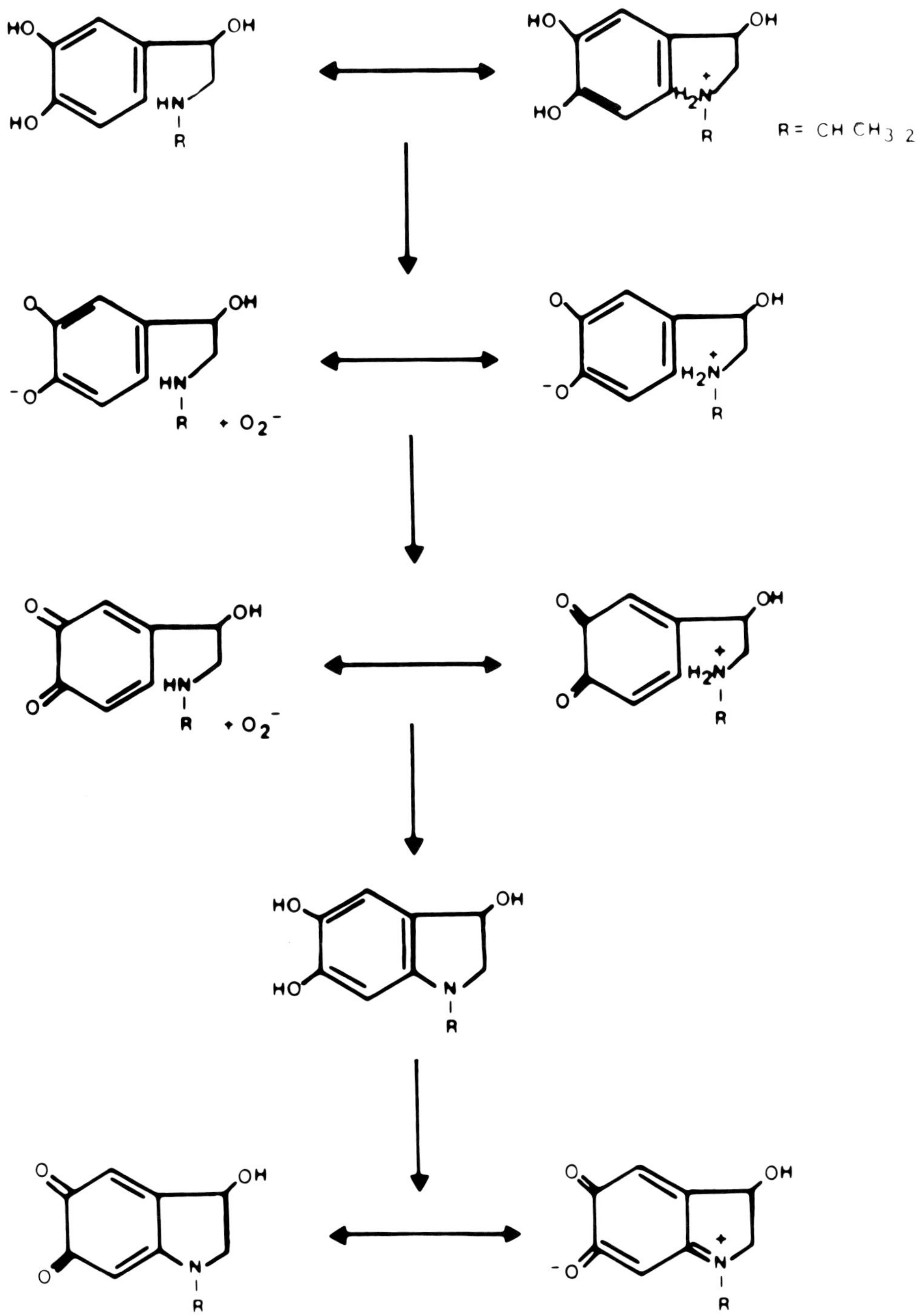

FIGURE 1. Oxidative pathways of catecholamines and related compounds.

oxygen to form hydrogen peroxide and fully oxidized products capable of covalent binding. Complexed transition metals such as those present in hemes and porphyrins also serve as pro-oxidants, although they act more slowly than do free metal ions.[35] Catecholamine solutions readily oxidize when incubated under physiological conditions with intact cells or membrane preparations. Solutions containing low concentrations of ISO (<100 nM) undergo oxidation in less than 10 sec at 37°C.[36]

Several reports have suggested that cardiac enzymes present in various subcellular compartments may catalyze the formation of aminochromes.[37-39] Green and Ritcher[39] have reported that the oxidation of adrenaline to adrenochrome is catalyzed by a cyanide-insensitive system present in heart and skeletal muscle. A cytochrome-indophenol oxidase system present in certain tissues has also been implicated in the formation of oxidative metabolites.[39] Catecholamine oxidase, an enzyme found in smooth muscle, catalyzes the oxidation of catecholamines to the respective aminochromes.[40] Hepatic fragments of endoplasmic reticulum are also capable of oxidizing catecholamines.[41]

Although several pathways for the oxidation of catecholamines have been proposed, levels of oxidative metabolites in vivo have never been demonstrated unambiguously. However, the presence of fluorescent adrenaline oxidation products associated with proteins in rabbit tissues has been reported.[27] Significant amounts of adrenaline oxidation products have been found in the liver, heart, and skeletal muscle of mice after X-ray radiation.[30] Aminochromes play an important role as intermediates in the in vivo formation of melanin pigments.[27] The adrenochrome-adrenaline system serves as a hydrogen carrier in mammalian skeletal muscle.[42] In view of their high chemical reactivity, aminochromes may have a transient existence in vivo. The formation of relatively stable addition products with molecules such as bisulfites and thiols could temporarily stabilize adrenochrome in vivo.[30]

Several oxidation intermediates of epinephrine metabolism appear to have the potential to cause severe peroxidative injury. A number of studies suggest that oxidative metabolites may have direct effects on the activity of membrane-bound enzymes. Takeo et al.[43] have suggested that adrenochrome specifically inhibits the activity of Na^+/K^+ ATPase in a dose-dependent manner. Other investigators have shown that an indole metabolite of epinephrine causes inhibition of actomyosin ATPase by binding to sulfhydryl groups associated with ATPase activity.[44,45]

Spontaneous peroxidative reactions are a normal occurrence in aerobic systems. Cells that utilize molecular oxygen generate highly reactive oxygen-free radicals. Free radicals are formed during intermediary metabolism and during the reduction of oxygen by enzymes of the electron transport system in mitochondria. These processes generate and leak superoxide anion and hydrogen peroxide. Enzymes that catalyze the incorporation of molecular oxygen to organic molecules, such as xanthine oxidase, aldehyde oxidase, monoamine oxidase, NADPH-cytochrome c reductase, NADPH cytochrome P-450 reductase, certain hydrolases, and several flavoproteins, generate superoxide.[46] Endogenous cellular antioxidants inactivate free radicals and protect aerobic cells from oxidant injury. Glutathione, glutathione reductase, glutathione peroxidase, catalase, tocopherol, retinol, ascorbic acid, and superoxide dismutase are key components of the antioxidant defense system.[46,47] Inhibition of these protective mechanisms results in enhanced sensitivity to free radical-induced cellular damage. Excessive production of free radicals may result in alterations in the biological activity of cellular macromolecules.

Oxidative by-products of catecholamines such as quinones and aminochromes may form covalent bonds with nucleophilic groups of macromolecules.[29] Such interactions may result in denaturation of molecules of vital importance for the structural and functional integrity of the cell. However, it must be considered that a correlation between covalent binding and cellular damage does not establish that the binding itself results in cytotoxicity. In addition to the generation of semiquinones, quinones, and aminochromes, the oxidation process results in the generation of hydrogen peroxide, superoxide, and hydroxyl radicals.[28] Excessive production of these free radicals may exceed the capacity of endogenous antioxidant mechanisms and render the cells susceptible to oxidative damage. Reducing agents such as sodium bisulfite and ascorbic acid prevent the oxidation of ISO.[33] If oxidative by-products of ISO are partly responsible for the production of toxicity, agents which prevent the oxidation of the parent compound should reduce or prevent the severity of the toxic response.

PROTECTIVE EFFECTS OF ASCORBIC ACID AND SODIUM BISULFITE

Accumulation of high calcium levels is the primary change observed after exposure to toxic concentrations of ISO. Calcium influx occurs as early as 1 to 2 min following ISO administration.[48,49] The pattern of calcium uptake induced by toxic concentrations of ISO in cultured myocardial cells shows a biphasic response.[48] The initial phase of calcium uptake is blocked selectively by agents which prevent calcium channel-mediated calcium transport (i.e., propranolol and verapamil). This phase is present when cardiac cells are exposed to nontoxic concentrations of ISO. Its occurrence is unaffected by ascorbic acid or sodium bisulfite, suggesting that antioxidants do not interfere with calcium channel-mediated events. In contrast, the second phase of calcium uptake is associated with the progressive accumulation of calcium and is only observed in the presence of toxic concentrations of ISO. This phase is not antagonized by propranolol or verapamil, but is significantly reduced by antioxidants. These observations suggest that the accumulation of calcium observed during the second phase is not related to channel-mediated calcium transport. The protective actions of antioxidants, but not verapamil or propranolol, during this phase suggest that toxic concentrations of ISO cause nonchannel mediated (nonspecific) calcium translocation across the sarcolemma. The excessive accumulation of calcium within myocardial cells may account for the loss of structural and functional integrity induced by toxic concentrations of ISO. Such an alteration would have three major consequences: (1) the entry of more calcium from the extracellular space into the myocyte; (2) alterations in mitochondrial, sarcoplasmic, and sarcolemmal function; and (3) the loss of critical intracellular components such as enzymes, cofactors, and ions.

Toxic concentrations of ISO cause mitochondrial injury, as reflected by an increase in mitochondrial membrane permeability (Figure 2). If ISO-induced calcium accumulation within myocardial cells results in mitochondrial injury, a reduction in the calcium availability to the cells should reduce the extent of mitochondrial injury. This concept is consistent with previous studies which show that the severity of ISO-induced myocardial necrosis is enhanced by modifications which promote calcium influx.[6] In our studies, the severity of ISO-induced mitochondrial injury is reduced under hypocalcemic (0.9 m*M*) but not normocalcemic (1.8 m*M*) conditions (Figure 3).

Yates et al.[15] have shown that propranolol and verapamil only partially protect the heart from adrenochrome-induced toxicity. Similar results have been obtained in our laboratory using cultured heart cells. Propranolol or verapamil do not completely prevent the mitochondrial damage induced by toxic concentrations of ISO (Figure 2). In contrast, antioxidants significantly reduce the extent and severity of mitochondrial injury.[50] These observations support the concept that ISO-induced mitochondrial injury is due to nonspecific translocation of calcium unrelated to calcium channels.

Lebedev et al.[51] have suggested that oxidative damage to membrane lipids results in increased calcium permeability. Primary products of lipid peroxidation have the ability to transport calcium across the lipid bilayer. Furthermore, oxidation products of ISO may interact with membrane lipids to cause a deterioration of calcium regulatory mechanisms. These changes, in turn, may result in an uncontrolled rise of calcium uptake into the cell. The ability of oxidative metabolites of catecholamines to cause peroxidative injury may be

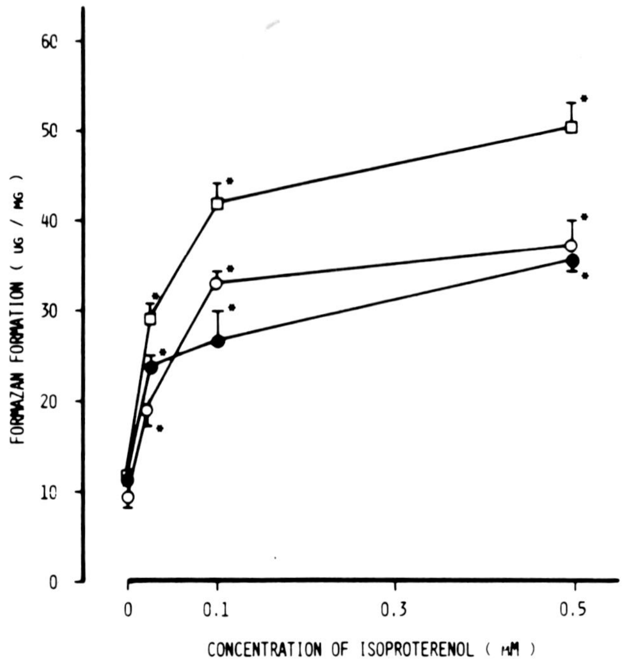

FIGURE 2. Mitochondrial membrane permeability of cultured heart myocytes exposed to ISO alone and in the presence of verapamil (1×10^{-5} *M*) or propranolol (1×10^{-6} *M* for 1.5 hr. Cultures concurrently exposed to ISO and verapamil or propranolol were compared to untreated controls and respective vehicle controls. Values represent the mean ± SEM; *, significantly different from their respective controls (p <0.01, n = 5); □, ISO alone; ●, ISO + propranolol; ○, ISO + verapamil.

influenced by the status of endogenous defense systems. Thus, an interaction between oxidation products of ISO and sulfhydryl donors normally present within myocardial cells may influence the extent and severity of cardiotoxicity. In this regard, a 20% reduction in the glutathione content of myocardial cells exposed to toxic concentrations of ISO has been observed.[50,52] This reduction might render other cellular macromolecules susceptible to oxidative damage. The extent of oxidative insult may be further aggravated by the formation of free radicals during the oxidation of ISO. Reducing agents can prevent the oxidation of catecholamines and reduce preformed quinones to their parent polyphenols. Additionally, agents such as sodium bisulfite can form an addition complex that deactivates oxidative metabolites of the parent catecholamine.

The concept that antioxidants may suppress the formation of oxidative metabolites is supported by previous studies in which (1) antioxidants suppressed free radical formation induced by ionizing radiation and carcinogenic compounds,[53] (2) α-tocopherol reduced the damage produced by hypoxia in isolated rabbit hearts,[54] and (3) ascorbic acid prevented acetaminophen hepatotoxicity.[55] These data suggest that antioxidants may prove useful in preventing various forms of toxic injury in which reactive intermediates may be involved.

CONCLUSIONS

Toxic concentrations of ISO cause a biphasic increase in myocyte calcium uptake. The initial phase may be a β-receptor component since it is antagonized by agents which interfere with channel-mediated calcium transport mechanisms. In contrast, the second phase of

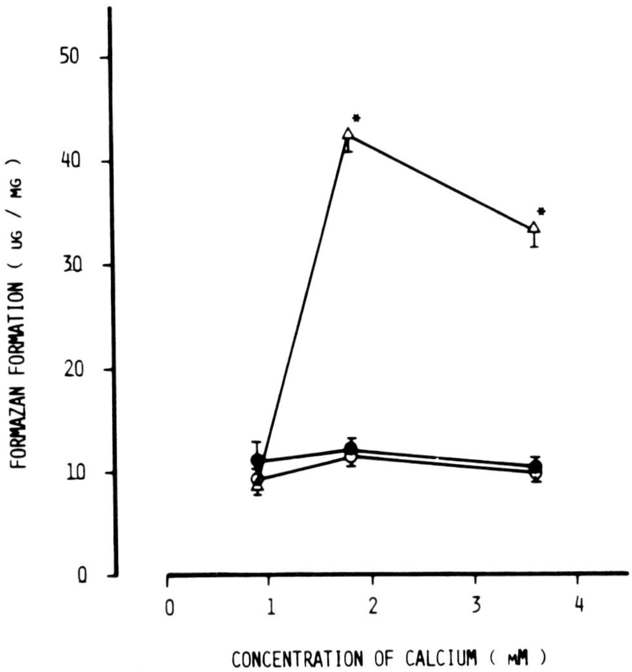

FIGURE 3. Effects of various concentrations of calcium on the mitochondrial permeability of cultured rat heart myocytes 1.5 hr after exposure to 2.4×10^{-7} or 1×10^{-4} *M* ISO. Values represent the mean ± SEM; *, significantly different from their respective controls ($p < 0.01$, n = 5); ○, untreated control; ●, 2.4×10^{-7} *M* ISO; △, 1×10^{-4} *M* ISO.

calcium uptake appears to be due to a nonspecific uptake process unrelated to calcium channels. This phase is antagonized by agents which prevent the oxidation of ISO, suggesting that oxidative damage to the plasma membrane may account for the altered calcium permeability. Oxidative by-products of ISO may interact with the sarcolemma to cause myocardial damage and loss of calcium homeostasis. Similar mechanisms may be involved in other forms of myocardial cell injury. Recently, a common causal pathway in the pathogenesis of catecholamine-induced toxicity and reperfusion injury has been proposed.[2] Since oxygen free radicals are thought to be involved in the pathogenesis of reperfusion injury, one can speculate that oxidative damage to myocardial cells results in the alterations of calcium homeostasis common to both forms of cellular injury.

REFERENCES

1. **Dhalla, N. S., Das, P. K., and Sharma, G. P.,** Subcellular basis of cardiac contractile failure, *J. Mol. Cell. Cardiol.,* 10, 363, 1978.
2. **Rona, G.,** Catecholamine cardiotoxicity, *J. Mol. Cell. Cardiol.,* 17, 291, 1985.
3. **Raab, W.,** Pathogenic significance of adrenaline and related substances in heart muscle, *Exp. Med. Surg.,* 1, 188, 1943.
4. **Rona, G., Chappel, C. I., Balazs, T., and Gaudry, R.,** An infarct-like myocardial lesion and other toxic manifestations produced by isoproterenol in rat, *Arch. Pathol.,* 67, 443, 1959.
5. **Godfraind, T., Ghysel-Burton, J., and Sturbois, X.,** Ionic changes evoked by isoprenaline in rat hearts in vivo and in vitro and their reduction by creatinol-*O*-phosphate, *Arzneim. Forsch.,* 29, 1468, 1979.

6. **Blaicklock, R. G., Hirsch, E. M., Dapson, S., Paino, B., and Lehr, D.,** Epinephrine-induced myocardial necrosis. Effects of aminophylline and adrenergic blockade, *Res. Commun. Chem. Pathol. Pharmacol.*, 34, 179, 1981.
7. **Dhalla, N. S., Ziegelhoffer, A., and Horrow, J. A. C.,** Regulatory role of membrane systems in heart function, *Can. J. Physiol. Pharamacol.*, 55, 1211, 1977.
8. **Nayler, W. G.,** The role of calcium in the ischemic myocardium, *Am. J. Pathol.*, 102, 262, 1981.
9. **Rona, G., Kahn, D. S., and Chappel, C. I.,** Studies on infarct-like myocardial necrosis produced by isoproterenol: a review, *Rev. Can. Biol.*, 22, 241, 1963.
10. **Chappel, C. I., Rona, G., Balazs, T., and Goudry, R.,** Comparison of cardiotoxic actions of certain sympathomimetic amines, *Can. J. Biochem. Physiol.*, 37, 35, 1959.
11. **Rona, G., Boutet, M., Huttner, I., and Peters, H.,** Pathogenesis of isoproterenol-induced myocardial alterations. Functional and morphological correlates, in *Recent Advances in Studies on Cardiac Structure and Metabolism,* Vol. 3, Dhalla, N., Ed., University Park Press, Baltimore, 1975, 507.
12. **Chappel, C. I., Rona, G., Balazs, T., and Gaudry, R.,** Comparison of cardiotoxic actions of certain sympathomimetic amines, *Can. J. Biochem.*, 37, 35, 1959.
13. **Yates, J. C. and Dhalla, N. S.,** Induction of necrosis and failure in the isolated perfused rat heart with oxidized isoproterenol, *J. Mol. Cell. Cardiol.*, 7, 807, 1975.
14. **Dhalla, N. S., Yates, J. C., Lee, S. L., and Singh, A.,** Functional and subcellular changes in the isolated rat heart perfused with oxidized isoproterenol, *J. Mol. Cell. Cardiol.*, 10, 31, 1978.
15. **Yates, J. C., Taam, G. M. L., Singal, P. K., Beamish, R. E., and Dhalla, N. S.,** Protection against adrenochrome-induced myocardial damage by various pharmacological interventions, *Br. J. Exp. Pathol.*, 61, 242, 1980.
16. **Singal, P. K., Dhillon, K. S., Beamish, R. E., Kapar, N., and Dhalla, N. S.,** Myocardial cell damage and cardiovascular changes due to i.v. infusion of adrenochrome in rats, *Br. J. Exp. Pathol.*, 63, 167, 1982.
17. **Yeager, J. C. and Iams, S. G.,** The hemodynamics of isoproterenol-induced cardiac failure in the rat, *Circ. Shock,* 8, 151, 1981.
18. **Balazs, T. and Ferrans, V. J.,** Cardiac lesions induced by chemicals, *Environ. Health Perspect.*, 20, 181, 1978.
19. **Kjekshus, J. K.,** Role of free fatty acids in catecholamine-induced cardiac necrosis, in *Recent Advances in Studies on Cardiac Structure and Metabolism,* Vol. 6, Fleckenstein, A. and Rona, G., Eds., University Park Press, Baltimore, 1975, 183.
20. **Mosinger, B., Stejskal, J., and Mosinger, B., Jr.,** Heart infarction-like effect induced by natural cath-echolamines in vitro, *Exp. Patho..*, 14, 157, 1977.
21. **Hoak, J. C., Warner, E. D., and Connor, W. E.,** New concept of levarterenol-induced acute myocardial necrosis, *Arch. Pathol.*, 87, 332, 1969.
22. **Joseph, X., Bloom, S., Pledger, G., and Balazs, T.,** Determinants of resistance to the cardiotoxicity of isoproterenol in rats, *Toxicol. Appl. Pharmacol.*, 69, 199, 1983.
23. **Balzs, T., Ohtake, S., and Noble, J. F.,** The development of resistance to the ischemic cardiopathic effect of isoproterenol, *Toxicol. Appl. Pharmacol.*, 21, 200, 1972.
24. **Harrison, W. H., Whisler, W. W., and Hill, B. J.,** Catecholamine oxidation and ionization properties indicated from the H^+ release, tritium exchange, and spectral changes which occur during ferricyanide oxidation, *Biochemistry,* 7, 3089, 1968.
25. **Borg, D. C.,** Transient free radical forms of hormones: EPR spectra from catechol and adrenochrome, *Proc. Natl. Acad. Sci.*, 53, 633, 1965.
26. **Graham, D. G., Tiffany, S. M., Bell, W. R., Jr., and Gutknecht, W. F.,** Autoxidation versus covalent binding of quinones as the mechanism of toxicity of dopamine, 6-hydroxydopamine, and related compounds toward C1300 neuroblastoma cells *in vitro, Mol. Pharmacol.*, 14, 644, 1978.
27. **Heacock, R. A.,** The chemistry of adrenochrome and related compounds, *Chem. Rev.*, 59, 181, 1959.
28. **Tse, D. C. S., McCreery, R. C., and Adams, R. N.,** Potential oxidative pathways of catecholamines, *J. Med. Chem.*, 19, 37, 1976.
29. **Kalyanaraman, B., Felix, C. C., and Sealy, R. C.,** Semiquinone anion radicals of catecholamines, catechol, estrogens, and their metal ion complexes, *Environ. Health Perspect.* 64, 185, 1985.
30. **Heacock, R. A. and Powell, W. S.,** Adrenochrome and related compounds, in *Progress in Medicinal Chemistry,* Vol. 9, West, G. B. and Ellis, G. P., Eds., Elsevier, New York, 1973, 275.
31. **Misra, H. P. and Fridovich, I.,** The role of superoxide anion in the autooxidation of epinephrine and a simple assay for superoxide dismutase, *J. Biol. Chem.*, 247, 3170, 1972.
32. **Bors, W., Michel, C., Saran, M., and Lengfelder, E.,** The involvement of oxygen radicals during the autoxidation of adrenalin, *Biochim. Biophys. Acta,* 540, 162, 1978.
33. **Graham, D. G.,** Oxidative pathways for catecholamines in the genesis of neuromelanin and cytotoxic quinones, *Mol. Pharmacol.*, 14, 633, 1978.

34. **Svengin, A. B. and Powis, G.,** Pulse radiolysis studies of antitumor quinones: radical lifetimes, reactivity with oxygen and one-electron reduction potentials, *Arch. Biochem. Biophys.*, 209, 119, 1981.
35. **Smith, A. K.,** Autooxidation and cell growth, *Art Sci. Tissue Culture,* 1, 1, 1982.
36. **Mahan, L. C. and Insel, P. A.,** Use of superoxide and catalase to protect catecholamines from oxidation in tissue culture studies, *Anal. Biochem.*, 136, 208, 1984.
37. **Valerino, D. M. and McCormack, J. J.,** Xanthine oxidase-mediated oxidation of epinephrine, *Biochem. Pharmacol.*, 20, 47, 1971.
38. **Gudbjarnason, S., Doell, B., and Oskarsdottir, G.,** Docosahexaenoic acid in cardiac metabolism and function, *Acta Biol. Med. Ger.*, 37, 777, 1976.
39. **Green, D. E., and Ritcher, D.,** Oxidation of epinephrine to adrenochrome in vitro, *Biochem. J.*, 31, 596, 1973.
40. **Inchiosa, M. A. and Rodriguez, I. B.,** Conditions affecting the purification and solution of a uterine epinephrine-oxidizing enzyme, *Biochem. Pharmacol.*, 18, 1883, 1969.
41. **Powis, G.,** Hepatic microsomal metabolism of epinephrine and adrenochrome by superoxide-dependent and -independent pathways, *Biochem. Pharamcol.*, 28, 83, 1979.
42. **Falk, J. E.,** The formation of hydrogen carriers by haematin-catalyzed peroxidations, *Biochemistry,* 44, 24, 1949.
43. **Takeo, S., Fliegel, L., Beamish, R. E., and Dhalla, N. S.,** Effects of adrenochrome on rat heart sarcolemmal ATPase activities, *Biochem. Pharmacol.*, 29, 559, 1979.
44. **Leonardi, R. G. and Inchiosa, M. A., Jr.,** Purification and substrate specificity of uterine catecholamine oxidase, *Biochem. Pharmacol.*, 21, 2804, 1972.
45. **Altiere, R. J. and Inchiosa, M. A., Jr.,** Concentration-dependent inhibition of myosin ATPase by an indole metabolite of epinephrine, *Life Sci.*, 26, 1523, 1980.
46. **Puglia, C. D. and Powell, S. R.,** Inhibition of cellular antioxidants: a possible mechanism of toxic cell injury, *Environ. Health Perspect.*, 57, 307, 1984.
47. **Salminen, A., Kainulainen, H., Arstilla, A. V., and Vihko, V.,** Vitamin E deficiency and the susceptibility of mouse cardiac and skeletal muscles, *Acta Physiol. Scand.*, 122, 565, 1984.
48. **Ramos, K., Combs, A. B., and Acosta, D.,** Role of calcium in isoproterenol cytotoxicity to cultured myocardial cells, *Biochem. Pharmacol.*, 33, 1989, 1984.
49. **Bloom, S. and Davis, D. L.,** Calcium as a mediator of isoproterenol-induced myocardial necrosis, *Am. J. Pathology,* 69, 459, 1972.
50. **Ramos, K., Combs, A. B., and Acosta, D.,** Cytotoxicity of isoproterenol to cultured heart cells: effects of antioxidants on modifying membrane damage, *Toxicol. Appl. Pharmacol.*, 70, 317, 1983.
51. **Lebedev, A. V., Levitsky, D. O., Loginov, V. A., and Smirnov, V. N.,** The effect of primary products of lipid peroxidation in the transmembrane transport of calcium ions, *J. Mol. Cell. Cardiol.*, 14, 99, 1982.
52. **Boor, P. J.,** Cardiac glutathione: diurnal rhythm and variation in drug-induced cardiomyopathy, *Res. Commun. Chem. Pathol. Pharmacol.*, 24, 27, 1979.
53. **Emanuel, N. B.,** Free radicals and the action of inhibitors of radical processes under pathological states and aging of living organisms and man, *Q. Rev. Biophys.*, 9, 283, 1976.
54. **Guarnieri, C., Ferrari, R., Visioli, O., Caldera, C. M., and Nayler, W. G.,** Effect of alpha-tocopherol on hypoxic-perfused and reoxygenated rabbit heart muscle, *J. Mol. Cell. Cardiol.*, 10, 893, 1978.
55. **Hargreaves, R. J., Evans, J. G., Pelling, D., and Butterworth, K. R.,** Studies on the effect of L-AA on acetaminophen-induced hepatotoxicity, *Toxicol. Appl. Pharmacol.*, 64, 380, 1982.

CHRONIC VITAMIN C DEFICIENCY, HYPERCHOLESTEROLEMIA, AND ATHEROSCLEROSIS

Emil Ginter

INTRODUCTION

A chronic vitamin C deficiency may provoke a number of metabolic disorders, increasing the risk of atherosclerosis. This subject was analyzed in detail in a review 10 years ago,[1] and more recent data on the role of the redox system of vitamin C in the prevention of disorders of lipid metabolism and atherogenesis are summarized in this chapter.

EPIDEMIOLOGY

The principal sources of vitamin C in human nutrition are vegetables and fruit. Population groups with a high consumption of these products have lower levels of serum cholesterol, particularly of cholesterol in the low-density lipoproteins (LDL), and show a lower rate of atherosclerotic complications than those consuming current European or North American diets. A significant indirect correlation between consumption of fruit and vegetables and mortality caused by ischemic heart disease has been repeatedly established.[1] In the U.S., there has been a steady fall in coronary mortality rates since 1968, and simultaneously the consumption of vegetables and fruits, particularly those rich in vitamin C, and the intake of synthetic ascorbic acid have risen during this period.[2,3] A steady fall in cerebrovascular mortality in Great Britain was accompanied by an enhanced consumption of vegetables and fruit.[4] These data, however, should be interpreted with caution as they need not necessarily reflect a causal relationship. Persons with a high consumption of vegetables and fruit consume less cholesterol and saturated animal fats. Moreover, vegetables and fruit contain, besides vitamin C, further substances with a hypocholesterolemic action, e.g., various components of dietary fiber, particularly pectin.

BIOCHEMICAL ASPECTS

The first data on a causal relationship between vitamin C, lipid metabolism, and atherosclerosis were obtained in experimental animals. Acute vitamin C deficiency (scurvy) provoked disorders of lipid metabolism in guinea pigs, while high doses of ascorbic acid, on the contrary, depressed plasma cholesterol levels and slowed down atherogenesis in rabbits fed a cholesterol diet.[1] The methodology of these studies was rightly criticized, because (1) metabolic disorders observed in scorbutic guinea pigs are often of a nonspecific nature, as scurvy is accompanied by food refusal, drop in body weight, and hemorrhages, and (2) the rabbit (and also the rat, mouse, hamster, cat, dog, and pig) is not a suitable laboratory animal for the study of the metabolic functions of vitamin C because, in contrast to man, it synthesizes ascorbate in the liver and is not dependent on a vitamin C supply from the diet. Vitamin C deficiency cannot be induced in these animals. On the other hand, guinea pigs, like humans and monkeys, do not express the gene for gulonolactone oxidase, a key enzyme in vitamin C biosynthesis, and are therefore extremely sensitive to vitamin C deficiency.

A characteristic feature of the prevailing nutritional situation of numerous population groups, often also in economically advanced countries, is a chronic low intake of vitamin C, especially during the winter and spring months when the offering of fresh vegetables and fruit is lower. Such a situation is simulated in the guinea pig by a model of chronic marginal vitamin C deficiency: the animals are fed an ascorbate-free diet supplemented with a small maintenance dose of ascorbic acid (1 mg/kg of body weight). Vitamin C levels in the blood and tissues of these animals are very low, only slightly higher than in scurvy. Thanks to the maintenance dose of ascorbic acid, deficient guinea pigs consume food normally, their appearance, behavior, and weight curves are normal, and practically no hemorrhages occur in their bodies. Long-term studies, lasting up to a year, have shown a marked endothelial proliferation, accumulation of lipids, and even atherosclerotic changes in the arteries of guinea pigs with chronic vitamin C deficiency.[1]

According to the definition given by a study group of the World Health Organization (WHO), "atherosclerosis is a variable combination of changes of the intima of arteries consisting of the focal accumulation of lipids, complex carbohydrates, blood and blood products, fibrous tissue and calcium deposits, and associated with medial changes." The mechanism of this process has not been precisely elucidated as yet, but the majority of authors ascribe the decisive role to two factors: (1) impairment of the integrity of the endothelial barrier in the arterial wall (endothelial injury may be caused by numerous factors, e.g., hypertension, extremely high LDL cholesterol levels, immunological factors, endotoxins, cigarette smoking, lipoperoxidation reactions, extreme nutritional imbalance, etc.) and (2) persistently high plasma lipid levels, particularly LDL cholesterol and possibly also triglycerides. Endothelial injury may lead to desquamation of endothelial cells. This exposes the underlying connective tissue to plasma lipoproteins and platelets which aggregate and release the content of their granules. Infiltration of platelet factors and lipoproteins leads to focal proliferation of arterial smooth-muscle cells, which begin to produce a large amount of collagen, elastin, and glycosaminoglycans. In the next phase, cholesterol from LDL is deposited both inside the cells and in their surrounding connective tissue matrix. If the LDL cholesterol concentration in plasma is permanently high and there is a permanent desquamation of the arterial endothelium, a complicated lesion ensues, containing newly formed connective tissue and cholesterol. This lesion may finally undergo calcification.[5]

A whole series of factors are responsible for the onset of atherosclerosis, hence it would not be warranted to assign the decisive role to any single nutritional factor. However, there exists well-founded evidence for inclusion of chronic vitamin C deficiency among these factors. The following sections of this chapter deal with several biochemical functions of ascorbic acid whose impairment, caused by vitamin C deficiency, may either directly or indirectly play a role in the atherogenesis.

Collagen Biosynthesis

Collagen is the major extracellular protein in connective tissues. The rate of collagen production may be of critical importance in maintaining the integrity of arterial endothelium. Collagen contains two unusual amino acids, hydroxyproline and hydroxylysine, that occur only rarely elsewhere in other animal proteins. Ascorbic acid participates in the hydroxylation of peptide-bound proline to hydroxyproline and lysine to hydroxylysine. Prolyl and lysyl hydroxylases could be regarded as mixed-function oxidases because they catalyze a reaction in which one atom of molecular oxygen is incorporated into the proline (or lysine) to form hydroxyproline (or hydroxylysine) and the other is incorporated into a cosubstrate. Uncoupled decarboxylation of α-ketoglutarate, which occurs slowly during proline hydroxylation, produces a reactive iron-oxo complex and free hydroxyl radicals. In the absence of reduced vitamin C, the Fe^{3+} becomes bound to the enzymes, forming the stable and inactive Fe^{3+}–S–enzyme complex.[6] In the presence of reduced vitamin C, Fe^{3+} becomes reduced to Fe^{2+}, which can readily dissociate, thus reactivating the enzyme. The same events explain the requirement for ascorbic acid in lysine hydroxylation.[6] Vitamin C, although not a direct participant in the hydroxylation reaction, functions as a rather specific protective agent to keep the enzyme active by reversing the oxidation of the enzyme-bound iron.

According to other authors, vitamin C may exert a more direct influence on collagen synthesis, possibly at the level of gene expression.[7] Vitamin C deficiency results in production of underhydroxylated collagen which is more sensitive to intracellular degradation.[8]

Although collagen synthesis in vascular walls is less sensitive to vitamin C deficiency than in other tissues, several studies show striking changes in the metabolism of arterial collagen in vitamin C deficiency.[1] In the thoracic aorta of guinea pigs with chronic vitamin C deficiency, a decline of type I and an increase of type III collagen was observed; both types of collagen proved to be slightly underhydroxylated.[9] By modifying the levels of ascorbate in the culture medium, the nature of the extracellular matrix produced by the aortic smooth muscle cells could be altered: ascorbate increases the amounts of insoluble collagen and sulfated proteoglycans and decreases the amount of insoluble elastin.[10,11] Changes in the composition of aortic structural proteins and glycosaminoglycans may significantly influence the integrity of aortic endothelium and activate platelets, causing the secretion of substances that may play a crucial role in atherogenesis by inducing the migration of smooth muscle cells to the site of the lesion. Scanning electron microscopy of the guinea pig aortic endothelium during development of vitamin C deficiency showed a disorganized endothelium, with desquamation; with a persisting vitamin C deficiency, platelet aggregation and fibrin deposition on the desquamated initial surface was observed.[12]

Carnitine Biosynthesis

Carnitine, β-hydroxy-γ(trimethylamino) butyric acid, plays an important role in the translocation of fatty acids across the mitochondrial membrane to facilitate fatty acid oxidation. In this way, carnitine performs an important function in the provision of energy for heart and skeletal muscle. In liver and kidney, carnitine is synthesized from L-methionine and L-lysine. The pathway from lysine to carnitine involves, among others, two hydroxylation reactions, each requiring vitamin C.[13]

The two hydroxylases involved in carnitine biosynthesis, 6-*N*-trimethyllysine hydroxylase and 4-*N*-trimethylaminobutyrate hydroxylase, are members of a class of α-ketoglutarate, Fe^{2+}, and ascorbate-requiring dioxygenases. The function of vitamin C in these reactions is about the same as in hydroxylation of proline and lysine during collagen biosynthesis.[6] Ascorbate is not an electron donor directly for a hydroxylation reaction, but it is for the reduction of Fe ions which take part in oxygen activation, probably by a radical mechanism. In vitamin C-deficient guinea pigs, carnitine biosynthesis is significantly depressed, although views differ as to which of the two hydroxylases is the more sensitive to this deficiency. Therefore, carnitine concentrations in the skeletal muscle, heart, liver, and kidneys of vitamin C-deficient animals, decline.[14,15] This decline is more marked in animals with a long-term vitamin C deficiency. A decrease of carnitine in muscles occurs not only during acute vitamin C deficiency, but also in apparently healthy animals with a suboptimal supply of ascorbic acid.[14]

Vitamin C deficiency induces a decrease in carnitine level in tissues for which fatty acids are the most important energy sources, i.e., in heart and skeletal muscle. Impaired fatty acid oxidation may take part in the development of hypertriglyceridemia, in signs and symptoms of fatigue, and in the fatal course of myocardial infarction in persons with a chronic vitamin C deficiency.

Cholesterol Transformation into Bile Acids

Current data leave little doubt that high levels of plasma cholesterol, especially of LDL cholesterol, are associated with an increased risk for premature coronary atherosclerosis. The plasma cholesterol concentration is the result of the complicated interaction of many processes, e.g., absorption of exogenous cholesterol from the diet, synthesis of endogenous cholesterol, distribution of cholesterol between plasma and tissues, fecal excretion of cholesterol in the form of neutral sterols, and transformation of cholesterol into bile acids. LDLs are the major plasma cholesterol-transporting lipoproteins. LDLs are large spherical particles with a core composed of nearly 1500 molecules of cholesterol esters; its detergent coat contains about 500 molecules of unesterified cholesterol, 800 molecules of phospholipids, and only 1 protein molecule (aproprotein B-100). This protein is recognized and bound by the LDL receptor, a glycoprotein that projects from the surface of cells. LDL enters the cell by receptor-mediated endocytosis; inside the cell it is degraded and its cholesterol is released for metabolic use, e.g., for construction of cell membranes, synthesis of bile acids, and steroid hormones, etc. Cholesterol accumulated in the cell controls the number of LDL receptors by a feedback mechanism.[16] Up to 3/4 of the total LDL receptors in the human and guinea pig body are located in the liver. Cholesterol transformation to its principal catabolic product, bile acids, takes place uniquely in the liver. The bile produced by the liver carries off bile acids and nonmetabolized cholesterol into the intestine. Consequently, the liver plays a central role in the control of plasma cholesterol level.

Vitamin C is indispensible for the key, rate-limiting reaction of cholesterol transformation into bile acid, i.e., for the 7α-hydroxylation of the cholesterol nucleus.[17] This reaction takes place in the smooth endoplasmic reticulum of the liver cell. The enzyme catalyzing this reaction (cholesterol 7α-hydroxylase or 7α-monooxygenase) belongs to a group of enzymes called microsomal hydroxylases or mixed-function oxygenases. These enzymes catalyze hydroxylation of endogenous and particularly of exogenous substances. In hydroxylations of this type, one atom of molecular oxygen enters the substrate and the other is reduced with NADPH to water. The exact role of vitamin C in these reactions is not known. Ascorbate may potentiate the activation of molecular oxygen in the cytochrome P-450 cycle.[18]

In the liver of vitamin C-deficient guinea pigs, the activity of cholesterol 7α-hydroxylase is markedly decreased.[19,20] Cytochrome P-450 isolated from vitamin C-deficient liver microsomes had a much lower capacity to 7α-hydroxylate cholesterol than a corresponding system isolated from the liver of normal guinea pigs.[19] Since 7α-hydroxylase limits the overall rate of bile acid synthesis, cholesterol degradation is significantly slowed down even in apparently healthy guinea pigs with marginal vitamin C deficiency.[1,17,20] Consequently, in vitamin C-deficient animals, cholesterol becomes accumulated in the liver. In the subsequent phase, the number of LDL receptors in the liver probably becomes decreased through feedback, for vitamin C-deficient guinea pigs have a prolonged half-life and a lowered fractional catabolic rate of LDL in blood plasma (Figure 1). In vitro tests have shown that vitamin C may influence the number of LDL receptors. Receptor-mediated catabolism of LDL was increased two- to threefold in arterial smooth muscle when the culture medium was supplemented with physiological concentrations of ascorbate.[21] In vitamin C-deficient guinea pigs, the LDL cholesterol level in plasma increases; later cholesterol accumulated in the aorta and atherosclerotic changes take place in the aorta, particularly in the coronary arteries.[1,12] Long-term vitamin C deficiency induces hypercholesterolemia even in animals as phylogenetically distant as the marmoset monkey and the rainbow trout.[22,23] In primates, however, this effect is less distinct than in guinea pigs. Acceleration of cholesterol biosynthesis does not contribute to the elevation of plasma cholesterol in vitamin C-deficient animals. Cholesterol synthesis and the activity of 3-hydroxy-3-methylglutaryl coenzyme A reductase, the rate-limiting enzyme of cholesterol biosynthesis, are unchanged or even reduced in vitamin C-deficient guinea pigs.[1,24,25]

Cholesterol degradation may be even more accelerated by a simultaneous administration of vitamin C and pectin which disrupts enterohepatic recirculation and enhances fecal excretion of bile acids.[26] The principal sources of vitamin C and pectin in human nutrition are vegetables and fruit. The synergism of vitamin C and pectin may play a role in reducing coronary mortality in people consuming large quantities of those products.

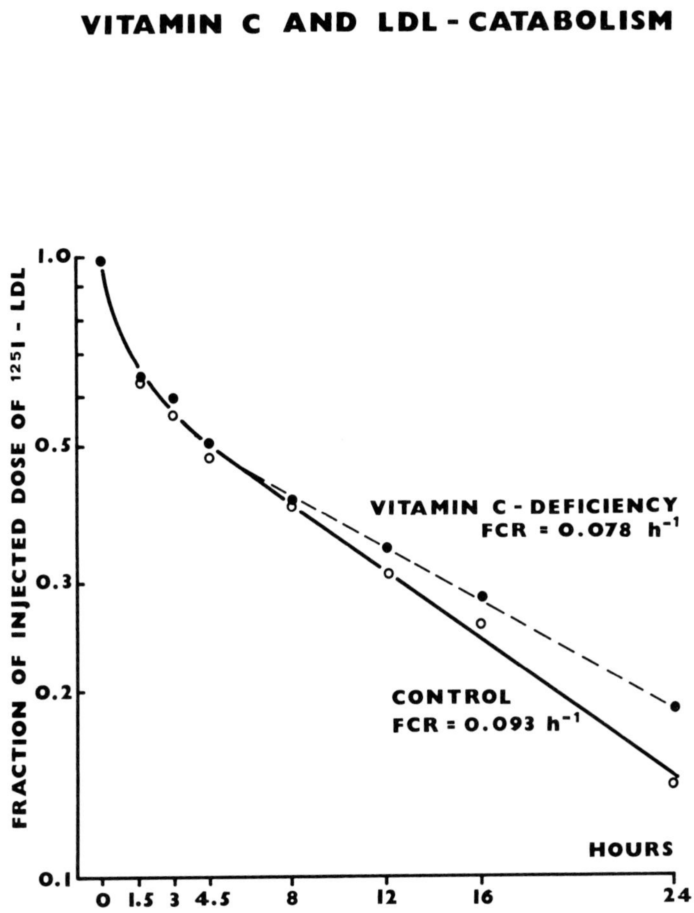

FIGURE 1. Plasma decay curves of labeled low-density lipoproteins (^{125}I-LDL) in control and vitamin C-deficient guinea pigs. In vitamin C-deficient animals, the halflife of the linear part of decay curve is prolonged (control, 11.1 hr; deficiency, 13.9 hr; $p < 0.01$) and the fractional catabolic rate (FCR) is significantly lowered ($p < 0.02$).

Prostaglandin Biosynthesis and Lipoperoxidation

Divergent data exist regarding the stimulating or inhibitory action of vitamin C on the synthesis of various prostaglandins in different tissues. Ascorbic acid caused in vitro a dose-dependent enhancement of the conversion of dihomo-γ-linolenic acid to prostaglandin E_1 by human platelets.[27] Prostaglandin E_1 inhibits platelet aggregation and, in agreement with this, very high doses of vitamin C inhibit both in vitro and in vivo platelet aggregation.[28] In experimental human vitamin C deficiency, on the other hand, platelet function was normal. Ascorbic acid increased the in vitro formation of prostacyclin (PGI_2) in the aortic wall by protecting the cyclooxygenase and PGI_2 synthase against the inhibitory activity of lipoperoxides.[29] Prostacyclin production by the vascular wall is diminished both in experimental and human atherosclerosis, probably due to lipoperoxidation. Administration of ascorbic acid to rabbits fed an atherogenic diet partially prevents the decreased PGI_2 formation by the aortic endothelium.[30] PGI_2 inhibits platelet aggregation, relaxes coronary arteries, and decreases blood pressure. An inverse association of serum vitamin C level and blood pressure was observed in male adults and ascorbate administration lowered systolic blood pressure in spontaneously hypertensive rats.[31]

Oxygen free radicals react with polyunsaturated fatty acids, giving rise to toxic endoperoxides. Evidence is accumulating that lipoperoxidation is involved in atherogenesis. LDLs contain about 800 molecules of phospholipids in each particle, i.e., around 1600 molecules of predominantly unsaturated fatty acids. Contact of LDL with arterial endothelial cells gives rise to lipoperoxides in the LDL coat. It is presumed that LDL particles thus altered become more atherogenic. Alongside fatty acid lipoperoxides, oxidized cholesterol derivates take part in the cytotoxic action of LDL on endothelial cells. Ingestion of oxidized fats or drugs and other xenobiotics that lead to fatty acid peroxidation will result in the appearance of oxidized fatty acids in lipoproteins. The content of lipoperoxides is substantially increased in the blood of atherosclerotic patients. Oxidized lipids are found in atherosclerotic lesions and in areas of inflammatory reactions. Ingestion of vitamin C depresses the quantity of lipoperoxides in human blood.[32] In guinea pigs injected with carbon tetrachloride, ascorbic acid significantly lowered in vivo lipid peroxidation.[33] Ascorbate prevents cardiotoxicity of isoproterenol in primary cultures of rat myocytes.[34]

All these effects are related to the ability of ascorbate or ascorbate radical to scavenge oxygen free radicals. Vitamin E, bound to membrane structures, is even more efficient. Several studies reported a synergistic protective effect of vitamin C and α-tocopherol against oxidant and free radical-mediated injuries. This problem is dealt with in more detail in another chapter of this book.[18]

VITAMIN C, LIPIDS, AND ATHEROSCLEROSIS: HUMAN STUDIES

Data on the effect of chronic vitamin C deficiency on lipid metabolism and atherosclerosis, practically unambiguous in experimental animals, are less concordant in human studies. This is due to several factors. To induce a manifest hypercholesterolemia in guinea pigs, the animals are to be kept in a chronic vitamin C deficiency for at least 2 months and evident atherosclerotic changes in coronary arteries occur in them only after 6 to 12 months which corresponds to 20 to 40% of their life span.[1] Such long-term studies on human are not possible. A short-term marginal vitamin C deficiency or acute scurvy does not alter serum cholesterol level in humans or may even lower it.[35] A spontaneous onset of scurvy, however, is usually related to malnutrition and overall metabolic breakdown. Vitamin C is but one of many nutritional factors affecting cholesterol metabolism and, hence, the degree of hypercholesterolemia induced by vitamin C deficiency in experimental animals depends on the composition of the experimental diets.[1] Long-term studies on outpatients do not permit control of the intake of nutrients capable of interacting with vitamin C and the results thus cannot be as unequivocal as in laboratory animals consuming an exactly defined diet.

A great part of the population of many countries is affected during the winter and spring months with a chronic marginal vitamin C deficiency. During this period, ascorbate level in the blood decreases, while that of plasma cholesterol rises. This is probably a causal relationship, for administration of 1 g of vitamin C per day for 1 year prevented an increase in serum cholesterol during the critical winter period.[36] When ascorbic acid was administered during the period of seasonal vitamin C deficiency to hypercholesterolemic persons in doses of 0.3 to 1.0 g daily, several independent studies reported a significant reduction of plasma total cholesterol.[37] This effect proved more striking in persons with high initial cholesterol levels. The prevalence of low vitamin C levels increases with age and there is a possibility that the age-related rise in serum cholesterol may result from vitamin C deficiency. A daily administration of 1 g ascorbic acid to inmates of a pensioners' home with very low intake of vitamin C from the diet (about 20 mg/day) significantly depressed the level of total cholesterol during the course of 1 year.[1] When ascorbic acid was given to elderly men, an increase of leukocyte vitamin C and decrease in LDL cholesterol were observed; no change in LDL occurred in similarly treated elderly women.[38]

In persons with high initial vitamin C status (serum ascorbate levels much above 1 mg/100 mℓ), ascorbate concentrations in tissues are close to the maximum steady-state levels and any further increase of vitamin C intake is meaningless. Reports on ineffectiveness of megadoses of vitamin C in such well-nourished individuals are irrelevant to the concept of chronic vitamin C deficiency as a risk factor of atherogenesis. In guinea pigs, an extremely high intake of ascorbic acid (2% in diet) exerts an adverse effect on cholesterol metabolism: it depresses the activity of cholesterol 7α-hydroxylase and enhances cholesterol level in blood serum and liver.[20,39]

The majority of human studies published in recent years report a significant drop of total serum cholesterol following long-term administration of ascorbic acid, particularly in persons with a low initial vitamin C status.[36,43] However, in patients with familiar hypercholesterolemia, administration of ascorbic acid over several weeks proved ineffective. Polygenic hypercholesterolemia could be normalized by a diet with a low content of cholesterol and saturated fatty acids and a high content of polyunsaturated fatty acids, vegetable-derived fiber and protein, tocopherols, and vitamin C.[44] In the case of familiar hypercholesterolemia conditioned by a disorder of LDL receptors, pharmacological intervention is necessary, but simultaneously a high vitamin C supply should be ensured, for effectiveness of bile acid sequestering agents (e.g., cholestyramin) declines with a deficiency of vitamin C.[45]

Much interest has been focused on high-density lipoproteins (HDL) following the observation that there exists an indirect correlation between HDL cholesterol and atherosclerosis, i.e., the higher the level of HDL cholesterol, the lower the incidence of coronary heart disease. It is possible that HDLs pick up cholesterol from arteries and other tissues and transport this cholesterol to other lipoproteins and to the liver for excretion from the body. In most but not in all cross-sectional studies, a significant positive correlation was found between HDL cholesterol level and plasma or leukocyte vitamin C concentrations.[38,46,47] However, the level of HDL cholesterol failed to be significantly altered in five persons with a marginal vitamin C deficiency.[35] In several studies, no effect of high doses of ascorbic acid was noted on HDL cholesterol level, but intake of vitamin C by these persons had been high also prior to the administration of ascorbic acid. On the other hand, administration of ascorbic acid to persons with disorders of lipoprotein metabolism moderately increased the HDL cholesterol level.[38,40] Compounds which induce hepatic microsomal cytochrome P-450 function (chlorinated hydrocarbon pesticides, some drugs, and ethylalcohol) may increase plasma HDL in man. HDL cholesterol and apoprotein A-I were directly proportional to liver cytochrome P-450 in subjects in which diagnostic liver biopsy was practiced.[46] Vitamin C is known to stimulate cytochrome P-450-dependent microsomal hydroxylations, thus probably influencing HDL synthesis.

OPTIMUM INTAKE OF VITAMIN C

There is as yet no answer to the question of optimum dietary intake of vitamin C for the human organism. Unfortunately, extreme divergences exist in this domain. The recommended doses range from 30 mg (WHO/FAC Commission, Medical Research Council of Great Britain), through 60 mg (U.S.), up to 10,000 mg/day. An optimum intake of vitamin C may be defined from a biochemical viewpoint as the dose ensuring maximum efficiency of ascorbate-dependent biochemical systems (synthesis of collagen and carnitine, cytochrome P-450-dependent microsomal hydroxylation of cholesterol and xenobiotics, etc.). Data obtained in experimental animals indicate that maximum steady-state levels of ascorbate in tissues are optimum for these functions.[49] The officially recommended doses do not ensure maximum steady-state levels in blood plasma and even less so in tissues. Therefore, the optimum human dose should be that which produces an ascorbate body pool which cannot be increased any futher by higher intakes of vitamin C. A pharmacokinetic study of ^{14}C-ascorbate turnover in healthy men estimated this value to be within the range of 100 mg/day.[50] Maximum body pool size found in this study, however, is lower than the maximum pool calculated from vitamin C levels in human tissues.[51] Hence, it is possible that the vitamin C dose ensuring maximum size of body pool is somewhat higher (approximately 150 mg/day). There exists, however, a great number of physiological and pathological situations that manifestly enhance vitamin C needs, e.g., high physical activity, pregnancy, mental stress, smoking, alcoholism, diabetes, infections, cancer, surgical intervention, etc.). In all probability, disorders of lipid metabolism and atherosclerosis, too, increase consumption of vitamin C. Experimental hypercholesterolemia has been shown to increase ^{14}C-ascorbate oxidation to $^{14}CO_2$ in guinea pigs by 30%.[52] Atherosclerotic patients not taking vitamin supplements often have a very low vitamin C status. Vitamin C levels in leukocytes in patients with coronary atherosclerosis were found to be significantly lower than in those without coronary artery disease, as demonstrated by coronary arteriography.[53] The apparent fall in leukocyte vitamin C after acute myocardial infarction is the result of the granulocytosis and vitamin C deficiency which persist for a long time after myocardial infarction.[54] Hence, it is probable that the officially recommended dose of 30 mg is suboptimal and that such an intake of vitamin C under stress situations correponds to a marginal vitamin C deficiency. For prevention of metabolic disorders related to atherosclerosis, the optimum daily intake of vitamin C is probably higher by one order of magnitude than the officially recommended doses.

CONCLUSIONS

Experimental, clinical, and epidemiological evidence is available to demonstrate that chronic marginal vitamin C deficiency plays a role in the disorders of lipoprotein metabolism, in the impairment of arterial endothelium integrity, and in the onset of atherosclerosis. The redox system of vitamin C is important for collagen synthesis (hydroxylation of proline and lysine), for carnitine synthesis (hydroxylation of trimethyllysine and trimethylaminobutyrate), and for cholesterol transformation into bile acids (7α-hydroxylation of cholesterol). It is probable that the redox system of vitamin C, through an interaction with oxygen free radicals, further affects metabolic areas, e.g., synthesis of prostaglandins and lipoperoxidation reactions. In experimental animals with a chronic vitamin C deficiency, hypercholesterolemia conditioned by a rise in LDL cholesterol level occurs, composition of collagen and glycosaminoglycans in arterial walls becomes altered, and, finally, morphological changes of arteries occur. The presence of hypercholesterolemia, hypertension, and atherosclerosis is more frequent in humans with a low intake of vitamin C. Optimum intake of ascorbic acid to prevent these disorders in humans is not precisely known, but probably exceeds tenfold the doses officially recommended and is in the range of some hundred milligrams of ascorbic acid per day.

REFERENCES

1. **Ginter, E.,** Marginal vitamin C deficiency, lipid metabolism, and atherogenesis, *Adv. Lipid Res.*, 16, 167, 1978.
2. **Ginter, E.,** Decline of coronary mortality in United States and vitamin C, *Am. J. Clin. Nutr.*, 32, 511, 1979.
3. **Verlangieri, A. J., Kapeghian, J. C., El-Dean, S., and Bush, M.,** Fruit and vegetable consumption and cardiovascular mortality, *Med. Hypotheses,* 16, 7, 1985.
4. **Acheson, R. M. and Williams, D. R. R.,** Does consumption of fruit and vegetables protect against stroke?, *Lancet,* 2, 1191, 1983.

5. **Ross, R. and Glomset, J. A.,** The pathogenesis of atherosclerosis, *N. Engl. J. Med.*, 295, 369, 1976.
6. **Myllyla, R., Majamaa, K., Gunzler, V., Hanauske-Abel, H. M., and Kivirikko, K. I.,** Ascorbate is consumed stoichiometrically in the uncoupled reactions catalyzed by prolyl 4-hydroxylase and lysyl hydroxylase, *J. Biol. Chem.*, 259, 5403, 1984.
7. **Murad, S., Grove, D., Lindberg, K. A., Reynolds, G., Sivarajah, A., and Pinnell, S. R.,** Regulation of collagen synthesis by ascorbic acid, *Proc. Natl. Acad. Sci. U.S.A.*, 78, 2879, 1981.
8. **Berg, R. A., Steinmann, B., Rennard, S. I., and Crystal, R. G.,** Ascorbate deficiency results in decreased collagen production: underhydroxylation of proline leads to increased intracellular degradation, *Arch. Biochem. Biophys.*, 226, 681, 1983.
9. **Ginter, E. and Adam, M.,** Characterization of the aortic collagens in guinea pigs with chronic vitamin C deficiency, *Atherosclerosis*, 46, 369, 1983.
10. **Faris, B., Ferrera, R., Toselli, P., Nambu, J., Gonnerman, W. A., and Franzblau, C.,** Effect of varying amounts of ascorbate on collagen, elastin and lysyl oxidase synthesis in aortic smooth muscle cell cultures, *Biochim. Biophys. Acta*, 797, 71, 1984.
11. **Barone, L. M., Faris, B., Chipman, S. D., Toselli, P., Oakes, B. W., and Franzblau, C.,** Alteration of the extracellular matrix of smooth muscle cells by ascorbate treatment, *Biochim. Biophys. Acta*, 840, 245, 1985.
12. **Fujinami, T.,** Experimental atherosclerosis with ascorbic acid deficiency as a model of endothelial injury hypothesis, *J. Jpn. Coll. Angiol.*, 20, 91, 1980.
13. **Hulse, J. D., Ellis, S. R., and Henderson, L. M.,** Carnitine biosynthesis, *J. Biol. Chem.*, 253, 1654, 1978.
14. **Hughes, R. E., Hurley, R. J., and Jones, E.,** Dietary ascorbic acid and muscle carnitine (β-OH-γ-(trimethylamino) butyric acid) in guinea-pigs, *Br. J. Nutr.*, 43, 385, 1980.
15. **Thoma, W. J. and Henderson, L. M.,** Effect of vitamin C deficiency on hydroxylation of trimethylaminobutyrate to carnitine in the guinea pig, *Biochim. Biophys. Acta*, 797, 136, 1984.
16. **Brown, M. S. and Goldstein, J. L.,** Lipoprotein receptors in the liver, *J. Clin. Invest.*, 72, 743, 1983.
17. **Ginter, E.,** Ascorbic acid in cholesterol and bile acid metabolism, *Ann. N.Y. Acad. Sci.*, 258, 410, 1975.
18. **Ginter, E.,** Interactions between vitamins C and E, and cytochrome P-450, in *Handbook of Free Radicals and Antioxidants in Biomedicine*, Vol. 2, Miquel, J., Quantanilha, A., and Weber, H., Eds., CRC Press, Boca Raton, Fla., 1988.
19. **Björkhem, I. and Kallner, A.,** Hepatic 7α-hydroxylation of cholesterol in ascorbate-deficient and ascorbate-supplemented guinea pigs, *J. Lipid Res.*, 17, 360, 1976.
20. **Holloway, D. E. and Rivers, J. M.,** Influence of chronic ascorbic acid deficiency and excessive ascorbic acid intake on bile acid metabolism and bile composition in the guinea pig, *J. Nutr.*, 111, 412, 1981.
21. **Aulinskas, T. H., van der Westhuyzen, D. R., and Coetzee, G. A.,** Ascorbate increases the number of low density lipoprotein receptors in cultured arterial smooth muscle cells, *Atherosclerosis*, 47, 159, 1983.
22. **John, T. M., George, J. C., Hilton, J. W., and Slinger, S. J.,** Influence of dietary ascorbic acid on plasma lipid levels in the rainbow trout, *Int. J. Vitam. Nutr. Res.*, 49, 400, 1979.
23. **McIntosh, G. H., Richmond, W., and Himsworth, R. L.,** Vitamin C deficiency and hypercholesterolemia in marmoset monkeys, *Nutr. Rep. Int.*, 23, 237, 1981.
24. **Holloway, D. E., Peterson, F. J., Prigge, W. F., and Gebhard, R. L.,** Influence of dietary ascorbic acid upon enzymes of sterol biosynthesis in the guinea pig, *Biochem. Biophys. Res. Commun.*, 102, 1283, 1981.
25. **Greene, Y. J., Harwood, H. J., and Stackpoole, P. W.,** Ascorbic acid regulation of 3-hydroxy-3-methylglutaryl coenzyme A reductase activity and cholesterol synthesis in guinea pig liver, *Biochim. Biophys. Acta*, 834, 134, 1985.
26. **Ginter, E., Kubec, F. J., Vozar, J., and Bobek, P.,** Natural hypocholesterolemic agent: pectin plus ascorbic acid, *Int. J. Vitam. Nutr. Res.*, 49, 406, 1979.
27. **Manku, M. S., Oka, M., and Horrobin, D. F.,** Differential regulation of the formation of prostaglandins and related substances from arachidonic acid and from dihomogammalinolenic acid. II. Effects of vitamin C, *Prostaglandins Med.*, 3, 129, 1979.
28. **Cordova, C., Musca, A., Violi, F., Perrone, A., and Alessandri, C.,** Influence of ascorbic acid on platelet aggregation in vitro and in vivo, *Atherosclerosis*, 41, 15, 1982.
29. **Beetens, J. R. and Herman, A. G.,** Vitamin C increases the formation of prostacyclin by aortic rings from various species and neutralizes the inhibitory effect of 15-hydroperoxyarachidonic acid, *Br. J. Pharmacol.*, 80, 249, 1983.
30. **Beetens, J. R., Coene, M. C., Verheyen, A., Zonnekeyn, L., and Herman, A. G.,** Influence of vitamin C on the metabolism of arachidonic acid and the development of aortic lesions during experimental atherosclerosis in rabbits, *Biomed. Biochim. Acta*, 43, S273, 1984.
31. **Yoshioka, M., Aoyama, K., and Matsushita, T.,** Effects of ascorbic acid on blood pressure and ascorbic acid metabolism in spontaneously hypertensive rats (SH rats), *Int. J. Vitam. Nutr. Res.*, 55, 301, 1985.

32. **Wartanovicz, M., Panczenko-Kresowska, B., Ziemlanski, S., Kowalska, M., and Okolska, G.,** The effect of α-tocopherol and ascorbic acid on the serum lipid peroxide level in elderly people, *Ann. Nutr. Metab.*, 28, 186, 1984.
33. **Kunert, K. J. and Tappel, A. L.,** The effect of vitamin C on in vivo lipid peroxidation in guinea pigs as measured by pentane and ethane production, *Lipids*, 18, 271, 1983.
34. **Ramos, K. and Acosta, D.,** Prevention by L(−) ascorbic acid of isoproterenol-induced cardiotoxicity in primary cultures of rat myocytes, *Toxicology*, 26, 81, 1983.
35. **Duane, W. C. and Hutton, S. W.,** Lack of effect of experimental ascorbic acid deficiency on bile acid metabolism, sterol balance, and biliary lipid composition in man, *J. Lipid Res.*, 24, 1186, 1983.
36. **Dobson, H. M., Muir, M. M., and Hume, R.,** The effect of ascorbic acid on the seasonal variations in serum cholesterol levels, *Scott. Med. J.*, 29, 176, 1984.
37. **Ginter, E.,** Pretreatment serum-cholesterol and response to ascorbic acid, *Lancet*, 2, 958, 1979.
38. **Ginter, E., Cerna, O., Budlovsky, J., Balaz, V., Hruba, F., Roch, V., and Sasko, E.,** Effect of ascorbic acid on plasma cholesterol in humans in a long-term experiment, *Int. J. Vitam. Nutr. Res.*, 47, 123, 1977.
39. **Ginter, E., Kosinova, A., Hudecova, A., and Mlynarcikova, U.,** Parabolic response of hepatic microsomal hydroxylating system and lipids to graded doses of ascorbic acid in guinea pigs on low and high α-tocopherol intake, *J. Nutr.*, 114, 485, 1984.
40. **Heine, H. and Norden, C.,** Vitamin C therapy in hyperlipoproteinemia, *Int. J. Vitam. Nutr. Res.*, Suppl. 19, 45, 1979.
41. **Bordia, A. K.,** The effect of vitamin C on blood lipids, fibrinolytic activity and platelet adhesiveness in patients with coronary artery disease, *Atherosclerosis*, 35, 181, 1980.
42. **Fidanza, A., Audisio, M., and Mastroiacovo, P.,** Vitamin C and cholesterol, *Int. J. Vitam. Nutr. Res.*, Suppl. 23, 153, 1982.
43. **Erden, F., Gulenc, S., Torun, M., Kocer, Z., Simsek, B., and Nebioglu, S.,** Ascorbic acid effect on some lipid fractions in human beings, *Acta Vitamol. Enzymatol.*, 7, 131, 1985.
44. **Choudhury, S., Jackson, P., Katan, M. B., Marenah, C. B., Cortese, C., Miller, N. E., and Lewis, B.,** A multifactorial diet in the management of hyperlipidaemia, *Atherosclerosis*, 50, 93, 1984.
45. **Ginter, E., Bobek, P., Ozdin, L., Kubec, F., and Vozar, J.,** Synergism between ascorbic acid and bile acid binding agents, *Adv. Physiol. Sci.*, 35, 67, 1981.
46. **Bates, C. J., Burr, M. K., and Leger, A. S. S.,** Vitamin C, high density lipoproteins and heart disease in elderly subjects, *Age Aging*, 8, 177, 1979.
47. **Burr, M. L., Bates, C. J., Sweetnam, P. M., and Barasi, M. E.,** Plasma ascorbate and HDL-cholesterol in women, *Hum. Nutr. Clin. Nutr.*, 36, 399, 1982.
48. **Luoma, P. V., Savolainen, M. J., Sotaniemi, E. A., Pelkonen, R. O., Arranto, A. J., and Ehnholm, C.,** Plasma high-density lipoproteins and liver lipids and proteins in man, *Acta Med. Scand.*, 214, 103, 1983.
49. **Ginter, E.,** Optimum intake of vitamin C for the human organism, *Nutr. Health*, 1, 66, 1982.
50. **Kallner, A., Hartmann, D., and Hornig, D.,** Steady-state turnover and body pool of ascorbic acid in man, *Am. J. Clin. Nutr.*, 32, 530, 1979.
51. **Ginter, E.,** Chronic marginal vitamin C deficiency: biochemistry and pathophysiology, *World Rev. Nutr. Diet.*, 33, 104, 1979.
52. **Ginter, E. and Zloch, Z.,** Raised ascorbic acid consumption in cholesterol-fed guinea pigs, *Int. J. Vitam. Nutr. Res.*, 42, 72, 1972.
53. **Ramirez, J. and Flowers, N. C.,** Leukocyte ascorbic acid and its relationship to coronary artery disease in man, *Am. J. Clin. Nutr.*, 33, 2079, 1980.
54. **Vallance, B. D., Hume, R., and Weyers, E.,** Reassessment of changes in leukocyte and serum ascorbic acid after acute myocardial infarction, *Br. Heart J.*, 40, 64, 1978.

THE ACTION OF FREE RADICAL SCAVENGERS AND ANTI-INFLAMMATORY DRUGS ON SYNOVIAL FLUID DEGRADATION INDUCED BY FREE RADICALS

P. Puig-Parellada

INTRODUCTION

The synovial fluid bathing the joint areas is considered a dialysate of blood plasma. In addition, hyaluronate (molecular weight, 1.6 to 3 million) which is secreted by synovial cells provides a synovial fluid with its characteristically viscous properties. In normal synovial fluid, the concentration of hyaluronate is about 3.5 mg/g of fluid. However, in inflammatory diseases such as rheumatoid arthritis, there is a decrease in viscosity of the synovial fluid. Such a decrease is one of the hallmarks of these diseases. Hyaluronic acid is by definition a polymer of alternating residues of glucuronic acid and *N*-acetyl-glucosamine. The implication of oxygen radicals in the decrease in viscosity of the synovial fluid is the main topic of this review.

DEPOLYMERIZATION OF HYALURONIC ACID IN "IN VITRO" SYSTEMS

Implication of Oxygen Free Radicals

In 1974 it was demonstrated by McCord[1] that enzymatically generated superoxide radicals rapidly degraded bovine synovial fluid and solutions of purified hyaluronic acid. The degradation was expressed by a marked reduction in viscosity and by an inability of the degraded synovial fluid to form a mucin clot characteristic of healthy synovial fluid. The addition of either catalase or superoxide dismutase (SOD) provided protection against degradation in both systems. This indicates that neither superoxide (O_2^-) nor hydrogen peroxide (H_2O_2) alone could effect depolymerization of the polysaccharide, but must be present simultaneously for the process to occur. The implication is that the actual depolymerizing species is generated secondarily by reaction between O_2^- and H_2O_2. The most likely candidate for this role is the reactive ·OH (hydroxyl radical). Additional support for ·OH radicals as a depolymerizing species comes from different lines of research as described in the following paragraphs.

Several effective scavengers of ·OH radicals, like mannitol, benzoate, dimethyl sulfoxide (DMSO), and ethanol, cause an almost complete preservation of the viscosity, indirectly implicating ·OH radicals in the synovial fluid depolymerization.[1-4] Another technique to induce a viscosity fall due to the production of free radicals is UV illumination of the synovial fluid. The production of oxy radicals was assayed by indirect methods.[5]

In this model, anti-inflammatory drugs like acetylsalicylic acid and indomethacin are inactive. In relation to indomethacin, the authors suggest the possible implication of singlet oxygen on which indomethacin seems to act specifically.[6] The findings with aspirin agree with previous results.[2,5]

Degradation of hyaluronic acid by the xanthine-xanthine oxidase system was strongly inhibited by SOD and catalase, whereas degradation induced by autoxidation of ferrous ions was weakly inhibited by catalase and unaffected by SOD. Both oxy radical-producing systems were inhibited by hydroxyl radical scavengers, suggesting that hydroxyl radical was the damaging species in both systems.[7]

Hydroxylation of salicylate to 2,3-dihydroxybenzoic acid has been used as an ·OH detecting system. Hydroxylation was stimulated by SOD, hydroxyl radical scavengers, and singlet oxygen quenchers.

It has been postulated also that ·OH radical production via the Fenton reaction cannot be assumed to occur in biological fluid or be used as an explanation for tissue-damaging reactions.[8] This conclusion was obtained by assaying the production of ·OH radicals by the formation of ethylene from methional. This technique has been discussed and some doubts were raised on the strict dependence of C_2H_4 production upon ·OH radicals.[9] According to the authors, their data suggest that the depolymerization observed in diluted synovial fluid may have been catalyzed by metal complexes added with the reagents or by a mechanism not involving metal ion catalyzed production of free ·OH radicals.[8]

Singlet oxygen was also implicated in the depolymerization of the synovial fluid by using a dye-sensitized photooxidation system[10] and by indirect methods, using specific scavengers.[2,3] It is rather interesting to point out that indomethacin reacts with singlet molecular oxygen.[6]

The observed decrease in viscosity of hyaluronic acid is related to a decrease in molecular weight of the polymer with the appearance of a polydisperse population of breakdown products. Some form of molecular interaction (aggregation) or conformational changes due to cross-linking takes place. Such structural alterations could have important consequences in terms of the altered physiological properties of the hyaluronate matrix.[6,11] Circular dichroism assays suggested that the generated free radicals randomly cleaved the hyaluronic acid side groups, causing a change in the tertiary structure, but not in the secondary structure.[10]

Iron Implication

Hydroxyl radicals are generated by a direct reaction between O_2^- and H_2O_2 (the Haber-Weiss reaction).

$$H_2O_2 + O_2^- \rightarrow O_2 + OH + OH^- \qquad (1)$$

This reaction does not proceed at a significant rate in biological systems.[12] However, addition of small amounts of iron salts to the superoxide-generating system causes the formation of ·OH radicals according to the following reactions:

$$Fe^{3+} + O_2^- \rightarrow Fe^{2+} + O_2 \qquad (2)$$

$$Fe^{2+} + H_2O_2 \rightarrow Fe^{3+} + \cdot OH + OH^- \qquad (3)$$

$$\text{Net:}\quad O_2^- + H_2O_2 \rightarrow O_2 + \cdot OH + OH^- \qquad (4)$$

The preceding reactions are defined as the ''iron-catalyzed Haber-Weiss reaction''. Reaction 2 suggests that iron (II) salts in aqueous solutions can react with O_2 to give O_2^-. O_2^- can undergo nonenzymatic dismutation and H_2O_2 can react with the remaining Fe^{2+} to produce ·OH radicals according to Reaction 3. The importance of iron in the degradation of hyaluronic acid has been demonstrated by the use of iron chelators[13] and by experiments which show that the degradation is induced by $FeSO_4$, but not $FeCl_3$.[14]

Degradation induced by Fe^{2+} salts can be inhibited by scavengers of the ·OH radicals such as formate, mannitol, and thiourea. Urea which reacts slowly with ·OH has little effect. Catalase is also a powerful inhibitor of degradation. SOD, on the contrary, accelerates the reaction implicating an increased supply of H_2O_2 which suggests that the concentration of H_2O_2 is the limiting factor in the production of ·OH.[14]

A fresh solution of ascorbic acid also promotes degradation of hyaluronic acid which is greatly increased by the presence of $FeSO_4$. Ascorbic acid reduces traces of Fe^{3+} present to Fe^{2+}.[14] Besides Fe^{2+}, the protein-bound iron can catalyze the Haber-Weiss reaction in biological systems. Experiments done with ferritine demonstrate its capacity to depolymerize synovial fluid in a xanthine-xanthine oxidase system. Apoferritin is also active. Both apoferritin and ferritin effects were iron dependent, since they were abolished by desferrioxamine which is known to form an iron complex which is unable to catalyze the Haber-Weiss reactions.[15]

Reference 15 discusses some of the possible mechanisms implicated in hydroxyl radical formation by ferritin and other protein-bound iron. Its implication in the Haber-Weiss reaction has been a matter of controversy.[16] The relevance of this data for the inflammatory process comes from the fact that at the onset of inflammation there is a rapid fall in serum iron. This iron can be found not only in its storage sites, but also in the rheumatoid synovial membranes[17] and lymph nodes.[18] The iron is present as ferritin molecules, frequently concentrated in complex cytoplasmic granules or lysosomes.

The concentration of transferrin is not increased in the synovial fluid, but is sufficient to bind about four times the amount of iron known to be present in synovial fluid.[19] Besides this protein-bound iron, it has been demonstrated, using as a biological system the bleomycin-dependent degradation of DNA, that macromolecular concentrations of free iron are present in the synovial fluid.[20] According to the authors, this iron is responsible for ·OH radical production which rapidly reacts with lipoproteins and other body fluid constituents such as hyaluronic acid.[21]

Iron-induced peroxidation of membrane phospholipids has been demonstrated with some iron chelators. Desferrioxamine, an iron chelator approved for clinical use, inhibits peroxidation at all concentrations tested and also inhibits the iron-catalyzed formation of hydroxyl radicals from superoxide[22] and, moreover, the superoxide itself.[23]

Ceruloplasmin and SOD are two enzymes with copper in their active sites. The levels of copper and ceruloplasmin in serum are significantly increased in arthritic patients, suggesting a possible protective role of both substances in inflammation.[17] The known action of copper in various chemical forms as anti-inflammatory drugs on animal models agrees with this data.[24]

The activities of SOD in the erythrocytes of patients with rheumatoid arthritis are not correlated with the erythrocyte copper concentration.[25] These results raise some important questions to be answered before a proper understanding can be gained of the possible implication of copper in rheumatoid arthritis. Like iron, copper stimulates phospholipid membrane peroxidation due to ·OH radical production, but the serum samples of patients with rheumatoid diseases offer less antioxidant protection than do normal control sera.[26] The data presented here on the implication of copper and iron in rheumatoid arthritis have been recently reviewed in several very interesting papers.[27-29]

Anti-Inflammatory Drugs

The actions of several anti-inflammatory drugs were evaluated as regards their capacity to inhibit the decrease in viscosity of the synovial fluid induced by the reaction between hypoxanthine and xanthine oxidase. The nonsteroidal anti-inflammatory drugs (indomethacin, niflimic acid, acetaminophen, imidazole mefenamic acid, and phenylbutazone) at 80 μM inhibit synovial fluid degradation. Acetylsalicylic acid is inactive. Acetaminophen and imidazole are more active in this system than as anti-inflammatory agents. These results indicate that the capacity of anti-inflammatory drugs to inhibit synovial fluid degradation is due to their ability to scavenge free radicals.[2]

It has been proposed that the bleaching of *p*-nitrosodimethylalanine can be used as specific technique for the detection of ·OH radicals generated in the oxidation of hypoxanthine by xanthine oxidase. This technique suggests that anti-inflammatory drugs act as ·OH radical scavengers.[30] Further research, however, has given rise to serious criticism of this technique.[30,31]

A more reliable way to implicate the ·OH radicals in the mechanism of action of anti-inflammatory drugs is to use as a model system the inactivation of bacteriophage T_2 by hydroxyl free radicals generated by pulse radiolysis experiments. The presence, in the system of anti-inflammatory drugs (indomethacin, flurbiprofen, D-penicillamine, and aspirin), strongly inhibited the inactivation of the bacteriophage, confirming that hydroxyl radicals react rapidly with these drugs.[32]

The superoxide anion generated from isolated macrophages, from a xanthine oxidase system, or from guinea pig peritoneal exudate was assayed using the spectrophotometric decrease in NADH absorption at 340 nm, in the presence of lactate dehydrogenase. The results obtained show that nonsteroidal anti-inflammatory drugs significantly inhibit the production of superoxide anion from macrophages and peritoneal exudates at physiological pH, but do not inhibit the superoxide anion generated in the xanthine oxidase system. This suggests that anti-inflammatory drugs do not work as O_2^- scavengers, but block O_2^- production by interfering with the cell membrane-associated mechanism producing superoxide anions.[33,34]

A recent study has verified the proposed actions of some anti-inflammatory drugs as radical scavengers and found that some of them influence enzyme activity when the xanthine-xanthine oxidase system is employed. Drugs like salicylic and acetylsalicylic acid and sulfasalazine inhibit the production of oxygen derived free radicals by xanthine oxidase. Sulfapiridine, 5-amino-salicylic acid, allopurinol, mannitol, glucuronic acid, paracetamol, ibuprofen, benoxaprofen, and gentisic acid exert their effect via scavenging of free radicals with very little effect on enzyme activity.[35] All the drugs tested were at millimolar concentration ranges which differ from the micromolar concentration range used by other authors.[2] This important difference in concentration permits us to consider, for instance, acetylsalicylic acid as an inhibitor of the enzyme[35] or as an inactive substance[2] in the same system.

Role of Polymorphonuclear Leukocytes (PMNLs)

In extracellular fluid, the possibility of oxy radical generation is apparent, provided there is a source of superoxide radical. The most likely physiological source of superoxide in extracellular fluid now appears to be metabolically activated PMNLs. Babior found that phagocytosing PMNLs produce and release O_2^- into the surrounding medium.[36] A number of arthritic conditions are characterized by large influxes of PMNLs in the synovial fluid. A strong correlation exists between decreased viscosity of the fluid, poor mucine clot formation, and the number of PMNLs present in the inflamed joint.[1]

The possible implication of PMNLs in synovial fluid depolymerization has been studied with human PMNLs suspended in a hyaluronic acid solution containing ferrous iron. Superoxide production was induced by addition of phorbol myristate acetate (PMA). Addition of mannitol and SOD to the PMNL system completely abolishes the effects of PMNLs on hyaluronic acid viscosity: this is characterized by a decrease in molecular weight of hyaluronic acid as shown by chromatographic analysis on Sepharose.[3,37]

The production of ·OH radicals during the PMNL phagocytic process has been demonstrated measuring the methane production from DMSO. Reactions producing ·OH (xanthine-xanthine oxidase) generated CH_4 from DMSO, whereas reactions yielding primarily O_2^- or H_2O_2 fail to produce CH_4. Neutrophils (PMNLs), monocytes, and alveolar macrophages also produce CH_4. Methane production from DMSO increases as the number of stimulated PMNLs is increased. Methane production is markedly decreased by potent ·OH scavengers and to a lesser degree by weaker ·OH scavengers. SOD or catalase also decreases CH_4 production.[9]

Hyaluronidase

The release of lysosomal hyaluronidase by the invading PMN could be one of the simplest ways to explain the depolymerization of hyaluronic acid. Its presence in pathological joint fluids, synovium, and PMNL leukocytes has been questioned by some authors[3] and considered an unsatisfactory explanation by others.[1] Lysosomal hyaluronidase is totally inactivated above pH 4.5[1] and the hyaluronidase activity found in synovial fluid does not correlate with inflammatory indexes.[38]

DEPOLYMERIZATION OF HYALURONIC ACID IN "IN VIVO" SYSTEMS

Oxygen Free Radicals and Arthritis

Deterioration of synovial fluid is a symptom which characterizes the inflammatory types of arthritis. Healthy synovial fluid contains a barely detectable amount of endogenous catalase (50 ng/mℓ or less), but does not contain enough endogenous SOD (1 μg/mℓ) to provide partial protection from superoxide-induced degradation. These values are at least two or three orders of magnitude less than intracellular concentrations of the enzymes. The endogenous SOD in the synovial fluid is the copper-zinc enzyme found in cytoplasm rather than the manganese enzyme found in mitochondria.[1]

The ability of synovial fluid from rheumatoid arthritic patients to scavenge O_2^- and H_2O_2 was examined. The synovial fluid contained no SOD activity, insufficient ceruloplasmin to scavenge any superoxide radical, and little if any catalase activity. The ascorbate concentrations in the plasma and synovial fluid were similar and too low for this molecule to have any significant O_2^- scavenging activity. It follows that synovial fluid from rheumatoid arthritic patients has no protection against O_2^- and H_2O_2 generated by active leukocytes. Since rheumatoid fluid has a high iron content, an interaction of O_2^- and H_2O_2 to generate ·OH radicals seems almost certain to occur in vivo.[39]

In a similar study, but with a different perspective,[19] no increase of SOD was observed; there was a slight increase of glutathione peroxidase and a marked increase of catalase in rheumatoid arthritic synovial fluid. The conclusion was similar to the previous one[39] in the sense that, although these enzymes are present in synovial fluid, their activity is insufficient to protect against oxygen free radicals and H_2O_2.

The importance of ceruloplasmin in the synovial fluid of rheumatoid arthritic patients has been denied by some authors[39,40] and considered by others to be an important protection against oxygen free radicals in serum and synovial fluid, both in control and rheumatoid arthritic patients, since it blocks the lipid peroxidation induced by the xanthine-xanthine oxidase system in rat liver microsomes. This conclusion was confirmed with gel filtration experiments, purified human ceruloplasmin, and the use of ceruloplasmin-free serum.[19]

In contrast to the work already mentioned,[19,39] a remarkable increase of SOD in rheumatoid arthritic patients was observed. The high SOD activity correlates both with the severity of the disease and with the inflammatory factors. These surprising results are explained by the authors, suggesting a nonspecific membrane alteration of the PMNLs and an inflow of this enzyme in synovial fluid. As for the synovial fluid, there are contradictory results when considering the protective enzyme activities in PMNLs and mononuclear cells.

The PMNLs of children with rheumatoid arthritis exhibited less SOD activity, but similar leukocyte glutathione peroxidase activity when compared with control values.[41,42] This decrease in SOD activity was observed only for cytoplasmic manganese enzyme in granulocytes of adults with rheumatoid arthritis.[43,44] Different results have been presented, showing that in PMNLs and mononuclear cells from patients with rheumatoid arthritis the SOD in mononuclear cells was twice that in PMNL and the SOD in both types of cells was significantly higher than in controls. These results are interpreted as suggesting the presence of immunologically hyperactive cells which in turn may lead to SOD induction.[45]

PMNLs from children with rheumatoid arthritis release more superoxide anion when stimulated with opsonized zymosan than control PMNLs.[41,42] With the chemiluminescence technique, it has been demonstrated that PMNLs from rheumatoid arthritic patients are unable to generate oxygen free radicals, implicating some specific antigen-antibody complexes which are not present in the sera of rheumatoid patients, but were present in the sera of lupus erythematosus patients.[46]

The SOD deficiency in granulocytes of children with rheumatoid arthritis[47,48] promotes the release of a sufficient amount of O_2^- by the PMNLs into the inflamed tissue to cause degradation of synovial fluid. Only the addition of SOD to the synovial fluid protects against such degradation. A higher release of O_2^- radicals in the chemiluminescence assay has also been observed in monocytes from patients with rheumatoid arthritis. No difference appears in granulocytes.[49]

Lipid Peroxidation and Arthritis

Lipid peroxides are generated both in the arachidonic acid metabolic pathway and as a consequence of the degradation products of free radical oxidation (peroxidation). In different models of experimental inflammation, a clear increase of lipid peroxides has been observed in serum and in different organs, especially the liver.[50,51] Liberation of lipid peroxides has also been observed in thermal injury. It has been suggested that there is a similarity between the burn toxin and lipid peroxides and that their release into the bloodstream is responsible for the cell damage of various organs such as spleen, kidney, and liver.[52] In carrageenan granuloma, an interrelation between an increase in lipid peroxidation and a reduction in drug metabolism has been demonstrated.[53]

In experimental models of arthritis, such as adjuvant arthritis in the rat, the unexpected finding was that the peroxidation of hepatic and splenic lipid is dramatically reduced. The levels of serum lipids are elevated.[54] In adjuvant arthritis in rats, the protective actions of SOD have been demonstrated on the different parameters assayed in sera and synovial fluid (severity, depression of albumin/globulin ratio, levels of acid phosphatase and B-glucuronidase, and TBA reactive substances). Similar results were obtained with vitamin E. These observations suggest that aggravation of adjuvant arthritis may be associated with lipid peroxidation due to superoxide and that SOD and antioxidants such as vitamin E may be beneficial for the treatment of arthritis.[55-57]

In patients with rheumatoid arthritis, higher lipid peroxides levels correlate with more active rheumatoid arthritis.[58] In synovial fluid and plasma, fluorescent degradation products of free radical oxidation were significantly elevated compared to normal controls. The elevation was largely derived from peroxidized polyunsaturated lipids. The levels declined after treatment with a variety of anti-inflammatory drugs.[59,60]

Consistent with lipid peroxidation in vivo is the presence of thiobarbituric acid (TBA) reactive material in serum and knee joint synovial fluid of rheumatoid patients. The amount of TBA reactive material in synovial fluid correlates with the concentrations of iron salts present, presumably because iron is an important catalyst of radical reactions in vivo. A correlation also exists between the amount of TBA reactive material present and the severity of the local inflammatory process, measured as knee score, leukocyte count, and synovial fluid C-reactive protein.[61]

The determination of lipid peroxidation as a parameter of peroxidation by O_2^- was measured in the serum and the synovial fluid of patients with rheumatoid arthritis and osteoarthritis. Both the lipid peroxidation levels and the SOD and catalase activity in the synovial fluid of the arthritic group were found to be higher than in the osteoarthritic group. No significant difference was observed in the serum of both groups. A positive correlation was found between the lipid peroxide values and the SOD activity in the synovial fluid of patients with rheumatoid arthritis.[62]

OXYGEN FREE RADICALS, PROTEOGLYCANS, AND COLLAGEN

In the connective tissue, three major macromolecules are implicated: hyaluronic acid, proteoglycans, and collagen. The first has already been considered. Not much information is available when looking for the implication of oxygen free radicals in the synthesis and degradation of proteoglycans and collagen.

In cultures of normal canine articular cartilage, anti-inflammatory drugs (fenoprofen, ibuprofen, and salicylates, but not indomethacin or sulindac) reversibly inhibit, at therapeutic doses, the net glucosaminoglycan synthesis in a concentration-dependent fashion.[63] Indomethacin is also inactive in experimental osteoarthritis in rabbits.[64]

With bovine articular cartilage in culture, the syntheses of hyaluronic acid and proteoglycans were inhibited after treatment with xanthine oxidase and hypoxanthine. Neither SOD nor specific chelators of iron are able to reverse the inhibition. Only catalase is active, which suggests that hydrogen peroxide is the active species. Inhibition also takes place when cartilage is treated with glucose oxidase which directly generates H_2O_2.[65]

Degradation of proteoglycan-hyaluronate aggregates occurs when incubated with oxygen-derived reactive species generated enzymatically (xanthine-xanthine oxidase). The effect was reversed by inclusion of SOD, catalase, or DETAPAC (an iron chelator) in the incubation, suggesting that hydroxyl radicals are the active species.[66]

The levels of collagenase and hyaluronic acid in cultured synovial cells are modified by the level of PGE_2 released into the medium. This action is probably related to the effects of PGE_2 on adenylate cyclase. Rabbit synovial fibroblast exposed to crystals of monosodium urate monohydrate induce syntheses and release of high levels of collagenase and PGE_2. Anti-inflammatory drugs affected collagenase and hyaluronic acid production indirectly by decreasing ambient PGE_2 levels.[67-70]

When acid-soluble calf skin collagen in dilute solutions is heated at 37°C the collagen will gel. This will not occur, however, when superoxide radicals are generated. The presence of SOD or catalase in the collagen solutions prevents the inhibition of gelatin. If SOD and catalase are both used simultaneously, a completely normal gelatin was found, implicating ·OH radicals in the process.[3]

Abnormal levels of lysosomal enzymes[71,72] and collagenase[72] are always present in the chronic inflammatory diseases.

The action of various anti-inflammatory drugs has been related to their ability to stabilize the lysosomal membrane and therefore inhibit indirectly the lysosomal degradation of proteoglycans and collagen. In in vitro experiments, this inhibition take places only at high concentration ($10^{-3}M$). The stabilizing effects of anti-inflammatory drugs on erythrocyte membrane have been used as a model system for lysosomal membrane stabilization.[73,74] It has been suggested that anti-inflammatory drugs inhibit red blood cell hemolysis by absorbing or destroying free radicals or peroxides.[73]

REFERENCES

1. **McCord, J. M.,** Free radicals and inflammation: protection of synovial fluid by superoxide dismutase, *Science,* 185, 529, 1974.
2. **Puig-Parellada, P. and Planas, J. M.,** Synovial fluid degradation induced by free radicals. In vitro action of several free radical scavengers and anti-inflammatory drugs, *Biochem. Pharamcol.,* 27, 535, 1978.
3. **Greenwald, R. A.,** Effect of oxygen-derived free radicals on connective tissue macromolecules, in *Biological and Clinical Aspects of Superoxide and Superoxide Dismutase,* Vol. 11B, Bannister, W. H. and Bannister, J. V., Eds., Elsevier/North-Holland, Amsterdam, 1980, 160.

4. **Del Maestro, R. F., Thaw, H. H., Bjork, J., Planker, M., and Arfors, K. E.,** Free radicals as mediators of tissue injury, *Acta Physiol. Scand.*, 492, 43, 1980.
5. **Ferraccioli, G. F., Ambanelli, U., Fietta, P., Giudicelli, N., and Giori, C.,** Decrease of osteoarthritic synovial fluid viscosity by means of u.v. illumination: a method to evaluate the free radical scavenging action of drugs, *Biochem. Pharmacol.*, 30, 1805, 1981.
6. **Bodanes, R. S. and Chan, P. C.,** Reaction of indomethacin with singlet molecular oxygen, *Biochem. Pharamacol.*, 29, 1337, 1980.
7. **Betts, W. H. and Cleland, L. G.,** Effect of metal chelators and anti-inflammatory drugs on the degradation of hyaluronic acid, *Arthritis Rheum.*, 25, 1469, 1982.
8. **Winterbourn, C. C.,** Hydroxyl radical production in body fluids. Roles of metal ions, ascorbate and superoxide, *Biochem. J.*, 198, 125, 1981.
9. **Repine, J. E., Eaton, J. W., Anders, M. W., Hoidal, J. R., and Fox, R. B.,** Generation of hydroxyl radical by enzymes, chemicals and human phagocytes in vitro, *J. Clin. Invest.*, 64, 1642, 1979.
10. **Andley, U. P. and Chakrabarti, B.,** Role of singlet oxygen in the degradation of hyaluronic acid, *Biochem. Biophys. Res. Commun.*, 115, 894, 1983.
11. **McNeill, J. D., Wiebkin, O. W., Betts, W. H., and Cleland, L. G.,** Depolymerisation products of hyaluronic acid after exposure to oxygen-derived free radicals, *Ann. Rheum. Dis.*, 44, 780, 1985.
12. **Koppenol, W. H., Butler, J., and van Leeuwen, J. W.,** The Haber-Weiss cycle, *Photochem. Photobiol.*, 28, 655, 1978.
13. **Halliwell, B.,** Superoxide-dependent formation of hydroxyl radicals in the presence of iron salts. Its role in degradation of hyaluronic acid by a superoxide-generating system, *FEBS Lett.*, 96, 238, 1978.
14. **Wong, S. F., Halliwell, B., Richmond, R., and Skowroneck, W. R.,** The role of superoxide and hydroxyl radicals in the degradation of hyaluronic acid induced by metal ions and by ascorbic acid, *J. Inorg. Biochem.*, 14, 127, 1981.
15. **Carlin, G. and Djursater, R.,** Xanthine oxidase induced depolymerization of hyaluronic acid in the presence of ferritin, *FEBS Lett.*, 177, 27, 1984.
16. **Baldwin, D. A., Jenny, E. R., and Aisen, P.,** The effect of human serum transferrin and milk lactoferrin on hydroxyl radical formation from superoxide and hydrogen peroxide, *J. Biol. Chem.*, 259, 13, 391, 1984.
17. **Muirden, K. D.,** Ferritin in synovial cells in patients with rheumatoid arthritis, *Ann. Rheum. Dis.*, 25, 387, 1966.
18. **Muirden, K. D.,** Lymph node iron in rheumatoid arthritis. Histology, ultrastructure, and chemical concentration, *Ann. Rheum. Dis.*, 29, 81, 1980.
19. **Biemond, P., Swaak, A. J. G., and Koster, J. F.,** Protective factors against oxygen free radicals and hydrogen peroxide in rheumatoid arthritis synovial fluid, *Arthritis Rheum.*, 27, 60, 1984.
20. **Gutteridge, J. M. C., Rowley, D. A., and Halliwell, B.,** Superoxide-dependent formation of hydroxyl radicals in the presence of iron salts. Detection of "free" iron in biological systems by using bleomycin-dependent degradation of DNA, *Biochem. J.*, 199, 263, 1981.
21. **Gutteridge, J. M. C., Rowley, D. A., and Halliwell, B.,** Superoxide-dependent formation of hydroxyl radicals and lipid peroxidation in the presence of iron salts. Detection of "catalytic" iron and anti-oxidant activity in extracellular fluids, *Biochem. J.*, 206, 605, 1982.
22. **Gutteridge, J. M. C., Richmond, R., and Halliwell, B.,** Inhibition of the iron-catalysed formation of hydroxyl radicals from superoxide and of lipid peroxidation by desferrioxamine, *Biochem. J.*, 184, 469, 1979.
23. **Sinaceur, J., Ribiere, C., Nordmann, J., and Nordmann, R.,** Desferrioxamine: a scavenger of superoxide radicals?, *Biochem. Pharmacol.*, 33, 1693, 1984.
24. **Milanino, R., Passarella, E., and Velo, G. P.,** Copper and the inflammatory process, *Adv. Inflamm. Res.*, 1, 281, 1979.
25. **Scudder, P., Stocks, J., and Dormandy, T. L.,** The relationship between erythrocyte superoxide dismutase activity and erythrocyte copper levels in normal subjects and in patients with rheumatoid arthritis, *Clin. Chim. Acta*, 69, 397, 1976.
26. **Gutteridge, J. M. C., Hill, C., and Blake, D. R.,** Copper stimulated phospholipid membrane peroxidation: antioxidant activity of serum and synovial fluid from patients with rheumatoid arthritis, *Clin. Chim. Acta*, 139, 85, 1984.
27. **Richardson, I.,** Salicylates, copper complexes, free radicals and arthritis, *J. Pharm. Pharamacol.*, 28, 666, 1976.
28. **Blake, D. R. and Lunerc, J.,** Copper, iron, free radicals and arthritis, *Br. J. Rheumatol.*, 24, 123, 1985.
29. **Blake, D. R., Hall, N. D., Bacon, P. A., Dieppe, P. A., Halliwell, B., and Gutteridge, J. M. C.,** The importance of iron in rheumatoid disease, *Lancet*, II, 1142, 1981.
30. **Puig-Parellada, P. and Planas, J. M.,** The possible significance of hydroxyl radicals in inflammation, in *Biological and Clinical Aspects of Superoxide and Superoxide Dismutase*, Vol. 11B, Bannister, W. H. and Bannister, J. V., Eds., Elsevier/North-Holland, Amsterdam, 1980, 141.

31. **Bors, W., Michel, C., and Saran, M.,** On the nature of biochemically generated hydroxyl radicals. Studies using the bleaching of *p*-nitrosodimethylaniline as a direct assay method, *Eur. J. Biochem.*, 95, 621, 1979.
32. **Hiller, K. O. and Wilson, R. L.,** Hydroxyl-free radicals and anti-inflammatory drugs: biological inactivation studies and reaction rate constants, *Biochem. Pharmacol.*, 32, 2109, 1983.
33. **Oyanagui, Y.,** Inhibition of superoxide anion production in macrophages by anti-inflammatory drugs, *Biochem. Pharmacol.*, 25, 1473, 1976.
34. **Oyanagui, Y.,** Inhibition of superoxide anion production in nonstimulated guinea pig peritoneal exudate cells by anti-inflammatory drugs, *Biochem. Pharmacol.*, 27, 777, 1978.
35. **Carlin, G., Djursäter, R., Smedegard, G., and Gerdin, B.,** Effect of anti-inflammatory drugs on xanthine oxidase and xanthine oxidase induced depolymerization of hyaluronic acid, *Agents Actions*, 16, 5, 1985.
36. **Babior, B. M.,** Oxygen-dependent microbial killing by phagocytes, *N. Engl. J. Med.*, 298, 659, 721, 1978.
37. **Greenwald, R. A. and Moy, W. W.,** Effect of oxygen-derived free radicals on hyaluronic acid, *Arthritis Rheum.*, 23, 455, 1980.
38. **Stephens, R. W., Sutherland, J., Ghosh, P., and Taylor, T. K. F.,** Human serum and synovial fluid hyaluronidase-bovine testicular hyaluronidase is not a valid substitute in drug evaluation studies, *Biochem. Pharmacol.*, 25, 1507, 1976.
39. **Blake, D. R., Hall, N. D., Treby, D. A., Halliwell, B., and Gutteridge, J. M. C.,** Protection against superoxide and hydrogen peroxide in synovial fluid from rheumatoid patients, *Clin. Sci.*, 61, 483, 1981.
40. **Al-Timimi, D. J. and Dormandy, T. L.,** The inhibition of lipid autoxidation by human caeruloplasmin, *Biochem. J.*, 168, 283, 1977.
41. **Rister, M.,** Superoxide anion and superoxide dismutase activity in arthritic conditions, *Agents Actions*, Suppl., 8, 137, 1981.
42. **Rister, M. and Bauermeister, K.,** Superoxide-dismutase and superoxide-radical-release in rheumatoid arthritis, *Klin. Wochenschr.*, 60, 561, 1982.
43. **Hemler, M. E., Cook, H. W., and Lands, W. E. M.,** Prostaglandin biosynthesis can be triggered by lipid peroxides, *Arch. Biochem. Biophys.*, 193, 340, 1979.
44. **Pasquier, C., Laoussadi, S., Sarfati, G., Raichvarg, D., and Amor, B.,** Superoxide dismutases in polymorphonuclear leukocytes from patients with ankylosing spondylitis or rheumatoid arthritis, *Clin. Exp. Rheum.*, 3, 123, 1985.
45. **Youssef, A. A. R. and Baron, D. N.,** Leukocyte superoxide dismutase in rheumatoid arthritis, *Ann. Rheum. Dis.*, 42, 558, 1983.
46. **D'Onofrio, C., Maly, F. E., Fischer, H., and Maas, D.,** Differential generation of chemiluminescence-detectable oxygen radicals by normal polymorphonuclear leukocytes challenged with sera from systemic lupus erythematosus and rheumatoid arthritis patients, *Klin. Wochenschr.*, 62, 710, 1984.
47. **Rister, M., Bauermeister, K., Gravert, U., and Gladtke, E.,** Superoxide-dismutase deficiency in rheumatoid arthritis, *Lancet*, I, 1094, 1978.
48. **Rister, M., Bauermeister, K., Gravert, U., and Gladtke, E.,** Superoxide dismutase and glutathione peroxidase in polymorphonuclear leukocytes, *Eur. J. Pediatr.*, 130, 127, 1979.
49. **Meske, S., Maly, F. E., Estefan, M., and Müller, W.,** Untersuchungen zur Sauerstoffradikalfreisetzung aus peripheren humanen Phagozyten (Granulozyten und Monozyten) bei Patienten mit chronischer Polyarthritis, *Z. Rheumatol.*, 44, 41, 1985.
50. **Sharma, S. C., Mukhtar, H., Sharma, S. K., and Murt, C. R. K.,** Lipid peroxide formation in experimental inflammation, *Biochem. Pharmacol.*, 21, 1210, 1972.
51. **Bragt, P. C., Schenkelaars, E. P. M., and Bonta, I. L.,** Dissociation between prostaglandin and malondialdehyde formation in exudate and increased levels of malondialdehyde in plasma and liver during granulomatous inflammation in the rat, *Prostaglandins Med.*, 2, 51, 1979.
52. **Nishigaki, I., Hagihara, M., Hiramatsu, M., Izawa, Y., and Yagi, K.,** Effect of thermal injury on lipid peroxide levels of rat, *Biochem. Med.*, 24, 185, 1980.
53. **Adolfs, M. J. P., Bonta, I. L., and Bragt, P. C.,** Increased lipid peroxidation and decreased heptatic aminopyrine metabolism during carrageenin-induced granulomatous inflammation in the rat, *Br. J. Pharmacol.*, 68, 123P, 1980.
54. **Robak, J.,** Adjuvant-induced and carrageenin-induced inflammation and lipid peroxidation in rat liver, spleen and lungs, *Biochem. Pharmacol.*, 27, 531, 1978.
55. **Yoshikawa, T., Tanaka, H., and Kondo, M.,** The increase of lipid peroxidation in rat adjuvant arthritis and its inhibition by superoxide dismutase, *Biochem. Med.*, 33, 320, 1985.
56. **Yoshikawa, T., Tanaka, H., and Kondo, M.,** Effect of vitamin E on adjuvant arthritis in rats, *Biochem. Med.*, 29, 227, 1983.
57. **Ohno, H., Takayanagi, N., Murase, J., and Iwata, H.,** Anti-inflammatory effect of Cu-Zn superoxide dismutase on carrageenin-induced arthritis in rabbits, *Nippon Yakurigaku Zasshi*, 82, 375, 1983.

58. **Muus, P., Bonta, I. L., and Den Oudsten, S. A.,** Plasma levels of malondialdehyde, a product of cyclo-oxygenase-dependent and independent lipid peroxidation in rheumatoid arthritis: a correlation with disease activity, *Prostaglandins Med.*, 2, 63, 1979.
59. **Lunec, J. and Dormandy, T. L.,** Fluorescent lipid peroxidation products in synovial fluid, *Clin. Sci.*, 56, 53, 1979.
60. **Lunec, J., Halloran, S. P., White, A. G., and Dormandy, T. L.,** Free-radical oxidation (peroxidation) products in serum and synovial fluid in rheumatoid arthritis, *J. Rheumatol.*, 8, 233, 1981.
61. **Rowley, D., Gutteridge, J. M. C., Blake, D., Farr, M., and Halliwell, B.,** Lipid peroxidation in rheumatoid arthritis: thiobarbituric acid-reactive material and catalytic iron salts in synovial fluid from rheumatoid patients, *Clin. Sci.*, 66, 691, 1984.
62. **Kaneda, H.,** A study on the lipid peroxide and its scavenging enzymes in rheumatoid arthritis, *Nippon Seikeigeka Gakkai Zasshi*, 56, 387, 1982.
63. **Palmoski, M. J. and Brandt, K. D.,** Effects of some nonsteroidal anti-inflammatory drugs on proteoglycan metabolism and organization in canine articular cartilage, *Arthritis Rheum.*, 23, 1010, 1980.
64. **Videman, T., Eronen, I., Friman, C., Ahtiala, K., and Mattila, M. J.,** Effects of indomethacin on glycosaminoglycan metabolism in the development of experimental osteoarthritis in rabbits, *Biochem. Pharmacol.*, 30, 2135, 1981.
65. **Bates, E. J., Lowther, D. A., and Johnson, C. C.,** Hyaluronic acid synthesis in articular cartilage: an inhibition by hydrogen peroxide, *Biochem. Biophys. Res. Commun.*, 132, 714, 1985.
66. **Bates, E. J., Harper, G. S., Lowther, D. A., and Preston, B. N.,** Effect of oxygen-derived reactive species on cartilage proteoglycan-hyaluronate aggregates, *Biochem. Int.*, 8, 629, 1984.
67. **Dayer, J. M., Roelke, M. S., and Krane, S. M.,** Effects of prostaglandin E_2 indomethacin, trifluoperazine and drugs affecting the cytoskeleton on collagenase production by cultures adherent rheumatoid synovial cells, *Biochem. Pharmacol.*, 33, 2893, 1984.
68. **Bocquet, J., Langris, M., Daireaux, M., Jouis, V., Pujol, J. P., Bailard, R., and Loyau, G.,** Mononuclear cell-mediated modulation of synovial cell metabolism. II. Increased hyaluronic acid synthesis by a monocyte cell factor (MCF), *Exp. Cell Res.*, 160, 9, 1985.
69. **McMillan, R. M., Vater, C. A., Hasselbacher, P., Hahn, J., and Harris, E. D.,** Induction of collagenase and prostaglandin synthesis in synovial fibroblasts treated with monosodium urate crystals, *J. Pharm. Pharmacol.*, 33, 382, 1981.
70. **Castor, C. W.,** Connective tissue activation. VII. Evidence supporting a role for prostaglandins and cyclic nucleotides, *J. Lab. Clin. Med.*, 85, 392, 1975.
71. **Galmiche, P. and Buneaux, J. J.,** Nouvelles recherches d'enzymologie synoviale, *Rev. Rhumatisme*, 43, 339, 1976.
72. **Thouveny, F., Thouveny, Y., Roux, H., and Recordier, A. M.,** Étude experimentale de la synthèse des enzymes lysosomiales par la synoviale: phosphatases acides, *Rev. Rhumatisme*, 43, 397, 1976.
73. **Fujihira, E. and Otomo, S.,** Stabilizing effect of anti-inflammatory drugs on erythrocyte membrane. II. On UV-induced canine erythrocyte hemolysis and lipid peroxidation, on UV-induced peroxide formation in extracted lipid, and on irradiated lipid-induced hemolysis, *Yakugaku Zasshi*, 90, 1355, 1970.
74. **Otomo, S. and Fujihira, E.,** Stabilizing effect of anti-inflammatory drugs on erythrocyte membrane. I. On hemolysis and lipid peroxidation induced by H_2O_2 in canine erythrocyte, *Yakugaku Zasshi*, 90, 1347, 1970.

α-TOCOPHEROL AND PLASMA LIPIDS IN MAN

John H. Kalbfleisch and Joseph J. Barboriak

INTRODUCTION

The prominent role of vascular diseases, especially atherosclerosis, as primary or contributing determinants of morbidity and mortality provides an opportunity for interventions aimed at modifying these variables and improving the quality of remaining life. Development of coronary artery disease (CAD) is associated with well-known risk factors; however, some of these factors, such as age, gender, and heredity, are unmodifiable. The results of a number of epidemiologic studies have shown that there are other risk factors that are amenable to modification.[1] These include high blood pressure, smoking, and abnormal blood lipid levels. This last variable is being investigated in depth and findings are available that indicate that elevated lipid levels, especially serum cholesterol, and very low levels of high-density lipoprotein cholesterol (HDL-C) may lead to a higher coronary heart disease morbidity and mortality.[2-4]

Among the many approaches attempting to reduce elevated blood lipid levels or aiming to increase the levels of HDL-C, the administration of α-tocopherol deserves special mention. The original observation reported by Herman et al. indicated a considerable increase in HDL-C levels following daily supplementation of the diet by 600 IU of α-tocopherol for a 4-week period.[5] These findings generated a great deal of interest since under normal conditions HDL-C levels are not easily altered by diet or pharmacologic treatment. The two factors which have been associated with increased HDL-C levels are regular prolonged physical activity and moderate alcohol intake. However, these methods may not be appropriate for application to the population at risk of developing CAD. Many of these subjects are physically not able to excercise to the extent necessary for a change in HDL-C and the recommendation of alcohol consumption is clearly associated with the danger of contributing to possible addiction. Thus, the relatively simple scheme of diet supplementation with α-tocopherol presents an apparently ideal means for modification of HDL-C levels. Since the increase in HDL-C levels is believed to be a protective factor against the development of atherosclerosis, Hermann's findings prompted a number of investigators to conduct further clinical trials and examine the α-tocopherol effects on HDL-C. In this presentation, most of the available reports on the effect of α-tocopherol on plasma HDL-C levels[6-19] as well as on the effect of α-tocopherol on other lipid classes have been reviewed.[20-23]

EFFECT OF α-TOCOPHEROL ON SERUM LIPIDS

Serum Total Cholesterol Levels

The results of studies that have measured total serum cholesterol before and after several weeks of α-tocopherol administration are summarized in Table 1. In this group of studies, the first report given by Harman in 1960 revealed no meaningful alteration of mean total cholesterol.[20] Among all studies, the observed changes in mean total cholesterol center about zero and most studies show changes of 15 mg/dℓ or less from the respective initial mean levels. Some of the studies included placebo-treated subjects and the results of these subsets are summarized in the lower part of Table 1. Placebo groups and α-tocopherol groups show a similar range of change in mean total cholesterol level. This similarity of changes in the α-tocopherol and placebo groups indicates an absence of an α-tocopherol influence on total cholesterol. While the total cholesterol level was essentially unaltered, α-tocopherol might

Table 1
EFFECT OF VITAMIN E ON SERUM TOTAL CHOLESTEROL LEVELS

Study subjects				Vitamin E		Initial	Final	Change	
M	F	Age	Type	Dose (IU)	Duration (weeks)	mean (mg/dℓ)	mean (mg/dℓ)	in mean (mg/dℓ)	Ref.
11	0	63.1	6 HS 5 CHD	600 mg	4	232.9	248.2	15.3	9
9	0	23 — 45	HS	300 mg	4	214.4	228.7	14.3	20
0	26	na	MD	600	3	177.2	191.2	14.0	10
7	8	26.0	HS	600	3	214.0s	225.2	11.2	15
0	13	42.4	HS	800	4	229.3	240.2	10.9	11
15	0	24.2	HS	400	3	164.7	175.2	10.5	17
13	26	39.4	HS	600	4	204.0	213.2	9.2	13
3	2	39.2	HL	800	6	332.4	341.2	8.8	6
15	0	36.1	HS	800	4	196.7	205.1	8.4	11
3	2	34.8	Low-HDL	600	5	184.6	193.0	8.4	5
4	1	34.2	HS	600	5	217.6	221.6	4.4	5
44	0	18 — 35	HS	600	4	205.2	209.5	4.3	22
0	46	18 — 35	HS	600	4	208.4	211.7	3.3	22
N = 15		30 — 60	HS	800	16	197.3	198.8	1.5	18
2	4	38.2	HS	800	5	204.3	205.8	1.5	6
7	2	35.0	HS	300	2	214.0	215.0	1.0	23
8	0	23 — 45	HS	600 mg	4	228.9	209.0	0.1	20
N = 6		31 — 55	HS	600	6	228.2	228.2	0.0	12
4	1	55.6	4 HL	400	4	297.4	297.4	0.0	14
6	3	38.0	HS	600	2	196.0	194.0	−2.0	23
7	0	41.0	RP	600	4	210.9	208.4	−2.5	16
0	10	47.0	HS	800	26	239.6	236.9	−2.7	19
20	0	44.0	HL	300	2	337.0	334.0	−3.0	23
0	13	44.5	HS	800	4	230.0	225.4	−4.6	19
14	12	42.0	HL	600	2	284.0	279.0	−5.0	23
29	0	58.4	CRP	800	4	235.5	228.5	−7.0	19
3	0	26.7	Low-HDL	800	4	215.0	207.3	−7.7	8
N = 15		30 — 60	HS	800	8	197.3	186.6	−10.7	18
26	0	61.7	CRP	800	26	242.8	231.5	−11.3	19
19	0	45.8	SCP	800	4	187.7	175.4	−12.3	11
11	3	49.2	HS	600	6	237.6	225.1	−12.5	7
1	3	25.0	HS	800	4	181.0	154.8	−26.2	8
0	5	na	HS	0	3	182.6	207.6	25.0	10
N = 15		30 — 60	HS	0	16	183.7	195.5	11.8	18
N = 15		30 — 60	HS	0	8	183.7	186.3	2.6	18
44	0	18 — 35	HS	0	4	206.6	207.5	0.9	22
0	12	56.7	HS	0	26	244.1	243.8	−0.3	19
17	0	59.2	CRP	0	26	238.8	235.1	−3.7	19
0	50	18 — 35	HS	0	4	217.3	212.1	−5.2	22
20	0	56.4	CRP	0	4	230.3	222.5	−7.8	19
0	3	25.0	HS	0	4	160.0	150.3	−9.7	8
0	16	49.8	HS	0	4	234.3	223.5	−10.8	19
10	5	26.0	HS	0	3	211.4s	200.3	−11.1	15
7	3	51.0	HS	0	6	236.2	225.1	−11.1	7

Note: HS = clinically normal healthy subjects: HL = hyperlipidemia: SCP = spinal cord patients: RP = renal patients: CHD = coronary heart disease: CRP = cardiac rehabilitation patients: MD = mammary dysplasia: low-HDL = subjects with low HDL-C at study start: N = total number of subjects: na = not available; and s = VLDL-C + LDL-C + HDL-C.

Table 2
EFFECT OF VITAMIN E ON SERUM TRIGLYCERIDE LEVELS

Study subjects				Vitamin E		Initial	Final	Change	
M	F	Age	Type	Dose (IU)	Duration (weeks)	mean (mg/dℓ)	mean (mg/dℓ)	in mean (mg/dℓ)	Ref.
4	1	55.6	4 HL	400	4	392.4	475.0	82.6	14
14	12	42.0	HL	600	2	252.0	306.0	54.0	23
11	0	63.1	6 HS 5 CHD	600mg	4	149.6	186.2	36.6	9
20	0	44.0	HL	300	2	737.0	767.0	30.0	23
0	10	47.0	HS	800	26	123.9	152.9	29.0	19
26	0	61.7	CRP	800	26	185.1	205.9	20.8	19
3	2	39.2	HL	800	6	112.0	125.4	13.4	6
0	46	18 — 35	HS	600	4	52.1	65.2	13.1	22
2	4	38.2	HS	800	6	89.2	99.5	10.3	15
0	13	44.5	HS	800	4	111.5	120.7	9.2	19
4	1	34.2	HS	600	5	82.2	86.4	4.2	5
7	2	35.0	HS	300	2	106.0	107.0	1.0	23
N = 15		30 — 60	HS	800	8	77.4	78.1	0.7	18
19	0	45.8	SCP	800	4	115.4	114.9	−0.5	11
15	0	24.2	HS	400	3	44.1	42.5	−1.6	17
0	13	36.1	HS	800	4	103.6	101.8	−1.8	11
44	0	18 — 35	HS	600	4	66.9	64.5	−2.6	22
1	3	25.0	HS	800	4	94.0	89.5	−4.5	8
29	0	58.4	CRP	800	4	175.4	168.7	−6.7	19
6	3	38.0	HS	600	2	136.0	128.0	−8.0	23
N = 15		30 — 60	HS	800	16	77.4	69.2	−8.2	18
7	0	41.0	RP	600	4	162.2	150.6	−11.6	16
13	26	39.4	HS	600	4	121.8	109.4	−12.4	13
15	0	36.1	HS	800	4	110.1	94.7	−15.4	11
11	3	49.2	HS	600	6	225.9	210.4	−15.5	7
3	2	34.8	Low-HDL	600	5	165.6	129.0	−36.6	5
3	0	26.7	Low-HDL	800	4	200.6	130.6	−70.0	8
0	50	18 — 35	HS	0	4	54.7	55.7	1.0	22
17	0	59.2	CRP	0	26	150.6	151.7	1.1	19
20	0	56.4	CRP	0	4	146.7	144.7	−2.0	19
0	12	56.7	HS	0	26	179.5	174.6	−4.9	19
7	3	51.0	HS	0	6	188.4	182.6	−5.8	7
0	3	25.0	HS	0	4	68.3	60.0	−8.3	8
44	0	18 — 35	HS	0	4	65.9	56.6	−9.3	22
N = 15		30 — 60	HS	0	8	91.3	87.4	−3.9	18
0	16	49.8	HS	0	4	158.6	144.8	−13.8	19
N = 15		30 — 60	HS	0	16	91.3	75.1	−16.2	18

Note: HS = clinically normal healthy subjects; HL = hyperlipidemia; SCP = spinal cord patients; RP = renal patients; CHD = coronary heart disease; CRP = cardiac rehabilitation patients; low-HDL = subjects with low HDL-C at start of study; and N = total number of subjects.

have produced a shift among the cholesterol levels of individual lipoproteins of which the total is composed. This is particularly important since the relative level of HDL-C is inversely related to the development of ischemic heart disease.

Serum Triglyceride Levels

High triglyceride levels are also believed to be a risk factor for cardiovascular disease. This is especially true for individuals with extensive coronary artery occlusion in whom elevated triglycerides raise blood viscosity and contribute to progression of the atherosclerotic process. A summary of results from studies that have examined the effect of α-tocopherol on this lipid is presented in Table 2. Inspection of Table 2 shows that, even in studies with hypertriglyceridemia subjects, α-tocopherol supplements did not reduce the elevated triglyceride levels.[6,14,21] Study groups with initially normal triglyceride levels exhibited changes from +36.6 to −36.6 mg/dℓ, with a average change of −1.5 mg/dℓ, clearly not evidence for a lowering effect. Control groups also showed a reduction in mean triglyceride level, indicating that supplementation with α-tocopherol is without any consistent effect on this lipid class.

EFFECT OF α-TOCOPHEROL ON LIPOPROTEINS

Serum High-Density Lipoprotein Cholesterol Levels

Cholesterol transport from body tissues to the liver is regulated by high-density lipoprotein (HDL); therefore, any mechanism that will increase the level of HDL-C would represent additional removal of some tissue cholesterol and thus retard the development of atherosclerosis and act as a protective factor in heart disease. Table 3 summarizes the effect of α-tocopherol supplementation on HDL-C as shown in the reviewed reports. The groups of subjects in Table 3 cover a wide age range, both sexes, and different health status. The initial reports of Herman et al. indicated a pronounced increase in HDL-C.[5,8] The subsequent studies prompted mostly by Herman's report could not reproduce the dramatic increase in HDL-C. Inspection of the reported results of the majority of studies shows that HDL-C levels were not markedly affected by α-tocopherol supplementation for periods up to 6 weeks. A few studies extended the treatment period up to 6 months; however, their results still indicated a small or no change in HDL-C level.[18,19] Changes in HDL-C in placebo-treated groups are also summarized in Table 3 and these changes are of the same order of magnitude as observed in the α-tocopherol-treated subjects. Because the total cholesterol level was essentially unaffected and the level of HDL-C did not change markedly, a major redistribution of the lipoprotein fractions as a result of α-tocopherol treatment appears unlikely.

Investigators have also studied the effect of different types of interventions on fractions of HDL-C. HDL can be separated into HDL2 and HDL3 by several methods and the cholesterol can be measured as HDL2-C and HDL3-C. The effect of α-tocopherol supplementation on HDL-C and HDL2-C and HDL3-C levels was examined in 1 study of 26 female patients with mammary dysplasia (Table 4).[10] This group of patients showed an increase in mean HDL-C (Table 3), little change in HDL2-C (Table 4), and a significant increase in HDL3-C (Table 4). The five healthy subjects in the same study did not show meaningful changes in these variables. Because of the selection of a special type of patient for study (women with mammary dysplasia), implication of similar effects for other populations would be unjustified.

Serum Low-Density Lipoprotein Cholesterol and Serum Very Low-Density Lipoprotein Cholesterol Levels

Elevated levels of plasma low-density lipoprotein (LDL) and very low-density lipoprotein (VLDL) have been shown to be associated with an increased incidence of atherosclerosis.[24] The cholesterol of LDL (LDL-C) and VLDL (VLDL-C) is taken up by body tissues and therefore higher levels of these lipoproteins have a unfavorable prognostic indication for atherosclerosis and related circulatory diseases. Few of the reviewed studies investigated the effect of α-tocopherol on VLDL-C and LDL-C. The results that are available are summarized in Table 4. No appreciable decrease in either VLDL-C or LDL-C is consistently reported in groups of subjects receiving α-tocopherol supplements. It is thus tentatively concluded that α-tocopherol does not influence the level of VLDL-C or LDL-C. This information and the aforementioned effects of α-tocopherol on HDL-C suggest that vitamin E does not modify a possible redistribution of cholesterol among the lipoprotein fractions.

Table 3
EFFECT OF VITAMIN-E ON SERUM HDL-C LEVELS

Study subjects				Vitamin E		Initial	Final	Change	
				Dose	Duration	mean	mean	in mean	
M	F	Age	Type	(IU)	(weeks)	(mg/dℓ)	(mg/dℓ)	(mg/dℓ)	Ref.
3	2	34.8	Low-HDL	600	5	16.8	61.2	44.4	5
4	1	34.2	HS	600	5	45.4	72.0	26.6	5
3	0	26.7	Low-HDL	800	4	31.7	47.7	16.0	8
0	26	na	MD	600	3	46.0	52.9	6.9	10
0	13	42.4	HS	800	4	65.9	71.4	5.5	11
19	0	45.8	SCP	800	4	41.4	43.3	2.2	11
2	4	38.2	HS	800	6	66.8	68.7	1.9	6
15	0	24.2	HS	400 mg	3	41.0	42.5	1.5	17
15	0	36.1	HS	800	4	60.6	61.7	1.1	11
13	26	39.4	HS	600	4	48.5	48.9	0.4	13
3	2	39.2	HL	800	6	45.6	45.8	0.2	6
7	0	41.0	RP	600	4	37.6	37.6	0.0	16
N = 15		30 — 60	HS	800	8	53.3	52.9	−0.4	18
N = 15		30 — 60	HS	800	16	53.3	52.4	−0.9	18
29	0	58.4	CRP	800	4	47.3	46.3	−1.0	19
4	1	55.6	4 HL	400	4	41.0	39.8	−1.2	14
11	3	49.2	HS	600	6	38.2	36.9	−1.3	7
N = 6		31 — 55	HS	600	6	62.6	61.1	−1.5	12
11	0	63.1	6 HS 5 CHD	600 mg	4	46.2	42.9	−3.3	9
0	10	47.0	HS	800	26	68.1	64.6	−3.5	19
7	8	26.0	HS	600	3	54.3e	50.4e	−3.9	15
26	0	61.7	CRP	800	26	46.8	42.3	−4.5	19
0	13	44.5	HS	800	4	66.8	62.0	−4.8	19
1	3	25.0	HS	800	4	59.5	45.8	−13.7	8
10	5	26.0	HS	0	3	50.1	52.8	2.7	15
0	12	56.7	HS	0	26	56.1	56.1	0.0	15
7	3	51.0	HS	0	6	31.7	31.6	−0.1	7
N = 15		30 — 60	HS	0	8	51.1	50.8	−0.3	18
17	0	59.2	CRP	0	26	46.8	45.8	−1.0	19
0	16	49.8	HS	0	4	60.2	58.1	−2.1	19
0	5	na	HS	0	3	49.2	48.0	−1.2	10
N = 15		30 — 60	HS	0	16	51.1	48.5	−2.6	18
20	0	56.4	CRP	0	4	48.5	44.4	−4.1	19
0	3	25.0	HS	0	4	70.2	54.9	−15.3	8

Note: HS = clinically normal healthy subjects; HL = hyperlipidemia; SCP = spinal cord patients; RP = renal patients; CHD = coronary heart disease; CRP = cardiac rehabilitation patients; MD = mammary dysplasia; low-HDL = subjects with low-HDL at start of study; N = total number of subjects; na = not available; and e = estimated from graph.

APOLIPOPROTEINS

Only approximately 25% of the HDL molecule consists of cholesterol (by weight). Some research efforts have searched for treatment effects on apolipoproteins, since these protein constituents account for approximately half (50%) of the HDLs in plasma. Additionally, recent studies in postcoronary artery by-pass patients have shown that changes in apolipoproteins may be better correlated with occlusion scores (a measure of atherosclerotic lesion severity) than HDL-C.

The main apolipoprotein of HDL is apolipoprotein A (90%) and this is divided into apolipoprotein A-I and apolipoprotein A-II. Apolipoprotein C and E account for the remainder of the apolipoproteins of HDL. Apolipoprotein B is a major constituent of LDL. To date,

Table 4
EFFECT OF VITAMIN E ON VLDL-C, LDL-C, AND HDL-C SUBCLASSES AND APOLIPOPROTEIN LEVELS

Study variate	Study subjects M	F	Age	Type	Vitamin E Dose (IU)	Vitamin E Duration (weeks)	Initial mean (mg/dℓ)	Final mean (mg/dℓ)	Change in mean (mg/dℓ)	Ref.
VLDL-C	4	1	55.6	4 HL	400	4	118.6	126.8	8.2	14
VLDL-C	7	8	26.0	HS	600	3	15.1e	16.3	1.2	15
VLDL-C	15	0	24.2	HS	400 mg	3	25.4	23.6	−1.8	17
VLDL-C	10	5	26.0	HS	0	3	19.0e	14.9	−4.1	15
LDL-C	7	8	26.0	HS	600	3	144.6e	158.5	13.9	15
LDL-C	15	0	24.2	HS	400 mg	3	99.0	110.6	11.6	17
LDL-C	N = 15		30 — 60	HS	800	16	128.6c	132.7	4.1	18
LDL-C	4	1	55.6	4 HL	400	4	137.4	130.0	−7.4	14
LDL-C	N = 15		30 — 60	HS	800	8	128.6c	118.1	−10.5	18
LDL-C	N = 15		30 — 60	HS	0	16	114.3c	131.9	17.6	18
LDL-C	N = 15		30 — 60	HS	0	8	114.3c	117.9	3.6	18
LDL-C	10	5	26.0	HS	0	3	142.3e	132.6	−9.7	15
VLDL + LDL	0	5	na	HS	600	3	136.3	163.5	26.7	10
VLDL + LDL	0	26	na	MD	600	3	131.6	129.4	−2.2	10
HDL2-C	0	26	na	MD	600	3	18.0	20.0	2.0	10
HDL2-C	0	5	na	HS	0	3	22.8	20.8	−2.0	10
HDL3-C	0	26	na	MD	600	3	20.7	26.0	5.3	10
HDL3-C	0	5	na	HS	0	3	22.3	23.3	1.0	10
Apo A-I	N = 6		30 — 55	HS	600	6	149.0	149.0	0.0	12
Apo A-II	N = 6		30 — 55	HS	600	6	49.0	52.0	3.0	12
Apo B	N = 6		30 — 55	HS	600	6	83.0	84.0	1.0	12

Note: HS = clinically normal healthy subjects; HL = hyperlipidemia; MD = mammary dysplasia; N = total number of subjects; na = not available; c = calculated from total cholesterol, HDL-C, and triglyceride; and e = estimated from graph.

there is only one reported study (with six subjects) with information on apolipoprotein A-I, A-II, and B following α-tocopherol supplementation.[12] As shown in Table 4, mean levels of A-I, A-II, and B showed little change after 6 weeks of treatment.

CLINICAL IMPLICATIONS

Clinical research commonly tests for the effect of a treatment or intervention by (1) detecting a change in a directly measured end-point variable such as morbidity, mortality, or quality of life or by (2) documenting a change in a variable that relates to (or is highly prognostic for) an end-point variable. The α-tocopherol studies have examined for effects on lipid classes such as cholesterol and triglycerides and subsets of lipoproteins, i.e., variables that are amenable to modification and can be expected to aid in the prevention or attenuation of cardiovascular disease. High levels of circulating lipids are a risk factor for atherosclerosis and myocardial infarction, while HDL-C is inversely related to the risk of ischemic heart disease.[12] A decrease in HDL is associated with an increase in atheromatous lesions; similarly an increase in LDL is associated with an increase in the incidence of atheroma.[15] Although it is difficult to compare the effects of α-tocopherol among the aforementioned individual studies (because of the different ages, lipid profiles, and health status of the study subjects), an overview of these studies suggests that α-tocopherol is not consistently effective as an agent for raising HDL-C, lowering serum total cholesterol levels, or lowering serum triglyceride levels. Additionally, there are only a few studies that have reported values for

LDL-C, VLDL-C, HDL-C subsets, and individual apolipoproteins. In these reports, the changes resulting from α-tocopherol supplementation are small. However, further investigation is necessary before more universal conclusions can be made.

The general lack of a consistent α-tocopherol effect on serum lipids and lipoproteins does not mean that vitamin E is not useful for modifying other biological processes which may be involved in the genesis of heart disease. A possible benefit of vitamin E would be in controlling processes which lead to or promote atherosclerosis and, in particular, the reduction of free redical formation (since this has been implicated as contributing to vascular damage).[23] The administration (or presence) of vitamin E is important in protecting cell membranes from the effects of lipid peroxidation, while its absence or deficit promotes cell membrane damage (by the uncoupling of phospholipids and protein necessary for the formation of the cell membrane).[23] A reduced formation of lesions in the arterial wall can be an effect of vitamin E that is independent of a direct effect on circulating cholesterol. Blood coagulability is potentially influenced since vitamin E has been reported to have a mild inhibitory action on platelet aggregation.[23] Although there are several processes that have been identified as possibly modified by α-tocopherol treatment, there are few or no clinical studies to confirm such activity in patients. Even if an effect of α-tocopherol on plasma lipids or another variable can be documented, one should realize that some of these changes may require a long time to be reflected in a change in morbidity or mortality.

This paper summarizes results of several studies that have investigated the α-tocopherol effects on serum lipids and lipoproteins. A review of these studies illustrates some common problems encountered in clinical studies. Several of the study groups had a very small number of subjects and consquently lack statistical precision. Other studies did not include a placebo group. Inclusion of a placebo-treated group in a randomized double-blind trial affords a means of controlling undesired or unavoidable kinds of bias. For example, participation in a study might be accompanied with a change in the usual lifestyle (diet, physical activity) in some subjects. If subjects are randomized to study groups, one of which is placebo, and if group sizes are adequate, then comparison of the treated groups with the placebo group tests for a treatment effect free of bias, if biasing factors are distributed randomly to all study groups. The study of subsets of patients requires extra care since the definition of a subset might create biasing effects if a placebo-treated group is not available for comparison. A subset of subjects formed by a restricted range of initial values (e.g., low HDL-C) could generate an additional effect (regression toward the mean) that becomes comingled with a (possible) real treatment effect. A placebo comparison group would help the investigator detect and account for such a problem. Lastly, the majority of studies were short term. One study was 16 weeks long and one was 26 weeks long.[18,19] Extrapolation of results observed over a short term to much longer studies could be misleading since the observed effects may be temporary and the presence of possible side effects due to a long-term treatment may remain unrecognized.

CONCLUSIONS

The impression of the initial favorable effect of α-tocopherol on HDL-C could not be reproduced in subsequent studies. One would expect that if such an effect is clearly demonstrable, then it should be observed in additional studies with subjects of a wide age range, varying health conditions, and, to some extent, different α-tocopherol dosage. The design of some of the reported studies was not ideal and it remains possible that better designed studies (double blind, placebo controlled) may demonstrate a beneficial effect of α-tocopherol. Nevertheless, it seems that if there is an effect of α-tocopherol on HDL-C, this effect is probably not as dramatic as suggested by the initial studies.

REFERENCES

1. **Levy, R. I.,** Cholesterol and cardiovascular disease: no longer whether, but rather when, in whom, and how?, *Circulation,* 72, 686, 1985.
2. **Gordon, T., Castelli, W. P., Hjortland, M. C., Kannel, W. B., and Dawber, T. R.,** High density lipo-protein as a protective factor against coronary heart disease, *Am. J. Med.,* 62, 707, 1977.
3. **Miller, G. J. and Miller, N. E.,** Plasma-high-density-lipoprotein concentration and the development of ischemic heart disease, *Lancet,* 1, 16, 1975.
4. **Carew, T. E., Kochinsky, T., Hayes, S. B., and Steinberg, D.,** A mechanism by which high-density lipoproteins may slow the atherogenic process, *Lancet,* 1, 1315, 1976.
5. **Herman, W. J., Ward, K., and Faucett, J.,** The effect of alpha-tocopherol on high-density lipoprotein cholesterol, *Am. J. Clin. Pathol.,* 72, 848, 1979.
6. **Hatam, L. and Kayden, H.,** The failure of alpha-tocopherol supplementation to alter the distribution of lipoprotein cholesterol in normal and hyperlipoproteinemic persons, *Am. J. Clin. Pathols.,* 76, 122, 1981.
7. **Herman, W. J.,** Letter to editor, *Am. J. Clin. Pathol.,* 76, 124, 1981.
8. **Herman, W. J.,** Letter to editor, *Am. J. Clin. Pathol.,* 76, 844, 1981.
9. **Schwartz, P. L. and Rutherford, I. M.,** The effect of tocopherol on high-density lipoprotein cholesterol, *Am. J. Clin. Pathol.,* 76, 843, 1981.
10. **Sundaram, G. S., London, R., Manimekala, S., Nair, P. P., and Goldstein, P.,** Alpha-tocopherol and serum lipoproteins, *Lipids,* 16, 223, 1981.
11. **Barboriak, J. J., El Ghatit, A. Z., Shetty, K. R., and Kalbfleisch, J. H.,** Vitamin E supplements and plasma high-density lipoprotein cholesterol, *Am. J. Clin. Pathol.,* 77, 371, 1982.
12. **Ehnholm, C., Huttunen, J. K., Kostiainen, E., Lukka, M., and Abo, K.,** Vitamin E does not influence plasma lipoprotein metabolism in healthy subjects with normal nutritional status, *Clin. Chem. Acta,* 121, 321, 1982.
13. **Howard, D. R., Rundell, C. A., and Batsakis, J. G.,** Vitamin E does not modify HDL-cholesterol, *Am. J. Clin. Pathol.,* 77, 86, 1982.
14. **Kesaniemi, Y. A. and Grundy, S. M.,** Lack of effect of tocopherol on plasma lipids and lipoprotein in man, *Am. J. Clin. Nutr.,* 36, 224, 1982.
15. **Bolster, M., Cole, M., Duggan, P. F., and Crowley, M.,** Lack of effect of ingestion of vitamin E on lipoprotein-cholesterol fractions in healthy adults, *Ir. Med. J.,* 76, 152, 1983.
16. **Chapkin, R. S., Haberstroh, B., Liu, T., and Holub, B. J.,** Effect of vitamin E supplementation on serum and high-density lipoprotein cholesterol in renal patients on maintenance hemodialysis, *Am. J. Clin. Nutr.,* 38, 353, 1983.
17. **Serfontein, W. J., Ubbink, J. B., and de Villiers, L. S.,** Further evidence on the effect of vitamin E on the cholesterol distribution in lipoprotein with special reference to HDL subfractions, *Am. J. Clin. Pathol.,* 79, 604, 1983.
18. **Stampfer, M. J., Willett, W., Castelli, W. P., Taylor, J. O., Fine, J., and Hennekens, C. H.,** Effect of vitamin E on lipids, *Am. J. Clin. Pathol.,* 79, 714, 1983.
19. **Kalbfleisch, J. H., Barboriak, J. J., Else, B. A., Hughes, C. V., and Tristani, F. E.,** Alpha-tocopherol supplements and high-density lipoprotein levels, *Br. J. Nutr.,* 55, 71, 1986.
20. **Harman, D.,** Vitamin E: effect on serum cholesterol and lipoproteins, *Circulation,* 22, 151, 1960.
21. **Farrell, M. P. and Bieri, J. G.,** Megavitamin E supplementation in man, *Am. J. Clin. Nutr.,* 28, 1381, 1975.
22. **Tsai, A. C., Kelley, J. J., Peng, B., and Cook, N.,** Study on the effect of megavitamin E supplementation in man, *Am. J. Clin. Nutr.,* 31, 831, 1978.
23. **Szczeklik, A., Gryglewski, R. J., Domagala, B., Dworski, R., and Basista, M.,** Dietary supplementation with vitamin E in hyperlipoproteinemias: effects on plasma lipid peroxides, antioxidant activity, prostacyclin generation and platelet aggregability, *Throm. Haemorrh.,* 54, 425, 1985.
24. **Castelli, W. P., Doyle, J. T., and Gordon, T.,** HDL cholesterol and other lipids in coronary heart disease: the cooperative lipoprotein phenotyping study, *Circulation,* 55, 767, 1977.

EFFECT OF VITAMIN E ON LOW-DENSITY LIPOPROTEIN METABOLISM IN THE ARTERIAL WALL

Yasushi Saito and Nobuhiro Morisaki

INTRODUCTION

Various compounds, including proteins, sugars, and lipids, are metabolized in the arterial wall[1] as in other tissues and various changes of their metabolism are observed in some pathological conditions that cause the risk of atherosclerosis.[2-8] Lipid accumulation, especially that of cholesterol ester, in the arterial wall is a major finding in atheromatous lesions. Lipid accumulation is caused by disturbances of the metabolism of lipids originating from low-density lipoprotein (LDL) incorporated into the cells.[9] Recently, not only native LDL but also denatured LDL has been examined as an atherogenic lipoprotein. Native LDL is metabolized in parenchymal cells, such as smooth muscle cells and endothelial cells, whereas denatured LDL, but not native LDL, is metabolized in the macrophages called scavenger cells.[10] Denatured LDL has not been found in human plasma in vitro, but it is prepared in vitro by chemical modifications such as acetylation or methylation.[11] Denatured LDL has a more negatively charged surface structure than native LDL and so moves further to the anode on electrophoresis. Native LDL binds to the LDL receptor and enters parenchymal cells where it is metabolized in lysosomes or microsomes (LDL pathway).[12] The binding of LDL is regulated by the amount of LDL incorporated into the cells. Excess incorporation of LDL into the cells decreases the LDL receptor activity (down regulation of the receptor). Because of this down regulation, excess uptake of LDL by the cells does not occur, and thus foam cell formation from parenchymal cells in the presence of native LDL has not been observed. However, there seems to be no down-regulation mechanism in the metabolism of denatured LDL in macrophages: large amounts of denatured LDL can be incorporated and cause lipid accumulation, because the uptake is beyond the capacity of the macrophages to metabolize the denatured LDL.

Recently, the formation of denatured LDL from LDL was reported to be mediated by endothelial cells in culture.[13] LDL that had been incubated with endothelial cells was mainly incorporated into macrophages, not into parenchymal cells, and the denaturation of LDL by the endothelial cells was suggested to be induced by peroxidation, because the value for thiobarbituric acid reactive substance (TBARS) of LDL treated with endothelial cells was increased and antioxidants prevented the denaturation of LDL.[14] Furthermore, the formation of oxidized LDL was demonstrated using a platelet aggregation system in vitro.[15] A certain level of lipid peroxides, measured as TBARS, was detected in LDL.[16] The correlation between the TBARS content of LDL and clinical atherosclerosis remains to be studied, but in the early 1950s, Glavind et al.[17] reported that the lipid peroxide value in the arterial wall increases with progression of atheroma. Thus, results suggest that lipid peroxidation is important in the metabolism of LDL in the arterial wall and in the formation of atheroma. Vitamin E is reported to have an antioxidant action in various tissues and to prevent initiation or progression of pathological conditions in some tissues.[18] Negative correlations between the vitamin E content and atherosclerosis have been observed in the arterial wall and blood.[19-21] In this chapter, LDL metabolism and its biological action are discussed in relation to lipid peroxidation and antioxidants in endothelial cells, smooth muscle cells, and macrophages, which are the main cells composing the arterial wall.

LIPID METABOLISM IN THE ARTERIAL WALL AND VITAMIN E

The major lipid accumulated in atheromatous lesions is cholesterol ester, which is mainly

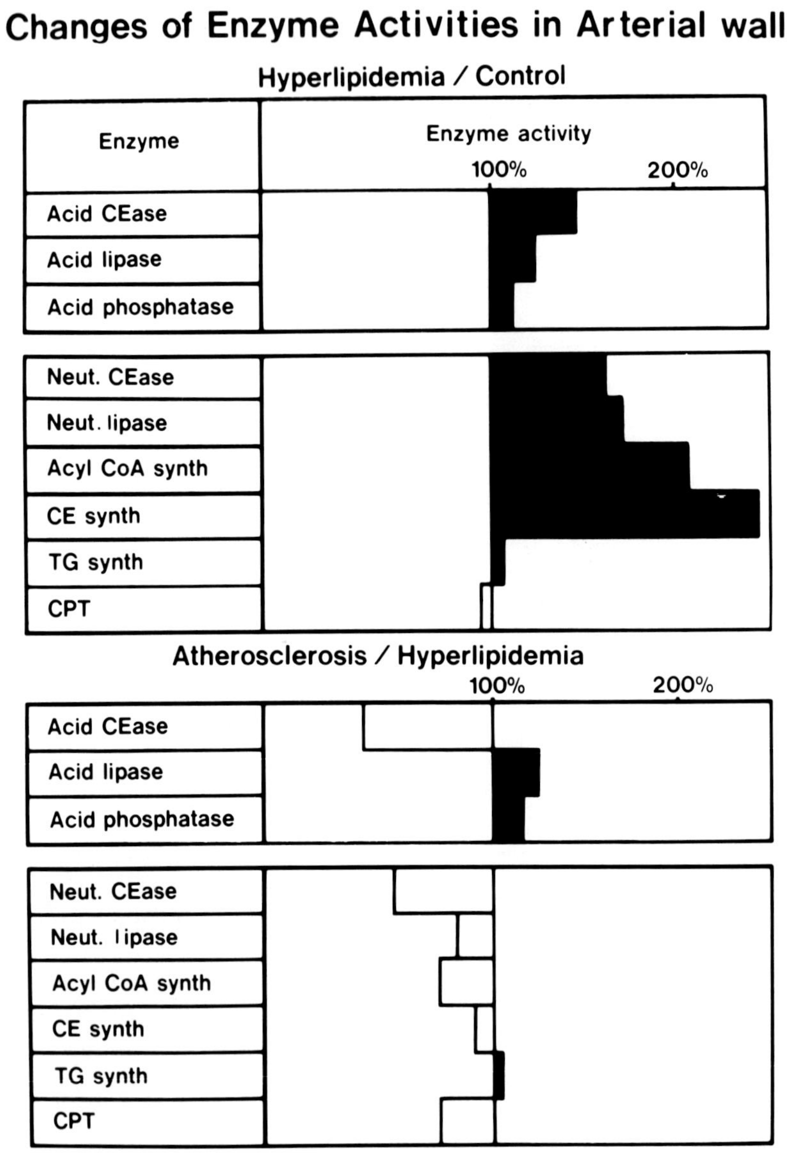

FIGURE 1. Changes of enzyme activities in the arterial wall; enzyme activities in rat aorta in hyperlipidemia (without atherosclerosis) as percentages of those in control aorta and of those in atherosclerosis as percentages of those in hyperlipidemia (without atherosclerosis). Closed and open columns show percentage increases and decreases, respectively. For details, see Reference 6, and for abbreviations, see the note to Table 1A.

derived from LDL. When the arterial wall is exposed to a high concentration of LDL, the metabolism of not only cholesterol ester, but also of phospholipid, FFA, and triglyceride increase.[6] The activities of the enzymes involved in these metabolic processes are also almost all increased in the experimental hypercholesterolemic rate arterial wall without atheromatous lesions, as shown in Figure 1. These increases are considered to be changes that tend to prevent lipid accumulation in the cells. However, all these enzyme activities except that of the cholesterol esterifying enzyme are lower in atheromatous lesions than in nonatheromatous lesions. The changes from nonatheromatous to atheromatous lesions due to hypercholesterolemia might be associated with change from a hypermetabolic to a hypometabolic state in

Table 1A
EFFECTS OF TOCOPHEROL DEFICIENCY ON ENZYME ACTIVITIES IN THE ARTERIAL WALL OF NORMOTENSIVE RATS ON NORMAL AND HIGH CHOLESTEROL DIETS

Enzyme activity (nmol/mg protein)	Normal diet		High cholesterol diet	
	TS	TD	TS	TD
Acid CEase	5.2 ± 0.5	1.1 ± 0.2*	7.2 ± 0.6	4.9 ± 1.2*
Neutral CEase	0.63 ± 0.33	0.560 ± 0.22	0.98 ± 0.7	0.77 ± 0.27
Acid lipase	14.2 ± 1.2	5.7 ± 0.9*	42.3 ± 1.2	15.4 ± 1.6*
Neutral lipase	11.9 ± 1.8	7.4 ± 1.1*	19.5 ± 1.3	16.7 ± 0.7*
Acyl-CoA synth	42.7 ± 4.4	24.9 ± 4.1*	62.2 ± 3.8	36.3 ± 2.9*
TG synth	1.62 ± 0.27	0.89 ± 0.09*	1.96 ± 0.14	1.77 ± 0.20
CPT	2.74 ± 0.86	13.79 ± 0.82	3.7 ± 1.3	7.4 ± 2.4*

Note: Values are means ±SD. TS, tocopherol-supplemented diet: TD, tocopherol-deficient diet: CEase, cholesterol esterase: TG, triglyceride: *, significant difference: synth, synthetase: CPT, cholinephosphotransferase.

From Shirai, K., Matsuoka, N., Morisaki, N., Murano, S., Sasaki, N., Shinomiya, M., Saito, Y., Kumagai, A., and Mizobuchi, M., *Artery*, 6, 484, 1980. With permission.

Table 1B
EFFECTS OF TOCOPHEROL DEFICIENCY ON ENZYME ACTIVITIES IN THE ARTERIAL WALL OF SPONTANEOUSLY HYPERTENSIVE RATS ON NORMAL AND HIGH CHOLESTEROL DIETS

Enzyme activity (nmol/mg protein)	Normal diet		High cholesterol diet	
	TS	TD	TS	TD
Acid CEase	9.1 ± 0.7	2.5 ± 1.4*	7.3 ± 2.3	2.6 ± 0.6*
Neutral CEase	3.2 ± 0.7	2.7 ± 0.2	3.3 ± 0.6	2.8 ± 0.9
Acid lipase	10.3 ± 1.9	13.0 ± 5.0	17.2 ± 2.7	12.7 ± 3.0*
Neutral lipase	13.4 ± 0.5	15.9 ± 8.8	20.1 ± 4.0	17.2 ± 4.8
Acyl-CoA synth	10.7 ± 3.0	9.0 ± 2.1	23.0 ± 9.0	16.9 ± 2.5
TG synth	2.18 ± 0.12	2.44 ± 0.23	3.61 ± 1.20	1.67 ± 0.25*
CPT	0.402 ± 0.063	0.285 ± 0.084	0.507 ± 0.134	0.436 ± 0.094

Note: See note for Table 1A.

From Murano, S., Shirai, K., Matsuoka, N., Morisaki, N., Sasaki, N., Shinomiya, M., Mizobuchi, M., Saito, Y., and Kumagai, A., *Tohoku J. Exp. Med.*, 136, 29, 1982. With permission.

the arterial wall. Therefore, it is important to elucidate the changes in the enzyme activities in conditions with risk of atherosclerosis, such as tocopherol deficiency, to clarify the correlation between lipid peroxidation and atherosclerosis. When rats were given a tocopherol-deficient diet for 16 weeks, no atheromatous lesions were observed, but TBARS increased in their arterial wall, and in other tissues.[4,5] The activities of the enzymes mentioned above in thse rats are shown in Tables 1A and B. The observed changes in their activities were similar to those in atherosclerotic lesions. These results suggest that metabolic changes in tocopherol-deficient rat arterial wall tend to cause formation of atheroma.

CAUSE OF DERANGEMENT OF ENZYME ACTIVITIES IN THE ARTERIAL WALL IN TOCOPHEROL-DEFICIENT RATS

The enzymes assayed in studies on this subject were mainly localized in particulate fractions isolated by subcellular fractionation, such as lysosomes, microsomes, and mitochondria.[3-8, 22-25] Activities could be demonstrated in the membranes only when the enzymes were located in those structures. Lipid peroxidation is thought to cause membrane fragility and increased release of enzymes from the membranes.[26] These results suggest that increase in lipid peroxide may result in decrease in the activities of these enzymes because of a change in the localization of the enzymes in the cells. Another possibility is that lipid peroxides denaturate the enzymes because they affect protein structure.[27,28] These results indicate that in tocopherol deficiency, metabolic disturbances in the arterial wall could occur not only by denaturation of LDL, but also by changes in enzymes. The following section reports results on the metabolism of cells isolated from the arterial wall.

LDL AND ENDOTHELIAL CELLS (ECs) (TABLES 2A AND B AND FIGURE 2)

As in other parenchymal cells, in ECs, LDL was shown to be metabolized through the LDL pathway.[29] However, in this section, only the interaction between LDL and ECs through lipid peroxidation is discussed.

Ross and Glomset proposed a classical ''response to injury'' hypothesis,[30,31] which was somewhat amended recently.[32] In both the classical and amended hypothesis, EC injury is important for initiation of atherosclerosis. when Ecs are injured by various factors, (1) the permeability of the vessels increases, resulting in a large influx of plasma components, and (2) platelets adhere and aggregate at injured loci, resulting in the secretion of many factors including platelet-derived growth factor[33] and chemotactic factors for smooth muscle cells such as 12-hydroxyeicosatetraenoic acid.[34] These responses to EC injury may lead to many atherosclerotic processes. The factors and mechanisms of EC injury are not well understood. Possible factors are mechanical injury, LDL, lipid peroxidation, immunologic complexes, toxins, viruses, and homocysteine.[30-32] Of these, LDL may be the most important under physiological conditions. LDL cytotoxicity was first observed in human umbilical ECs[35,36] and human arterial smooth muscle cells (SMCs).[36] LDL cytotoxicity is inhibited by high-density lipoprotein (HDL).[36,38] The protective effect of HDL on LDL cytotoxicity has also been observed with bovine ECs.[39] The mechanism of HDL protection is unknown. LDL cytotoxicity is not species-specific or cell-specific (Table 2A) and is not mediated by LDL receptor-mediated uptake of LDL.[41] However, native LDL itself is not toxic to ECs.[40,43-45] ECs are sensitive to peroxidative attack, and only oxidized LDL, demonstrated by its reactivity to thiobarbituric acid, is toxic to cultured ECs.[45] Addition of general radical scavengers, such as vitamin E, butylated hydroxytoluene, and glutathione, or superoxide dismutase, which selectively scavenges the superoxide anion, to the serum during isolation of LDL prevented the cytotoxicity of LDL to ECs. Treatment with catalase, which scavenges hydrogen peroxide, was also partially effective in preventing LDL cytotoxicity,[43] and the chelating agent EDTA is also reported to be effective.[40,43,44] This effect of EDTA was attributed to its action in chelating divalent cations, which are necessary for cell free lipid peroxidation. These compounds were effective only when cells were treated with them before isolation of LDL.[43] Thus, the cytoxicity of LDL was exerted by oxidized LDL formed during isolation of LDL and the superoxide anion or hydrogen peroxide was involved in this oxidation of LDL. There are reports that lipid peroxides or oxidized sterols, which could be present in plasma LDL, are toxic to the artery. Injection of oxidative derivatives of cholesterol, such as 25-hydroxycholesterol and cholestane-3β,5α,6β-triol induced EC injury

Table 2A
MODIFICATION OF LDL IN CELL-FREE SYSTEMS AND BIOLOGICAL EFFECTS OF THE MODIFICATION (RELATED TO LIPID PEROXIDATION ONLY)[a]

Method of modification	Target cells	Biological effect	Inhibitors	Ref.
During isolation (?)	huv EC	Toxicity		35
During isolation (?)	huv EC, human aortic SMC	Toxicity	HDL	36
During isolation (?)	huv EC	Toxicity		37
During isolation (?)	huv EC	Toxicity	HDL	38
During isolation (?)	Bovine aortic EC	Toxicity	HDL	39
Malondialdehyde (MDA) glutaraldehyde	Human blood monocytes	Lipid metabolism		15
During isolation	huv EC, human aortic SMC	Toxicity	EDTA	40
During isolation (?)				41
During isolation (?)	Human fibroblasts	Toxicity	HDL	42
During isolation	Human fibroblasts	Toxicity	Vitamin E, BHT, EDTA, SOD, catalase	43
During isolation	Human fibroblasts	Toxicity	EDTA, glutathione HDL	44
During isolation	huv EC, human fibroblasts	Toxicity	BHT, SOD	45
Cu^{2+}	Mouse peritoneal Mϕ	Lipid metabolism	Probucol	46

[a] EC, endothelial cells; huv, human umbilical vein; SMC, smooth muscle cell; Mϕ, macrophages; HDL, high-density lipoprotein; EDTA, ethylenediamine tetraacetic acid; BHT, butylated hydroxytoluene; BHA, butylated hydroxyanisole; and SOD, superoxide dismutase.

in vivo.[54,55] These compounds were also toxic to aortic SMCs in vitro.[56,57] Injection of the lipid peroxide of linoleic acid also injured ECs in vivo.[58] It is still unknown whether substantial amounts of oxidized LDL are present in the plasma. LDL contains cholesterol and phospholipids, which could both be oxidized. Actually vitamin E in the plasma is mainly present in LDL,[59] and so LDL is usually protected from peroxidation attack. However, oxidized LDL (lipid peroxides and oxidized sterols) could be present under certain conditions.[16,58,60]

There are reports that ECs themselves may form oxidized LDL (EC-modified LDL).[13,14,46-49,51,52] However, it is unknown whether LDL modified by ECs is toxic to ECs themselves. The biological significance of LDL modified by ECs will be discussed in relation to macrophages.

Not only oxidized LDL in the plasma, but also that in the arterial wall must be important in injury of ECs. Glavind et al.[17] reported that the lipid peroxide content of the arterial wall

Table 2B
MODIFICATION OF LDL BY CELLS AND BIOLOGICAL EFFECTS OF THE MODIFICATION[a]

Cell line	Target cells	Biological effect	Inhibitors	Ref.
Aggregated platelets	Human blood monocytes	Lipid metabolism		15
Rabbit aortic EC	Mouse peritoneal Mϕ	Lipid metabolism		13,47
huvEC, rabbit aortic EC, swine, guinea pig SMC	Mϕ (J774 cells) mouse peritoneal Mϕ	Lipid metabolism		48
huvEC, rabbit aortic EC	Mouse peritoneal Mϕ	Lipid metabolism	Vitamin E, BHT, EDTA	14
huvEC, bovine aortic SMC	Human fibroblasts	Toxicity	BHT	49
Human aortic SMC (+Fe or Cu)	Human blood monocytes	Lipid metabolism	BHT	50
Rabbit aortic EC			Phospholipase inhibitor (*p*-bromophenacyl bromide)	51
Rabbit and bovine aortic EC	Mouse peritoneal Mϕ	Motility		52
Rabbit aortic EC	Mouse peritoneal Mϕ	Lipid metabolism	Probucol	46
Human aortic SMC, monkey aortic SMC (+Fe or Cu)	Human blood monocytes	Lipid metabolism	BHT, BHA, SOD, desferrioxamine, diethylene triamine pentacetic acid	53

[a] EC, endothelial cells; huv, human umbilical vein; SMC, smooth muscle cells; Mϕ macrophages; HDL, high-density lipoprotein; EDTA, ethylenediamine tetraacetic acid; BHT, butylated hydroxytoluene; BHA, butylated hydroxyanisole; and SOD, superoxide dismutase.

was positively correlated with the severity of atherosclerosis. The nature of these lipid peroxides is unknown, but they are most likely to be oxidation products of LDL. Recently much attention has been paid to the role of macrophages in development of atherosclerosis.[61] Macrophages derived from monocytes accumulate in the intima of the arterial walls in atheromatous lesions, especially under hypercholesterolemic conditions.[32,62] Macrophages may be converted to foam cells[10,32,62] and they secrete macrophage-derived growth factor for smooth muscle cells.[63] Moreover, native LDL may be oxidized by incubation with macrophages.[76] The superoxide anion produced by macrophages may be important in this oxidation. Thus, oxidized LDL in both the plasma and arterial wall could injure ECs in vivo and initiate atherosclerotic processes.

Oxidized LDL affects not only EC but also eicosanoid metabolism. The major prostaglandin synthesized by ECs is PGI_2. PGI_2 is synthesized from arachidonic acid by cyclooxygenase and prostacyclin synthetase, and prostacyclin synthetase is strongly inhibited by oxidized LDL.[16,64] Since PGI_2 is a potent inhibitor of platelet aggregation,[65] oxidized LDL also accelerates thrombosis by inhibiting PGI_2 biosynthesis in ECs. All the above results suggest that antioxidants such as vitamin E could prevent atherosclerosis by inhibiting formation of oxidized LDL.

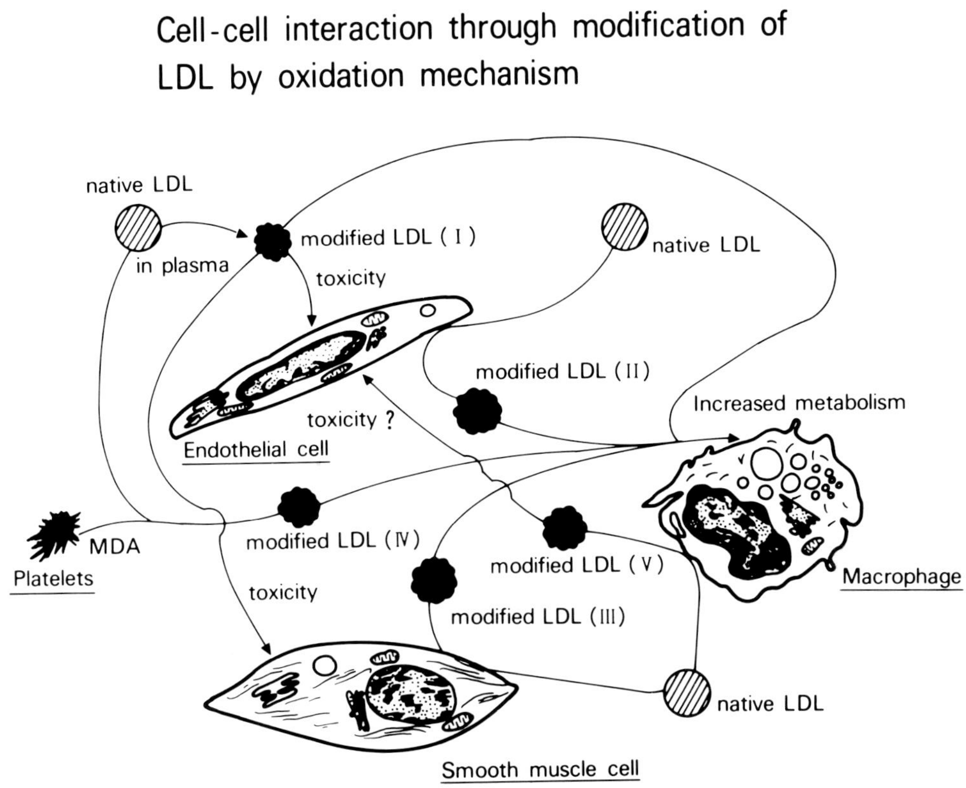

FIGURE 2. Cell-cell interactions through oxidative modification of LDL. For details, see text and Tables 2A and B.

LDL AND SMCs (TABLES 2A AND B AND FIGURE 2)

Like macrophages, SMCs may also be converted to foam cells[66] and they proliferate in the intima of atheromatous lesions. Foam cells mainly accumulate cholesterol ester intracellularly. The LDL pathway in cultured SMCs has been studied extensively, but the mechanism of foam cell formation is not yet known. Foam cells are not formed on treatment of SMCs with native LDL in vitro, because of down regulation of the LDL receptor,[12] and it is still unknown whether oxidized LDL is related to their formation. As mentioned above, (1) cholesterol esterase activity (especially acidic) is greatly decreased in vitamin E deficiency (Table 1A and B), and (2) preincubation of vitamin E with LDL or arterial wall increased acidic cholesterol esterase activity with LDL-cholesterol ester as a substrate (Figure 3). These results indicate that cholesterol esterase in SMCs is sensitive to peroxidative attack and that tocopherol deficiency may lead to cholesterol ester accumulation in SMCs. Coetzee et al.[67] reported that antioxidants including vitamin E increased the catabolism of ^{125}I-LDL in chloroquine-treated human skin fibroblasts. These results suggest that vitamin E stimulates LDL metabolism in lysosomes by protecting them from peroxidative attack. The effects of lipid peroxides on LDL receptor protein, receptor-mediated internalization of LDL, and regulation of the LDL receptor number (recycling and synthesis) are unknown. Oxidized LDL is reported to be cytotoxic to SMCs as it is to ECs.[36,40] Since necrosis is commonly observed in atheromatous lesions, this cytotoxicity might be related to atheromatous processes.

SMCs proliferate in the intima of atheromatous lesions on stimulation by various factors such as platelet derived growth factor,[33] macrophage derived growth factor,[63] endothelial cell derived growth factor,[68] epidermal growth factor, and transforming growth factor from

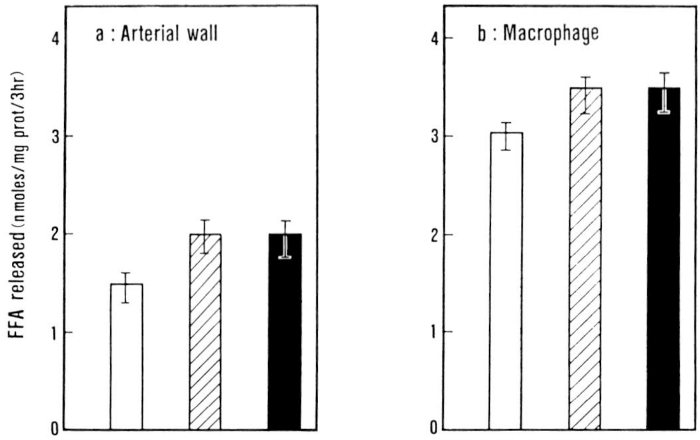

FIGURE 3. Effects of vitamin E on cholesterol esterase activity in rat aorta and peritoneal macrophages. (a) LDL or aortic homogenates were preincubated with 50 μM vitamin E and then acid cholesterol esterase activity was assayed as reported;[22] □, control; ▨, preincubated with LDL; and ■, preincubated with aortic homogenate. (b) Acetylated LDL or macrophage homogenates were preincubated with 50 μM vitamin E and then acid cholesterol esterase activity was assayed as reported;[22] □, control; ▨, preincubated with acetylated LDL; and ■, preincubated with macrophages.

platelets.[69] Moreover, nonpeptide factors are also important for regulation of SMC proliferation. Acyclic lipid peroxides, which are synthesized enzymatically in SMCs from polyunsaturated fatty acids including arachidonic acid, inhibit SMC proliferation without being toxic: only when SMCs are rapidly proliferating. Moreover, antioxidants inhibit acyclic lipid peroxidation and stimulate SMC proliferation. Therefore, some oxidant stress is necessary for regulation of cell proliferation and too high contents of antioxidants may stimulate excess proliferation of SMCs. This oxidant stress hypothesis is reviewed elsewhere.[70]

LDL AND MACROPHAGES (TABLES 2A AND B AND FIGURE 2)

Macrophages are one type of precursor of foam cells. Unlike SMCs, macrophages are readily converted to foam cells or lipid loaded cells in vitro. Macrophages do not metabolize native LDL.[10] Macrophages have a scavenger pathway and metabolize denatured LDLs such as acetylated LDL, LDL-dextran complex,[71] β-migrating very low-density lipoprotein,[72] and modified LDL, which is formed during incubation of native LDL with divalent cations such as Cu^{2+} and Fe^{2+}, or under more physiological conditions with human umbilical vein ECs. The modified LDL formed by ECs was suggested to be oxidized LDL because it reacted with thiobarbituric acid and its formation was prevented by antioxidants such as vitamin E, butylated hydroxytolene, and butylated hydroxyanisole.[14,49,50,53] However, not all ECs modify LDL (bovine aortic ECs do not modify LDL)[49] and other cells besides ECs can modify LDL (e.g., bovine,[49] swine,[48] guinea pig,[48] human,[50,53] and monkey[53] aortic SMCs). There are reports that divalent cations such as Cu^{2+} and Fe^{2+} are involved in the modification of LDL[50,53] and that chelating agents of divalent cations prevent the formation of modified LDL.[14,53] This modification is suggested to be mediated by superoxide anion generated by the cells.[43,53] Activation of phospholipase A_2 coupled with oxygen-centered radicals was suggested to be important for modification of LDL.[51] Whether this modification occurs intracellularly through endocytosis and retroendocytosis of LDL or on the cell surface is unknown. Modified LDL causes accumulation of a large amount of cholesterol ester in macrophages[13-15, 46-48,50,53] and also inhibits the motility of macrophages.[52] Modification of

LDL that leads to increase in its electrophoretic mobility increases the degradation of LDL in macrophages.[14] The electrophoretic mobility of LDL is increased by its cell-mediated oxidation, but not by its cell-free oxidation.[49] However, as LDL oxidized in a cell-free system has the same effect on macrophages,[15,46] the change in electrophoretic mobility may not be related to the increased metabolism of modified LDL and inhibition of chemotactic activity. LDL modified by ECs has the same receptor as acetylated LDL,[48] but acetylated LDL does not inhibit the chemotactic activity of macrophages.[52] Moreover, other types of modified LDL that are formed during incubation of native LDL with malondialdehyde, glutaraldehyde, and aggregated platelets, which produce malondialdehyde, also increased cholesterol ester accumulation in macrophages,[15] but the metabolism of these modified LDLs is not mediated by the acetylated LDL receptor.[15] The lipid-lowering drug and antioxidant probucol was reported to inhibit lipid accumulation by preventing oxidation of LDL by rabbit ECs.[46] These results indicate that antioxidants such as vitamin E can inhibit foam cell formation by preventing oxidation of LDL. However, studies are required on whether these modifications can occur in vivo because many antioxidative systems are present in vivo. One report showed that lipoproteins containing apo-B isolated from the arterial walls have increased electrophoretic mobility, suggesting the presence of modified LDL.[73]

Antioxidants such as vitamin E increased cholesterol esterase activity in macrophages with acetylated LDL-cholesterol ester as substrate when they were preincubated with acetylated LDL or macrophages (Figure 3). Their effect might be the same as that in SMCs where they stimulate hydrolysis of LDL-cholesterol ester. Moreover, similar treatment with vitamin E decreased cholesterol ester hydrolysis-dependent cholesterol esterification.[74] Thus, in macrophages also vitamin E interferes with lipoprotein metabolism either by inhibiting LDL oxidation or by affecting enzymes related to lipid metabolism.

CONCLUSIONS

Lipid accumulation, which is consistently observed in atheromatous lesions, is caused by disturbances in the metabolism of not only lipid, but also other substances. Lipid peroxidation affects the characteristics of LDL and causes lipid accumulation in macrophages and cytotoxicity to ECs or SMCs. The cell-cell interactions due to modification of LDL are summarized in Figure 2. By damaging enzymes that are important in the metabolism of lipids, lipid peroxidation also causes metabolic changes, resulting in increased lipid accumulation. Vitamin E is known to prevent lipid peroxidation. Clinical studies have shown that the plasma levels of lipid peroxides are increased in some pathological conditions related to atherosclerosis. The above results suggest that vitamin E could prevent the formation of atheromatous lesions.

REFERENCES

1. **Kirk, J. E.,** *Enzymes of the Arterial Wall,* Acadmic Press, New York, 1969.
2. **St. Clair, R. W.,** Metabolism of the arterial wall and atherosclerosis, in *Atherosclerosis Reviews,* Vol. 1, Paoletti, R. and Gotto, A. M., Jr., Eds., Raven Press, New York, 1976, 61.
3. **Shinomiya, M., Matsuoka, N., Shirai, K., Morisaki, N., Sasaki, N., Murano, S., Saito, Y., and Kumagai, A.,** Effect of pantethine on cholesterol ester metabolism in rat arterial wall, *Atherosclerosis,* 36, 75, 1980.
4. **Shirai, K., Matsuoka, N., Morisaki, N., Murano, S., Sasaki, N., Shinomiya, M., Saito, Y., Kumagai, A., and Mizobuchi, M.,** Effects of tocopherol deficiency on lipid metabolism in the arterial wall of rats on normal and high cholesterol diets, *Artery,* 6, 484, 1980.

5. **Murano, S., Shirai, K., Matsuoka, N., Morisaki, N., Sasaki, N., Shinomiya, M., Mizobuchi, M., Saito, Y., and Kumagai, A.,** Effect of tocopherol deficiency on lipid metabolism in arterial wall of spontaneously hypertensive rats on normal and high cholesterol diets, *Tohoku J. Exp. Med.*, 136, 29, 1982.
6. **Morisaki, N., Murano, S., Shinomiya, M., Sasaki, N., Shirai, K., Matsuoka, N., Mizobuchi, M., Akikusa, B., Saito, Y., and Kumagai, A.,** Lipid metabolism in arteriosclerotic arterial wall of rats, *Atherosclerosis*, 43, 51, 1982.
7. **Morisaki, N., Shinomiya, M., Mizobuchi, M., Fujiyama, Y., Sasaki, N., Shirai, K., Matsuoka, N., Saito, Y., and Kumagai, A.,** Lipid metabolism in the aorta and the brain microvessels of rabbits on high cholesterol diet, *Artery*, 11, 225, 1982.
8. **Morisaki, N., Fujiyama, Y., Matsuoka, N., Saito, Y., and Kumagai, A.,** Role of aging in lipid metabolism in the aorta of rats, *Gerontology*, 30, 13, 1984.
9. **Brown, M. S. and Goldstein, J. L.,** Receptor-mediated control of metabolism, *Science*, 191, 150, 1976.
10. **Goldstein, J. L., Ho, Y. K., Basu, S. K., and Brown, M. S.,** Binding site on macrophages that mediates uptake and degradation of acetylated low density lipoprotein, producing massive cholesterol deposition, *Proc. Natl. Acad. Sci. U.S.A.*, 76, 333, 1979.
11. **Basu, S. K., Goldstein, J. L., Anderson, R. G. W., and Brown, M. S.,** Degradation of cationized low density lipoprotein and regulation of cholesterol metabolism in homozygous familial hypercholesterolemia fibroblasts, *Proc. Natl. Acad. Sci. U.S.A.*, 73, 3178, 1976.
12. **Goldstein, J. L. and Brown, M. S.,** The LDL receptor defect in familial hypercholesterolemia, in *The Medical Clinics of North America*, Vol. 66(2), W. B. Saunders, Philadelphia, 1982, 335.
13. **Henriksen, T., Mahoney, E. M., and Steinberg, D.,** Enhanced macrophage degradation of low density lipoprotein previously incubated with cultured endothelial cells: recognition by receptors for acetylated low density lipoproteins, *Proc. Natl. Acad. Sci. U.S.A.*, 78, 6499, 1981.
14. **Steinbrecher, U. P., Parthasarathy, S., Leake, D. S., Witztum, J. L., and Steinberg, D.,** Modification of low density lipoprotein by endothelial cells involves lipid peroxidation and degradation of low density lipoprotein phospholipids, *Proc. Natl. Acad. Sci. U.S.A.*, 81, 3883, 1984.
15. **Fogelman, A. M., Shecher, I., Seager, J., Hokom, M., Child, J. S., and Edwards, P. A.,** Malondialdehyde alteration of low density lipoproteins leads to cholesteryl ester accumulation in human monocyte-macrophages, *Proc. Natl. Acad. Sci. U.S.A.*, 77, 2214, 1980.
16. **Szczeklik, A. and Gryglewski, R. J.,** Low density lipoproteins (LDL) are carriers for lipid peroxides and inhibit prostacyclin (PGI_2) biosynthesis in arteries, *Artery*, 7, 488, 1980.
17. **Glavind, J., Hartmann, S., Clemunessen, J., Jessen, K. E., and Dam, H.,** Studies on the role of lipid peroxides in human pathology. II. The presence of peroxidized lipids in the atherosclerotic aorta, *Acta Pathol. Microbiol. Scand.*, 30, 1, 1952.
18. **Lubin, B. and Machlin, L. J., Eds.,** Vitamin E; biochemical, hematological, and clinical aspects, *Ann. N.Y. Acad. Sci.*, 393, 1982.
19. **Iwakami, M.,** Peroxides as a factor of atherosclerosis, *Nagoya J. Med. Sci.*, 28, 50, 1965.
20. **Fukumoto, S.,** Effects of vitamin E on lipid metabolism and on experimental atherosclerosis, *Taishitsu Igaku Kenkyuujo Rep.*, 16, 1, 1965 (in Japanese).
21. **Ito, T.,** Tocopherol, non-protein SH and metals in the human aorta, *Jpn. Circ. J.*, 33, 25, 1969.
22. **Shinomiya, M., Matsuoka, N., Shirai, K., Saito, Y., and Kumagai, A.,** Studies on cholesterol esterase in rat arterial wall, *Atherosclerosis*, 33, 343, 1979.
23. **Morisaki, N., Matsuoka, N., Shirai, K., Saito, Y., and Kumagai, A.,** Studies on acyl-CoA synthetase in rat arterial wall, *Atherosclerosis*, 37, 439, 1980.
24. **Shirai, K., Matsuoka, N., Saito, Y., and Kumagai, A.,** Effect of phospholipids on lipase activity in rat arterial wall homogenate, *Tohoku J. Exp. Med.*, 138, 131, 1982.
25. **Sasaki, N., Matsuoka, N., Shirai, K., Saito, Y., and Kumagai, A.,** Studies on CDP-choline: 1,2-diacylglycerol choline phosphotransferase activity in rat arterial wall, *Atherosclerosis*, 43, 185, 1982.
26. **Roubal, W. T. and Tappel, A. L.,** Polymerization of proteins induced by free radical lipid peroxidation, *Arch. Biophys.*, 113, 150, 1966.
27. **Little, C. and O'Brien, P. J.,** The effectiveness of a lipid peroxide in oxidizing protein and non-protein thiols, *Biochem. J.*, 106, 419, 1968.
28. **Lewis, S. E. and Wills, E. D.,** The destruction of -SH groups of proteins and amino acids by peroxides of unsaturated fatty acids, *Biochem. Pharmacol.*, 11, 901, 1962.
29. **Reckless, J. P. D., Weinstein, D. B., and Steinberg, D.,** Lipoprotein and cholesterol metabolism in rabbit endothelial cells in culture, *Biochim. Biophys. Acta*, 529, 475, 1978.
30. **Ross, R. and Glomset, J. A.,** The pathogenesis of atherosclerosis (first of two parts), *N. Engl. J. Med.*, 295, 369, 1976.
31. **Ross, R. and Glomset, J. A.,** The pathogenesis of atherosclerosis (second of two parts), *N. Engl. J. Med.*, 295, 420, 1976.
32. **Ross, R.,** The pathogenesis of atherosclerosis — recent progress, *N. Engl. J. Med.*, 314, 488, 1986.

33. **Ross, R., Glomset, J., Kariya, B., and Harker, L.,** A platelet-dependent serum factor that stimulates the proliferation of arterial smooth muscle cells in vitro, *Proc. Natl. Acad. Sci. U.S.A.*, 71, 1207, 1974.
34. **Nakao, J., Ooyama, T., Ito, H., Chang, W., and Murota, S.,** Comparative effect of lipoxygenase products of arachidonic acid on rat aortic smooth muscle cell migration, *Atherosclerosis*, 44, 339, 1982.
35. **Henriksen, T., Evensen, S. A., and Carlander, B.,** Injury of endothelial cells in culture induced by low density lipoproteins, *Eur. J. Clin. Invest.*, 7, 243, 1977.
36. **Hessler, J. R., Robertson, A. L., and Chisolm, G. M., III,** LDL-induced cytotoxicity and its inhibition by HDL in human vascular smooth muscle and endothelial cells in culture, *Atherosclerosis*, 32, 213, 1979.
37. **Henriksen, T., Evensen, S. A., and Carlander, B.,** Injury to human endothelial cells in culture induced by low density lipoproteins, *Scand. J. Clin. Lab. Invest.*, 39, 361, 1979.
38. **Henriksen, T., Evensen, S. A., and Carlander, B.,** Injury to cultured endothelial cells induced by low density lipoproteins: protection by high density lipoproteins, *Scand. J. Clin. Lab. Invest.*, 39, 369, 1979.
39. **Tauber, J. P., Cheng, J., and Gospodarowitcz, D.,** Effect of high and low density lipoproteins on proliferation of cultured bovine vascular endothelial cells, *J. Clin. Invest.*, 66, 696, 1980.
40. **Chisolm, G. M., Hessler, J. R., Morel, D. W., and Lewis, L. J.,** In vitro autooxidation renders low density lipoprotein toxic to cells in culture, *Arteriosclerosis*, 1, 359a, 1981.
41. **Børsum, T., Henriksen, T., Carlander, B., and Reisvaag, A.,** Injury to human cells in culture induced by low density lipoprotein: an effect independent of receptor binding and endocytic uptake of low density lipoprotein, *Scand. J. Clin. Lab. Invest.*, 42, 75, 1982.
42. **Evensen, S. A., Nilsen, E., and Galdal, K. S.,** Injury to cultured human fibroblasts induced by low density lipoproteins: potentiating and protective factors, *Scand. J. Clin. Lab. Invest.*, 42, 285, 1982.
43. **Morel, D. W., Hessler, J. R., and Chisolm, G. M.,** Low density lipoprotein cytotoxicity induced by free radical peroxidation of lipid, *J. Lipid Res.*, 24, 1070, 1983.
44. **Hessler, J. R., Morel, D. W., Lewis, L. J., and Chisolm, G. M.,** Lipoprotein oxidation and lipoprotein-induced cytotoxicity, *Arteriosclerosis*, 3, 215, 1983.
45. **Evensen, S., Galdal, K. S., and Nilsen, E.,** LDL-induced cytotoxicity and its inhibition by anti-oxidant treatment in cultured human endothelial cells and fibroblasts, *Atherosclerosis*, 49, 23, 1983.
46. **Parthasarathy, S., Young, S. G., Witztum, J. L., Pittman, R. C., and Steinberg, D.,** Probucol inhibits oxidative modification of low density lipoprotein, *J. Clin. Invest.*, 77, 641, 1986.
47. **Henriksen, T., Mahoney, E. M., and Steinberg, D.,** Interaction of plasma lipoproteins with endothelial cells, *Ann. N.Y. Acad. Sci.*, 401, 102, 1982.
48. **Henriksen, T., Mahoney, E. M., and Steinberg, D.,** Enhanced macrophage degradation of biologically modified low density lipoprotein, *Arteriosclerosis*, 3, 149, 1983.
49. **Morel, D. W., Di Corleto, P. E., and Chisolm, G. M.,** Endothelial and smooth muscle cells alter low density lipoprotein in vitro by free radical oxidation, *Arteriosclerosis*, 4, 357, 1984.
50. **Heinecke, J. W., Rosen, H., and Chait, A.,** Iron and copper promote modification of low density lipoprotein by human arterial smooth muscle cells in culture, *J. Clin. Invest.*, 74, 1890, 1984.
51. **Parthasarathy, S., Steinbrecher, V. P., Barnett, J., Witztum, J. L., and Steinberg, D.,** Essential role of phospholipase A_2 activity in endothelial cell-induced modification of low density lipoprotein, *Proc. Natl. Acad. Sci. U.S.A.*, 82, 3000, 1985.
52. **Quinn, M. T., Parthasarathy, S., and Steinberg, D.,** Endothelial cell-derived chemotactic activity for mouse peritoneal macrophages and the effects of modified forms of low density lipoproteins, *Proc. Natl. Acad. Sci. U.S.A.*, 82, 5949, 1985.
53. **Heineche, J. W., Baker, L., Rosen, H., and Chait, A.,** Superoxide-mediated modification of low density lipoprotein by arterial smooth muscle cells, *J. Clin. Invest.*, 77, 757, 1986.
54. **Imai, H., Werthessen, N. T., Lequesne, P. W., Soloway, A. H., and Kanisawa, M.,** Angiotoxicity of oxygenated sterols and possible precursors, *Science*, 207, 651, 1980.
55. **Peng, S. K., Taylor, C. B., Hill, J. C., and Morin, R. J.,** Cholesterol oxidation derivatives and arterial endothelial damage, *Atherosclerosis*, 54, 121, 1985.
56. **Peng, S. K., Taylor, C. B., Tham, P., Werthessen, N. T., and Mikkelson, B.,** Effect of auto-oxidation products from cholesterol on aortic smooth muscle cells, *Arch. Pathol. Lab. Med.*, 102, 57, 1978.
57. **Peng, S. K., Tham, P., Taylor, C. B., and Mikkelson, B.,** Cytotoxicity of oxidation derivatives of cholesterol on cultured aortic smooth muscle cells and their effect on cholesterol biosynthesis, *Am. J. Clin. Nutr.*, 32, 1033, 1979.
58. **Yagi, K.,** Assay for serum lipid peroxide level and its clinical significance, in *Lipid Peroxides in Biology and Medicine*, Yagi, K., Ed., Academic Press, New York, 1982, 223.
59. **McCormick, E. C., Cornwell, D. G., and Brown, J. B.,** Studies on the distribution of tocopherol in human serum lipoproteins, *J. Lipid Res.*, 1, 221, 1960.
60. **Lee, D. M.,** Malondialdehyde formation in stored plasma, *Biochem. Biophys. Res. Commun.*, 95, 1663, 1980.
61. **Gerrity, R. G.,** The role of the monocyte in atherogenesis. I. The transition of blood-borne monocytes into foam cells in fatty lesions, *Am. J. Pathol.*, 103, 181, 1981.

62. **Walker, L. N. and Bowyer, D. E.,** Endothelial healing in the rabbit aorta and the effect of risk factors for atherosclerosis, *Arteriosclerosis,* 4, 479, 1984.
63. **Leibovich, S. J. and Ross, R.,** A macrophage-dependent factor that stimulated the proliferation of fibroblasts in vitro, *Am. J. Pathol.,* 84, 501, 1976.
64. **Turk, J., Wyche, A., and Needleman, P.,** Inactivation of vascular prostacyclin synthetase by platelet lipoxygenase products, *Biochem. Biophys. Res. Commun.,* 95, 1628, 1980.
65. **Moncada, S., Gryglewski, R. J., Bunting, S., and Vane, J. R.,** An enzyme isolated from arteries transforms prostaglandin endoperoxides to an unstable substance that inhibits platelet aggregation, *Nature (London),* 263, 663, 1976.
66. **Imai, H., Lee, K. T., Pastori, S., Panlilio, E., Florentin, R., and Thomas, W. A.,** Atherosclerosis in rabbits, architectural and subcellular alterations of smooth muscle cells of aortas in response to hyperlipidemia, *Exp. Mol. Pathol.,* 5, 273, 1966.
67. **Coetzee, G. A., Stein, O., and Stein, Y.,** Modulation by sodium ascorbate of the effect of chloroquine on low density lipoprotein retention and degradation in cultured human skin fibroblasts, *Atherosclerosis,* 32, 277, 1979.
68. **Gajdusek, C., DiCorleto, P., Ross, R., and Schwartz, S. M.,** An endothelial cell-derived growth factor, *J. Cell Biol.,* 85, 467, 1980.
69. **Todaro, G. J., DeLarco, J. E., and Fryling, C. M.,** Sarcoma growth factor and other transforming peptides produced by human cells: interaction with membrane receptors, *Fed. Proc.,* 41, 2996, 1982.
70. **Cornwell, D. G. and Morisaki, N.,** Fatty acid paradoxes in the control of cell proliferation: prostaglandins, lipid peroxides and co-oxidation reactions, in *Free Radicals in Biology,* Vol. 6, Pryor, W. A., Ed., Academic Press, New York, 1984, 182.
71. **Basu, S. K., Brown, M. S., Ho, Y. K., and Goldstein, J. L.,** Degradation of low density lipoprotein dextran sulfate complexes associated with deposition of cholesteryl esters in mouse macrophages, *J. Biol. Chem.,* 254, 7141, 1979.
72. **Goldstein, J. L., Ho, Y. K., and Brown, M. S.,** Cholesteryl ester accumulation in macrophages resulting from receptor-mediated uptake and degradation of hypercholesterolemic canine β-very low density lipoproteins, *J. Biol. Chem.,* 255, 1839, 1980.
73. **Hoff, H. G. and Gaubatz, J. W.,** Isolation, purification and characterization of a lipoprotein containing apoB from the human aorta, *Atherosclerosis,* 42, 273, 1982.
74. **Saito, Y., Shinomiya, M., and Kumagai, A.,** Effect of tocopherol on low-density lipoprotein cholesterol ester metabolism in the arterial wall, *Ann. N.Y. Acad. Sci.,* 393, 183, 1982.
75. **Goto, Y.,** Lipid peroxides as a cause of vascular diseases, in *Lipid Peroxides in Biology and Medicine,* Yagi, K., Ed., Academic Press, New York, 1982, 295.
76. **Hiramatsu, K.,** personal communication.

SERUM α-TOCOPHEROL, COENZYME Q, AND TBA-REACTIVE SUBSTANCE IN ACUTE MYOCARDIAL DAMAGE AND STROKE

Masayoshi Kibata

INTRODUCTION

Glavind reported in 1952[1] that lipid peroxides accumulate in atherosclerotic lesions in proportion to the severity of the process. Since then, many workers have suggested that lipid peroxides are associated with the initiation and development of atherosclerosis. On the other hand, it has been demonstrated clinically and experimentally that vitamin E, a natural antioxidant, plays a role in the inhibition of lipid peroxides in our body and inhibits development of atherosclerosis. Kibata et al. reported earlier[2] that serum vitamin E levels correlate inversely with serum peroxide levels, which in Japan are determined routinely by Yagi's fluorometric method[3-5] and are designated as the thiobarbituric acid (TBA) value (Figure 1).

Coenzyme Q_{10} is a member of the electron transport system of mitochondria which is receiving increasing attention in the therapay for cerebral, coronary, and renal artery ischemias in the acute stage, as its antioxidant activity is being elucidated. In this chapter, studies on the use of coenzyme Q_{10}, in comparison with vitamin E, in acute vascular accidents, acute myocardial infarction (AMI), and stroke (cerebral vascular accident, CVA) are presented. Further, research on an experimental model of AMI, namely, the myocardial necrosis that is induced in the rat by injection of a large dose of isoproterenol, is summarized.

METHODS

Blood is collected as soon as possible from CVA or AMI patients admitted to the emergency rooms of hospitals. In addition, some cases are monitored by collecting blood samples every other day during the first week and then once a week for an additional 3 weeks. The sera are stored at $-20°C$ for up to 3 weeks before conducting chemical determinations. The TBA value is determined by Yagi's method. Vitamin E is measured as α-tocopherol by the method described by Abe and Katsui.[5] Cholesterol and triglycerides are measured enzymatically. Coenzyme Q_{10} is assayed by high performance liquid chromatography as described by Abe and Katsui.[5] Lipoproteins are separated by the ultracentrifugation method into the very low-density fraction (VLDL, less than 1.006), low-density fraction (LDL, 1.006 to 1.063), and high-density fraction (HDL, 1.063 to 1.21).

Male Wistar rats, weighing 200 g, are injected with 80 mg/kg of isoproterenol once or twice at an interval of 12 hr, in order to induce myocardial necrosis (by Rona's method).[6] TBA and vitamin E in the blood and in the tissues of heart and liver are assayed before injection and at 12, 24, 48, and 76 hr after injection.

One group of rats was fed a diet containing 1% of α-tocopherol nicotinate for 1 month and compared with another group fed the control diet. Another group of rats was injected with daily dosages of 30 mg/kg of α-tocopherol acetate for 7 days prior to the isoproterenol injection. In order to obtain lipoprotein fractions, serum samples from three animals were pooled to attain one sample.

RESULTS AND DISCUSSION

Vitamin E dropped quickly (Table 1) and TBA increased within 1 to 2 days from the onset of a vascular accident. Administration of 400 or 600 mg/day of vitamin E in 31 stroke

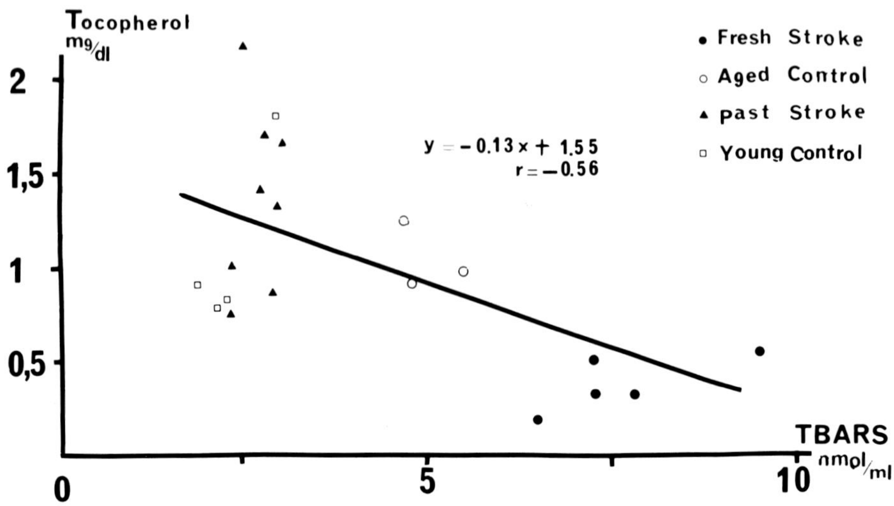

FIGURE 1. Correlation between serum tocopherol levels and serum TBARS.

Table 1
SERUM VITAMIN E LEVELS (μg/mℓ) AND PERCENT DISTRIBUTION IN LIPOPROTEINS OF CONTROL, CVA, AND AMI

	Control (n = 14)	CVA (n = 16)	AMI (n = 13)
		minimum in 1st week	
Vitamin E	12.66 ± 3.47	7.23 ± 2.32	8.05 ± 2.79
		% distribution	
VLDL	16 ± 12	20 ± 11	20 ± 8
LDL	45 ± 11	45 ± 13	51 ± 10
HDL	37 ± 10	35 ± 8	29 ± 9

patients in the acute phase prevented the elevation of TBA (Figures 2 and 3). These findings led to the assumption that vitamin E migrated to the lesions and was utilized to counteract the lipid peroxidation process. The level of Q_{10} correlated directly ($p < 0.01$) with serum vitamin E and inversely with cholesterol, showed its minimum level at 2 or 3 days after the accident, and remained at low levels for 1 or 2 weeks thereafter. Vitamin E showed a rapid decrease in the early stage, whereas coenzyme Q_{10} decreased gradually (Table 2; Figure 4)

According to our data, coenzyme Q_{10} increased in the LDL and decreased in the HDL fraction and vitamin E increased in VLDL and decreased in HDL in patients with CVA or AMI. The decrease of vitamin E in HDL was observed as early as within 2 days, but coenzyme Q_{10} in HDL took 1 to 2 weeks to reach its lowest value (Figure 5). The HDL cholesterol level, as measured by the precipitation method, showed a similar curve and the lowest level was attained at 1 week (Figure 6).

TBA in rat heart tissue increased slightly 12 hr after isoproterenol injection, and vitamin E declined to 0.12 μg/mg protein (Figure 7), while liver TBA showed a sharp increase and returned to normal level within 48 hr, together with a decrease in vitamin E (Figure 8). TBA increased more slowly in blood than in liver and took more than 72 hr to return to its normal range (Figure 9). These observations suggest that a large amount of lipid peroxide produced in the liver is released into the bloodstream, leading to an increase in TBA levels. The mechanims which suddenly induced a large production of lipid peroxide was assumed

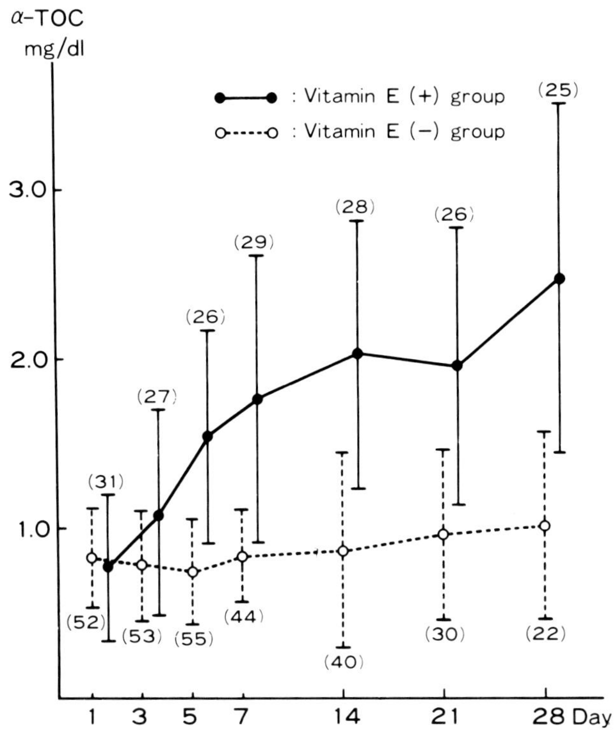

FIGURE 2. Mean values of serum α-tocopherol after the onset of stroke. The solid line represents the group to which 400 or 600 mg/day of α-tocopherol acetate was given for 28 days. The dotted line represents the group without α-tocopherol supplementation.

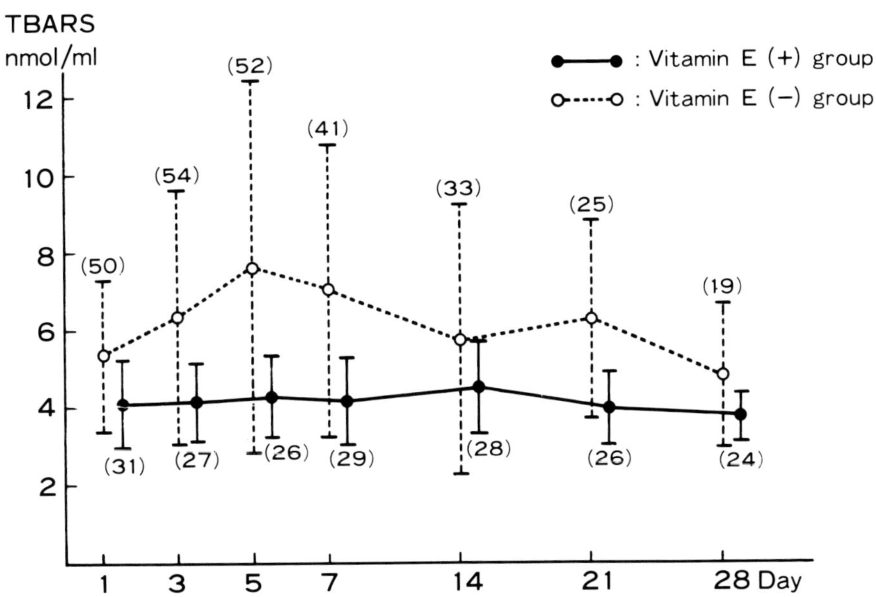

FIGURE 3. Mean values of serum TBARS of α-tocopherol-treated (solid line) and untreated (dotted line) patients.

Table 2
SERUM COENZYME Q_{10} LEVELS (ng/mℓ) OF CONTROL AND OF PATIENTS WITH CVA AND AMI

	CVA (n = 16)		AMI (n = 18)	
Control (n = 21)	Day of attack	Minimum in 1st week	Day of attack	Minimum in 1st week
840.64 ± 242.23	699.13 ± 389.26	563.55 ± 204.05	800.20 ± 409.88	576.78 ± 239.91

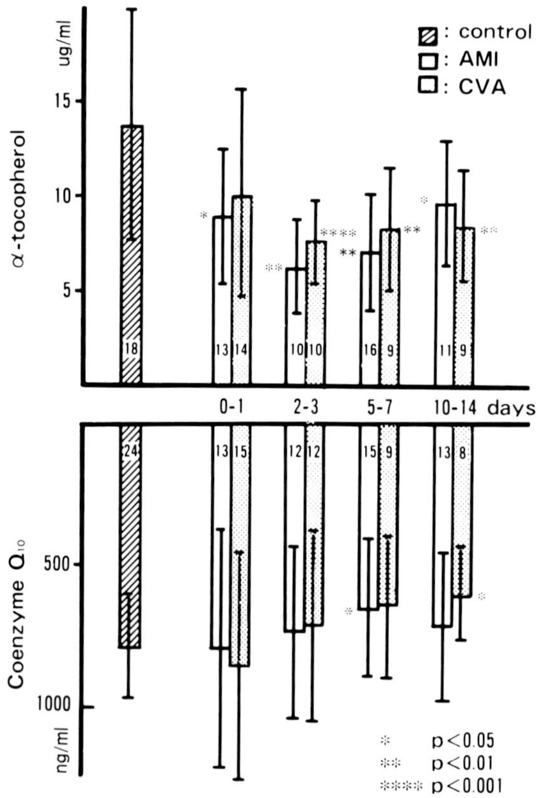

FIGURE 4. Values of serum α-tocopherol and Q_{10} after the onset of CVA and AMI.

to be damage to liver cells due to the ischemia caused by the isoproter nol-induced myocardial necrosis (IPMN) shock, although a direct effect of isoproterenol is probable.

In comparison with control rats, the vitamin E level in the heart, liver, and blood of rats fed α-tocopherol nicotinate showed increases, and TBA decreased correspondingly (Figure 10). Isoproterenol injection elevated the TBA levels in serum liver, but only minor changes were noted in rats fed α-tocopherol nicotinate. The coenzyme Q levels in various tissues were unaffected by α-tocopherol feeding (Figure 11). The level of coenzyme Q_9 decreased significantly only in the heart tissue of the α-tocopherol-fed and isoproterenol-injected group and showed small increases in the liver and serum.

The fact that a high percentage of vitamin E and TBA was noted in HDL, in contrast to coenzyme Q_9 (which was mostly found in LDL), suggests that coenzyme Q_9 is predominantly located in the core of VLDL and LDL, together with cholesterol, whereas vitamin E and

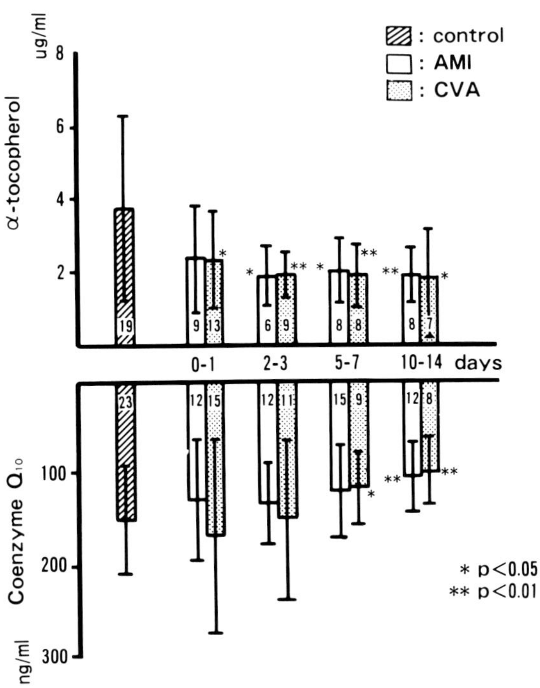

FIGURE 5. Values of HDL α-tocopherol and coenzyme Q_{10} after the onset of CVA and AMI.

TBA exist in the surface area of the lipoprotein structure, and that the latter move readily to HDL via the VLDL-HDL pathway (Figure 12). When the production of liver TBA increases, vitamin E is consumed in the liver at a higher rate, resulting in a decrease in the level of vitamin E in VLDL and also in HDL. On the other hand, the increased liver TBA is released into the bloodstream, as a component of VLDL, and is transferred to LDL together with coenzyme Q_9, because the VLDL-LDL pathway is stimulated (Figure 13).

ACKNOWLEDGMENT

The author is grateful to Drs. H. Uehara, T. Fuchimoto, K. Shoji, M. Shimono, K. Sakurai, M. Minami, T., Ishioka, and Y. Fukada for their dedicated cooperation in collecting the blood samples and technical assistance.

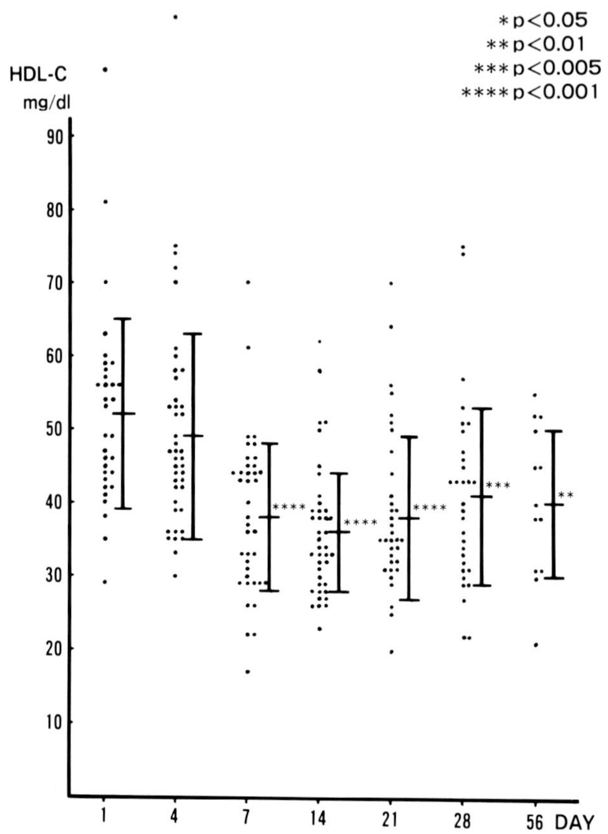

FIGURE 6. Values of HDL cholesterol after the onset of AMI. These data were determined by the precipitation method.

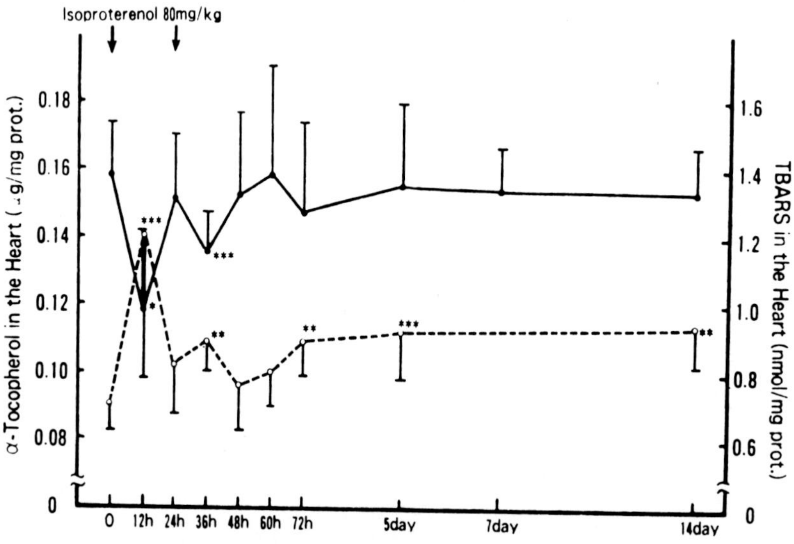

FIGURE 7. Changes in α-tocopherol and TBARS in rat heart after isoproterenol injection (mean ± SE). Each group contained four to five rats; α-tocopherol, closed circle, TBARS, open circle; *p, less than 0.05; **p, less than 0.02; ***p, less than 0.01.

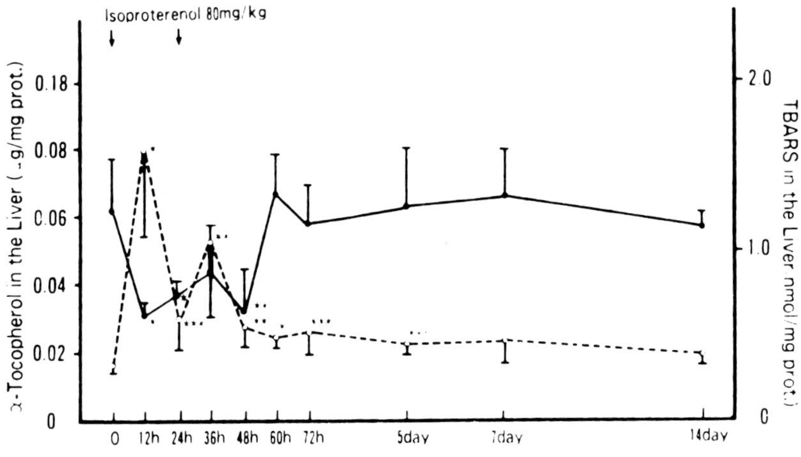

FIGURE 8. Changes in α-tocopherol and TBARS in rat liver after isoproterenol injection (mean ± SE).

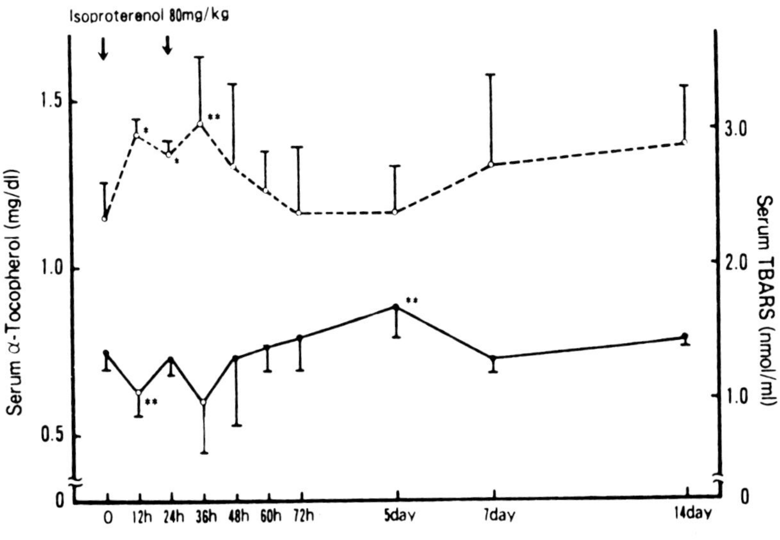

FIGURE 9. Changes in α-tocopherol and TBARS in rat serum after isoproterenol injection (mean ± SE).

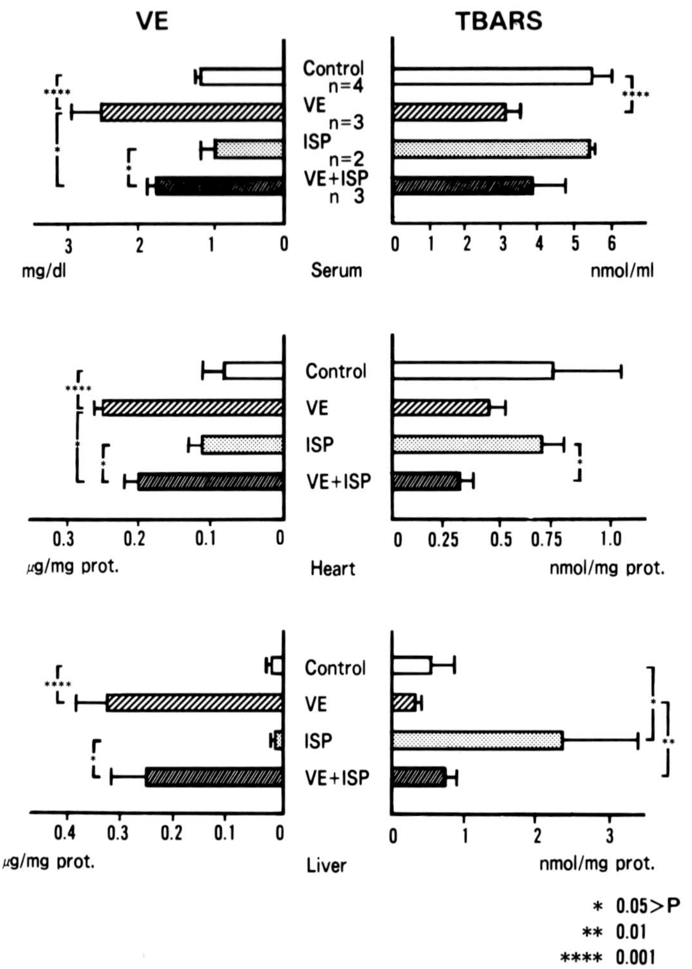

FIGURE 10. Effects of α-tocopherol feeding for 1 month on vitamin E and TBARS in serum, heart, and liver of control and isoproterenol-injected rats; control, control diet, placebo injected; ISP, control diet, isoproterenol injected; VE, α-tocopherol injected; VE + ISP, α-tocopherol supplemented diet, isoproterenol injected; **p*, less than 0.05; ***p*, less than 0.01; and ****p*, less than 0.001.

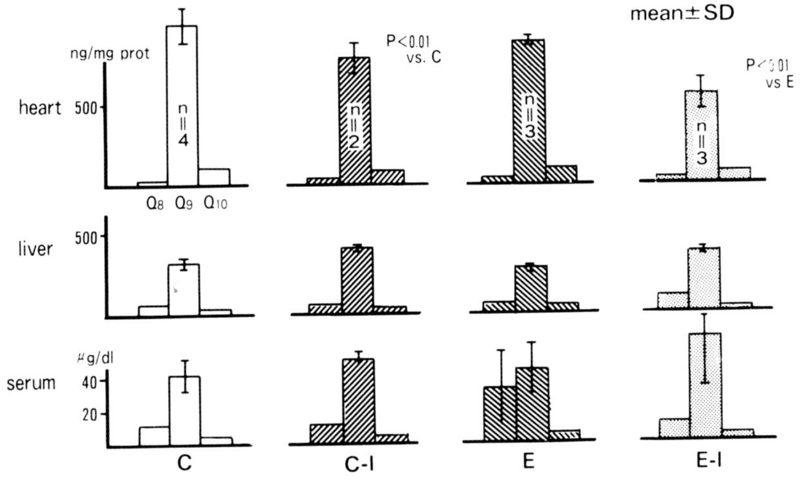

FIGURE 11. Levels of coenzyme Q_8, Q_9, and Q_{10} in heart, liver, and serum of rats of group C (fed control diet and placebo injected), C-I (fed diet without added tocopherol and isoproterenol injected), E (fed diet with added α-tocopherol nicotinate, for 1 month, and placebo injected), and E-I (fed diet with α-tocopherol and isoproterenol injected).

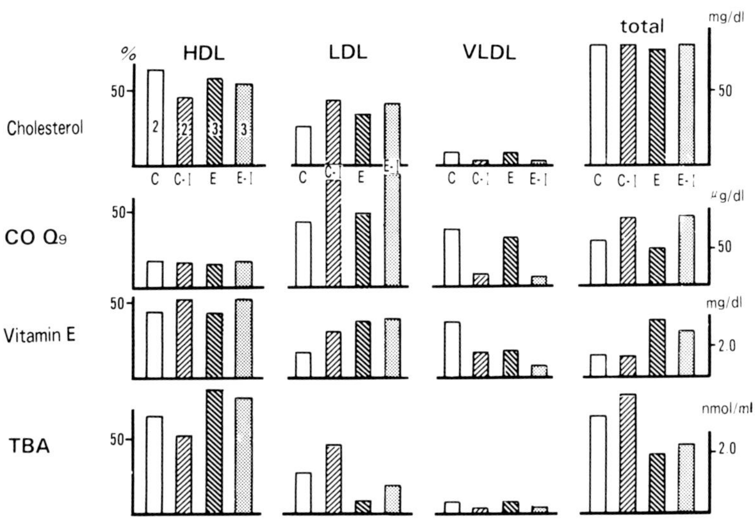

FIGURE 12. Percent of total cholesterol, coenzyme Q_9, vitamin E, and TBA in each lipoprotein; C, placebo injected for 7 days and saline injected on the day of the experiment; C-I, placebo injected for 7 days followed by isoproterenol; E, α-tocopherol nicotinate (30 mg/kg/day) injected for 7 days; E-I, days of treatment with α-tocopherol nicotinate and isoproterenol.

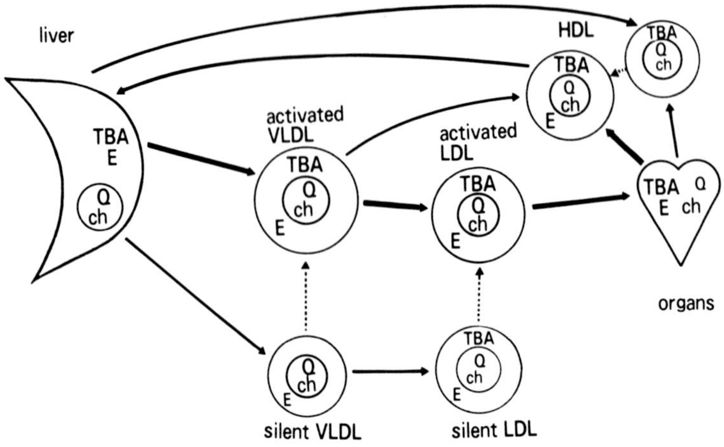

FIGURE 13. A speculative scheme, which shows the flows of TBA, vitamin E, coenzyme Q, and cholesterol between lipoprotein fractions in the acute stages of CVA and AMI.

REFERENCES

1. **Glavind, J., Hartman, S., Clemmesen, J., Jessen, K. E., and Dam, H.,** Studies on the role of lipoperoxides in human pathology, *Acta Pathol. Microb. Scand.*, 30, 1, 1952.
2. **Kibata, M., Schimizu, Y., Miyake, K., Shimono, M., Shoji, K., Miyahara, K., Fuchimoto, T., and Nasu, Y.,** Alpha-tocopherol and TBA reactive substance (TBARS) in serum of the stroke at acute stage, *Igakunoayumi*, 101, 591, 1977 (in Japanese).
3. **Yagi, K.,** A simple fluorometric assay for lipoperoxide in blood plasma, *Biochem Med.*, 15, 212, 1976.
4. **Kibata, M., Shimizu, Y., Miyake, K., Shoji, K., Miyahara, K., Nasu, Y., and Kimura, I.,** Studies on vitamin E in lipid metabolism, *Tocopherol, Oxygen and Biomembranes*, de Duve, C. and Hayaishi, O., Eds., Elsevier/North-Holland, Amsterdam, 1978, 283.
5. **Abe, K. and Katsui, G.,** Fluorometric determination of tocopherol in serum, *Eiyo To Shokuryo*, 28, 277, 1975 (in Japanese).
6. **Rona, G., Chappel, C. I., Balazs, T., and Gaudry, R.,** An infarct-like myocardial lesion and other toxic manifestation produced by isoproterenol in the rat, *Arch. Pathol.*, 67, 443, 1959.

SELENIUM AND LIPID PEROXIDATION IN HUMAN AND EXPERIMENTAL CARDIOVASCULAR DISEASE

Hans-Joachim Peters and Heinz Koehler

INTRODUCTION

It has been suggested that increased lipid peroxidation can cause degeneration of organs or tissues[1-7] and that lipid peroxides formed at the primary site are transferred through the circulation to other organs or tissues and provoke damage by propagating lipid peroxidation.[8,9] Lipid peroxidation occurs in tissues when antioxidant protective systems are deficient. Among the protective systems are the biological antioxidants, selenium (Se), Se-dependent glutathione peroxidase (GPx), and vitamin E.[10-25] As an integral part of the enzyme GPx, Se plays an important role in preventing oxidative damage of tissues by lipid peroxides. In animals, Se deficiency is well established and is reflected in low Se levels in biological fluids and tissues. Se has been recognized as playing an essential role in human beings, mainly in the prevention of a number of diseases. Thus, in man, a low Se status has been associated with various pathological conditions, among others, cardiovascular diseases (CVD).[26-56] Keshan disease, an endemic cardiomyopathy in China (Szechwan Province), was prevalent in rural areas located in a region where the Se content was low in foods. Multifocal necrosis and fibrous replacement of the myocardium, characterized by its histopathological features, were present in most cases, although it is not specific for this disease.

Intervention studies with sodium selenite in several thousand children, compared with control subjects, for 4 successive years revealed that Keshan disease is Se responsive. The incidence rate among Se-treated children was only 0 to 2.2 per 1000, while it was 9.5 to 13.5 per 1000 subjects in the placebo group. The difference is statistically significant.[57-60] Further, in a recent study, Moore et al.[61] report a significant inverse relationship between total Se and severity of coronary arteriosclerosis.

In 1957 Mills[62] discovered GPx, an enzyme which catalyzes the formation of oxidized glutathione disulfide (GSSG) from hydroperoxides (ROOH) and reduced glutathione: ROOH $+ 2\ \mathrm{GSH} \xrightarrow{\mathrm{GPx}} \mathrm{ROH} + \mathrm{GSSG} + H_2O$. GPx shows high activity in liver and erythrocytes, moderate activity in heart and lung, and low activity in muscle. This enzyme is specific for its hydrogen donor, GSH, and nonspecific for hydroperoxides.[63,64] The antihypoxic effect of vitamin E and its derivatives is due to the preservation of an elevated content of ubiquinone and α-tocopherol and regeneration of the NADH-ubiquinone-reductase system functioning in myocardial mitochondria.[65] The product of lipid peroxidation malondialdeyde in the range of 10 to 30 m*M* induces hemolysis of human erythrocytes in a concentration-dependent fashion. This hemolytic effect is markedly reduced in the presence of vitamin E and sodium selenite.[66] The above supports the view that Se deficiency and lipid peroxidation play an important role both in mitochondrial injury and in the development of CVD.

ANIMAL EXPERIMENTS

The laboratory animals were 23 domestic pigs, weighing from 25 to 40 kg. Following orotracheal intubation, the muscle-relaxed animals were operated under nitrous oxide/oxygen anesthesia (N_2O/O_2) with addition of halothane and ether. After thoracotomy, flow transducers were placed around the aorta abdominalis, the arteria carotis communis, and the right and left coronary arteries. Pressure in the left ventricle was measured with a catheter-tip manometer, and in the right ventricle, pressure was measured with a fluid-filled catheter. Myocardial infarction was induced by complete ligation of the ramus interventricularis

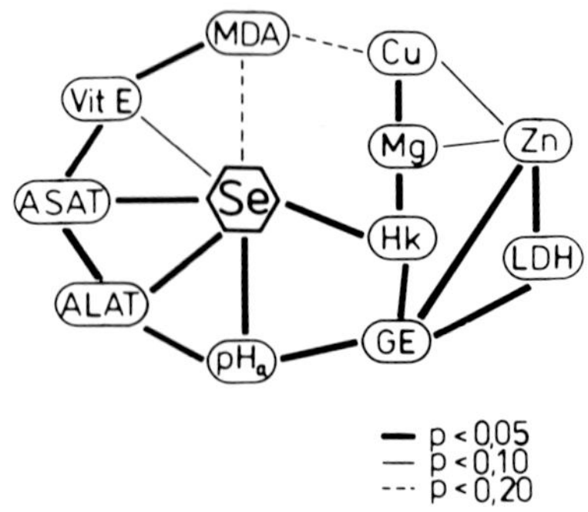

FIGURE 1. Parameters with significant correlation coefficients.

anterior at the passage between the middle and distal thirds. Effects on the cardiac rhythm and hemodynamics were evaluated electrocardiographically and by recording blood pressure. At 30 min before setting the ligature, sodium selenite was infused in concentrations of 1 to 3 mg/kg of body weight. Serum samples were taken before Se application and at 15 min after the end of infusion. First of all, we determined Se, vitamin E, malondialdehyde (MDA), total protein (GE), aspartate aminotransferase (ASAT), alanine aminotransferase (ALAT), lactatedehydrogenase (LDH), pH value (pHa), and hematocrit (Hk). Se determinations were carried out using hydride atomic absorption,[84] vitamin E with bathophenanthroline,[85] and MDA demonstration with the thiobarbituric acid reaction.[86] We found no relationships between Se, MDA, and vitamin E, but, by interpretation of the hemodynamic signs, we could see a well-defined cardioprotective effect of sodium selenite (Figure 1). The hemodynamic action consisted of an increase of heart rate, maximum aortic flow rate, cardiac output, minute volume of the heart and coronary flow, as well as a decrease in total peripheral resistance. The antiarrhythmic effect of Se was manifested in reduced occurrence of ventricular extrasystoles and flutter and disappearance of "late potentials" in the peri-infarction area.

In muscle tissue of the infarcted and noninfarcted areas of the pig heart muscle tissue (left ventricle, 100-μm slices, deep frozen), we determined Se by laser atomic absorption (Figure 2).[87] The highest Se concentrations were found in the endocardium. On the other hand, Se was not detectable in the epicardium.

ARTERIOSCLEROSIS IN HUMAN PATIENTS

MDA, vitamin E, GPx, and Se were analyzed in the blood of 96 male patients aged 45 to 60 years with arteriosclerotic obstructive disease in the leg. In addition, 32 of these subjects had manifestations of coronary or cerebral arteriosclerosis. In the parameters mentioned, no significant differences were found between patients with general arteriosclerosis and localized arteriosclerosis of the leg. Patients with arteriosclerosis and hyperlipoproteinemia (HLP) had higher values of MDA than patients without HIP (Table 1). No correlations were found between MDA and total cholesterol, triglycerides, vitamin E, GPx, and Se. In a group of hospitalized patients with the clinical diagnosis of acute myocardial infarction (AMI), compared with a healthy group, we analyzed Se serum concentrations. In the group with AMI, the Se values were significantly lower than in the control group (Table 2). Moreover, there was a positive correlation between Se and the rate of mortality. In our opinion, the above data justify further research to gain a better understanding of the relationship between Se, lipoperoxides, antioxidants, and cardiovascular disease.

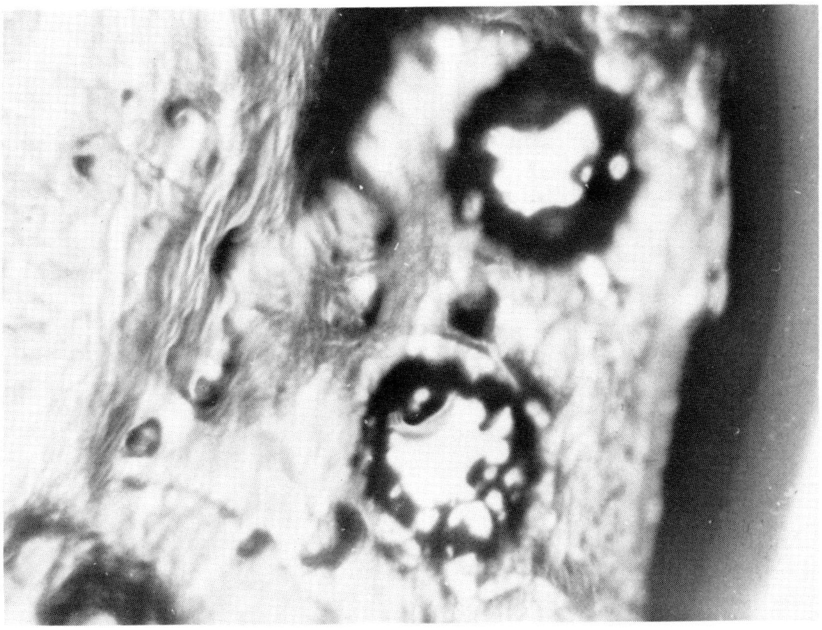

FIGURE 2. Laser shots in slices of myocardium.

Table 1
MDA (μMOL/ℓ) CONCENTRATIONS IN SERUM OF PATIENTS WITH AND WITHOUT HLP

	With HLP	Without HLP
General arteriosclerosis	1.89 ± 0.59	1.82 ± 0.84
Arteriosclerosis of the leg	2.49 ± 0.92	1.80 ± 0.80

Table 2
SERUM SE CONCENTRATIONS (μMOL/ℓ)

	Control Group (n = 93)	Group with AMI (n = 54)
$\bar{x}$	981	670
SD	209	266

REFERENCES

1. **Glavind, J., Hartmann, S., Clemmesen, J., Jessen, K. E., and Dam, H.,** Studies on the role of lipoperoxides in human pathology. II. The presence of peroxidized lipids in the atherosclerotic aorta, *Acta Pathol.*, 30, 1, 1952.
2. **Haugaard, N.,** Cellular mechanisms of oxygen toxicity, *Physiol. Rev.*, 48, 311, 1968.

3. **Maseki, M., Nishigaki, J., Hagihara, M., Tomoda, Y., and Yagi, K.,** Lipid peroxide levels and lipid content of serum lipoprotein fractions of pregnant subjects with or without pre-eclampsia, *Clin. Chim. Acta,* 115, 155, 1981.
4. **Di Luzio, N. R.,** Antioxidants, lipid peroxidation and chemical-induced liver injury, *Fed. Proc. Fed. Am. Soc. Exp. Biol.,* 32, 1875, 1973.
5. **Hiramitsu, T., Hasegawa, Y., Hirata, K., Nishigaki, J., and Yagi, K.,** Formation of lipoperoxide in the retina of rabbit exposed to high concentration of oxygen, *Experientia,* 32, 622, 1976.
6. **Plaa, G. L. and Witschi, H.,** Chemicals, drugs and lipid peroxidation, *Annu. Rev. Pharmacol.,* 16, 125, 1976.
7. **Chance, B., Sies, H., and Boveris, A.,** Hydroperoxide metabolism in mammalian organs, *Physiol. Rev.,* 59, 527, 1979.
8. **Ueda, K., Yoshioka, T., Takehara, Y., and Abe, K.,** Lipoperoxides, vitamin E, and activities of superoxide dismutase, glutathione peroxidase, and catalase in regenerating rat liver, *Biochem. Int.,* 1, 663, 1983.
9. **Szczeklik, A., Gryglewski, R. J., Domagala, B., Dworski, R., and Basista, M.,** Dietary supplementation with vitamin E in hyperlipoproteinemias: effects on plasma lipid peroxides, antioxidant activity, prostacyclin generation and platelet aggregability, *Thromb. Haemost.,* 54, 425, 1985.
10. **Tappel, A. L. and Dillard, C. J.,** In vivo lipid peroxidation: measurement via exhaled pentane and protection by vitamin E, *Fed. Proc.,* 40, 174, 1981.
11. **Harrison, J. H. and Conrad, H. R.,** Selenium content and glutathione peroxidase activity in tissues of the dairy cow after short-term feeding, *J. Dairy Sci.,* 67, 2464, 1984.
12. **Marubayashi, S., Dohi, K., Ochi, K., and Kawasaki, T.,** Role of free radicals in ischaemic rat liver cell injury: prevention of damage by alpha-tocopherol administration, *Surgery,* 99, 184, 1986.
13. **Thomson, C. D., Ong, L. K., and Robinson, M. F.,** Effects of supplementation with high-selenium wheat bread on selenium, glutathione peroxidase and related enzymes in blood components of New Zealand residents, *Am. J. Clin. Nutr.,* 41, 1015, 1985.
14. **Peters, H.-J., Voigt, H., Koehler, H., Vorberg, B., and Reuter, W.,** Lipid peroxides and antioxidants in the blood of patients with arteriosclerotic circulatory disorders, *Z. Gesamte Inn. Med. Ihre Grenzgeb.,* 40, 276, 1985.
15. **Julicher, R. H. M. and Sterrenberg, L.,** Evidence for lipid peroxidation during the calcium paradox in vitamin E deficient rat heart, *Naunyn Schmiedebergs Arch. Pharmakol.,* 326, 87, 1984.
16. **Jain, S. K.,** Vitamin E and stabilization of membrane lipid organization in red blood cells with peroxidative damage, *Biomed. Biochim. Acta,* 42, 43, 1983.
17. **Doni, M. G. and Falanga, A.,** The effect of vitamin E or selenium on the oxidant-antioxidant balance in rats, *Br. J. Exp. Pathol.,* 65, 75, 1984.
18. **Hill, K. E. and Burk, R. F.,** Influence of vitamin E and selenium on glutathione-dependent protection against microsomal lipid peroxidation, *Biochem. Pharmacol.,* 33, 1065, 1984.
19. **Nakayama, T., Kodama, M., and Nagata, C.,** Free radical formation in DNA by lipid peroxidation, *Agric. Biol. Chem.,* 48, 517, 1984.
20. **Diplock, A. T.,** The role of vitamin E in biological membranes, *Ciba Found. Symp.,* 101, 45, 1983.
21. **Ohki, K., Takamura, T., and Nozawa, Y.,** Effect of alpha-tocopherol on lipid peroxidation and acyl chain mobility of liver microsomes from vitamin E-deficient rat, *J. Nutr. Sci. Vitaminol.,* 30, 221, 1984.
22. **Blasig, I. E., Blasig, R., and Loewe, H.,** Myocardial lipid peroxidation during isoproterenol-induced blood flow reduction in rat myocardium, *Biomed. Biochem. Acta,* 43, 171, 1984.
23. **Mikaelyan, E. M., Melik-Agaeva, E. A., and Mikhitaryan, V. G.,** Regulation of the lipid peroxidation by antioxidants in erythrocyte membranes in stress, *Zh. Eksp. Klin. Med.,* 24, 217, 1984.
24. **Siddons, R. C. and Mills, C. F.,** Glutathione peroxidase activity and erythrocyte stability in calves differing in selenium and vitamin E status, *Br. J. Nutr.,* 46, 345, 1981.
25. **Bieri, J. G., Corash, L., and van Hubbard, S.,** Medical uses of vitamin E, *N. Engl. J. Med.,* 308, 1063, 1983.
26. **Verlinden, M., van Sprundel, M., van der Auwera, J. C., and Eylenbosch, W. J.,** The selenium status of Belgian population gorups. II. Newborns, children, and the aged, *Biol. Trace Elem. Res.,* 5, 103, 1983.
27. **Peters, H.-J., Koehler, H., Duck, H.-J., Guenther, K. R., and Pankau, H.,** Cardioprotective function of trace element selenium in experimental myocardial infarction, *Dtsch. Gesundheitswes.,* 37, 187, 1983.
28. **Pankau, H., Schauer, J., Koehler, H., Peters, H.-J., Duck, H.-J., and Guenther, K. R.,** Effects of sodium selenite in hemodynamic animal experiments, *Selenium Tellurium Abstr.,* 23(63373), 1982.
29. **Peters, H.-J., Koehler, H., Duck, H.-J., Englemann, L., Zwanziger, H., Vorberg, B., Pohle, B., Pankau, H., and Haustein, B.,** Selenium and its clinical-pathobiochemical importance in myocardial infarction, *Wiss. Z. Karl Marx Univ. Leipzig Math. Naturwiss. Reihe,* 35, 195, 1986.
30. **Voigt, H., Reuter, W., Peters, H.-J., Koehler, H., and Herrmann, W.,** Relations between the degree of severity of arteriosclerosis and parameters of trace elements and lipids, *Z. Gesamte Hyg. Ihre Grenzgeb.,* 30, 321, 1984.

31. **Valentine, J. L., Kang, H. K., Dang, P.-M., and Schluchter, M.,** Selenium concentrations and glutathione peroxidase activities in a population exposed to selenium via drinking water, *J. Toxicol. Environ. Health,* 6, 731, 1980.
32. **Miettinen, T. A., Alfthan, G., Huttunen, J. K., Pikkarainen, J., Naukkarinen, V., Mattila, S., and Kumlin, T.,** Serum selenium concentration related to myocardial infarction and fatty acid content of serum lipids, *Br. Med. J.,* 287, 517, 1983.
33. **Salonen, J. T., Alfthan, G., Huttunen, J. K., Pikkarainen, J., and Puska, P.,** Association between cardiovascular death and myocardial infarction in a matched-pair longitudinal study, *Lancet, ii,* 175, 1982.
34. **Shamberger, R. J., Gunsch, M. S., Willis, C. E., and McCormack, L. J.,** Selenium and heart disease. II. Selenium and other trace metal intakes and heart disease in 25 countries, in *Trace Substances in Environmental Health,* Vol. 12, Hemphill, D. D., Ed., University of Missouri, Columbia, 1978, 48.
35. **Levander, O. A., Sutherland, B., Morris, V. C., and King, J. C.,** Selenium metabolism in human nutrition, in *Selenium in Biology and Medicine,* Spallholz, J. E., Martin, J. L., and Ganther, H. E., Eds., AVI Publishing, Westport, Conn., 1981.
36. **Diplock, A. T.,** Metabolic and functional defects in selenium deficiency, *Philos. Trans. R. Soc. London Ser. B,* 294, 105, 1981.
37. **Jaakola, K., Tummavuori, J., Pirinen, A., Kurkela, P., Tolonen, M., and Arstila, U.,** Selenium levels in whole blood of Finnish volunteers before and during organic and inorganic selenium supplementation, *Scand. J. Clin. Lab. Invest.,* 43, 473, 1983.
38. **Agostoni, A., Gerli, G. C., Beretta, L., Palazzini, G., Buso, G. P., Xusheng, H., and Moschini, G.,** Erythrocyte antioxidant enzymes and selenium serum levels in an Andean population, *Clin. Chim. Acta,* 133, 153, 1983.
39. **Huguet, C., Chappuis, Ph., Peynet, J., Thuillier, F., Legrand, A., and Rousselet, F.,** Intérêt dans le diagnostic biologique des cardiomyopathies non obstructives, *Ann. Biol. Clin. (Paris),* 41, 277, 1983.
40. **Facchinetti, T., Delaini, F., Salmona, M., Donati, M. B., Feuerstein, S., and Wendel, A.,** The influence of selenium intake on chronic adriamycin toxicity and lipid peroxidation in rats, *Toxicol. Lett.,* 15, 301, 1983.
41. **Mancinella, A.,** Il selenio nell economia generale degli organismi viventi: sindromi da intossicazione e da carenza, impieglu attuali, prospettive terapeutiche, *Clin. Ter.,* 10L, 529, 1982.
42. **Shamberger, R. J.,** Selenium in the drinking water and cardiovascular disease, *J. Environ. Pathol. Toxicol.,* 4(2), 305, 1980.
43. **Anke, M., Partschefeld, M., Kursa, J., and Kroupova, V.,** Selenium-deficient myopathy, *Wiss. Z. Friedrich Schiller Univ. Jena Math. Naturwiss. Reihe,* 32, 809, 1983.
44. **Levander, O. A.,** Recent developments in selenium nutrition, in *Nutrition Update,* Vol. 1, Weininger, J. and Briggs, C., Eds., John Wiley & Sons, New York, 1983, 147.
45. **Virtamo, J. and Huttunen, J. K.,** Selenium in human disease, *Ann. Clin. Res.,* 17, 87, 1985.
46. **Oster, O., Prellwitz, W., Kasper, W., and Meinertz, T.,** Congestive cardiomyopathy and selenium deficiency, *Clin. Chim. Acta,* 128, 125, 1983.
47. **Wang, Y. X., Böcker, K., and Reuter, H.,** Selenium and myocardial infarction: glutathione peroxidase in platelets, *Klin. Wochenscher.,* 59, 817, 1981.
48. **Virtamo, J., Valkeila, E., Alfthan, G., Punsar, S., Huttunen, J. K., and Karvonen, M. J.,** Serum selenium and the risk of coronary heart disease and stroke, *Am. J. Epidemiol.,* 122, 276, 1985.
49. **Kok, F. J., Hofmann, A., Vandenbroucke, J., and Valkenburg, H. A.,** Selenium and cardiovascular disease, *Int. J. Epidemiol.,* 14, 335, 1985.
50. **Tang, F.-Z., Guo, J., Zhao, W.-G., Tang, S.-Y., and Xie, Y.-J.,** Anti-arrythmic effects of sodium selenite, *Acta Pharamacol. Sin.,* 4, 244, 1983.
51. **Goldman, I. S. and Kantrowitz, N. E.,** Cardiomyopathy associated with selenium deficiency, *N. Engl. J. Med.,* 305, 701, 1981.
52. **Ellis, N., Lloyd, B., Lloyd, R. S., and Clayton, B. E.,** Selenium and vitamin E in relation to risk factors for coronary heart disease, *J. Clin. Pathol.,* 37, 200, 1984.
53. **Gomanova, M. I., Shemjakin, F. M., Kudrin, A. N., Nikolayev, S. M., and Chermysheva, L. F.,** The determination of selenium in the blood and heart of rats during experimental myocardial infarction, *Farmacija,* 26, 56, 1977.
54. **Nikolayev, S. M., Kudrin, A. N., and Kaktursky, L. V.,** Medico-prophylactic influence of sodium selenite on the course of experimental myocardial infarction, *Farmakol. Toksikol.,* 5, 571, 1976.
55. **van Vleet, J. F., Ferrans, V. J., and Weirich, W. E.,** Cardiac disease induced by adriamycin administration in dogs and an evaluation of vitamin E and selenium as cardioprotectants, *Am. J. Pathol.,* 99, 13, 1980.
56. **Pankau, H., Duck, H.-J., Koehler, H., Peters, H.-J.,** Systolic and diastolic coronary arteries blood flow profiles during selenium application, *Selenium Tellurium Abstr.,* 25(73971), 1984.
57. **Ge, K., Xue, A., Bai, J., and Wang, S.,** Keshan disease — an endemic cardiomyopathy in China, *Virchows Arch. A,* 401, 1, 1983.

58. **Gu, B.,** Pathology of Keshan disease, *Chin. Med. J.*, 96, 251, 1983.
59. **Chen, X., Yang, G., Chen, J., Chen, X., Wen, Z., and Ge, K.,** Studies on the relations of selenium and Keshan disease, *Biol. Trace Elem. Res.*, 2, 91, 1980.
60. Keshan Disease Research Group of the Chinese Academy of Medical Sciences, Observations on effect of sodium selenite in prevention of Keshan disease, *Chin. Med. J.*, 92, 471, 1979.
61. **Moore, J. A., Noiva, R., and Wells, I. C.,** Selenium concentrations in plasma of patients with arteriographically defined coronary arteriosclerosis, *Clin. Chem. (N.Y.)*, 30, 1171, 1984.
62. **Mills, G. C.,** Hemoglobin catabolism. I. Glutathione peroxidase, an erythrocyte enzyme which protects hemoglobin from oxidative breakdown, *J. Biol. Chem.*, 229, 189, 1957.
63. **Chow, C. K. and Tappel, A. L.,** An enzymatic protective mechanism vs. lipid peroxidation damage to lungs of oxygen exposed rats, *Lipids*, 7, 518, 1972.
64. **Flohé, L., Guenzler, W. A., and Schock, H. H.,** Glutathione peroxidase: a selenoenzyme, *FEBS Lett.*, 32, 132, 1973.
65. **Donchenko, G. V., Khmelevskii, and Yu, V.,** Redox processes in rat myocardium in acute hypoxia, *Fiziol. Zh. (Kiev)*, 30, 180, 1984.
66. **Pan, H. and Feng, L.,** Effect of malondialdehyde on erythrocytes, *Sheng Wu Hua Hsueh Yu Sheng Wu Wu Li Jinzhan*, 56, 34, 1984.
67. **Ottolenghi, A.,** Interaction of ascorbic acid and mitochondrial lipids, *Arch. Biochem. Biophys.*, 79, 355, 1959.
68. **Meerson, F. Z., Kagan, V. E., Kozlov, Yu. P., Belkina, L. M., and Arkhipenko, Yu. V.,** The role of lipid peroxidation in pathogenesis of ischemic damage and the antioxidant protection of the heart, *Basic Res. Cardiol.*, 77, 465, 1982.
69. **Roth, E., Torok, B., Zsoldos, T., and Matkovics, B.,** Lipid peroxidation and scavenger mechanism in experimentally induced heart infarcts, *Basic Res. Cardiol.*, 80, 530, 1985.
70. **Meerson, F. Z. and Manukhina, E. B.,** Disturbance of contractile function of vena portae in stress and infarction and its pervention, *Basic Res. Cardiol.*, 80, 407, 1985.
71. **Savov, V. M., Didenko, V. V., Dosmagambetova, R. S., Meerson, F. Z., and Kagan, V. E.,** Lipid peroxidation in the myocardium in experimental infarct, *Biol. Nauki (Moscow)*, 5, 30, 1985.
72. **Lankin, V. Z., Tikhaze, A. K., Rakita, D. R., Afonskaia, N. I., and Ruda, M. I.,** Utilization of active forms of oxygen and lipoperoxides in the blood of patients with myocardial infarction, *Ter. Arkh.*, 57, 58, 1985.
73. **Hess, M. L. and Manson, N. H.,** The oxygen free radical system and myocardial dysfunction, *Adv. Myocardiol.*, 5, 177, 1985.
74. **Rao, P. S., Cohen, M. V., and Mueller, H. S.,** Production of free radicals and lipid peroxidase in early experimental myocardial ischemia, *J. Mol. Cell Cardiol.*, 15, 713, 1983.
75. **Rao, P. S. and Mueller, H. S.,** Lipid peroxidation and acute myocardial ischemia, *Adv. Exp. Med. Biol.*, 161, 347, 1983.
76. **Doni, M. G., Falanga, A., Delaini, F., Vicenzi, E., Tomasiak, M., and Donati, M. B.,** The effect of vitamin E or selenium on the oxidant-antioxidant balance in rats, *Br. J. Exp. Pathol.*, 65, 75, 1984.
77. **Fukuzawa, K., Takase, S., and Tsukatani, H.,** The effect of concentration on the antioxidant effectiveness of alphatocopherol in lipid peroxidation induced by superoxide free radicals, *Arch. Biochem. Biophys.*, 240, 117, 1985.
78. **Summerfield, F. W. and Tappel, A. L.,** Effect of dietary polyunsaturated fats and vitamin E on aging and peroxidative damage to DNA, *Arch. Biochem. Biophys.*, 233, 408, 1984.
79. **Quintanilha, A. T.,** Effects of physical exercise and/or vitamin E on tissue oxidative metabolism, *Biochem. Soc. Trans.*, 12, 403, 1984.
80. **Wartanowicz, M., Panczenko-Kresovska, B., Ziemlanski, S., Kowalska, M., and Okolska, G.,** The effect of alpha-tocopherol and ascorbic acid on the serum lipid peroxide level in elderly people, *Ann. Nutr. Metab.*, 28, 186, 1984.
81. **Diplock, A. T.,** Vitamin E, selenium and free radicals, *Med. Biol.*, 62, 78, 1984.
82. **Jackson, M. J., Jones, D. A., and Edwards, R. H.,** Techniques for studying free radical damage in muscular dystrophy, *Med. Biol.*, 62, 135, 1984.
83. **Nevé, J., Vertongen, F., and Molle, L.,** Selenium deficiency, *Clin. Endocrinol. Metab.*, 14, 629, 1985.
84. **Peters, H.-J. and Dittrich, K.,** Determination of selenium in biological fluids by hybride atomic absorption, *Z. Med. Labor. Diagn.*, 22, 237, 1981.
85. **Peters, H.-J., Vorberg, B., and Koehler, H.,** A modified micromethod for vitamin E determination in serum, *Z. Gesamte Hyg. Ihre Grenzgeb.*, 29, 441, 1983.
86. **Yagi, K.,** A simple fluorometric assay for lipoperoxide in blood plasma, *Biochem. Med.*, 15, 212, 1976.
87. **Dittrich, K. and Wennrich, R.,** Laser vaporization in atomic spectroscopy, *Prog. Anal. Atom. Spectrosc.*, 7, 315, 1984.

THE NEUROPATHOLOGY OF CHRONIC VITAMIN E DEFICIENCY IN HUMANS AND OTHER MAMMALS

James S. Nelson

EXPERIMENTAL VITAMIN E DEFICIENCY

Ringsted[1] described a neurological syndrome consisting of ataxia, bilateral symmetrical muscle atrophy, and paralysis in adult rats maintained on a vitamin E-deficient diet for prolonged periods. Subsequently, Einarson and Ringsted[2] investigated the neuropathological basis of the syndrome. Lesions were found in the spinal cord, peripheral nerves, and skeletal muscles. The earliest alteration involved the dorsal roots and dorsal columns, especially the fasciculus gracilis. The process began at the lumbar level and gradually progressed rostrally, sometimes as high as the lower cervical segments. Axons and myelin sheaths were lost in approximately equal numbers. These changes were accompanied by fibrillary astrocytosis and lipid phagocyte formation. Somewhat later, myopathic changes were evident in skeletal muscles. During the later stages of the process, degenerative changes in the anterior horns, including the neuronal accumulation of lipofuscin or ceroid pigment and, possibly, nerve cell loss with mild astrocytosis, were observed. In contrast, Wolf and Pappenheimer[3] found no significant neuropathologic changes in rats maintained up to 365 days on a vitamin E-deficient diet. In a later publication,[4] however, they noted the presence of lipopigment in central nervous system (CNS) neurons of vitamin E-deficient rats and absence of the pigment in rats receiving protective doses of wheat germ oil or tocopherol. The accumulation of lipopigment was not associated with nerve cell loss or neurological distrubances. The investigations of Wolf and Pappenheimer[3,4] were stimulated in part by the growing use, in the late 1930s, of vitamin E as a treatment of amyotrophic lateral sclerosis and related motor neuron disorders. They provided early, substantial experimental evidence against an etiologic or pathogenetic relationship between tocopherol deficiency and amyotrophic lateral sclerosis. Later, in a critical review of his studies, Einarson[5] acknowledged that he had probably overemphasized the significance of the anterior horn cell alterations. He reaffirmed the occurrence of degeneration in the dorsal columns of the spinal cord. It is this lesion, rather than the controversial and dubious anterior horn cell degeneration, that has been shown to be the characteristic neuropathological lesion in the vitamin E-deficient rat.

The regular occurrence of concomitant loss of axons and myelin sheaths, especially in the rostral segments of the dorsal columns of rats with chronic vitamin E deficiency, has now been demonstrated by several groups of investigators, including Luttrell and Mason,[6] Malamud et al.,[7] and Pentschew and Schwarz.[8] The latter also called attention to the presence of swollen, dystrophic preterminal and terminal axons (spheroids) in the dorsal columns, gracile and cuneate nuclei, dorsal horns. Clarke's column, the sensory nucleus of the fifth nerve, and the nucleus of the tractus solitarius. Subsequently these observations have been confirmed by Machlin et al.[9] (Figures 1 and 2). More recently, Towfighi[10] has described similar, dystrophic changes in peripheral sensory axons from vitamin E-deficient rats. Ultrastructural studies have shown that the axonal swelling results from local accumulation of smooth membrane complexes, neurofilaments, mitochondria, and collections of electron-dense structures descriptively identified as microbodies or multivesicular or multigranular bodies[11,12] (Figure 3). The nervous system lesions are commonly accompanied by a primary necrotizing myopathy affecting the skeletal muscles (Figure 4).

A limited amount of information was derived from the initial neuropathologic studies of vitamin E deficiency in primates. Einarson and Telford[13] examined four vitamin E-deficient rhesus monkeys and noted relatively minor degenerative changes confined to nerve cell

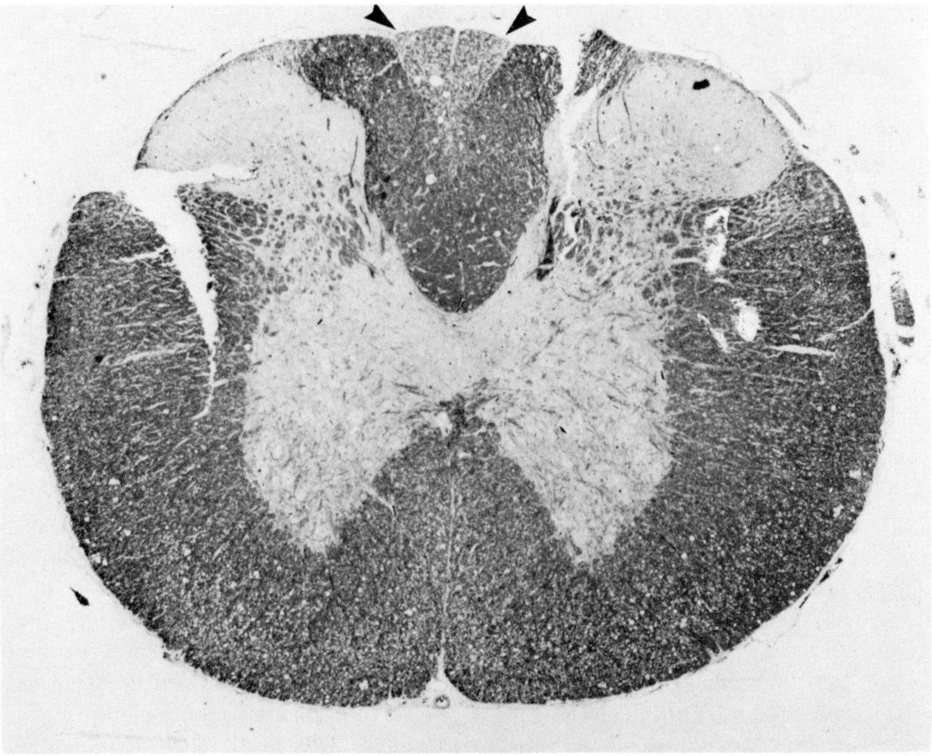

FIGURE 1. Spinal cord, vitamin E-deficient rat; loss of axons and myelin sheaths in fasciculus gracilis (arrows). (Loyez stain; magnification × 8.5.) (Reproduced from author's paper in Ciba Foundation Symposium 101, *Biology of Vitamin E*, Pitman, London, 1983, 92. With permission.)

bodies such as accumulation of lipopigment and neuronophagia. Axons in the CNS were not affected, and the peripheral nerves were not commented upon. Vitamin E deficiency had been induced by feeding the monkeys a deficient diet. Three of the animals had been maintained on the regimen for 6 to 17 months, and the fourth monkey had been maintained for almost 3 years before examination. No neuropathologic findings were reported by Hayes[14] in his studies of experimental vitamin E deficiency in New and Old World monkeys. The absence of findings in this case may be the consequence of a restricted neuropathologic examination.

Subsequently, Nelson et al.[15] demonstrated a consistent group of progressive central and peripheral nervous system lesions in seven rhesus monkeys maintained on a vitamin E-deficient diet for 30 to 33 months. The principal neuropathologic alteration was loss of sensory axions in the posterior columns, sensory roots, and peripheral nerves (Figures 5 and 6). Morphologic and morphometric studies indicated that the distal segments of the axons were affected most severely and large-caliber myelinated fibers were selectively involved. Swollen, dystrophic axons (spheroids) occurred infrequently. Degeneration and phagocytosis of small numbers of neuronal perikarya were observed in the dorsal root ganglia and the anterior horns. The number of affected neurons was not proportional to the number of affected axons. Accumulation of lipopigment was evident in neuronal perikarya and CNS endothelial cells. The nervous system lesions were usually accompanied by a chronic necrotizing myopathy. The differences between this study and those previously reported may be related to the length of time the monkeys were maintained on the vitamin E-deficient diet and to the extent of the neuropathologic examination.

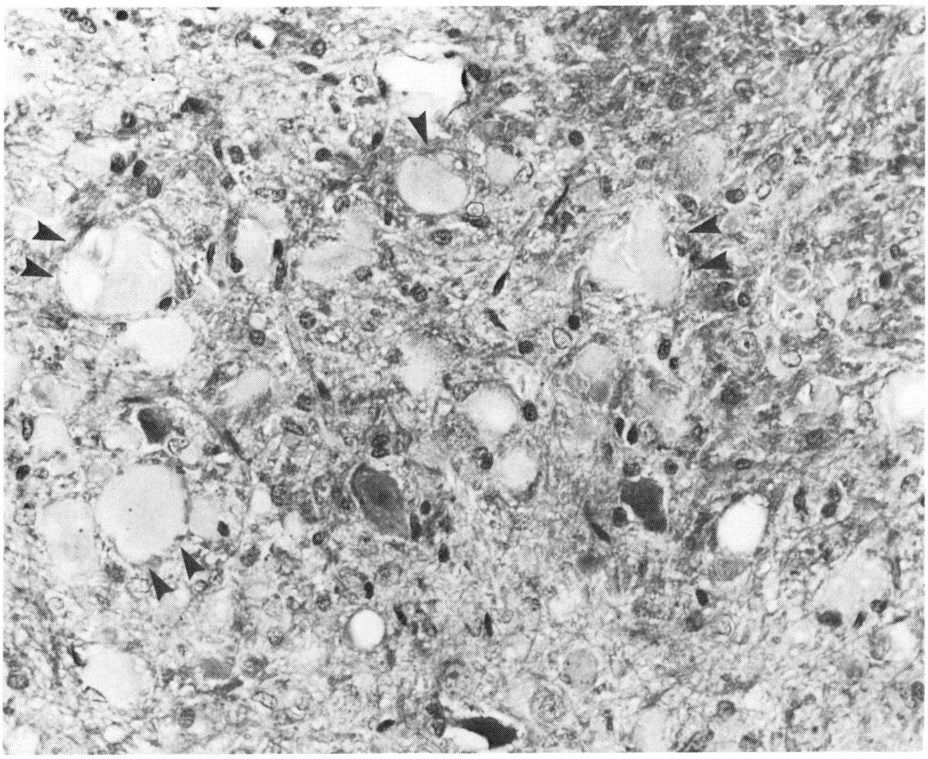

FIGURE 2. Nucleus gracilis, vitamin E-deficient rat; numerous dystrophic axons (arrows). (Luxol fast blue-PAS; magnification × 250.)

The neuropathology of chronic vitamin E deficiency has not been widely studied in other subhuman mammals. Dystrophic axons were reported in the gracile and cuneate nuclei vitamin E - deficient by Hayes et al.[16] Recently, vitamin E deficiency has been implicated in the pathogenesis of subependymal hemorrhages in fetal hamsters.[17]

VITAMIN E DEFICIENCY IN HUMANS

Dystrophic axons similar in type and distribution to those observed in vitamin E-deficient rats were noted at autopsy by Sung[18] and Sung and Stadlan[19] in children and young adults with cystic fibrosis or congenital biliary atresia. Although serum levels of vitamin E had not been measured, the authors pointed out that, in all likelihood, the patients were vitamin E deficient because both cystic fibrosis and biliary atresia are characteristically associated with malabsorption of fat-soluble vitamins, including vitamin E. Very low serum levels of vitamin E (secondary to intestinal malabsorption) and a progressive spinocerebellar disorder are regularly observed in patients with abetalipoproteinemia. The studies of Muller et al.[20] provided initial evidence of an etiological link between the vitamin E deficiency and the spinocerebellar syndrome. Further, Muller et al.[20] indicated the feasibility of treating the neurological disorder with large doses of vitamin E. These observations form the basis of recent investigations demonstrating that humans with lipid malabsorption disorders and chronic vitamin E deficiency can and do develop a progressive, crippling neurological disorder which may be treated or prevented by giving vitamin E.

Congenital Biliary Atresia

Neurologic[21] and neuropathologic[22] studies involving ten children, 3.5 to 15 years of age

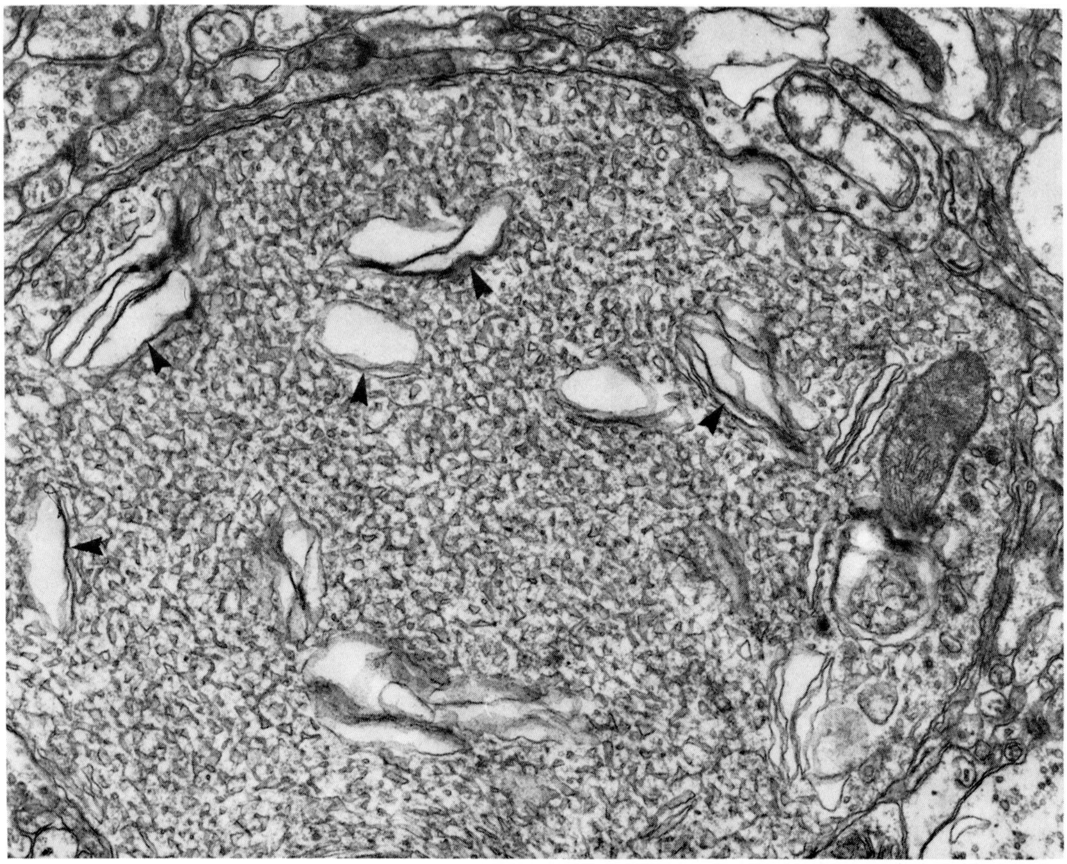

FIGURE 3. Nucleus gracilis, vitamin-E deficient rat; ultrastructural appearance of dystrophic axon. A network of smooth membrane profiles with characteristic clefts (arrows) fills and distends the axon. (Magnification × 25,000.)

with biliary atresia have been reported. The six children on whom neuropathological studies were done died between 3.5 and 12 years of age. Serum vitamin E concentrations were measured in four of the six affected children. Levels below 0.3 mg/dℓ were noted regularly. Two children developed a progressive neurological disorder, with loss of deep tendon reflexes, disturbances in vibration sense, proprioception, light touch sensation, and gait; dysmetria; and occasionally abnormal extraocular movements and distal motor weakness. Identical neurological disturbances were observed in the four other living patients with congenital biliary atresia and low serum levels of vitamin E and in the four children reported by Guggenheim et al.[23] In the latter report, treatment with vitamin E improved neurological function.

Similar neuropathologic changes were found in each of the six patients, although the extent of the examination varied according to the tissue available. In two patients a detailed examination of both the central and peripheral nervous system was carried out. Numerous dystrophic axons and astrocytosis were evident in the major sensory relay nuclei of the caudal medulla and spinal cord (Figure 7). There was concomitant loss of axons and myelin sheaths from the posterior columns of the spinal cord (Figure 8). Axonal loss was more severe in the rostral segments of the posterior columns. With age, the brain stem and spinal cord lesions increased in severity. In the older patients, there was mild loss of axons and myelin sheaths from the lateral columns in the regions of the dorsal and ventral spinocerebellar tracts. Loss of axons and myelin sheaths was evident in the posterior roots and there was

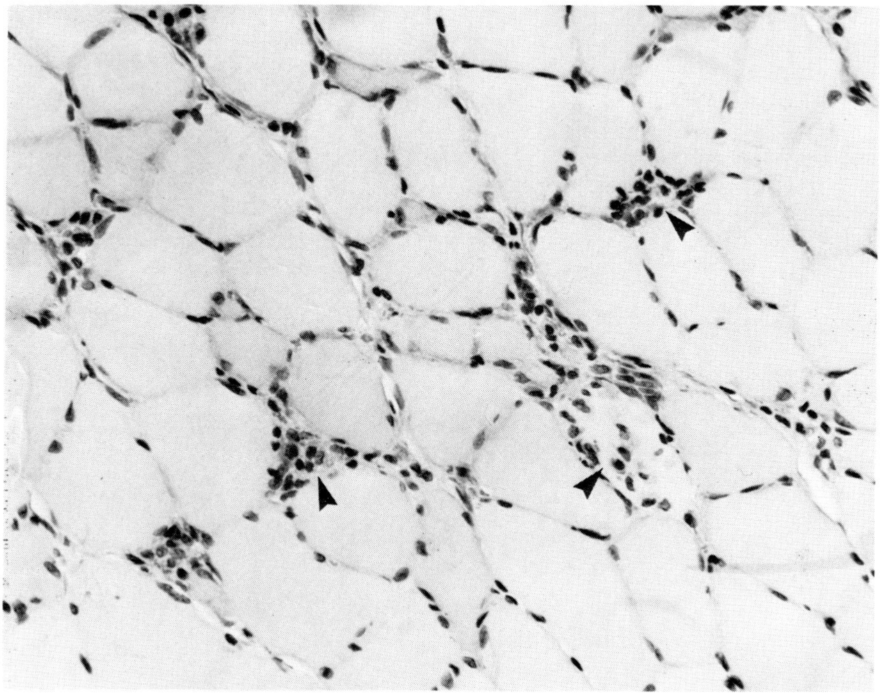

FIGURE 4. Skeletal muscle, vitamin E-deficient rat; necrotizing myopathy with active necrosis and phagocytosis (arrows) of skeletal muscle fibers. (Hematoxylin-eosin; magnification × 250.) (Reproduced from author's paper in Ciba Foundation Symposium 101, *Biology of Vitamin E,* Pitman, London, 1983, 92. With permission.)

mild depletion of neurons in the dorsal root ganglia. The losses in these structures, however, were not proportional to the extent of axonal loss in the rostral segments of the cord. Foci of astrocytosis and parenchymal loss were present in the oculomotor nuclei of two children with abnormal extraocular movements. Lipopigment occurred in large neurons throughout the central and peripheral nervous system. With age, the number of neurons with lipopigment accumulation increased. In the peripheral nervous system, major nerve trunks were intact. Occasional degenerating axons were found in the sural nerves of two patients (Figure 9). Fiber spectrum analysis of the sural nerves of these two patients demonstrated, a mild loss of large-caliber myelinated axons. Necrotizing myopathy was not observed. Random variation of muscle fiber diameter and lipopigment accumulation occurred regularly. In one of the older children, groups of small-diameter muscle fibers reminiscent of neurogenic atrophy were seen in some muscles.

Recently, pathologic studies of nerve and muscle biopsies from children and young adults with cholestatic liver disease, vitamin E deficiency, and characteristic neurologic findings have been reported with increasing frequency.[23-29] Peripheral nerve lesions have included loss of large caliber myelinated fibers, axonal degeneration and dystrophy, alterations in myelin, and abnormal Schwann cell inclusions. In muscle biopsies, evidence of denervation atrophy and accumulation of lipopigment have been noted. Saito et al.[30] have reported the autopsy findings in a patient with familial intrahepatic cholestasis. The neuropathologic findings were similar to those described previously in the cases of biliary atresia, except for greater degrees of nerve cell loss and more widespread axonal dystrophy in the former. The question of whether the variations on the basic theme of distal axonopathy and axonal dystrophy described in these reports reflect the spectrum of neuropathology associated with

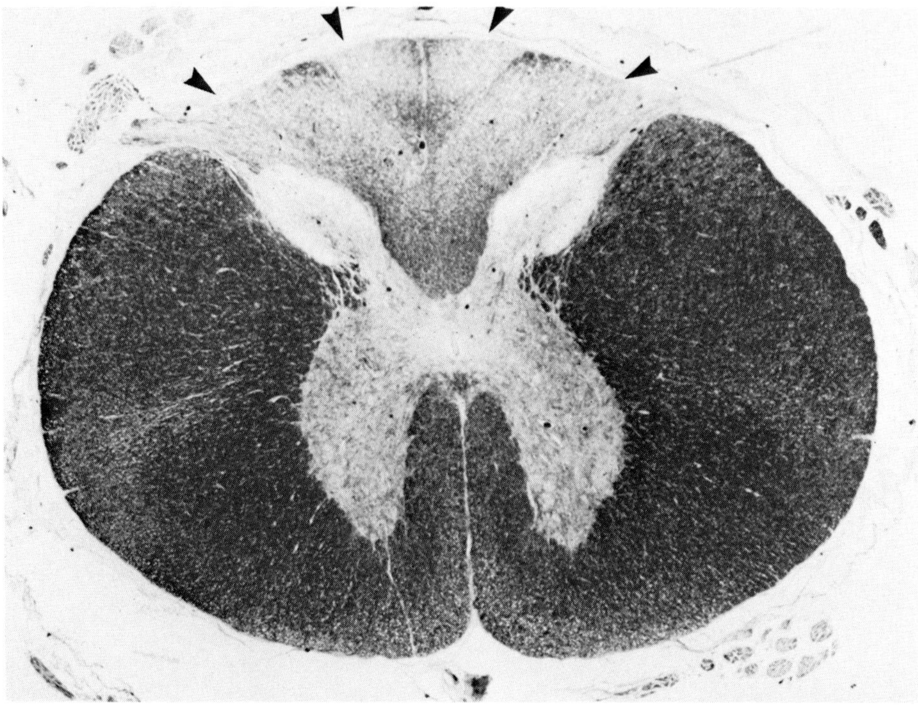

FIGURE 5. Spinal cord, vitamin E-deficient monkey. The loss of axons and myelin sheaths from the posterior columns is indicated by arrows. (Loyez stain; magnification × 8.5.) (Reproduced from author's paper in Ciba Foundation Symposium 101, *Biology of Vitamin E*, Pitman, London, 1983, 92. With permission.).

vitamin E deficiency or are the results of other factors is unresolved. Its solution will require further study.

Abetalipoproteinemia

Serum vitamin E levels are extremely low or undetectable in abetalipoproteinemia (Bassen-Kornzweig syndrome), an autosomal recessive disorder characterized by absence of detectable plasma apoprotein B.[31] This protein is an essential component of all the low-density lipoproteins, including chylomicrons as well as the low-density (LDL) and very low-density VLDL) types. Consequently, these substances are absent from affected patients and the capacity for intestinal absorption and plasma transport of vitamin E is seriously impaired. A neurologic syndrome similar to that seen in the cases of vitamin E deficiency and biliary atresia develops within the first two decades of life, often by 10 years of age.[32] Treatment of these patients with vitamin E controls or prevents development of the neurologic disorder.[33]

A limited amount of information is available concerning the neuropathologic findings in abetalipoproteinemia. In earlier autopsy studies, demyelination of the posterior columns, spinocerebellar tracts, and peripheral nerves were described.[34,35] Abnormalities observed in peripheral nerves include patchy demyelination, selective loss of large-caliber myelinated axons, and varying degrees of paranodal demyelination,.[36-39] The neuropathology of variants of abetalipoproteinemia, such as hypobetalipoproteinemia, is not available.

We have reported post-mortem neuropathologic studies on two patients with abetalipoproteinemia.[40] Similar findings were present in both cases. There was marked loss of axons and myelin sheaths in the posterior roots, posterior columns, and gracile and cuneate nuclei. Only a few dystrophic axons were present. Loss of axons and myelin sheaths also occurred in the spinocerebellar tracts and peripheral nerves. Random skeletal muscle fiber atrophy

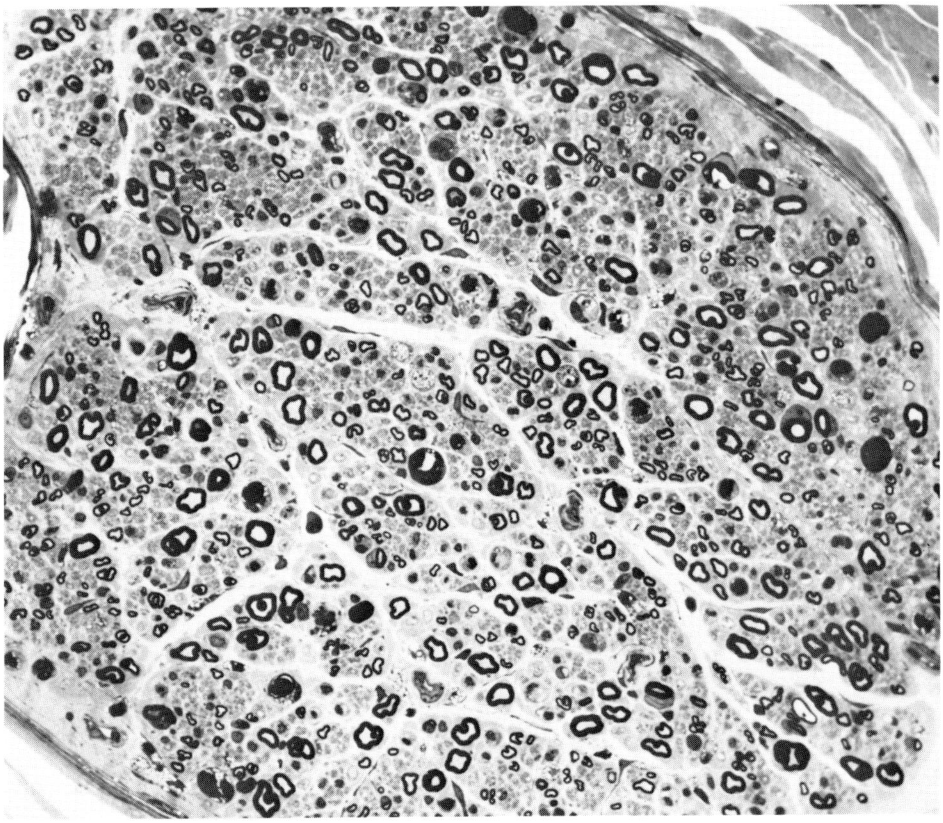

FIGURE 6. Sural nerve, vitamin E-deficient monkey. Axonal loss is moderate in amount. Degenerating axons and macrophages are present; 1.5-μm plastic section. (Alkaline toluidine blue; magnification × 250.) (Reproduced from author's paper in Ciba Foundation Symposium 101, *Biology of Vitamin E*, Pitman, London, 1983, 92. With permission.)

was evident without necrosis. Lipopigment accumulation involved both neurons and skeletal muscle fibers.

Acquired Intestinal Malabsorption

The occurrence of vitamin E deficiency with progressive neurologic symptoms in acquired forms of lipid malabsorption has now been amply demonstrated.[41] Neuropathologic findings have been reported in three of these cases.[42-44] The findings described by Weder et al.[43] and Brin et al.[44] closely resemble those in the biliary atresia cases. The pathologic studies on the sural nerve and deltoid muscle by Bertoni et al.[42] were carried out after 18 and 124 days, respectively, of vitamin E therapy. Focal loss of myelinated axons and excessive numbers of residual bodies in 5 to 10% of Schwann cells were noted in the sural nerve biopsy. Mild variability of fiber diameter, fiber degeneration and regeneration, periodic acid-schiff (PAS)-positive diastase-resistant sarcoplasmic granules, and histochemical evidence of abnormal mitochondrial function were observed in the muscle biopsy. The effect of vitamin E treatment on the morphologic findings in either the nerve or muscle biopsy is uncertain.

Isolated Vitamin E Deficiency

A few cases of vitamin E deficiency and neurologic disorder in the absence of generalized intestinal fat malabsorption have been recognized.[45] Nerve and muscle biopsy findings in

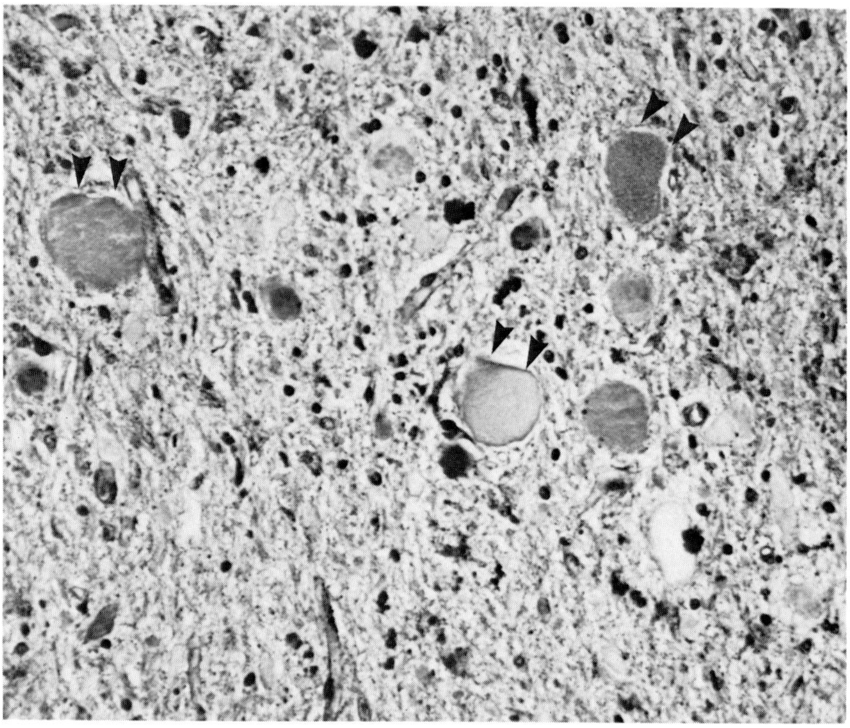

FIGURE 7. Nucleus gracilis; patient with biliary atresia, vitamin E deficiency, and progressive neurologic disorder. Dystrophic axons are (arrows) similar to those in Figure 2. (Luxol fast blue-PAS; magnification × 250.)

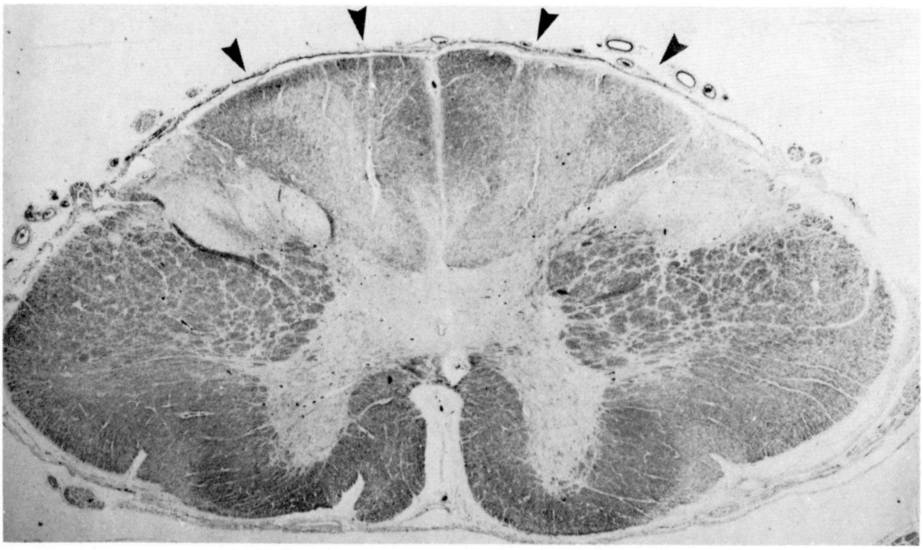

FIGURE 8. Spinal cord; patient with biliary atresia, vitamin E deficiency, and progressive neurologic disorder; loss of axons and myelin sheaths from posterior columns (arrows). Compare with Figures 1 and 5. (Loyez stain; magnification × 8.5.) (Reproduced from author's paper in Ciba Foundation Symposium 101, *Biology of Vitamin E,* Pitman, London, 1983, 92. With permission.)

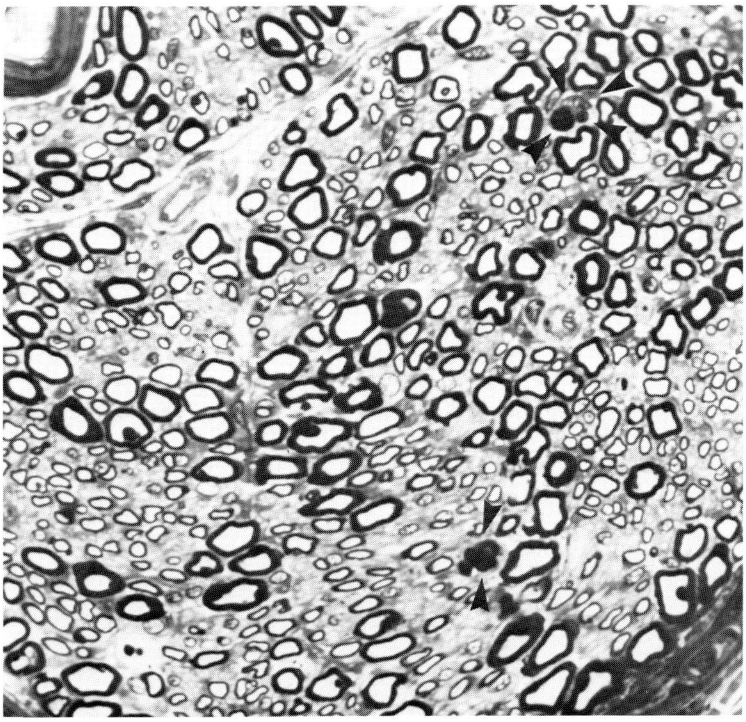

FIGURE 9. Sural nerve; patient with biliary atresia, vitamin E deficiency, and progressive neurologic disorder; occasional degeneration axons (arrows); 1.5-μm plastic section. (Alkaline toluidine blue; magnification × 250.)

one case have been reported.[46] Lipopigment was present in both Schwann cells and skeletal muscle fibers. Variation in muscle fiber diameter and loss of myelinated fibers more severely affecting those of large caliber were also noted.

Cystic Fibrosis

A progressive spinocerebellar disorder and chronic vitamin E deficiency have been observed in cases of cystic fibrosis.[47-49] Correlative neuropathological studies involving such cases have not been reported. Lipopigment was present in the nerve and muscle biopsy from the case reported by Bye et al.[49] Degenerative changes were absent in the nerve, however, evidence of denervation and regeneration was observed in the muscle biopsy. Post-mortem studies by Sung[18] and Geller et al.[50] support the view that neuropathologic findings in cases of cystic fibrosis with vitamin E deficiency and spinocerebellar syndrome are very likely to be similar to those seen in the biliary atresia cases. Unfortunately, the neurologic status of the patients reported by Sung[18] and Geller et al.[50] is unknown. Consequently, the clinical significance of the lesions reported by these authors is unclear.

CONCLUSIONS

A morphologically similar, distal axonopathy chiefly affecting large-caliber myelinated sensory axons develops regularly in rats, monkeys, and humans with prolonged, chronic vitamin E deficiency. The dorsal or posterior columns of the spinal cord are most severely affected. The numbers of dystrophic axons which occur in the major sensory relay nuclei of the medulla and spinal cord may vary from species to species. Axonal degeneration in the peripheral nerves is less severe than in the spinal cord. Except for accumulations of lipopigment, nerve cell bodies are usually minimally affected. In humans a characteristic spinocerebellar disorder develops which is readily correlated with the neuropathologic lesions. In rats and monkeys, vitamin E deficiency results in a chronic necrotizing myopathy with obvious necrosis and phagocytosis of muscle fibers. Skeletal muscle necrosis as well as clinical evidence of myopathy is minimal in vitamin E-deficient humans. Lipopigment within muscle fibers and changes suggestive of denervation atrophy are the principal alterations reported. The pathogenesis of the neural and skeletal muscle disorders has not been fully delineated. Clinical and laboratory studies demonstrating a favorable response of these disturbances to treatment with vitamin E indicate clearly, however, that deficiency of this vitamin is directly involved in their evolution.

ACKNOWLEDGMENTS

The author thanks Ms. Sandy Brickey and Mrs. Barbara Cogswell for their assistance in preparing this manuscript.

REFERENCES

1. **Ringsted, A.,** A preliminary note on the appearance of paresis in adult rats suffering from chronic avitaminosis E, *Biochem. J.*, 29, 788, 1935.
2. **Einarson, L. and Ringsted, A.,** *Effect of Chronic Vitamin E Deficiency on the Nervous System and the Skeletal Musculature in Adult Rats. A Neurotropic Factor in Wheat Germ Oil,* Oxford University Press, London, 1938.
3. **Wolf, A. and Pappenheimer, A. M.,** Central nervous system in vitamin E deficient rats, *Arch. Neurol. Psychiatry,* 48, 538, 1942.
4. **Wolf, A. and Pappenheimer, A. M.,** Occurrence and distribution of acid fast pigment in the central nervous system, *J. Neuropathol. Exp. Neurol.*, 4, 402, 1945.
5. **Einarson, L.,** Criticizing review of the concepts of the neuromuscular lesions in experimental vitamin E deficiency, preferably in adult rats, *Acta Psychiatr. Scand. Suppl.*, 78, 9, 1952.
6. **Luttrell, C. N. and Mason, K. E.,** Vitamin E deficiency, dietary fat, and spinal cord lesions in the rat, *Ann. N.Y. Acad. Sci.*, 52, 113, 1949.
7. **Malamud, N., Nelson, M. M., and Evans, H. M.,** The effect of chronic vitamin E deficiency on the nervous system in the rat, *Ann. N.Y. Acad. Sci.*, 52, 135, 1949.
8. **Pentschew, A. and Schwarz, K.,** Systemic axonal dystrophy in vitamin E deficient adult rats, *Acta Neuropathol.*, 1, 313, 1962.
9. **Machlin, L. J., Filipski, R., Nelson, J., Horn, L. R., and Brin, M.,** Effects of a prolonged vitamin E deficiency in the rat, *J. Nutr.*, 107, 1200, 1977.
10. **Towfighi, J.,** Effects of chronic vitamin E deficiency on the nervous system of the rat, *Acta Neuropathol.*, 54, 261, 1981.
11. **Lampert, P., Blumberg, J. M., and Pentschew, A.,** An electron microscopic study of dystrophic axons in the gracile and cuneate nuclei of vitamin E deficient rats, *J. Neuropathol. Exp. Neurol.*, 23, 60, 1964.
12. **Schochet, S.,** Mitochondrial changes in axonal dystrophy produced by vitamin E deficiency, *Acta Neuropathol. Suppl.* 5, 54, 1971.
13. **Einarson, L. and Telford, I. R.,** Effect of vitamin E deficiency on the central nervous system in various laboratory animals, *Biol. Skrifter, K. Dan. Vidensk. Selskab.*, 11(2), 1, 1960.
14. **Hayes, K. C.,** Pathophysiology of vitamin E deficiency in monkeys, *Am. J. Clin. Nutr.*, 27, 1130, 1974.
15. **Nelson, J. S., Fitch, C. D., Fischer, V. W., Broun, G. O., and Chou, A. C.,** Progressive neuropathologic lesions in vitamin E deficient rhesus monkeys, *J. Neuropathol. Exp. Neurol.*, 40, 166, 1981.
16. **Hayes, K. C., Nielsen, S. W., and Rousseau, J. E.,** Vitamin E deficiency and fat stress in the dog, *J. Nutr.*, 99, 196, 1969.
17. **Keeler, R. F. and Young, S.,** Role of vitamin E in the etiology of spontaneous hemorrhagic necrosis of the central nervous system of fetal hamsters, *Teratology,* 20, 127, 1979.
18. **Sung, J. H.,** Neuroaxonal dystrophy in mucoviscidosis, *J. Neuropathol. Exp. Neurol.*, 23, 567, 1964.
19. **Sung, J. H. and Stadlan, E. M.,** Neuroaxonal dystrophy in congenital biliary atresia, *J. Neuropathol. Exp. Neurol.*, 25, 341, 1966.

20. **Muller, D. P. R., Lloyd, J. K., and Bird, A. C.,** Long term management of abetalipoproteinemia. Possible role for vitamin E, *Arch. Dis. Child.*, 52, 209, 1977.
21. **Rosenblum, J. L., Keating, J. P., Prensky, A. L., and Nelson, J. S.,** A progressive neurologic syndrome in children with chronic liver disease, *N. Engl. J. Med.*, 304, 503, 1981.
22. **Nelson, J. S.,** Neuropathological studies of chronic vitamin E deficiency in mammals including humans, in *Biology of Vitamin E,* CIBA Foundation Symp. No. 101, Chairman: A. T. Diplock, Pitman, London, 1983, 92.
23. **Guggenheim, M. A., Ringel, S. P., Silverman, A., and Grabert, B. E.,** Progressive neuromuscular disease in children with chronic cholestasis and vitamin E deficiency: diagnosis and treatment with alpha tocopherol, *J. Pediatr.*, 100, 51, 1982.
24. **Tomasi, L. G.,** Reversibility of human myopathy caused by vitamin E deficiency, *Neurology,* 29, 1182, 1979.
25. **Alvarez, F., Landrieu, P., Laget, P., Lemonnier, F., Odieore, M., and Alagille, D.,** Nervous and ocular disorders in children with cholestasis and vitamin A and E deficiency, *Hepatology,* 3, 410, 1983.
26. **Neville, H. E., Ringel, S. P., Guggenheim, M. A., Wehling, C. A., and Starcevich, J. M.,** Ultrastructural and histochemical abnormalities of skeletal muscle in patients with chronic vitamin E deficiency, *Neurology,* 33, 483, 1983.
27. **Werlin, S. L., Harb, J. M., Swick, H., and Blank, E.,** Neuromuscular dysfunction and ultrastructural pathology in children with chronic cholestasis and vitamin E deficiency, *Ann. Neurol.*, 13, 291, 1983.
28. **Sokol, R. J., Heubi, J. E., Iannaccone, S. T., Bove, K. E., and Balistreri, W. F.,** Vitamin E deficiency with normal serum vitamin E concentrations in children with chronic cholestasis, *N. Engl. J. Med.*, 310, 1209, 1984.
29. **Landrieu, P., Selva, J., Alvarez, F., Ropert, A., and Metral, S.,** Peripheral nerve involvement in children with chronic cholestasis and vitamin E deficiency. A clinical, electrophysiological, and morphological study, *Neuropediatrics,* 16, 194, 1985.
30. **Saito, K., Yokoyama, T., Okaniwa, M., and Kamoshita, S.,** Neuropathology of chronic vitamin E deficiency in fatal familial cholestasis, *Acta Neuropathol.*, 58, 187, 1982.
31. **Herbert, P. N., Gotto, A. M., and Fredrickson, D. S.,** Familial lipoprotein deficiency, in *The Metabolic Basis of Inherited Diseases,* Stanbury, J. B., Wyngaarden, J. B., and Fredrickson, D. S., Eds., McGraw-Hill, New York, 1978, 544.
32. **Muller, D. P. R., Lloyd, J. K., and Wolff, O. H.,** Vitamin E and neurologic function, *Lancet,* i, 225, 1983.
33. **Muller, D. P. R. and Lloyd, J. K.,** Effect of large oral doses of vitamin E on the neurological sequelae of patients with abetalipoproteinemia, *Ann. N.Y. Acad. Sci.*, 393, 133, 1982.
34. **Sobrevilla, L. A., Goodman, M. L., and Kane, C. A.,** Demyelinating central nervous system disease, macular atrophy and acanthocytosis (Bassen-Kornzweig syndrome), *Am. J. Med.*, 37, 821, 1964.
35. **Dische, M. R. and Porro, R. S.,** The cardiac lesions in Bassen-Kornzweig syndrome, *Am. J. Med.*, 49, 568, 1970.
36. **Schwartz, J. R., Rowland, L. P., Eder, H., Marks, P. A., Osserman, E. F., Hirschberg, E., and Anderson, H.,** Bassen-Kornzweig syndrome: deficiency of serum B-lipoprotein, *Arch. Neurol.*, 8, 108, 1963.
37. **Yuill, G. M., Scholz, G., and Lascelles, R. G.,** Abetalipoproteinemia. A case report with pathological studies, *Postgrad. Med. J.*, 52, 713, 1976.
38. **Miller, R. G., Davis, C. J. F., Illingworth, D. R., and Bradley, W.,** The neuropathy of abetalipoproteinemia, *Neurology,* 30, 1286, 1980.
39. **Wichman, A., Buchthal, F., and Gregg, R. E.,** Peripheral nerve disease in abetalipoproteinemia, *Neurology,* 34(Suppl. 1), 182, 1984.
40. **Brin, M. F., Nelson, J. S., Roberts, W. C., Marquardt, M. D., Suswankosai, P., and Petito, C. K.,** Neuropathology of abetalipoproteinemia: a possible complication of the tocopherol (vitamin E) deficient state, *Neurology,* 33(Suppl. 2), 142, 1983.
41. **Satya-Murti, S., Howard, L., Krokel, G., and Wolf, B.,** The spectrum of neurologic disorder from vitamin E deficiency, *Neurology,* 36, 917, 1986.
42. **Bertoni, J. M., Abraham, F. A., Falls, H. F., and Itabashi, H.,** Small bowel resection with vitamin E deficiency and progressive spinocerebellar syndrome, *Neurology,* 34, 1046, 1984.
43. **Weder, B., Meinberg, O., Wildi, E., and Meier, C.,** Neurologic disorder of vitamin E deficiency in acquired intestinal malabsorption, *Neurology,* 34, 1561, 1984.
44. **Brin, M. F., Fetell, M. R., Green, P. H. A., Kayden, H. J., Hays, A. P., Behrens, M. M., and Baker, H.,** Blind loop syndrome, vitamin E malabsorption, and spinocerebellar degeneration, *Neurology,* 35, 338, 1985.
45. **Harding, A. E., Matthews, S., Jones, S., Ellis, C. J. K., Booth, I. W., and Muller, D. P. R.,** Spinocerebellar degeneration associated with a selective defect of vitamin E absorption, *N. Engl. J. Med.*, 313, 32, 1985.

46. **Burck, U., Goebel, H. H., Kuhlendahl, H. D., Meier, C., and Goebel, K. M.,** Neuromyopathy and vitamin E deficiency in man, *Neuropediatrics,* 12, 267, 1981.
47. **Elias, E., Muller, D. P. R., and Scott, J.,** Association of spinocerebellar disorders with cystic fibrosis or chronic childhood cholestasis and very low serum vitamin E, *Lancet,* ii, 1319, 1981.
48. **Willison, H. J., Muller, D. P. R., Matthews, S., Jones, S., Kriss, A., Stead, R. J., Hodson, M. E., and Harding, A. E.,** A study of the relationship between neurological function and serum vitamin E concentrations in patients with cystic fibrosis, *J. Neurol. Neurosurg. Psychiatry,* 48, 1097, 1985.
49. **Bye, A. M. E., Muller, D. P. R., Wilson, J., Wright, V. M., and Mearns, M. B.,** Symptomatic vitamin E deficiency in cystic fibrosis, *Arch. Dis. Child,* 60, 162, 1985.
50. **Geller, A., Gilles, F., and Schwachman, H.,** Degeneration of fasciculus gracilis in cystic fibrosis, *Neurology,* 27, 185, 1977.

SYNAPTIC PLASTICITY IN THE CEREBELLAR GLOMERULI OF RATS: EFFECTS OF AGING AND VITAMIN E DEFICIENCY

Carlo Bertoni-Freddari

INTRODUCTION

Numerous hypotheses have been proposed to explain the mechanisms of aging,[1] but full understanding of the primary causes involved in the aging process is still lacking. Recently, many gerontologists using different experimental approaches have focused attention on the plasma membrane functions and/or structure to seek possible defects at cellular membrane level responsible for age-related impairments.[2-4] The membrane hypothesis of aging[5-9] attributes a primary role in cellular aging of postmitotic cells to the free radical-induced deterioration of the plasma membrane structure, which results in a gradual loss of passive permeability of the cell membrane for potassium (and probably also for water). These alterations lead to a considerable condensation of the intracellular colloids through the increase of potassium concentration and loss of intracellular water, which results in a decrease of practically all kinds of enzyme activities, an impairment of protein synthesis, and a decrease in the rate of degradation of damaged components. This hypothesis is supported by numerous experimental observations in various cellular models.[7,8] Although various biological systems have been used to demonstrate membrane changes occurring in old organisms,[10] one of the most important models is represented by the nervous system. Neurons, being postmitotic, are as old as the organism which they belong to; therefore, true aging alterations can easily be evidenced in this type of cells. Synaptic junctions are highly differentiated areas of the neuronal membrane. Because of their peculiar function in the transmission of the nervous impulse, they are very specific, plastic, dynamic, metabolically active, and vulnerable structures. The purpose of the present chapter is to discuss age-related changes of rat synaptic junctions and the alterations caused in young animals by a dietary treatment favoring an overproduction of free radicals, such as vitamin E deficiency.

THE CONCEPT OF SYNAPTIC PLASTICITY

The nervous system has a great capability to adapt its structure and function to internal and environmental stimuli, assuring the best response. During the last decade, it has been widely accepted that this adaptive ability is maintained throughout the life span of the organism and that it involves all the structures of the nervous tissue.[11] The term "neuronal plasticity" served to define this dynamic condition of the nerve cells and it was used in a broad sense, including hypertrophy of nerve cell somas and increased activity of a particular kind of enzyme.[12] More recently, it has been found that single structures as well as particular aspects of their functions of the central nervous system (CNS) can display their own plasticity[11] which appears to be of great importance in understanding the cellular mechanisms involved in the adaptive capabilities of the nervous system. Because of their peculiar function, synaptic junctions are currently investigated with regard to their adaptive changes under different experimental conditions and the term "synaptic plasticity" has become of common use to indicate this dynamic state of the junctional areas between the neurons.[12,13] As mentioned above, with improvement of investigative methodologies, various aspects of the synaptic plasticity have been taken into account. For example, biochemical plasticity means that molecules, like neurotransmitter substances, enzyme activities, receptor binding etc., may be actively modulated by several factors at synaptic endings.[14] The physiologist, when speaking of plasticity at synaptic regions, refers to variations of spike amplitudes or electric postsynaptic potentials which take place in the response to defined stimuli during neurotransmitter-induced discharge.[15] From a morphological point of view, synaptic plasticity means the ultrastructural modifications of synaptic membranes to adapt architecture of the synaptic contacts to the changing requirements.[16]

EFFECT OF AGING ON SYNAPTIC PLASTICITY IN THE CEREBELLAR CORTEX

Most of the areas of the brain, if not all, appear to be seriously affected by aging.[16] Cell loss, metabolic changes, lipofuscin accumulation, and a decreased number of synaptic contacts have been found in various zones of the CNS, like the hippocampus and cerebral cortex, assumed to be involved in learning and memory formation and/or retrieval.[16,17] Less attention has been paid by gerontologists to the cerebellum, which is involved in motor coordination, gait, and posture and it is because of the ordered arrangement of its cellular components and fibers represents a very reliable and sensitive anatomical model for quantitative morphological studies.[18]

The cerebellar model is advantageous also from a methodological point of view. Namely, when performing quantitative morphological studies on some other parts of the nervous system, it is difficult to sample the same area in various animals, if there is no precise anatomical structure which can be used as an orientation reference. In this respect, the cerebellum offers a good chance to overcome the problem: very discrete and separated areas, the so-called glomeruli, can easily be visualized in the granular layer of the cerebellar cortex and may well serve as an anatomical model for quantitative morphological studies.[19] The structure of the glomeruli is fairly well known: they can be considered as cell-free islands where the mossy fiber coming from the periphery swells up and makes synaptic contacts with the finger-like dendrites of the surrounding granular cells. With regard to the kind of neurotransmitters used by the synapses in this zone, it was originally thought that most of the junctions were cholinergic;[20] however, from the biochemical demonstration of certain enzymes involved in transmitter synthesis and degradation in this area of the CNS, it has been concluded that the inputs coming to this area are mainly GABAergic and serotoninergic.[21]

When performing quantitative morphological studies, a necessary prerequisite is a good staining of the structures to be investigated in order to facilitate and objectify their identification. The ethanolic phosphotungstic acid (E-PTA) technique, introduced in 1968 by Bloom and Aghajanian,[22] is currently used to stain preferentially synaptic junctions and has been proved to be very useful in visualizing the developmental maturation of synaptic connectivity.[23,24] As reported by Bloom and Aghajanian,[22] E-PTA has the advantage of staining selectively the synaptic paramembranous material, whereas all other membranous intracellular organelles are electron-lucent. Many morphological features of the synaptic appositions can therefore easily be measured. In our laboratory, we adapted the E-PTA staining procedure to investigate quantitatively the morphological aspect of synaptic plasticity in the cerebellar glomeruli of young (3 months), adult (11 months), and old (28 months) female Wistar rats (Figure 1).[19] Three parameters were investigated by means of quantitative morphometry: the numerical density of the synapses (Nv), i.e., the number of contacts per unit volume of tissue; the surface density of the synaptic contact zones (Sv), i.e., the total surface contact area per unit volume of tissue; and the average area ($\overline{S}$) of the single synapse which gives a measure of the size of the synapses.

Our quantitative studies revealed a notable biological importance of these parameters. Namely, it has been convincingly demonstrated that the number (Nv) and size ($\overline{S}$) of the synapses are in an inverse correlation, aiming to maintain constant the total synaptic surface contact area (Sv) among the dendritic network while changing external stimulations throughout the life span of the organism.[25-27]

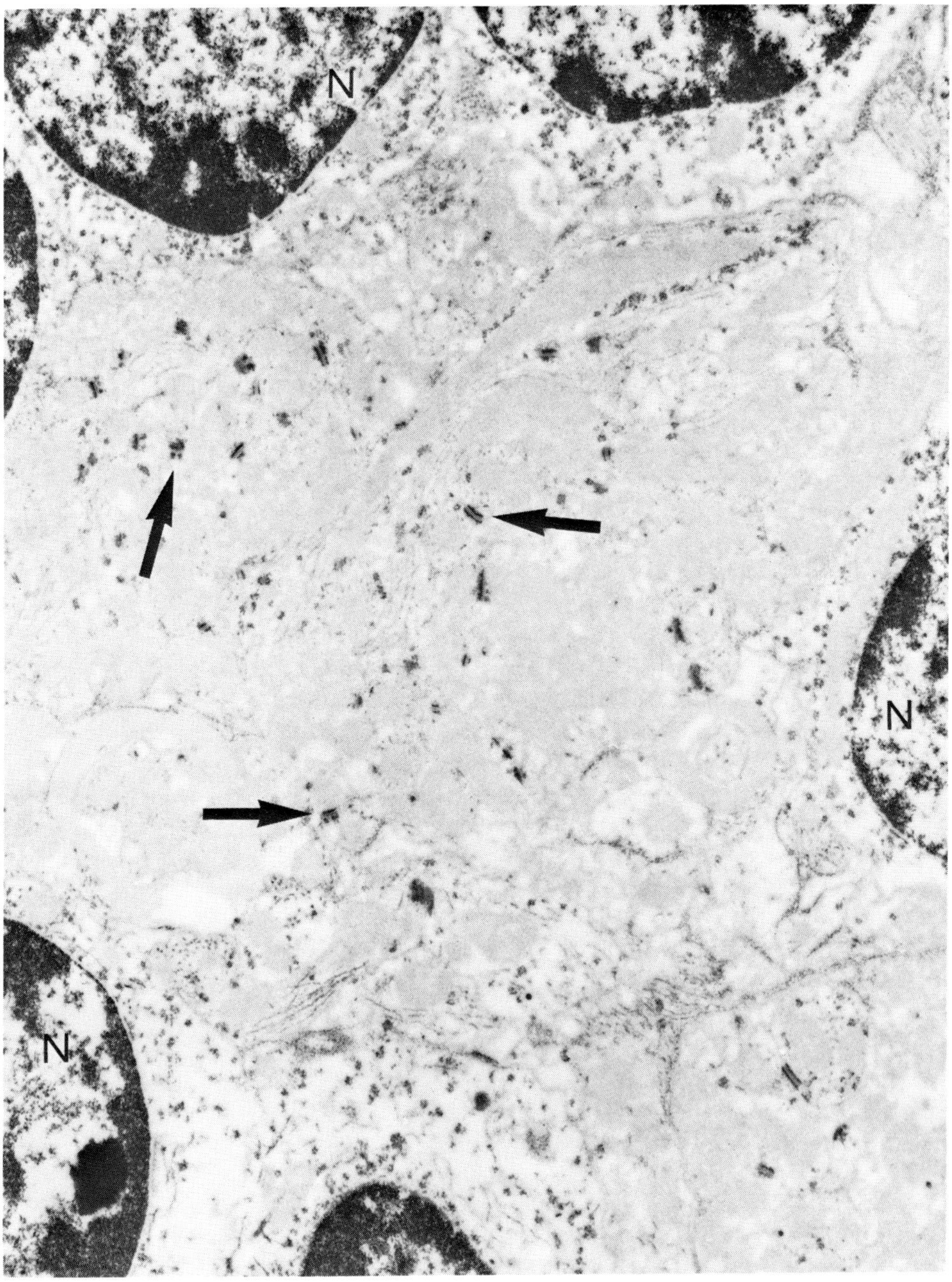

FIGURE 1. Electron microscope picture of rat cerebellar glomerulus stained by means of the E-PTA method; N, nuclei of granular cells; arrows, synaptic junctions. (Magnification × 5000.)

Table 1 reports the data of Nv, $\overline{S}$, and Sv obtained in the three groups of rats studied.[28] Interesting conclusions can be drawn either by analyzing each parameter in the different age groups or by taking into account the three parameters altogether per age group. Namely, a significant increase of the number of contacts between young and adult ages was found, whereas, when young and adult groups were compared with old rats, a marked decrease was shown. $\overline{S}$, i.e., the size of the synapses,[27] appears to be the same in young and old rats, but is substantially decreased in the adult animals. Considering the data of Sv, i.e.,

Table 1
NUMERICAL DENSITY (Nv), AVERAGE SIZE ($\overline{S}$), AND SURFACE DENSITY (Sv) OF THE SYNAPSES IN THE CEREBELLAR GLOMERULI OF YOUNG, ADULT, AND OLD RATS

Age (months)	Nv (number of synapses/ μm^3)	$\overline{S}$ (μm^2)	Sv ($\mu m^2/\mu m^3$)
3	1.65 ± 0.06	0.0841 ± 0.0044	0.1199 ± 0.0039
11	1.85 ± 0.04	0.0706 ± 0.0024	0.1213 ± 0.0021
28	1.39 ± 0.07	0.0820 ± 0.0038	0.0988 ± 0.0031

Statistical significance (Student's *t* test):

(Nv):	3/11	$p < 0.05$	($\overline{S}$): 3/11	$p < 0.02$	(Sv):	3/11	N.S.
	3/28	$p < 0.001$	3/28	N.S.		3/28	$p < 0.0001$
	11/28	$p < 0.0001$	11/28	$p < 0.02$		11/28	$p < 0.0001$

Note: N.S. = not significant.

the final result of the inverse relationship between Nv and $\overline{S}$, no significant difference exists between young and adult animals. On the contrary, in old rats the total surface area of the synaptic contact zones is markedly reduced.

By analyzing Nv, $\overline{S}$, and Sv altogether per age group, one can obtain quantitative information about the morphological aspect of synaptic plasticity during defined periods of life: young animals are characterized by a high number of synaptic contacts of large size which provide a high value of total synaptic surface contact area. During adulthood, although the surface area of the synaptic contact zones remains constant, consistent rearrangements seem to take place at nerve endings, i.e., the number of contacts increases and the size of the single synapse decreases. In aged animals, morphological adaptations of the synaptic appositions were observed, too: their number decreases while their size increases, resulting in a reduced synaptic contact surface area among the dendritic network.

In conclusion from the data reported in Table 1, it can easily be inferred that synaptic remodeling is an ongoing process which occurs from development to aging and results in different morphological patterns of the synaptic junctions at various ages. The possible causes/effects leading to these patterns, which probably also involve the location of the contacts,[29,30] should be found in the functional connectivity requirements of each period of life. In other words, a redundant number of synapses of large size during development and maturation may well represent the great potentiality of the young CNS to eliminate the scarcely used synapses and to reinforce and stabilize those contacts with increased activity.[31] In adulthood, the smaller size, the increased number of junctions, and the unchanged value of the total surface area of the synaptic contact zones, as compared to young animals, seem to indicate that the total synaptic connectivity is achieved very early during development and that, as the individual becomes adult, synaptic appositions undergo a sort of fragmentation which leads to a finer distribution of the contacts and, as a consequence, to a more detailed control of a wider domain. The latter assumption seems to be supported by much experimental data reporting improved behavioral, ultrastructural, biochemical, and physiological performances (such as learning and memory) of the adult CNS.[32-34] During aging, when synaptic circuits are established by means of a low number of enlarged contacts providing a reduced total synaptic contact surface area,[27] it appears that nerve cells are not able to efficiently maintain (probably due to an impaired energy metabolism)[35] the high number of contacts built through development and maturation in adulthood and therefore some contacts (the less used or the newly formed and not reinforced ones) are reabsorbed. However, as it is often observed in several living systems, the reabsorbed material appears to be not completely degraded to its basic components, but is recycled and used to enlarge and stabilize the persisting contacts. This kind of "compensative synaptogenesis" has been observed by us[19,27] and other authors in the hippocampus[36] of old animals and also appears to take place in several areas of the CNS and in different experimental conditions (e.g., undernutrition and lesions) affecting the normal functioning of synaptic turnover. Although speculative, there is a great deal of supporting data for this hypothesis. Buell and Coleman[37,38] reported morphological evidence of increasing dendritic sprouting in the old normal human brain, as compared to brains from either adult or demented individuals, and interpreted these data in terms of reactive capacity of the nonpathologic old brain to occupy (with newly formed dendritic branches) the vacated places left by the dendrites of dead neurons in order to carry out a sort of compensation for the lost functions. Our data regarding the increased size of synapses in old CNS[19,27] may well represent the "synaptic aspect" of the reactive process demonstrated by these authors. Barnes and McNaughton[15] and Landfield et al.[39] found that the persisting large synaptic contacts in the hippocampus of old rats are much more powerful, with regard to their electric postsynaptic potential, than the smaller synapses of adult animals. This finding clearly indicated that these junctions have been reinforced and are able to fire at a slower rate, but with a greater degree of strength. Cotman et al.,[40] Hoff et al.,[41] and Scheff et al.[42] demonstrated that lesion-induced synaptogenesis is markedly delayed in old rats and that, because of the reduced metabolism of old CNS, the removal of debris of the lesioned fibers is impaired and the reactive synaptogenesis slows down. Finally, it is well known, since beginning of research on aging of the brain, that old people remember well what happened during their adolescence, whereas they show clear deficits in the so called short-term memory. This known, macroscopic behavior, together with the reported involvement of the synaptic junctions in memory formation[43] may well indicate that the establishment and reinforcement of new contacts are impaired during aging.

The observed ultrastructural alterations of the synaptic membranes obviously must have a common basis: it is the author's belief that the best explanation for these phenomena is offered by the membrane hypothesis of aging.[5-9] As mentioned before, this hypothesis assumes that the membranous structures are primary targets of the free radical attacks, due to the relatively more compact structure as compared to the cytosol. As a matter of fact, lipid peroxidation[2] and most probably also the direct, radical-induced damage of proteins increase with advancing age; meanwhile the capacity of the defensive mechanisms against such damages including the rate of replacement of the components[2,7,8] decreases. Although several drugs and toxins are also capable of free radical production,[2] most of the harmful oxygen free radicals derive from the normal metabolic processes, i.e., from respiration.[46] In other words, the living systems are practically autodestructive, since the cellular macromolecules are continuously exposed to the hazard of free radical attacks and are more or less vulnerable according to their chemical composition. In this respect, biological plasma membranes appear to be preferential sites of free radical reactions because of their content in unsaturated fatty acids which are part of the structure of their phospholipids. This is particularly valid for the synaptic junctions which, as clearly demonstrated also by the increased binding of OsO_4 in sections stained for conventional electron microscopy, show a higher content of unsaturated lipids than any other part of the neuronal membrane. Therefore, we believe that free radical-mediated injuries may play a crucial role in damaging the structure of the plasma membranes especially in the postmitotic nerve cells.[16,28] The replacement of the damaged components is slowed down during aging due to the overall decrease of the rates of synthetic processes.[9] On the other hand, the increasing condensation of the old cells favors the intermolecular cross-linking efficiency of the OH· free radicals formed in the cells by means of heterolysis of hydrogen peroxide by transient metals (as in the Fenton reaction).[7,8] This situation results in a relative shift of the proportion of harmful radicals formed and quenched in a unit of time, i.e., in spite of the decreased oxygen consumption in old age, the oxyradicals will destruct an ever-increasing portion of the organic macromolecules in the living cells. In other words, one should not expect any special process in old age which is not present also in the young individuals, but the quantitative physicochemical relationships are shifted toward a less favorable situation in which the living system loses its capacity for replacing the damaged components. In order to verify the involvement of cellular defense mechanisms against free radical attacks in the preservation of the integrity of plasma membrane structure, experiments in young rats fed a diet completely devoid of the biological antioxidant vitamin E for 10 months have been carried out.

EFFECT OF VITAMIN E DEFICIENCY ON THE PLASTICITY OF CEREBELLAR SYNAPTIC JUNCTIONS

Vitamin E (α-tocopherol) is a known lipid-soluble substance very abundant in oils of vegetable seeds, especially those of wheat, soyabean, cotton, sunflower, and corn; margarine from hydrogenated vegetable oils is one of the best food sources of α-tocopherol. The early discovery by Evans and Bishop (1922),[47] that pregnant laboratory animals, fed on a rancid diet, reabsorbed their fetuses, led to the proposal that this phenomenon was due to the absence of a lipid-soluble compound which had been destroyed by oxidative processes in the rancid lard. Evans called this compound vitamin E and nowadays it is known that α-tocopherol is a biological lipid antioxidant which is very important in maintaining of foodstuffs and avoiding free radical effects originating from normal metabolic processes in the living cells. With advancing knowledge, a more defined biological role has been proposed for vitamin E, and two main different hypotheses are currently investigated to clarify the mechanism of action of α-tocopherol. The first and most acknowledged is the antioxidant theory which holds that vitamin E is able to inhibit the peroxidation of unsaturated fatty acids brought about mainly by molecular oxygen.[48,49] Support for this hypothesis is mainly provided by (1) the action of synthetic antioxidants which, although unrelated to vitamin E, are able to substitute α-tocopherol in preventing deficiency symptoms[50,51] and (2) by in vitro experiments which clearly show inhibition by vitamin E of autoxidation of unsaturated lipids.[52] Moreover, in isolated cellular organelles, such as microsomes and mitochondria, it has also been found that α-tocopherol can help to quench free radical attacks by acting as a cofactor in enzyme-dependent lipid peroxidation and in biological oxidation or oxidative phosphorylation.[53] The second and very intriguing hypothesis about the mechanism of action of α-tocopherol was developed by Lucy and Diplock.[54,55] According to these authors, the molecules of vitamin E are bound to the biological membranes by means of specific physicochemical interactions between the peculiar phytyl side chain of the vitamin and the arachidonyl residues of membrane phospholipids: this binding allows the vitamin E molecules to exert a stabilizing effect upon the structure of the biological membranes, reducing in this way the sites available for free radical attacks. A strong support for this hypothesis was provided by data from experiments performed in a tissue culture system where it was shown that the uptake of radioactive cholesterol and its content, as well as the content of radioactive arachidonyl residues of membrane phospholipids, were significantly increased in these cells by the presence of α-tocopherol in the culture medium.[56,57] Therefore these results demonstrate an effect of vitamin E which does not involve its antioxidant function, but rather sustains its action on the architecture of the membrane by controlling the profiles of cholesterol and unsaturated phospholipids. As can be easily inferred, these two proposed mechanisms of action of vitamin E do not exclude each other, may be valid at the same time, and, what is more, both point to a protective role of α-tocopherol for biological membranes.

Young female Wistar rats of 1 month of age were fed for 10 months with a vitamin E-deficient diet of the composition reported in Table 2. This long period of vitamin E deficiency

Table 2
COMPOSITION OF THE VITAMIN E-DEFICIENT DIET

Maize starch and sucrose	58%
Extracted casein	18%
Torula yeast	10%
Melted filtered fat	8%
Osborne-Mendel salts	5%
Vitamin supplement lacking vitamin E	1%

was necessary to induce exhaustion of the organism's stores of this vitamin according to the data reported by other authors.[58] Loss of weight from the age of 4 months and progressive loss of hair were the macroscopic signs of this diet.[59]

Figure 2 (A and B) reports the results of Nv and Sv that we obtained in young rats fed with the vitamin E-deficient diet as compared to age-matched controls and other age groups. With regard to age variations, Nv increases between 3 and 11 months of age, remains constant throughout the adulthood period (11 to 18 months), and decreases from 18 up to 28 months. The vitamin E-deficient group showed a substantial (20%) reduction of the number of contacts as compared to animals of the same age, and its value was similar to the one found in the 23-month-old group. In Figure 2B, the results of Sv are reported: a constancy from 3 up to 18 months of age and a progressive decrease from this age on were demonstrated for this parameter. The vitamin E-deficient animals showed a 25% reduction as compared to their littermates.

The distribution of $\overline{S}$ in young (3 months), adult (11 months), vitamin E-deficient (11 months), and old (28 months) groups is shown in Figure 3. Young and old animals showed almost the same wide distribution of synaptic size, whereas adult rats displayed a synaptic population of smaller size distributed over a narrower range. In vitamin E-deficient animals, the size of the synaptic contacts appears to be distributed over the same range as shown by their littermates, but the majority of $\overline{S}$ (about 60%) is of even smaller size. Considering the mean of $\overline{S}$ for all the groups investigated, we found a significant decrease in the 11-month-old rats when compared to 3- and 28-month-old groups and no difference between 3 and 28 months of age. Vitamin E deficiency did not affect the mean of $\overline{S}$: no difference was found between the vitamin E-deficient rats and the other groups investigated. In conclusion, Nv and Sv were markedly reduced by α-tocopherol deficiency, while $\overline{S}$ appeared not to be affected with regard to its mean value although it was substantially modified in percentage distribution.

Interpreting these data in terms of synaptic plasticity, the reasons for the present alterations can find a reliable explanation in the proposed mechanism of action of α-tocopherol. Synaptic ultrastructural features, as mentioned above, can be modified by several environmental and experimental conditions even in the adult nervous system.[40] Taking into account our present findings, it appears that vitamin E deficiency may well be one of the factors able to influence the dynamic morphology of synaptic membranes, since our dietary treatment resulted in ultrastructural alterations very similar to the ones found in old rats. The only difference we found was the lack of a sort of "compensative synaptogenesis" for the recovery of the lost contact area brought about by the reduced number of synapses.

A common denominator may be responsible for these alterations, i.e., the increased peroxidative damage which occurs at cell membranes during aging and in situations (e.g., vitamin E deficiency) when protection against free radical attacks is lacking. Synaptic junctions appear to be very vulnerable to peroxidative processes because of the molecular composition of their membranes and, in particular, of the high content in polyunsaturated fatty acids of their phospholipids.[60] Since an ongoing turnover in the basic membrane con-

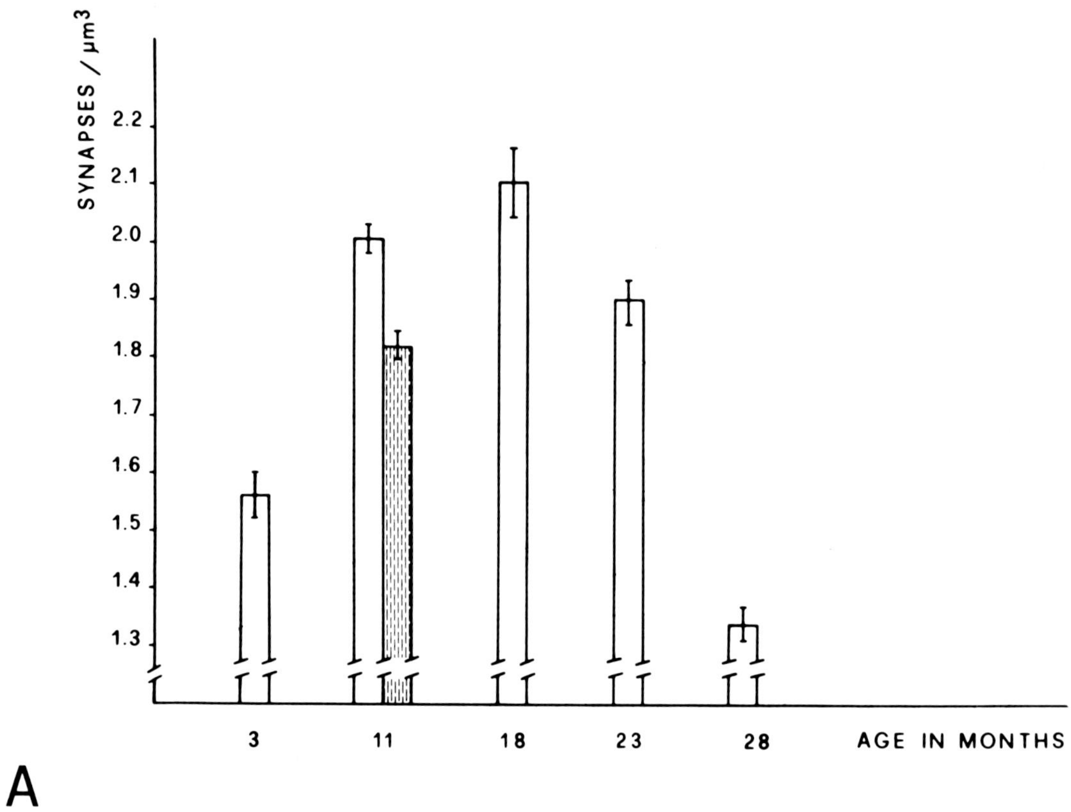

FIGURE 2. (A) Numerical density of the synapses (Nv), mean ± SEM; open columns, control rats of different ages, dashed column, vitamin E-deficient group.

Statistical significance (Student's *t* test):

3/11	$p < 0.0001$	11/18	N.S.	18/23	$p < 0.002$
3/18	$p < 0.0001$	11/23	$p < 0.002$	18/28	$p < 0.0001$
3/23	$p < 0.0001$	11/28	$p < 0.0001$	23/28	$p < 0.0001$
3/28	$p < 0.005$				

Vitamin E-deficient rats/control group of the same age ($p < 0.0001$).

(B) Surface density of the synaptic contact zones (Sv), mean ± SEM; ●, control animals of different ages; ■, vitamin E-deficient rats; N.S., not significant.

Statistical significance (Student's *t* test):

3/11	N.S.	11/18	N.S.	18/23	$p < 0.0001$
3/18	N.S.	11/23	$p < 0.0001$	18/28	$p < 0.0001$
3/23	$p < 0.0001$	11/28	$p < 0.0001$	23/28	$p < 0.001$
3/28	$p < 0.0001$				

Vitamin E-deficient rats/control group of the same age ($p < 0.0001$); N.S., not significant.

(From Bertoni-Freddari, C., Giuli, C., and Pieri, C., *Mech. Ageing Dev.*, 24, 225, 1984. With permission.)

stituents, brought about mainly by the axoplasmic flow, takes place at nerve endings,[61] several factors can modulate the molecular composition of synaptic membranes and, as a consequence, synaptic plasticity. Vitamin E may well be one such modulating agent, and this fact is strongly supported by the growing body of data reporting its direct or mediated protection of plasma membrane structure. As a matter of fact, besides its antioxidant effect

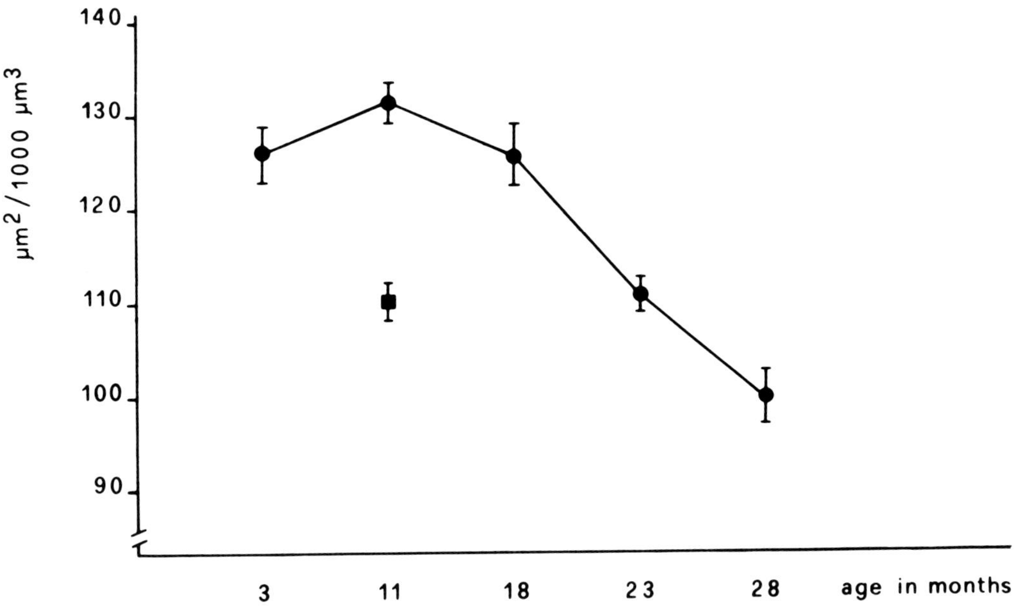

B

FIGURE 2 (continued)

on polyunsatured lipids[48,49] and the stabilizing role on plasma membrane, recent advances in understanding the biochemical mechanisms involved in α-tocopherol action point to a more specific protection of this compound upon CNS structures.[62-64] In this context, it has been found that synaptosomes and gray matter from cerebellar cortex show a very high content of α-tocopherol.[65,66] In addition, Burton reported that the amount of vitamin E within membranes is independent from the α-tocopherol blood concentration; rather it is a function of the polyunsaturated fatty acids composition and of the functional protection requirements of a particular zone of the cellular membrane.[67] Thus, it is understandable why so high an amount of vitamin E was found at synaptic endings.

Although well-differentiated areas of the neuronal membrane, synapses are properly supplied with basic constitutents to substitute and/or repair those worn out during the transmission of the nervous impulse.[68] It has been demonstrated that during aging the turnover the basic synaptic constituents (e.g., phospholipids), mediated by axoplasmic flow, is slowed down;[69,70] vitamin E deficiency, on the contrary, appeared to increase the rate of transport of these compounds, supporting the idea that an accelerated turnover occurs at nerve endings when protective mechanisms are not sufficiently efficient.[71] This latest assumption finds convincing support in the recent data reported by Niki et al.,[72] demonstrating that vitamin E is able to inhibit, in vitro, the oxidation of phosphatidylcholine — a phospholipid which has been found to be very abundant (42.4%) within rat synaptic membranes.[60] Apart from — or in addition to — these actions of α-tocopherol upon synaptic membranes, it is worthwhile to mention another interesting point which can further explain the results obtained by us. Vatassery et al.[65,66] showed that mice cerebellar content of vitamin E is very low and after a short period of feeding animals with a diet lacking this compound, the cerebellar uptake of labeled α-tocopherol is very much increased. These authors inferred that cerebellum is much more sensitive to vitamin E deficiency than any other part of the brain and, at least for EPTA-stained synaptic junctions, our present findings appear to confirm their findings.

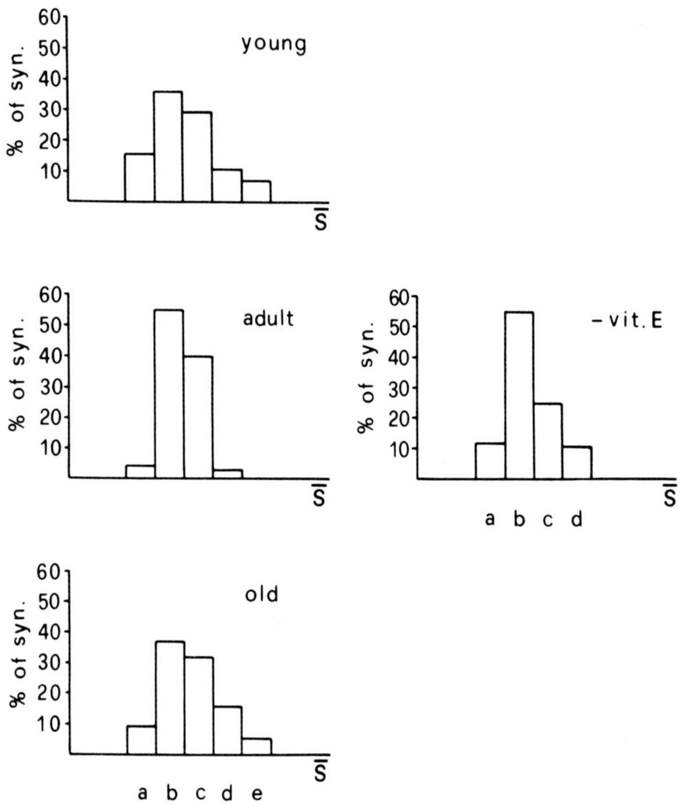

FIGURE 3. Percentage distribution of the average area of the single synaptic contact ($\bar{S}$). (a) <0.04 μm^2; (b) 0.04 to 0.08 μm^2; (c) 0.08 to 0.12 μm^2; (d) 0.12 to 0.16 μm^2; (e) >0.16 μm^2. Means ± SEM of young, adult, and old control groups are reported in Table 1. Mean SEM of the vitamin E-deficient animals is 0.0759 ± 0.0028. No significant difference was found comparing the Vitamin E-deficient group with 3-, 11-, and 28-month-old rats.

CONCLUSION

Aging has been defined by several investigators as a time-related loss of the capacity of the cell to maintain homeostasis — a fact which has been proven to be particularly true for cell membranes, especially in those cells, like neurons, which have the same age as the organism to which they belong. Primary causes of the aging process, as defined above, have been claimed to be attacks of free radicals originating from normal metabolic processes within the cells. For aerobic organisms, like human beings, oxygen is necessary for survival, but if improperly controlled, it is also the main source of free radicals.[73] Living cells, through repeated evolutive events, have improved their defense mechanisms to inactivate free radicals, and several experimental data suggest that the medium life span of a single species may be determined by the balance between the rate of free radical production and the efficiency of protective mechanisms; therefore an impairment in the mechanisms and/or compounds involved in free radical quenching may represent a potential threat to longevity. In this respect, our data, considering as a biological membrane model the synaptic junctions of the cerebellar cortex, support the hypothesis that plasma membrane structure of postmitotic nerve cells is a very vulnerable site for free radical attacks and that vitamin E may be an efficient substance to preserve the membrane functional integrity at synaptic endings.

ACKNOWLEDGMENT

The author is very much indebted to Dr. Imre Zs.-Nagy, Scientific Coordinator of the F. Verzar International Laboratory for Experimental Gerontology (VILEG), for encouragement and critical reading of this manuscript.

REFERENCES

1. **Finch, C. E. and Schneider, E. L., Eds.,** *Handbook of Biology of Aging,* Van Nostrand-Reinhold, New York, 1985.
2. **Hegner, D.,** Age-dependence of molecular and functional changes in biological membrane properties, *Mech. Ageing Dev.,* 14, 101, 1980.
3. **Sun, A. Y. and Sun, G. Y.,** Neurochemical aspects of the membrane hypothesis of aging, *Interdiscip. Top. Gerontol.,* 15, 34, 1979.
4. **Packer, L., Deamer, D. W., and Heath, R. L.,** Regulation and deterioration of structure in membranes, *Adv. Gerontol. Res.,* 2, 77, 1967.
5. **Zs.-Nagy, I.,** A membrane hypothesis of aging, *J. Theor. Biol.,* 75, 189, 1978.
6. **Zs.-Nagy, I.,** The role of membrane structure and function in cellular aging: a review, *Mech. Ageing Dev.,* 9, 237, 1979.
7. **Zs.-Nagy, I.,** Aging of the cellular membrane: basic principles and pharmacological interventions, *Geriatrika,* 1(8), 102, 1985.
8. **Zs.-Nagy, I.,** An attempt to answer the questions of theoretical gerontology on the basis of the membrane hypothesis of aging, in *Age Pigments: Biological Markers of Aging and Environmental Stress?,* by Aloj Totaro, E., Pisanti, F. A., and Glees, P., Eds., VNU Science Press, Utrecht, in press.
9. **Zs.-Nagy, I. and Semsei, I.,** Centrophenoxine increases the rates of total and mRNA synthesis in the brain cortex of old rats: an explanation of its action in terms of the membrane hypothesis of aging, *Exp. Gerontol.,* 19, 171, 1984.
10. **Grinna, L. S.,** Changes in cell membranes during aging, *Gerontology,* 23, 452, 1977.
11. **Cotman, C. W., Ed.,** *Neuronal Plasticity,* Raven Press, New York, 1978.
12. **Cotman, C. W. and Nieto-Sampedro, M.,** Cell biology of synaptic plasticity, *Science,* 225, 1287, 1984.
13. **Eccles, J. C.,** Synaptic plasticity, *Naturwissenschaften,* 66, 147, 1979.
14. **Black, I. B., Adler, J. E., Dreyfus, C. F., Jonakait, G. M., Katz, D. M., La Gamma, E. F., and Markey, K. M.,** Nuerotransmitter plasticity at the molecular level, *Science,* 225, 1266, 1984.
15. **Barnes, C. A. and McNaughton, B.,** Physiological compensation for loss of afferent synapses in rat hippocampal granule cells during senescence, *J. Physiol.,* 309, 473, 1980.
16. **Cotman, C. W. and Holets, V. R.,** Structural changes at synapses with age: plasticity and regeneration, in *Handbook of Biology of Aging,* Finch, C. E. and Schneider, E. L., Eds., Van Nostrand-Reinhold, New York, 1985, 617.
17. **Brizzee, K. R. and Ordy, J. M.,** Age pigments, cell loss and hyppocampal function, *Mech. Ageing Dev.,* 9, 143, 1979.
18. **Woodward, D. J., Hoffer, B. J., and Altman, J.,** Physiological and pharmacological properties of Purkinje cells in rat cerebellum degranulated by postnatal X-irradiation, *J. Neurobiol.,* 5, 283, 1974.
19. **Bertoni-Freddari, C. and Giuli, C.,** A quantitative morphometric study of synapses of rat cerebellar glomeruli during aging, *Mech. Ageing Dev.,* 12, 127, 1980.
20. **Israel, M. and Whittaker, V. P.,** The isolation of mossy fibre endings from the granular layer of the cerebellar cortex, *Experientia,* 21, 325, 1965.
21. **Balazs, R., Hajos, F., Johnson, A. L., Reynierse, G. L. A., Tapia, R., and Wilkin, G.,** Subcellular fractionation of rat cerebellum: an electron microscopic and biochemical investigation. III. Isolation of large fragments of the cerebellar glomeruli, *Brain Res.,* 86, 17, 1975.
22. **Bloom, F. E. and Aghajanian, G. K.,** Fine structural and cytochemical analysis of the staining of synaptic junctions with phosphotungstic acid, *J. Ultrastruct. Res.,* 22, 362, 1968.
23. **Bloom, F. E.,** The formation of synaptic junctions in developing rat brain, in *Structure and Function of synapses,* Pappas, G. D. and Purpura, D. P., Eds., Raven Press, New York, 1972, 101.
24. **Dyson, S. E. and Jones, D. G.,** The morphological categorization of developing synaptic junctions, *Cell Tissue Res.,* 167, 363, 1976.
25. **Bertoni-Freddari, C., Giuli, C., and Pieri, C.,** The effect of acute and chronic centrophenoxine treatment on the synaptic plasticity of old rats, *Arch. Gerontol. Geriatr.,* 1, 365, 1982.

26. **Bertoni-Freddari, C., Mervis, R. F., Giuli, C., and Pieri, C.,** Chronic dietary choline modulates synaptic plasticity in the cerebellar glomeruli of aging mice, *Mech. Ageing Dev.*, 30, 1, 1985.
27. **Bertoni-Freddari, C., Giuli, C., Pieri, C., and Paci, D.,** Quantitative investigation of the morphological plasticity of synaptic junctions in rat dentate gyrus during aging, *Brain Res.*, 366, 187, 1986.
28. **Bertoni-Freddari, C., Giuli, C., and Pieri, C.,** Effect of chronic vitamin E deficiency on the synapses of cerebellar glomeruli in young rats, *Mech. Ageing Dev.*, 24, 225, 1984.
29. **Greenough, W. T.,** Development and memory: the synaptic connection, in *Brain and Learning*, Teyler, T., Ed., Greylock, Conn., 1978, 127.
30. **Greenough, W. T. and Green, J. S.,** Experience and the changing brain, in *Aging: Biology and Behaviour*, McGauch, J. L., March, J. G., and Kiesler, S. B., Eds., Academic Press, New York, 1981, 159.
31. **Purves, D. and Lichtman, J. V.,** Elimination of synapses in the developing nervous system, *Science*, 210, 153, 1980.
32. **Greenough, W. T., West, R. W., and Devgood, T. J.,** Subsynaptic plate perforations: changes with age and experience in the rat, *Science*, 202, 1096, 1978.
33. **Dyson, S. E. and Jones, D. G.,** Synaptic remodelling during development and maturation: junction differentiation and splitting as a mechanism for modifying connectivity, *Dev. Brain. Res.*, 13, 125, 1984.
34. **Carlin, R. K. and Siekevitz, P.,** Plasticity in the central nervous system: do synapses divide?, *Proc. Natl. Acad. Sci. U.S.A.*, 80, 3517, 1983.
35. **Smith, C. B.,** Aging changes in cerebral energy metabolism, *Trends Neurosci.*, 6, 203, 1984.
36. **Cotman, C. W. and Scheff, S. W.,** Compensatory synapse growth in aged animals after neuronal death, *Mech. Ageing Dev.*, 9, 103, 1979.
37. **Buell, S. J. and Coleman, P. D.,** Dendritic growth in the aged human brain and failure of growth in senile dementia, *Science*, 206, 854, 1979.
38. **Buell, S. J. and Coleman, P. D.,** Quantitative evidence for selective dendritic growth in normal human aging but not in senile dementia, *Brain Res.*, 214, 23, 1981.
39. **Landfield, P. W., McGaugh, J. L., and Lynch, G.,** Impaired synaptic potentiation processes in the hippocampus of aged, memory-deficient rats, *Brain Res.*, 150, 85, 1978.
40. **Cotman, C. W., Nieto-Sampedro, M., and Harris, E. W.,** Synapse replacement in the nervous system of adult vertebrates, *Physiol. Rev.*, 61, 684, 1981.
41. **Hoff, S. F., Scheff, S. W., Benardo, L. S., and Cotman, C. W.,** Lesion induced synaptogenesis in the dentate gyrus of aged rats. I. Loss and reacquisition of normal synaptic density, *J. Comp. Neurol.*, 20, 246, 1982.
42. **Scheff, S. W., Benardo, L. S., and Cotman, C. W.,** Decline in reactive fiber growth in the dentate gyrus of aged rats compared to young adult rats following entorhinal cortex removal, *Brain Res.*, 199, 21, 1980.
43. **Lynch, G. and Baudry, M.,** The biochemistry of memory: a new and specific hypothesis, *Science*, 224, 1057, 1984.
44. **Nagy, K. and Zs.-Nagy, I.,** Alterations in the molecular weight distribution of proteins in rat brain synaptosomes during aging and centrophenoxine treatment, *Mech. Ageing Dev.*, 28, 171, 1984.
45. **Zs.-Nagy, I. and Floyd, R. A.,** Hydroxyl free radical reactions with amino acids and proteins studied by electron spin resonance spectroscopy and spin trapping, *Biochim. Biophys. Acta*, 790, 238, 1984.
46. **Halliwell, B.,** Biochemical mechanism accounting for the action of oxygen on living organisms: the key role of superoxide dismutase, *Cell. Biol. Int. Rep.*, 2, 113, 1978.
47. **Evans, H. and Bishop, K. S.,** On the existence of a hitherto unrecognised dietary factor essential for reproduction, *Science*, 56, 650, 1922.
48. **Tappel, A. L.,** Vitamin E as the biological lipid antioxidant, *Vitam. Horm. (N.Y.)*, 20, 493, 1962.
49. **Tappel, A. L.,** Vitamin E and free radical peroxidation of lipids, *Ann. N.Y. Acad. Sci.*, 203, 12, 1972.
50. **Di Luzio, N. R.,** Antioxidants, lipid peroxidation and chemical induced liver injury, *Fed. Proc. Fed. Am. Soc. Exp. Biol.*, 32, 1875, 1973.
51. **Leibovitz, B. E. and Siegel, B. V.,** Aspects of free radical reactions in biological systems: aging, *J. Gerontol.*, 35, 45, 1980.
52. **Packer, L. and Smith, J. R.,** Extension of the lifespan of cultured normal human diploid cells by Vitamin E, *Proc. Natl. Acad. Sci. U.S.A.*, 71, 4763, 1974.
53. **Molenaar, I., Vos, J., and Hommes, F. A.,** Effect of Vitamin E deficiency on cellular membranes, *Vitam. Horm. (N.Y.)*, 30, 45, 1972.
54. **Lucy, J. A.,** Functional and structural aspects of biological membranes a suggested structural role for Vitamin E in the control of membrane permeability and stability, *Ann. N.Y. Acad. Sci.*, 203, 4, 1972.
55. **Diplock, A. T. and Lucy, J. A.,** The biochemical modes of action of vitamin E and selenium: a hypothesis, *FEBS Lett.*, 29, 205, 1973.
56. **Giasuddin, A. S. M. and Diplock, A. T.,** The influence of Vitamin E on the growth and plasma membrane permeability of mouse fibroblasts in culture, *Arch. Biochem. Biophys.*, 196, 270, 1979.
57. **Giasuddin, A. S. M. and Diplock, A. T.,** The influence of vitamin E on membrane lipids of mouse fibroblasts in culture, *Arch. Biochem. Biophys.*, 210, 348, 1981.

58. **Bieri, J. G. and Farrel, P. M.,** Vitamin E, *Vitam. Horm. N.Y.*, 34, 31, 1976.
59. **Bertoni-Freddari, C., Giuli, C., Lustyik, G., and Zs.-Nagy, I.,** In vivo effects of vitamin E deficiency on the interacellular monovalent electrolyte concentration in brain and liver on rat. An energy dispersive X-ray microanalytic study, *Mech. Ageing Dev.*, 16, 169, 1981.
60. **Breckenridge, W. C., Gombos, G., and Morgan, I. G.,** The lipid composition of adult rat brain synaptosomal plasma membranes, *Biochim. Biophys. Acta,* 266, 695, 1972.
61. **Grafstein, B. and Forman, D. L.,** Intracellular transport in neurons, *Physiol. Rev.*, 60, 1167, 1979.
62. **Nelson, J. S., Fitch, C. D., Fischer, V. W., Broun, G. O., and Chou, A. C.,** Progressive neuropathologic lesions in vitamin E deficient rhesus monkeys, *J. Neuropathol. Exp. Neurol.*, 40, 166, 1981.
63. **Muller, D. P. R., Lloyd, J. K., and Wolff, O. H.,** Vitamin E and neurological function, *Lancet,* January 29, 225, 1983.
64. **Towfighi, J.,** Effects of chronic Vitamin E deficiency on the nervous system of the rat, *Acta Neuropathol.*, 54, 261, 1981.
65. **Vatassery, G. T., Angerhofer, C. K., and Peterson, F. J.,** Vitamin E concentrations in the brains and some selected peripheral tissues of selenium-deficient and vitamin E-deficient mice, *J. Neurochem.*, 42, 554, 1984.
66. **Vatassery, G. T., Angerhofer, C. K., Knox, C. A., and Desmukh, D. S.,** Concentrations of vitamin E in various neuroanatomical regions and subcellular fractions, and the uptake of vitamin E by specific areas of rat brain, *Biochim. Biophys. Acta,* 792, 118, 1984.
67. **Burton, G. W., Joyce, A., and Ingold, K. V.,** Is vitamin E the only lipid-soluble, chain-breaking antioxidant in human blood plasma and erythrocyte membranes?, *Arch. Biochem. Biophys.*, 221, 281, 1983.
68. **Koenig, H. L., Di Giambernardino, L., and Bennet, G.,** Renewal of proteins and glycoproteins of synaptic constituents by means of axonal transport, *Brain, Res.*, 62, 413, 1973.
69. **Geinisman, Y., Bondareff, W., and Telser, A.,** Diminished axonal transport of glycoproteins in the senescent rat brain, *Mech. Ageing Dev.*, 6, 363, 1977.
70. **Stromska, D. P. and Ochs, S.,** Axoplasmic transport in aged rats, *Exp. Neurol.*, 77, 215, 1982.
71. **Wood, P. L. and Boegman, R. J.,** Increased rate of rapid axonal transport in vitamin E deficient rats, *Brain Res.*, 84, 325, 1975.
72. **Niki, E., Kawakami, A., Yamamoto, Y., and Kamika, Y.,** Oxidation of lipids. VIII. Synergistic inhibition of oxidation of phosphotidylcholine liposome in aqueous dispersion by vitamin E and vitamin C, *Bull. Chem. Soc. Jpn.*, 58, 171, 1985.
73. **Halliwell, B. and Gutteridge, J. M. C.,** Oxygen radicals and the nervous system, *Trends Neurosci.*, January, 1985, p. 22.

α-TOCOPHEROL AS A POTENTIAL MODIFIER OF THE TOXICITY OF ANTHRACYCLINE ANTIBIOTICS

Dorota Soltysiak-Pawluczuk and Yeu-Ming Wang

INTRODUCTION

During the last decade, much work has been done to understand the complex mechanism of anthracycline antibiotics-induced toxicities and the biochemical mechanism of protection. Cardiotoxicity and, likely, carcinogenicity are side effects of anthracycline antibiotics. Cyclical oxidation and reduction of anticancer anthracycline antibiotics by membrane-bound and cytosolic enzymes generates oxygen radical intermediates that damage cells. Normally, free radicals formed under physiological conditions are effectively removed by protective mechanisms located close to the sites of their formation. Their increased formation during biotransformation of anthracycline antibiotics might, however, result in a cellular imbalance, especially in tissues with a limited protective system. Biological membranes are involved in the mechanisms of action of the anthracycline antibiotics. α-Tocopherol (vitamin E), a free radical scavenger and a membrane-stabilizing agent, has been shown to reduce the drug-induced toxicities. This review is focused mainly, therefore, on the interaction of anthracycline antibiotics and α-tocopherol at the membrane level.

PHYSIOLOGICAL FUNCTION OF α-TOCOPHEROL

α-Tocopherol as an Antioxidant

α-Tocopherol is an integral component of the membranous antioxidant defense system.[1] The presence of specific binding sites for vitamin E in membranes of erythrocytes[2] and adrenocortical cells[3] suggests that one of the physiologically relevant actions of the vitamin is associated with its binding to specific sites in the membrane. The low-density lipoprotein fraction has been demonstrated to function as a transport system for α-tocopherol to cells from plasma through a high-affinity receptor mechanism,[4] whereas a cytosolic binding protein may be involved in intracellular α-tocopherol transfer between subcellular membranes.[5]

α-Tocopherol has an amphipathic structure with a polar head group and an apolar tail. It is located in the lipid bilayer in a phytyl tail-to-phytyl tail arrangement with the phenolic hydroxyl group near the lipid-water interface.[6] α-Tocopherol exists predominantly in an unionized hydroxyl form when localized in the model membranes at physiological pH values.[7] The chroman ring with a hydroxyl residue is required for the antioxidative action of α-tocopherol. Neither α-tocopheryl acetate nor α-tocopheryl quinone, an oxidized metabolite of α-tocopherol, inhibits lipid peroxidation in vitro.[8] The phytyl side chain of α-tocopherol does not influence its antioxidant activity, but enhances the retention of α-tocopherol in membranes.[9]

α-Tocopherol is capable of scavenging oxygen-derived free radicals in vitro[10,11] as well as lipid peroxy radicals in membrane models and in living systems.[8,12,13] Among the tocopherols and other model compounds, α-tocopherol reduced the rate and extent of the formation of the lipid peroxyl radicals in the most effective way in vitro,[11] and 1 mol of α-tocopherol completely prevented peroxidation of 100 mol of polyunsaturated fatty acids in liposomes.[8] α-Tocopherol has been proved to be the main lipophilic, chain-breaking antioxidant in human plasma and erythrocyte membranes,[12] as well as in rat liver and Novikoff tumor samples.[14]

The subcellular fractions from different rat tissues are characterized by differing α-tocopherol content, with the highest reported in mitochondria and microsomes.[15] The highest level of microsomal α-tocopherol was found in lung and heart.[16] Studies of microsomes isolated from different animal tissues established the quantitative relationship between α-tocopherol content and protection against lipid peroxidation in vitro,[16] raising the possibility that the relatively high α-tocopherol content in rat heart microsomes may be related to their resistance to lipid peroxidation.[16,17] Experiments with liposomes containing various amounts of α-tocopherol demonstrated that, depending on the effectiveness of oxidative stress, a critical concentration of antioxidant is needed to inhibit lipid peroxidation.[18] Mitochondrial, microsomal, and nuclear membranes from liver and muscles of vitamin E-deficient rats showed higher sensitivity to oxidative damage than did those of controls, the nuclear membrane being the most sensitive to vitamin E deficiency.[19] Enhanced lipid peroxidation in microsomal membranes was also observed after vitamin E-deficient animals were exposed to radiation.[20]

The reaction of the hydroxyl group of α-tocopherol with free radicals produces α-tocopherol radicals, as was demonstrated recently during peroxidation of phosphatidylcholine liposomes.[21] The mechanism of α-tocopherol regeneration from α-tocopherol radicals has not been explained yet, but its existence could compensate for the very low ratio, estimated to be 1:1000, of α-tocopherol to polyunsaturated fatty acids in membranes.[1] The reduction of α-tocopherol radicals by ascorbate in solutions[13] and liposomes[8,21,22] has a high rate constant, which suggests rapid interaction between these two compounds. The relevance of such experiments to the situation in vivo is questionable because the two vitamins are present in different cellular phases, but the recent finding that α-tocopherol is oriented in membrane with its phenoxyl moiety close to the aqueous phase makes this statement more probable.[6]

Glutathione (GSH), another cellular reducing agent present in cytosol in higher concentration than ascorbate, might also be involved in α-tocopherol regeneration. GSH depletion of mouse and rat liver in vivo greatly increased spontaneous lipid peroxidation.[23] The inhibition of lipid peroxidation by GSH was vitamin E-dependent, being observed only in microsomes obtained from rats given vitamin E supplement and not in those from the deficient groups.[24] The recent demonstration in microsomal membranes of a GSH-dependent protein that inhibits lipid peroxidation by free radical scavenging mechanisms[25,26] supported the participation of GSH in the regeneration reaction and maintenance of tissue α-tocopherol level.

α-Tocopherol as a Membrane-Structure Stabilizer

Biological membranes are organized as a fluid mosaic with dynamic assembly of membranous components. The membrane fluidity determines the lateral mobility and consequent interactions of membrane components important for the activity of membrane-bound enzymes, receptors, and transport systems. Cholesterol and phospholipids are among the physiological modulators of membrane fluidity.[27]

As an essential structural component of membrane, α-tocopherol may influence membrane integrity and function.[28] The fact that not only α-tocopherol, but also tocopheryl acetate, which does not have antioxidant activity, can prevent the oxidative hemolysis of erythrocytes by interacting with the erythrocyte membrane supports this suggestion.[29] The physicochemical interaction between α-tocopherol and unsaturated fatty acid has been proposed to have a stabilizing effect on biological membranes.[28] α-Tocopherol-unsaturated fatty acid complexes can be formed experimentally through hydrophobic and hydrogen-bonding interactions[18,30] in model and natural membrane systems.[31] α-Tocopherol may influence the composition of membrane phospholipids in a specific way without affecting neutral lipids.For instance, α-tocopherol significantly enhanced the arachidonate content of membrane phospholipids in cultured fibroblasts.[32] In addition, the specific loss of arachidonic acid was observed in the hepatic mitochondrial and microsomal membranes of ducklings deficient in vitamin E.[33]

α-Tocopherol increased membrane fluidity in erythrocytes[34] and in liposomes prepared from membrane lipids of platelets,[35] whereas its opposite effect was observed in intestinal brush-border membranes.[36] α-Tocopherol inclusion into liposomes was found to increase permeability for ascorbate,[31] but decreased that of glucose and chromate ions.[37] In cultured fibroblasts, α-tocopherol stimulated membrane glucose transport, with optimal glucose transport of fibroblasts occurring when the medium contained α-tocopherol, linoleic acid, and cholesterol.[38]

In vitamin E deficiency, the activity of glyceraldehyde-3-phosphate dehydrogenase of ghosts was lower compared to that of control.[39] This alteration involved only the membrane enzyme and not its soluble form in cytosol, which suggested that this modification may result from an alteration of surrounding lipids. Vitamin E deficiency also influenced the ultrastructure of liver mitochondrial and microsomal membranes.[33] Thus, lower vitamin E content might not only reduce the antioxidant efficiency of membranes, but destabilize the membrane structure, leading to greater exposure of polyunsaturated fatty acids of membrane lipids to peroxidation.[40]

Potential Implication of α-Tocopherol in Gene Activity Regulation

α-Tocopherol binds to nuclei[41] and may influence RNA and protein synthesis.[42] The nuclear α-tocopherol-binding receptor identified as a nonhistone chromosomal protein is suspected of mediating potential α-tocopherol involvement in the regulation of gene expression.[41]

Particularly interesting is the effect of α-tocopherol on xanthine oxidase and, probably, quinone reductase. Xanthine oxidase can generate free radical intermediates during metabolism of anthracyclines,[43] and α-tocopherol has been suggested to be a corepressor of xanthine oxidase biosynthesis. The marked increase in enzyme activity observed in α-tocopherol deficiency is correlated with elevated synthesis of the enzyme protein.[44] In agreement with this finding is a recent observation of significant suppression of rat liver xanthine oxidase activity after high dietary vitamin E supplementation.[45]

In contrast, quinone reductase is involved in anthracycline detoxification. The preliminary observation of the elevation of quinone reductase activity in liver of rats fed a diet high in vitamin E may result from the direct effect of α-tocopherol on enzyme biosynthesis.[45] The suggestion is attractive in view of the fact that the induction of quinone reductase contributes to the protective mechanism of action of some anticarcinogenic compounds.[46]

THE MECHANISM OF ACTION OF ANTHRACYCLINE ANTIBIOTICS

Daunorubicin (daunomycin) and its analogue doxorubicin (adriamycin, ^{14}OH daunorubicin) isolated from *Streptomyces* species are the most frequently used anticancer drugs of a broad class of anthracycline antibiotics. Both agents have a four-ring anthracycline structure linked by a glycosidic bond to an amino sugar, daunosamine. Doxorubicin is active against a broad spectrum of solid tumors, whereas daunorubicin is mainly effective against acute leukemias.[47]

The complex mechanism of the cytotoxicity of the anthracycline antibiotics has not yet been explained. The intercalation of the drug into DNA leads to changes in nucleic acid conformation and subsequent inhibition of nucleic acids and protein synthesis.[47] Mediated by topoisomerase, these drugs induce strand breaks in DNA. They also produce free radical intermediates and subsequent reactive oxygen species. Some recent evidence implicates cellular membranes as the sites responsible for the cytotoxic effects of the anthracyclines.

Metabolism of Anthracycline Antibiotics

Anthracycline antibiotics are the substrates for cytoplasmic aldo-keto reductases requiring NADPH as a cofactor. Reduction of a side chain carbonyl group yielded alcohol metabolites that retained some cytotoxic activities, but had lowered ability to interact with membrane lipids.[48]

The quinone structure of anthracycline antibiotics makes them good electron acceptors for membrane-bound and cytosolic flavoproteins acting as one-electron or two-electron transfer enzymes. The direct two-electron reduction of anthracycline antibiotics by NAD(P)H-quinone oxidoreductase (DT-diaphorase) to the stable hydroquinones may be regarded as a detoxification process.[49] On the other hand, the one-electron reduction of anthracyclines proceeds by a free radical mechanism.[43,50] This enzymatic one-electron reduction of a stable quinone form of anthracycline results in the formation of a reactive semiquinone radical that rapidly reduces molecular oxygen to superoxide radical. The regenerated anthracycline antibiotic molecule participates again in the cyclic process of drug reduction and oxidation with continuous production of oxygen radicals.[48]

Mitochondrial,[51,52] microsomal,[48,50,53] and nuclear[54] electron transport chains, together with cytosolic enzymes,[43] are the sites of the activation of anthracycline antibiotics to reactive radical intermediates. No information is available yet on the biotransformation of the anthracycline antibiotics by plasma membranes. However, the presence of such membrane enzymes as NADH dehydrogenase, NAD(P)H dehydrogenase, and xanthine oxidase,[55] which can reduce anthracycline antibiotics in an univalent way, suggests free radical generation at this cellular site.

Anthracycline Antibiotics as Membrane-Active Antitumor Agents

Anthracycline antibiotics interact with membrane proteins and phospholipids. Daunorubicin may be attached photochemically by a covalent reaction to the surface protein of sarcoma 180 cells.[56] The specific interaction of anthracycline antibiotics with phospholipids was demonstrated in models[57-59] and in biological membranes.[58] Doxorubicin molecules with the positively charged amino group of the sugar moiety form electrostatic complexes with such negatively charged phospholipid components as cardiolipin or phosphatidic acid,[58] with a particularly high affinity to cardiolipin.[59]

The interaction of doxorubicin with membrane components influences their organization in the membrane. ^{31}P magnetic resonance measurements demonstrated the structural consequences of doxorubicin-cardiolipin interaction on the rearrangement of model and mitochondrial membranes.[56] The cytostatic concentration of anthracyclines induced changes in the rate of agglutination mediated by lectin in sarcoma 180 cells,[59] suggesting that the drug altered the dynamic mobility of lectin receptors. Electron microscopy revealed that the surface ultrastructure of hepatocytes was modulated by doxorubicin and that the cells, after initial loss of microvilli, exhibited numerous blebs and holes on the surface.[60] Likewise, alterations in electron spin resonance of spin-labeled ghost membranes and mastocytoma cells[61] indicated that the subtle conformational changes of membrane proteins were caused by an interaction with anthracyclines.

A rapid, dose-dependent decrease in membrane fluidity of sarcoma 180 cells was the earliest noticed action of doxorubicin, before its effects on nucleic acid were observed.[62] Alterations in membrane fluidity modify membrane permeability and membrane-bound enzyme activities. Such significant changes in membrane permeability and transport by daunorubicin were observed in HeLa cells,[63] dog heart sarcolemmal vesicles,[64] and in sarcoma 180 cells.[65] In addition, doxorubicin inhibited transmembrane NADH dehydrogenase in hepatoma cells and in virus-transformed liver cells.[66]

Plasma membranes are involved in cellular resistance to anthracyclines. A reduced uptake of anthracyclines in resistant cells has been associated with the presence of specific glycoprotein (P180) in the plasma membrane.[67] Moreover, the amount of daunorubicin that could be protoincorporated to surface proteins increased proportionally to the drug sensitivity of sarcoma 180 cell lines.[56] In these cells, progressive changes in membrane fluidity correlated with the expression of drug resistance.[68]

The direct interaction of doxorubicin with plasma membranes may be linked to the cytotoxicity of the drug. After being bound to membrane-impermeable agarose beads[69] or cross-linked polymers.[70] which limited drug action to the cell surface, the drug still retained its cytotoxic activity.

Cytotoxic activity of anthracycline antibiotics may result directly from semiquinone radical forms of the drugs or indirectly from a secondarily generated, toxic, oxygen-derived free radical species. Superoxide anion radicals are regarded as the primary source of oxidative damage in cells.[71] Their production by anthracyclines has been demonstrated in submitochondrial particles,[51] microsomes,[50] and nuclei.[54] Subsequent formation of a hydrogen peroxyl radical is also found in submitochondrial particles.[72]

Free radicals generated through redox cycling of anthracyclines were demonstrated to initiate lipid peroxidation in microsomes[53] and in mitochondria.[73] Such peroxidative damage of membrane structural integrity may be subsequently followed by membrane dysfunction.

CARDIOTOXICITY OF ANTHRACYCLINE ANTIBIOTICS

Chronic administration of anthracycline antibiotics results in a dose-dependent cardiotoxicity. Significant pathological changes in cardiac tissue have been noted in humans as well as in mouse,[73] rat,[74] and rabbit.[75] Microscopic examination of cardiac tissue revealed the disruption of mitochondrial and sarcoplasmic reticulum membranes.[76] These morphological alterations seemed to be correlated with impairment by anthracyclines of some cardiac membrane functions such as mitochondrial oxidative phosphorylation[77] and calcium translocation in the mitochondria[78] and sarcolemmal membrane.[64]

Free Radical Generation in Cardiac Tissue and the Cardiac Defense System

The cardiac mitochondrial NADH dehydrogenase[51,52,79] and microsomal NADPH-cytochrome P-450 oxidoreductase[80] were demonstrated to reduce anthracycline to semiquinone forms. The generation of superoxide and hydroxyl radicals as well as hydrogen perioxide was stimulated in a concentration-dependent manner by doxorubicin in both subcellular fractions, but the rate of these reactions was much higher in cardiac submitochondrial particles than in microsomes. Since the heart muscle is rich in mitochondria, this subcellular fraction may be the major site responsible for redox cycling of anthracyclines, and the impairment of mitochondrial function may be the primary factor in anthracycline cardiotoxicity.[51]

The limited biochemical protective system of the heart makes this tissue particularly susceptible to doxorubicin-induced peroxidative damage. In mouse heart, compared with mouse liver, the activity of catalase was 150 times lower and that of superoxide dismutase was 4 times lower.[81] The diminished catalase content suggests that in cardiac tissue GSH peroxidase participates mainly in hydrogen peroxide detoxification. However, 72 hr after a single dose of doxorubicin, a depression of GSH peroxidase activity further impaired the cardiac defense system against peroxides and led to a decrease in hydrogen peroxide and lipid hydroperoxide detoxification.[81]

Cardiac endogenous nonprotein sulfhydryl groups are involved in doxorubicin-induced cardiotoxicity. Doxorubicin administration resulted in lowering of GSH levels in heart and in other tissues.[75,82] This effect of anthracycline can be most harmful in heart because of the cardiac-specific low GSH level.[82] In addition, doxorubicin inhibited the activity of sulfhydryl-dependent NADP-linked isocitrate dehydrogenase[83,84] and malic enzyme[84] in heart homogenates. The content of these NADPH-generating enzymes is very high in cardiac tissue. NADPH is needed for GSH regeneration by glutathione reductase. Thus, the inhibition of these enzymes could be amplified in heart because of the low cardiac GSH level and because of reduced provision of NADPH for GSH generation. The involvement of sulfhydryl groups in anthracycline cardiotoxicity is also supported by the finding that the process can be significantly reduced by concurrent administration of sulfhydryl-containing compounds such as cysteamine,[82] GSH,[85] or *N*-acetylcysteine.[86] These compounds increase the intracellular level of free sulfhydryl groups, an augmentation of an inadequate chemical defense system in the heart that might allow detoxification of doxorubicin-generated free radicals.

Lipid Peroxidation in Cardiac Tissue Induced by Anthracyclines

Stimulation of NAD(P)H-dependent lipid peroxidation by anthracyclines was observed in cardiac microsomes[80,87] and in cardiac mitochondria.[73,87] The doxorubicin-induced peroxidation of liver mitochondrial phospholipids in vitro may be inhibited by GSH,[79] but a low GSH level in the heart could enhance induction of the lipid peroxidation process.[88] Thus, the observed anthracycline-induced damage of cardiac mitochondria and its membrane functions might result partly from drug-induced lipid peroxidation process in the heart.[89]

High cardiolipin content in mitochondria, together with the high affinity of anthracyclines to cardiolipin, increases their interaction with mitochondrial membrane. Such drug fixation in the membrane structure not only disturbs membrane integrity, but also increases the possibility of free radical generation during drug metabolism by membrane enzymes. Mitochondrial membrane proteins such as inorganic phosphate and Ca^{2+} carriers as well as cytochrome c oxidase require cardiolipin for full activity, and they were shown to be inhibited by doxorubicin-cardiolipin complex formation.[60,90] *N*-Acetyldoxorubicin, which does not bind to cardiolipin, does not inhibit mitochondrial respiration.[51]

MODULATION OF ANTHRACYCLINE-INDUCED CARDIOTOXICITY BY α-TOCOPHEROL

Vitamin E given in very high doses several days before doxorubicin administration diminished doxorubicin-induced cardiotoxicity in mice[73] and rabbits.[80] After four consecutive daily i.m. injections of vitamin E, α-tocopherol levels were raised in rabbit cardiac tissue and serum.[80] In rabbit erythrocytes in vitro, α-tocopherol prevented doxorubicin-induced oxidation of GSH, and no hemolysis was observed.[80] Cardiac tissue levels of GSH also remained unchanged in vitamin E-treated rabbits 24 hr after doxorubicin injection.[81] In addition, light and electron microscopic study of these animals revealed none of the pathological changes usually induced by anthracycline administration in rabbit cardiac tissue, nor did electrocardiograms show the changes typical for doxorubicin-caused cardiotoxicity.[82] Histological evidence showed that pretreatment with α-tocopherol prevents doxorubicin-induced cardiomyopathy in mice as well.[76] The reduced cardiac toxicity was correlated with diminished lipid peroxidation in cardiac tissue.[76] Similarly, α-tocopherol inhibited both endogenous and doxorubicin-augmented peroxidation of liver mitochondrial phospholipids in vitro.[79] Finding the metabolic profile of doxorubicin essentially unchanged after α-tocopherol pretreatment suggested that the protective effect of α-tocopherol may depend on maintenance of pharmacologic rather than physiologic tissue concentrations.[91]

CARCINOGENIC ACTION OF ANTHRACYCLINE ANTIBIOTICS IN MAMMARY GLAND

Doxorubicin and daunomycin induce cancer in rat mammary gland exclusively. The first observation, 25 years ago,[92] of a high incidence of mammary tumors in rats after sublethal single doses of doxorubicin or daunorubicin was confirmed more recently.[93,94] Mammary cells are an actively proliferating cellular population whose high susceptibility to carcinogenesis arises from the presence of large proliferative compartments in the mammary gland.[95] This fact implicates cellular DNA as the main target of the carcinogenic activity of anthracyclines. The frequencies and types of DNA breaks induced by anthracycline antibiotics in isolated mammary and liver cells have revealed a significantly lower quantity of DNA single-strand breaks in hepatocytes. Doxorubicin uptake by these two types of cells was similar, which indicated that the specific toxic effect of the drug was not caused by its preferential accumulation in mammary cells.[94]

Some biochemical characteristics of mammary cells suggest that they might be more exposed to free radicals generated during anthracycline metabolism than hepatocytes. ^{3}H-daunorubicin was eliminated very slowly from mammary cells and, compared with hepatocytes, quinone reductase activity in mammary cells is lower and xanthine oxidase activity is higher.[96] This means that mammary cells are less able to detoxify anthracycline by two-electron reduction by quinone reductase and may generate more anthracycline semiquinone radicals by xanthine oxidase than do hepatocytes. Moreover, the cytosolic defense system of mammary cells is less efficient than that of hepatocytes. GSH level was found to be two times lower and GSH peroxidase activity was found to be ten times lower in mammary cells compared with hepatocytes.[94] Therefore, although the concentration of anthracycline antibiotics in mammary tissue and surrounding fat was lowest after an i.v. injection of the drug, the low drug clearance and specific metabolic characteristics of mammary cells increase the possibility of a higher accumulation of daunorubicin semiquinone free radicals and, subsequently, oxygen-active species. This fact seemed to be correlated with more DNA lesions in this tissue than in others and may be implicated in anthracycline antibiotics-induced malignant transformation.

Changes in membrane fluidity may influence the mobility of the receptor involved in regulating cell proliferation. By altering the physical properties of membranes, anthracycline can induce redistribution of membranous receptors. Doxorubicin is known to modulate the expression of growth-regulatory receptors at the cell surface of actively growing HeLa and 3T3 cells,[97] a mechanism that may alter the cellular response to surface receptors important for normal growth control.[62]

PROTECTIVE EFFECT OF α-TOCOPHEROL AGAINST THE CARCINOGENIC ACTION OF ANTHRACYCLINE ANTIBIOTICS

α-Tocopherol has been shown to have an anticarcinogenic effect;[98] i.p. administration of α-tocopherol for 4 days before daunorubicin injection reduced the occurrence of daunorubicin-induced mammary tumors in Sprague-Dawley rats,[94] and dietary supplements of vitamin E were also shown to suppress chemically induced mammary gland carcinogenesis.[99] The role of α-tocopherol as a protective agent against the carcinogenic effect of anthracyclines has not yet been elucidated; it may interfere with drug side effects in several ways.

α-Tocopherol may protect DNA from daunorubicin-generated free radicals either directly, because of its antioxidant property, or by modulating enzymatic systems involved in daunorubicin metabolism. Reduced glutathione has been shown to protect DNA against chemically induced damage.[100] Thus, it is conceivable that α-tocopherol prevents depletion of intracellular levels of reduced glutathione by daunorubicin. Mammary cells isolated from rats maintained on vitamin E-supplemented diets were more resistant to daunorubicin-induced single-strand DNA breaks than controls.[96] Dietary α-tocopherol has also been shown to reduce chemically induced peroxidative damage to rat brain DNA.[101]

The plasma membrane may be another cellular site of α-tocopherol protection against the carcinogenic action of anthracyclines. Doxorubicin is known to modulate rearrangement of surface membrane receptors for growth regulatory factors.[97] This anthracycline effect may be modified by α-tocopherol, since α-tocopherol has been shown to be required for receptor function.[102] The potential relationship between anthracycline mammary tumor induction and α-tocopherol protection needs further study.

CONCLUSION

At present it is difficult to decide which of the many anthracycline-induced cellular modifications are critical for cell type-specific anthracycline toxicity. Numerous reports provide evidence that membranes may be the primary cellular sites of cytotoxic effects of the drugs. High membrane content of polyunsaturated fatty acids, especially in subcellular membranes, and the presence of membrane-bound enzymatic systems that can generate reactive species during anthracycline metabolism make membranes very susceptible to oxidative damage. The selective toxicity of anthracyclines on certain tissues seems to be correlated not only with differences in the distribution and capacity of enzymes that activate anthracyclines to free-radical intermediates, but also with cellular defense systems against oxidative damage induced by the drugs. Heart and mammary glands are known to have limited intracellular protective systems in comparison with those of other tissues.

Mammalian cells require α-tocopherol for normal function. A wide variety of symptoms observed in experimentally produced vitamin E deficiency in animals indicate tissue-specific susceptibility to damage. The protection by α-tocopherol supplementation against the toxic effects of anthracyclines may arise not only from its antioxidant properties, but also from its effects on membrane structure. Suppression by α-tocopherol of the molecular mobility of phospholipids may protect them both against lipid peroxidation initiated by free radicals and enzymatic hydrolysis leading to the release of arachidonate and its metabolites. Although α-tocopherol seems to protect against membrane-damaging processes induced by anthracyclines, the tissue-specific site that contributes to the toxicity of the drug as well as the exceptional susceptibility of this site require further exploration.

ACKNOWLEDGMENTS

We would like to thank Ms. Rosalyn Zenon for typing this manuscript and Mrs. Lore Feldman for the editorial comments. Our research results presented in this manuscript were supported by Grant CA-35363 from the National Cancer Institute, Bethesda, Md., and a grant from Hoffmann-La Roche, Inc., Nutley, N.J.

REFERENCES

1. **Tappel, A. L.,** Vitamin E and selenium protection from in vivo lipid peroxidation, *Ann. N.Y. Acad. Sci.*, 355, 18, 1980.
2. **Wimalasena, J., Davis, M., and Kitabichi, A. E.,** Characterization and solubilization of the specific binding sites for D-α-tocopherol from human erythrocyte membranes, *Biochem. Pharmacol.*, 31, 3455, 1982.
3. **Kitabchi, A. E., Wimalasena, J., and Barker, J.,** Specific receptor sites for α-tocopherol in purified isolated adrenocortical cell membrane, *Biochem. Biophys. Res. Commun.*, 96, 1739, 1980.
4. **Traber, M. G. and Kayden, H. J.,** Vitamin E is delivered to cells via the high affinity receptor for low-density lipoprotein, *Am. J. Clin. Nutr.*, 40, 747, 1984.
5. **Murphy, D. J. and Mavis, R. D.,** Membrane transfer of α-tocopherol. Influence of soluble α-tocopherol-binding factors from the liver, lung, heart, and brain of the rat, *J. Biol. Chem.*, 256, 10, 464, 1981.
6. **Perly, B., Smith, I. C. P., Hughes, L., Burton, G. W., and Ingold, K. U.,** Estimation of the location of natural α-tocopherol in lipid bilayers by ^{13}C-NMR spectroscopy, *Biochim. Biophys. Acta*, 819, 131, 1985.
7. **Drummond, C. J. and Grieser, F.,** The ionization behaviour of DL-α-tocopherol (vitamin E) in model membranes: micelles and vesicles, *Biochim. Biophys. Acta*, 16, 275, 1985.
8. **Fukuzawa, K., Chida, H., Tokumura, A., and Tsukatani, H.,** Antioxidative effect of α-tocopherol incorporation into lecithin liposomes on ascorbic acid-Fe^{2+}-induced lipid peroxidation, *Arch. Biochem. Biophys.*, 206, 173, 1981.

9. **Niki, E., Kawakami, A., Saito, M., Yamamoto, Y., Tsuchiya, J., and Kamiya, Y.,** Effect of phytyl side chain of vitamin E on its antioxidant activity, *J. Biol. Chem.*, 260, 2191, 1985.
10. **Fukuzawa, K. and Gebicki, J. M.,** Oxidation of α-tocopherol in micelles and liposomes by the hydroxyl, perhydroxyl, and superoxide free radicals, *Arch. Biochem. Biophys.*, 226, 242, 1983.
11. **Burton, G. W., Cheeseman, K. H., Doba, T., Ingold, K. U., and Slater, T. F.,** Vitamin E as an antioxidant *in vitro* and *in vivo*, *Ciba Found. Symp.*, 101, 4, 1983.
12. **Burton, G. W., Joyce, A., and Ingold, K. U.,** Is vitamin E the only lipid-soluble, chain-breaking antioxidant in human blood plasma and erythrocyte membranes?, *Arch. Biochem. Biophys.*, 22, 281, 1983.
13. **Niki, E., Saito, T., Kawakami, A., and Kamiya, Y.,** Inhibition of oxidation of methyl lineoleate in solution by vitamin E and vitamin C, *J. Biol. Chem.*, 259, 4177, 1984.
14. **Cheeseman, K. H., Collins, M., Proudfoot, K., Slater, T. F., Burton, G. W., Webb, A. C., and Ingold, K. U.,** Studies on lipid peroxidation in normal and tumour tissues, *Biochem. J.*, 235, 507, 1986.
15. **Taylor, S. L., Lamden, M. P., and Tappel, A. L.,** Sensitive fluorometric method for tissue tocopherol analysis, *Lipids*, 11, 530, 1976.
16. **Kornbrust, D. J. and Mavis, R. D.,** Relative susceptibility of microsomes from lung, heart, liver, kidney, brain and testes to lipid peroxidation: correlation with vitamin E content, *Lipids*, 15, 315, 1980.
17. **Mimnaugh, E. G., Trush, M. A., Ginsburg, E., and Gram, T. E.,** Differential effects of anthracycline drugs on rat heart and liver NADPH-dependent lipid peroxidation, *Cancer Res.*, 42, 3574, 1982.
18. **Fukuzawa, K., Takase, S., and Tsukatani, H.,** The effect of concentration on the antioxidant effectiveness of α-tocopherol in lipid peroxidation induced by superoxide free radicals, *Arch. Biochem. Biophys.*, 240, 117, 1985.
19. **Quintanilha, A. T. and Packer, L.,** Vitamin E, physical exercise and tissue oxidative damage, *Ciba Found. Symp.*, 101, 56, 1983.
20. **Konings, A. W. T. and Drijver, E. B.,** Radiation effects on membranes. I. Vitamin E deficiency and lipid peroxidation, *Radiat. Res.*, 80, 494, 1979.
21. **Scarpa, M., Rigo, A., Maiorino, M., Ursini, F., and Gregolin, C.,** Formation of α-tocopherol radical and recycling of α-tocopherol by ascorbate during peroxidation of phosphatidylcholine liposomes. An electron paramagnetic resonance study, *Biochim. Biophys. Acta*, 801, 215, 1984.
22. **Doba, T., Burton, G. W., and Ingold, K. U.,** Antioxidant and coantioxidant activity of vitamin C. The effect of vitamin C, either alone or in the presence of vitamin E or a water-soluble vitamin E analogue, upon the peroxidation of aqueous multilamellar phospholipid liposomes, *Biochim. Biophys. Acta*, 835, 298, 1985.
23. **Younes, M. and Siegers, C. P.,** Lipid peroxidation as a consequence of glutathione depletion in rat and mouse liver, *Res. Commun. Chem. Pathol. Pharmacol.*, 27, 119, 1980.
24. **Reddy, C. C., Scholz, R. W., Thomas, C. E., and Massaro, E. J.,** Vitamin E dependent reduced glutathione inhibition of rat liver microsomal lipid peroxidation, *Life Sci.*, 31, 571, 1982.
25. **Burk, R. F.,** Glutathione-dependent protection by rat liver microsomal protein against lipid peroxidation, *Biochim. Biophys. Acta*, 757, 21, 1983.
26. **McCay, P. B., Lai, E. K., Powell, S. R., and Breuggemann, G.,** Vitamin E functions as an electron shuttle for glutathione-dependent free radical reductase activity in biological membranes, *Fed. Proc. Fed. Am. Soc. Exp. Biol.*, 45, 451, 1985.
27. **Shinitzky, M.,** Membrane fluidity in malignancy, adversative and recuperative, *Biochim. Biophys. Acta*, 738, 251, 1984.
28. **Diplock, A. T. and Lucy, J. A.,** The biochemical modes of action of vitamin E and selenium: a hypothesis, *FEBS Lett.*, 29, 205, 1973.
29. **Mino, M. and Sugita, K.,** The membrane action of α-tocopherol upon oxidative damage in erythrocytes, in *Tocopherol, Oxygen and Biomembranes*, de Duve, C. and Hayaishi, O., Eds., Elsevier/North-Holland, New York, 1978, 71.
30. **Srivastava, S., Phadke, R. S., Govil, G., and Rao, C. N. R.,** Fluidity, permeability and antioxidant behaviour of model membranes incorporated with α-tocopherol and vitamin E acetate, *Biochim. Biophys. Acta*, 734, 353, 1983.
31. **Erin, A. N., Spirin, M. M., Tabidze, L. V., and Kagan, V. E.,** Formation of α-tocopherol complexes with fatty acids. A hypothetical mechanism of stabilization of biomembranes by vitamin E, *Biochim. Biophys. Acta*, 774, 96, 1984.
32. **Giasuddin, A. S. M. and Diplock, A. T.,** The influence of vitamin E on membrane lipids of mouse fibroblasts in culture, *Arch. Biochem. Biophys.*, 210, 348, 1981.
33. **Molennar, I., Hulstaert, C. E., and Hardonk, M. J.,** Role in function and ultrastructure of cellular membranes, in *Vitamin E. A Comprehensive Treatise*, Machlin, L. J., Ed., Marcel Dekker, New York, 1980, 319.
34. **Viti, V., Cicero, R., Callari, D., Guidoni, L., Billiteri, A., and Sichel, G.,** Effect of lipophilic vitamins on the erythrocyte membrane, *FEBS Lett.*, 158, 36, 1983.

35. **Steiner, M.,** Vitamin E changes the membrane fluidity of human platelets, *Biochim. Biophys. Acta,* 640, 100, 1981.
36. **Ohyashiki, T., Ushiro, H., and Mohri, T.,** Effects of α-tocopherol on the lipid peroxidation and fluidity of porcine intestinal brush-border membranes, *Biochim. Biophys. Acta,* 858, 294, 1986.
37. **Lucy, J. A.,** Structural interactions between vitamin E and polyunsaturated phospholipids, in *Tocopherol, Oxygen and Biomembranes,* de Duve, C. and Hayaishi, O., Eds., Elsevier/North-Holland, New York, 1982, 109.
38. **Giasuddin, A. S. M. and Diplock, A. T.,** The influence of vitamin E and selenium on the growth and plasma membrane permeability of mouse fibroblasts in culture, *Arch. Biochem. Biophys.,* 196, 270, 1979.
39. **Shapiro, S. S. and Mott, D. J.,** Alterations of enzymes in the red blood cell membranes in vitamin E deficiency, *Ann. N.Y., Acad. Sci.,* 393, 263, 1982.
40. **McCay, P. B., Poyer, J. L., Pfeifer, P. M., May, H. E., and Gilliam, J. M.,** A function for tocopherol. Stabilization of the microsomal membrane from radical attack during TPNH-dependent oxidations, *Lipids,* 6, 297, 1970.
41. **Nair, P. P., Patnaik, R. N., and Hauswirth, J. W.,** Cellular transport and binding of D-α-tocopherol, in *Tocopherol, Oxygen and Biomembranes,* de Duve, C. and Hayaishi, O., Eds., Elsevier/North-Holland, New York, 1978, 83.
42. **Catignani, G. L.,** Biochemical function. Role in nucleic acid and protein metabolism, in *Vitamin E. A Comprehensive Treatise,* Machlin, L. J., Ed., Marcel Dekker, New York, 1980, 318.
43. **Pan, S. S., Pedersen, L., and Bachur, N. R.,** Comparative flavoprotein catalysis of anthracycline antibiotic. Reductive cleavage and oxygen consumption, *Mol. Pharmacol.,* 19, 184, 1981.
44. **Catignani, G. L., Chytil, F., and Darby, W. J.,** Vitamin E deficiency: immunochemical evidence for increased accumulation of liver xanthine oxidase, *Proc. Natl. Acad. Sci. U.S.A.,* 71, 1966, 1974.
45. **Purewal, M. and Wang, Y.-M.,** Effect of dietary vitamin E (E) on the activities of rat hepatic quinone reductase, xanthine oxidase, and glutathione (GSH)-S-transferase, *Fed. Proc.,* 45(Abstr.), 840, 1985.
46. **Benson, A. M., Hunkeler, M. J., and Talalay, P.,** Increase of NAD(P)H:quinone reductase by dietary antioxidants: possible role in protection against carcinogenesis and toxicity, *Proc. Natl. Acad. Sci. U.S.A.,* 77, 5216, 1980.
47. **Myers, C. E.,** Anthracyclines, in *Principles of Cancer Treatment,* Chabner, B., Ed., W. B. Saunders, Philadelphia, 182, 416.
48. **Bachur, N. R.,** Anthracycline antibiotic pharmacology and metabolism, *Cancer Treat. Rep.,* 63, 817, 1979.
49. **Lind, C., Hochstein, P., and Ernsten, L.,** DT-diaphorase as a quinone reductase: a cellular device against semiquinone and superoxide radical formation, *Arch. Biochem. Biophys.,* 216, 178, 1982.
50. **Bachur, N. R., Gordon, S. L., and Gee, M. V.,** A general mechanism for microsomal activation of quinone anticancer agents to free radicals, *Cancer Res.,* 38, 1745, 1978.
51. **Davies, K. J. A. and Doroshow, J. H.,** Redox cycling of anthracyclines by cardiac mitochondria, *J. Biol. Chem.,* 261, 3060, 1986.
52. **Doroshow, J. H.,** Anthracycline antibiotic-stimulated superoxide, hydrogen peroxide, and hydroxyl radical production by NADH dehydrogenase, *Cancer Res.,* 43, 4543, 1983.
53. **Goodman, J. and Hochstein, P.,** Generation of free radicals and lipid peroxidation by redox cycling of adriamycin and daunomycin, *Biochem. Biophys. Res. Commun.,* 77, 797, 1977.
54. **Bachur, N. R., Gee, M. V., and Friedman, R. D.,** Nuclear catalyzed antibiotic free radical formation, *Cancer Res.,* 42, 1078, 1982.
55. **Low, H. and Crane, F. L.,** Redox function in plasma membranes, *Biochim. Biophys. Acta,* 515, 141, 1978.
56. **Yee, G., Carey, M., and Tritton, T. R.,** Photoaffinity labeling of the sarcoma 180 cell surface by daunomycin, *Cancer Res.,* 44, 1898, 1984.
57. **Henry, N., Fantine, E. O., Bolard, J., and Garnier-Suillerot, A.,** Interaction of adriamycin with negatively charged model membranes: evidence of two types of binding sites, *Biochemistry,* 24, 7085, 1985.
58. **Nicolay, K., van der Neut, R., Fok, J. J., and de Kruijff, B.,** Effects of adriamycin on lipid polymorphism in cardiolipin-containing model and mitochrondrial membranes, *Biochim. Biophys. Acta,* 819, 55, 1985.
59. **Goormaghtigh, E., Chatelain, P., Caspers, J., and Ruysschaert, J. M.,** Evidence of a specific complex between adriamycin and negatively charged phospholipids, *Biochim. Biophys. Acta,* 597, 1, 1980.
60. **Rogers, K. E., Carr, B. J., and Tookes, Z. A.,** Cell surface-mediated cytotoxicity of polymer-bound adriamycin against drug-resistant hepatocytes, *Cancer Res.,* 43, 2741, 1983.
61. **Sinha, B. K. and Chignell, C. F.,** Interaction of antitumor drugs with human erythrocyte ghost membranes and mastocytoma P815: a spin label study, *Biochem. Biophys. Res., Commun.,* 86, 1051, 1979.
62. **Murphree, S. A., Tritton, T. R., Smith, P. L., and Sartorelli, A. C.,** Adriamycin-induced changes in the surface membrane of sarcoma 180 ascites cells, *Biochim. Biophys. Acta,* 649, 317, 1981.

63. **Prosperi, E., Croce, A. C., Bottiroli, G., and Supino, R.,** Influence of daunorubicin on membrane permeability properties: detection by means of intracellular accumulation and efflux of fluorescein, *Chem. Biol. Interact.*, 54, 271, 1985.
64. **Caroni, P., Villani, F., and Carafoli, E.,** The cardiotoxic antibiotic doxorubicin inhibits the Na^+/Ca^{2+} exchange of dog heart sarcolemmal vesicles, *FEBS Lett.*, 130, 184, 1981.
65. **Choudhury, S. R., Deb, J. K., Choudhury, K., and Neogy, R. K.,** Inhibition of uridine transport through sarcoma-180 cell membrane by anthracycline antibiotics, *Biochem. Pharmacol.*, 31, 1811, 1982.
66. **Sun, I. L., Crane, F. L., Chou, J. Y., Low, H., and Grebing, C.,** Transformed liver cells have modified transplasma membrane redox activity which is sensitive to adriamycin, *Biochem. Biophys. Res. Commun.*, 116, 210, 1983.
67. **Center, M. S.,** Mechanisms regulating cell resistance to adriamycin. Evidence that drug accumulation in resistant cells is modulated by phosphorylation of a plasma membrane glycoprotein, *Biochem. Pharmacol.*, 34, 1471, 1985.
68. **Siegfried, J. A., Kennedy, K. A., Sartorelli, A. C., and Tritton, T. R.,** The role of membranes in the mechanism of action of the antineoplastic agent adriamycin, *J. Biol. Chem.*, 258, 339, 1983.
69. **Tritton, T. R. and Yee, G.,** The anticancer agent adriamycin can be actively cytotoxic without entering cells, *Science*, 217, 248, 1982.
70. **Wingard, L. B., Jr., Tritton, T. R., and Egler, K. A.,** Cell surface effects of adriamycin and carminomycin immobilized on cross-linked polyvinyl alcohol, *Cancer Res.*, 45, 3529, 1985.
71. **Fridovich, I.,** Biological effects of the superoxide radical, *Arch. Biochem. Biophys.*, 247, 1, 1986.
72. **Doroshow, J. H. and Davies, K. J. A.,** Redox cycling of anthracyclines by cardiac mitochondria. II. Formation of superoxide anion, hydrogen peroxide, and hydroxyl radical, *J. Biol. Chem.*, 261, 3068, 1986.
73. **Myers, C. E.,** Adriamycin: the role of lipid peroxidation in cardiac toxicity and tumor response, *Science*, 191, 165, 1977.
74. **Sonnevels, P.,** Effects of α-tocopherol on the cardiotoxicity of adriamycin in the rat, *Cancer Treat. Res.*, 62, 1033, 1978.
75. **Wang, Y.-M., Madanat, F. F., Kimball, J. C., Gleiser, C. A., Ali, M. K., Kaufman, M. W., and van Eys, J.,** Effect of vitamin E against adriamycin-induced toxicity in rabbits, *Cancer Res.*, 40, 1022, 1980.
76. **Ferrans, V. J.,** Overview of cardiac pathology in relation to anthracycline cardiotoxicity, *Cancer Treat. Rep.*, 62, 955, 1978.
77. **Mailer, K. and Petering, D. H.,** Inhibition of oxidative phosphorylation in tumor cells and mitochondria by daunomycin and adriamycin, *Biochem. Pharmacol.*, 25, 2085, 1976.
78. **Moore, L., Landon, E. J., and Cooney, D. A.,** Inhibition of the cardiac mitochondrial calcium pump by adriamycin in vitro, *Biochem. Med.*, 18, 131, 1977.
79. **Mimnaugh, E. G., Trush, M. A., Bhatnagar, M., and Gram, T. E.,** Enhancement of reactive oxygen-dependent mitochondrial membrane lipid peroxidation by the anticancer drug adriamycin, *Biochem. Pharmacol.*, 34, 847, 1985.
80. **Mimnaugh, E. G., Gram, T. E., and Trush, M. A.,** Stimulation of mouse heart and liver microsomal lipid peroxidation by anthracycline anticancer drugs: characterization and effects of reactive oxygen scavengers, *J. Pharmacol. Exp. Ther.*, 226, 806, 1983.
81. **Doroshow, J. H., Locker, G. Y., and Myers, C. E.,** Enzymatic defenses of the mouse heart against reactive oxygen metabolites. Alterations produced by doxorubicin, *J. Clin. Invest.*, 65, 128, 1980.
82. **Olson, R. D., MacDonald, J. S., van Boxtel, C. J., Boerth, R. C., Harbison, R. D., Slonim, A. E., Freeman, R. W., and Oates, J. A.,** Regulatory role of glutathione and soluble sulfhydryl groups in the toxicity of adriamycin, *J. Pharmacol. Exp. Ther.*, 215, 450, 1980.
83. **Fabregat, I., Satrustegui, J., and Machado, A.,** Interaction with protein SH groups could be involved in adriamycin cardiotoxicity, *Biochem. Med.*, 32, 289, 1984.
84. **Yasumi, M., Minaga, T., Nakamura, K., Kizu, A., and Ijichi, H.,** Inhibition of cardiac NADPH-linked isocitrate dehydrogenase by adriamycin, *Biochem. Biophys. Res. Commun.*, 93, 631, 1980.
85. **Yoda, Y., Nakazawa, M., Abe, T., and Kawakami, Z.,** Prevention of doxorubicin myocardial toxicity in mice by reduced glutathione, *Cancer Res.*, 46, 2551, 1986.
86. **Doroshow, J. H., Locker, G. Y., Ifrim, I., and Myers, C. E.,** Prevention of doxorubicin cardiac toxicity in the mouse by *N*-acetylcysteine, *J. Clin. Invest.*, 68, 1053, 1981.
87. **Kharasch, E. D. and Novak, R. F.,** Inhibitory effects of anthracenedione antineoplastic agents on hepatic and cardiac lipid peroxidation, *J. Pharmacol. Exp. Ther.*, 226, 500, 1983.
88. **Meredith, M. J. and Reed, D. J.,** Depletion in vitro of mitochondrial glutathione in rat hepatocytes and enhancement of lipid peroxidation by adriamycin and 1,3-Bis(2-chloroethyl)-1-nitrosourea (BSNU), *Biochem. Pharmacol.*, 32, 1383, 1983.
89. **Narabayashi, H., Takeshige, K., and Minakami, S.,** Alteration of innermembrane components and damage to electron-transfer activities of bovine heart submitochondrial particles induced by NADPH-dependent lipid peroxidation, *Biochem. J.*, 202, 97, 1982.

90. **Goormaghtigh, E. and Ruysschaert, J. M.,** Anthracycline glycoside-membrane interactions, *Biochim. Biophys. Acta,* 779, 271, 1984.
91. **Mimnaugh, E. G., Siddik, Z. H., Drew, R., Sikic, B. I., and Gram, T. E.,** The effects of α-tocopherol on the toxicity, disposition, and metabolism of adriamycin in mice, *Toxicol. Appl. Pharmacol.,* 49, 119, 1979.
92. **Bertazolli, C., Chieli, T., and Solcia, E.,** Different incidence of breast carcinomas or fibroadenomas in daunorubicin or adriamycin treated rats, *Experientia,* 27, 1209, 1971.
93. **Buciarelli, E.,** Mammary tumor induction in male and female Sprague-Dawley rats by adriamycin and daunomycin, *J. Natl. Cancer Inst.,* 66, 81, 1981.
94. **Wang, Y.-M. and Howell, S. K.,** α-tocopherol as a potential modifier of daunorubicin-induced mammary tumors in rats, *Ann. N.Y. Acad. Sci. U.S.A.,* 393, 186, 1982.
95. **Russo, J. and Russo, I. H.,** Influence of differentiation and cell kinetics on the susceptibility of the rat mammary gland to carcinogenesis, *Cancer Res.,* 1980, 2677, 40.
96. **Wang, Y.-M.,** Mechanisms and prevention of anticancer agent-induced cancer: interaction of vitamin E and daunorubicin, in *Vitamins and Cancer — Human Cancer Prevention by Vitamins and Micronutrients,* Meyskens, F. L. and Prasad, K. N., Eds., Humana Press, Campton, N.J., 1985, 213.
97. **Zuckier, G. and Tritton, T. R.,** Adriamycin causes up regulation of epidermal growth factor receptors in actively growing cells, *Exp. Cell Res.,* 148, 155, 1983.
98. **Ames, B. N.,** Dietary carcinogens and anticarcinogenesis. Oxygen radicals and degenerative diseases, *Science,* 221, 1256, 1983.
99. **Harman, D.,** Dimethylbenzanthracene induced cancer: inhibitory effect of dietary vitamin E, *Clin. Res.,* 17, 125, 1969.
100. **Chan, J. Y. H., Stout, D. L., and Becker, F. F.,** Protective role of thiols in carcinogen-induced DNA damage in rat liver, *Carcinogenesis,* 7, 1621, 1986.
101. **Summerfield, F. W. and Tappel, A. L.,** Vitamin E protects against methyl ethyl ketone peroxide-induced peroxidative damage to rat brain DNA, *Mutat. Res.,* 126, 113, 1984.
102. **Kilian, P. L., Dunlap, C. R., and Mueller, P.,** Reconstitution of acetylcholine receptor from *Torpedo californica* with highly purified phospholipids: effect of α-tocopherol, phylloquinone, and other terpenoid quinones, *Biochem. Biophys. Res. Commun.,* 93, 409, 1980.

ANTIOXIDANT THERAPY IN NEURONAL CEROID LIPOFUSCINOSIS (NCL)

Tuomas Westermarck, Pirkko Santavuori, and Hannu Heiskala

INTRODUCTION

The neuronal ceroid lipofuscinoses (NCLs) are a group of recessively inherited neurodegenerative lysosomal storage diseases, the incidence of which is particularly high in the Nordic countries. In Finland, the national average is ten times the world-wide average of 1 in 100,000.[1] Characteristics of the diseases are widespread deposits of ceroid and lipofuscin pigments in the body, visual failure, and progressive cerebral degeneration, causing mental retardation, gait disturbances, etc.[2] Depending upon chronological, clinical, electrophysiological, and neuropathological features, the NCL group can be subdivided in to the infantile form (INCL, Santavuori-Haltia disease), the late infantile form (LINCL, Jansky-Bielschowsky disease), the juvenile form (JNCL, Spielmeyer-Sjögren disease), and the adult form (Kuf disease).[1-3] These progressive diseases cause death at ages under 14, 6 to 22, and 15 to 35 years, respectively.[1-3] The adult form is not discussed here.

The pathogenesis of NCL is unknown. There are several hypotheses, including abnormal utilization of dolichols in glycoprotein synthesis, perhaps in enzymes involved in the degradation or intracellular recycling of lysosomal membrane; inherited defects in lysosomal protein catabolism; or destruction of cells and organelles by lipid peroxidation and other forms of oxygen radical damage.[4-11] Despite the fact that peroxidation is most probably a secondary phenomenon (the diseased tissues peroxidize more rapidly than normal tissues and cytotoxic end products of lipid peroxidation cause secondary damage), this hypothesis offers a therapeutic rationale for alleviation of neurodegeneration.[12]

DEVELOPMENT OF ANTIOXIDANT THERAPY

The use of antioxidants and transition metal chelators or other substances that inhibit oxidative reactions such as those promoted by oxygen radicals is an important approach to the treatment of NCL (Table 1). One of the most essential lipoperoxidation counteracting enzymes is the selenium-containing glutathione peroxidase (GSH-Px)[13] which in the presence of glutathione (GSH) can reduce fatty acid hydroperoxides to more inert hydroxy acids (LOH):

$$\mathrm{LOOH} + 2\,\mathrm{GSH} \rightarrow \mathrm{LOH} + \mathrm{GSSG} + \mathrm{H_2O}\ (1)^{13} \qquad (1)$$

However, it is unable to carry out this reaction when peroxides are present in membrane phospholipids. In vivo GSH-Px may prevent polar hydroxy acids from being inserted into membranes.

Nonbiologically significant decreases in superoxide dismutase or H_2O_2-removing enzymes have been observed.[14] Nevertheless, two independent reports from the Nordic countries showed that erythrocyte GSH-Px activity was lowered in NCL patients, and in one study it was shown that sodium selenite supplementation could reverse this.[14-16] Germain and Arneson[17] reported that the GSH-Px activity in neurons of animals may be increased by sodium selenite supplementation. When ^{75}Se-selenite was injected parenterally into rats, ^{75}Se was found in brain.[13,18] Mouse neuroblastoma cells can take up selenite, and, presumably, convert it into selenocysteine, since GSH-Px activity rose 30-fold in response to an addition of exogenous selenite, and selenocysteine is very likely to be present at the active site of GSH-Px.[17,19] Selenium cysteine cannot easily penetrate the blood-brain barrier, while selenium methionine

Table 1
PRESENT WESTERMARCK-SANTAVUORI THERAPY COMPARED TO THE PREVIOUSLY USED ZEMAN FORMULA

Substance	Previous oral dose[a]	Present oral dose[a]
Ascorbic acid	0.008—0.025 g/kg BW	None
D-α-Tocopherol	0.02—0.06 g/kg BW	0.02—0.06 g/kg BW
Methionine	0.008—0.025 g/kg BW	None
BHT	0.5 mg/kg BW	None
Pyridoxine	None	0.5—3 mg/kg BW
Riboflavin	None	0.03—0.3 mg/kg BW
Sodium selenite	None	0.075—0.1 mg/kg BW

[a] BW = body weight.

Table 2
THE MONITORING OF THE WESTERMARCK-SANTAVUORI TREATMENT

Substance	Optimal serum/blood level
Sodium selenite	S-Se, 2.5—4.5 μmol/ℓ
D-α-Tocopherol	S-vitamin E, 40—80 μmol/ℓ
Pyridoxine	AC,[a] 1.2—1.7
Riboflavin	AC,[b] 1.1—1.7

[a] E-ASAT (+ P-5-P)/E-ASAT.
[b] E-GR (+ FAD)/E-Gr.

can, but is unable to do it in the presence of phenylalanine and methionine.[20] Se-methionine is incorporated in body proteins, where it is not available for GSH-Px synthesis until these proteins turn over.[21] Sunde and Hoekstra[21] have asserted that selenium in an inorganic form may serve as the immediate precursor for selenium into GSH-Px. For these reasons, we are using sodium selenite (Tables 1 and 2).

It seemed logical to combine vitamin E with selenium because vitamin E may control the formation of lipid hydroperoxides through its antioxidant function, possibly by entering into a structural relation with membrane phospholipids.[22,23] Administration of vitamin E has been reported to decrease the neurological symptoms seen in patients with abnormal fat absorption (including a long-term study of patients with abetalipoproteinemia, strongly suggesting that vitamin E is essential for a normal neurological structure and function). Moreover, vitamin E protects monkeys against peroxidative destruction characterized by loss of cone outer segments, macular degeneration, and accumulation of lipofuscin and ceroid in vitamin E-deficient monkeys.[24,25] Vitamin E levels, however, are very low only in some advanced and hospitalized NCL patients, an observaton shared by Würtz-Jørgensen.[26,27]

In animals, both an excess and a deficiency of dietary vitamin E have been shown to significantly depress GSH-Px activity in liver and plasma.[28,29] Machlin and Gabriel showed that if a significant increase in tocopherol levels of tissues with slow accumulation rate, such as muscle and brain, is desired, a prolonged treatment at high levels of intake is necessary.[30] This is the rationale for the present use of a rather high dosage and frequent monitoring of vitamin E concentrations (Tables 1 and 2).

The combination of methionine and vitamin E apparently facilitates the excretion of selenium.[31] Large doses of L-methionine have been reported to induce depletion of glutathione in mice.[32] Studies on diets of man have shown that methionine is one of the most toxic of all nutritional amino acids.[33] Thus, the supplementation of methionine to NCL patients has been discontinued.

The rationale for vitamin B_6 treatment is that administration of vitamin B_6 may increase formation of cystein — a precursor of glutathione — through vitamin B_6-dependent enzymes, cystathione synthetase and cystathionase.[34,35] Riboflavin is needed for activation of vitamin B_6 and as a coenzyme of glutathione reductase, and there are reports of an effect of riboflavin on lipoperoxidation.[36-38]

Vitamins B_2 and B_6, like vitamin C, may have both anti- and pro-oxidant effects in vivo.[39,40] A potential source of H_2O_2 is through autoxidation of ascorbic acid, a reaction catalyzed by copper, iron, quinones, or by a photosensitizer such as riboflavin.[40] Since there is some evidence of raised iron content in certain cells and cerebrospinal fluid (CSF) in NCL patients, vitamin C is no longer included in the antioxidant regimen.

In a study concerning our juvenile patients, the riboflavin or pyridoxine values (measured by an activation coefficient method) usually did not differ in pretreated patients form those in controls; the treatment, however, increased vitamin activity significantly.[41-44] Because of the pro-oxidant potential, the dosage of vitamins B_2 and B_6 has been monitored by activation coefficient methods, aiming at rather high, but not too high, concentrations of vitamins (Table 2).[41,42]

Administration of phenolic antioxidants, such as 3,5-tert-butyl-4-hydroxytoluene (BHT) to mice and rats has an influence upon the incidence of mammary tumorigenesis induced by 7,12-dimethylbenz (*a*) anthracene (DMBA).[45] As a consequence of their peroxidative effect, polyunsaturated fatty acids are promotors of carcinogenesis; dietary BHT supplements are effective in reducing the incidence of tumors.[45] However, BHT supplementation of the basal diets of rats has been shown to increase the levels of both adrenal cholesterol and cholesterylester and to reduce body weight.[45] A similar response in NCL patients would be undesirable, and consequently BHT is not included in the newer treatment.

Transition metal chelators such as desferrioxamine have also been considered as antioxidants because the rates of lipid peroxidation and superoxide-dependent formation of hydroxyl radicals (OH·) are greatly accelerated in the presence of suitable catalysts such as iron and copper.[46] Zeman[47] showed high concentrations of these metals in ceroid lipofuscin, while we have been unable to confirm this.[48] In a pilot study, Gutteridge et al.[10] found an elevated content of nonprotein bound iron in the CSF of NCL patients, which we have been unable to confirm.[49] There is, moreover, some evidence that NCL patients have 2 decreased antioxidant protection against iron-dependent hydroxyl radical formation in vitro,[10,11] increased intestinal absorption of iron,[50] and increased levels of aluminium in the CSF,[48] which may enhance the iron-dependent damage to membranes.[51] In our pilot study with desferrioxamine, we were unable to show either definite clinical response or consistent changes in nonprotein bound iron in the CSF.[52] Further investigation is needed to establish whether these mechanisms have a real importance in patients with NCL.

CLINICAL RESPONSE

Spielmeyer-Sjögren's disease (JNCL) is the only form responding to antioxidant therapy.[53] JNCL begins with gradual loss of vision at the age of 4 to 8 years and is followed by progressive dementia, impairment of speech and motor functions, epilepsy, and death by the age of 20 years. The diagnosis is principally based on the presence of vacuolated lymphocytes, typical ophthalmologic findings, and rectal biopsy with characteristic electron-microscopic accumulations.

The treatments were initiated in 1970, following the suggestion by Zeman and Coll that antioxidants, such as tocopherol, ascorbic acid, methionine, and BHT, could counteract the rate of progression of JNCL (Table 1). In their open, nonrandomized trial, Santavuori and Moren found that in treated patients IQs were higher, the neurologic signs were less marked, and epilepsy was less frequent than in untreated patients.[54] The therapy did not benefit vision and was unsuccessful in advanced cases. These preliminary findings contributed to further development of antioxidant therapy as described above.

A recent study was performed on 125 JNCL patients, 49 of whom were treated with antioxidants.[53] We found a definite retardation of the progression of JNCL in about half the patients receiving antioxidants. The quality of the response was better with the Westermarck-Santavuori formula. This became more obvious in the follow-up studies performed in the spring of 1986. During the 5-year period, 1981 to 1986, six patients (between 20 and 26 years of age) had shown only slight deterioration or none at all. This differs strikingly from the fast progression of the disorder usually seen after the age of 18 years. In four of these six patients, mentioned above, the originally used vitamins C and E, methionine, and BHT therapy had been changed (since 1977 to 1981) to the Westermarck-Santavuori formula because a deterioration had been noticed. In two of the patients the clinical response was exceptionally good; and the progression of the disease seemed to have been halted. These patients were males aged 22 and 23 who have no motor dysfunction and are mentally normal (with academic achievements). Altogether, four children have been treated from early preclinical stage and two have been treated from late preclinical stage onward. Preclinical patients are children with no clinical signs, but definite histological confirmation of the disease. The follow-up time was 3 years in one case and between 5 and 8 years in the other cases. The results were less encouraging than expected. Only one girl was definitely better at the age of eight than her elder affected siblings had been. However, a prolonged follow-up time is needed. The initiation of antioxidant therapy between 8 and 10 years of age has produced the best responses; similarly, the response was better in males. This finding is in harmony with the hypothesis of the secondary nature of peroxidation in the pathogenesis of NCL since preclinical disease may not cause as much secondary peroxidation as a clinical disease.[12]

The side effects have been few. A boy aged 13 years developed partly brown- and white-spotted nails (serum selenium, 644 ng/g). Certain patients have reported nausea when taking sodium selenite on an empty stomach or with certain fruits. Vitamins E and B_2 and B_6 have been well tolerated, but there was one male patient with reversible muscular dystony coincident with administration of high doses of vitamin E. Norppa et al.[55] studied nine NCL patients receiving sodium selenite (Na_2SeO_3) injections or tablets (0.005 to 0.5 mg Se/kg, daily) for 1 to 13.5 months; the supplementation had no detectable effect on their peripheral lymphocyte chromosomes in vivo.

As stated before, the more aggressively deteriorating forms of NCL have not benefited from the treatment. We have experience from ten INCL patients, three of them treated at a preclinical stage. The only notable response has been an improvement of blood circulation. There is evidence that hypoxia as well as hyperoxia stimulate lipid peroxidation, while antioxidants decrease the ischemic deterioration of the tissue.[56] Our experience from LINCL is limited to three patients treated at an early preclinical or clinical stage with similarly poor results.[40] No reports are available from other countries concerning antioxidant therapy of INCL or LINCL.

DISCUSSION

While the basic metabolic defect in NCL is not established, the treatment can only be directed toward secondary and highly complex reactions involved with pro-oxidant and antioxidant mechanisms. The current evidence indicates that the childhood forms of NCL may have different phenotypes and there are several clinical descriptions of the various types.[57-59] There is also accumulating knowledge of biochemical variants. For example, according to Siakotos,[60] only one sixth of the JNCL patients have phospholipid hydroperoxide glutathione peroxidase selenoenzyme in their brains. This form may be the most sensitive to selenium treatment. These facts are in agreement with the observed heterogeneity of clinical response to antioxidant treatment.

To differentiate the effects of the currently used antioxidants, we have started a randomized study, initiating the treatment with a single antioxidant and performing crossovers; at this stage it is too early to draw any conclusions. It is very difficult to organize a randomized and double-blind study in NCL patients, since there is a great heterogeneity among the patients; there may be several biochemical phenotypes of the unknown primary defect as well as differences in antioxidative enzymes, making the different treatment groups most probably unequal. Further, the disease has a slow progression and therefore the effects of antioxidant therapy interact with those of the changing environments and rehabilitation procedures to which the patients have been exposed. Moreover, there is a scarce number of patients and, finally, the impact of a neurodegenerative disease on the family has to be considered; the family needs all the support available, and in this situation a double-blind study seems impractical.

CONCLUSIONS

The data suggest that our antioxidant treatment has a limited but definite effect on Spielmeyer-Sjögren disease. Since the basic etiopathogenesis remains unknown, this treatment is the only pharmacological way to help these patients (beside symptom-oriented measures such as anticonvulsive treatment). Most important, our limited experience (on six patients) suggests that antioxidant treatment does not lengthen the severe end stage of NCL patients. On the contrary, it may offer a little better quality of life for that age period.

REFERENCES

1. **Santavuori, P.,** Neuronal ceroid lipofuscinoses in childhood, in *Neurolipidoses: Donnees Actuelles. Recent Progress in Neurolipidoses and Allied Disorders,* Fondation Marcel Merieux, Lyon, 1984, 95.
2. **Zeman, W., Donahue, S., Dyken, P., and Green, J.,** The neuronal ceroid-lipofuscinoses (Batten-Vogt syndrome), in *Handbook of Clinical Neurology,* Vol. 10, Vinken, P. J. and Bruyn, G. W., Eds., North-Holland, Amsterdam, 1970, 588.
3. **Santavuori, P., Haltia, M., Rapola, J., and Raitta, C.,** Infantile type of so-called neuronal ceroid-lipofuscinosis. I. A clinical study of 15 patients, *J. Neurol. Sci.,* 18, 257, 1973.
4. **Wolfe, L. S., Ng Ying Kin, N. M. K., Palo, J., and Haltia, M.,** Dolichols in brain and urinary sediment in neuronal ceroid-lipofuscinosis, *Neurology,* 33, 103, 1983.
5. **Wolfe, L. S., Palo, J., Santavuori, P., Andermann, F., Andermann, E., Jacob, J. C., and Kolodny, E.,** Urinary sediment dolichols in the diagnosis of neuronal ceroid-lipofuscinosis, *Ann. Neurol.,* 19, 270, 1986.
6. **Krusius, T., Viitala, J., Palo, J., and Maury, C. P. J.,** Enrichment of high mannose-type glycans in nervous tissue glycoproteins in neuronal ceroid-lipofuscinosis, *J. Neurol. Sci.,* 72, 1, 1986.
7. **Palmer, D. N., Husbands, D. R., Winter, P. J., Blunt, J. W., and Jolly, R. D.,** Ceroid lipofuscinosis in sheep. I. Bis(monoacylglycero)phosphate, dolichol, ubiquinone, phospholipids, fatty acids, and fluorescence in liver lipopigment lipids, *J. Biol. Chem.,* 261, 1766, 1986.
8. **Palmer, D. N., Barns, G., Husbands, D. R., and Jolly, R. D.,** Ceroid lipofuscinosis in sheep. II. The major component of the lipopigment in liver, kidney, pancreas, and brain is low molecular weight protein, *J. Biol. Chem.,* 261, 1773, 1986.
9. **Armstrong, D. and Koppang, N.,** Ceroid lipofuscinoses: a model for ageing, in *Age Pigments,* Sohal, R. S., Ed., Elsevier/North-Holland, Amsterdam, 1981, 355.

10. **Gutteridge, J. M. C., Rowley, D. A., Halliwell, B., and Westermarck, T.,** Increased non-protein-bound iron and decreased protection against superoxide-radical damage in cerebrospinal fluid from patients with neuronal ceroid lipofuscinoses, *Lancet,* 2, 459, 1982.
11. **Gutteridge, J. M. C., Westermarck, T., and Santavuori, P.,** Iron and oxygen radicals in tissue damage: implications for the neuronal ceroid lipofuscinoses, *Acta Neurol. Scand.,* 68, 365, 1983.
12. **Halliwell, B. and Gutteridge, J. M. C.,** Lipid peroxidation, oxygen radical cell damage, and antioxidant therapy, *Lancet,* 1, 1396, 1984.
13. **Prohaska, J. R.,** Neurochemical aspects of selenium, in *Neurobiology of the Trace Elements,* Vol. 1, Dreosti, I. E. and Smith, R. M., Eds., Humana Press, Clifton N.J., 1983, 245.
14. **Marklund, S., Santavuori, P., and Westermarck, T.,** Superoxide dismutase, catalase and glutathione peroxidase in infantile, late infantile and juvenile ceroid lipofuscinosis, *Clin. Chim. Acta,* 145, 267, 1985.
15. **Westermarck, T. and Sandholm, M.,** Decreased erythrocyte glutathione peroxidase activity in neuronal ceroid lipofuscinosis (NCL) — corrected with selenium supplementation, *Acta Pharmacol. Toxicol.,* 40, 70, 1977.
16. **Jensen, C. E., Shukla, V. K. S., Gissel-Nielsen, G., and Clausen, J.,** Biochemical abnormalities in Batten's syndrome, *Scand. J. Clin. Lab. Invest.,* 38, 309, 1978.
17. **Germain, G. S. and Arneson, R. M.,** Selenium induced glutathione peroxidase activity in mouse neuroblastoma cells, *Biochem. Biophys. Res. Commun.,* 79, 119, 1977.
18. **Prohaska, J. R. and Ganther, H. E.,** Selenium and glutathione peroxidase in developing rat brain, *J. Neurochem.,* 27, 1379, 1976.
19. **Epp, O., Ladenstein, R., and Wendel, A.,** The refined structure of the selenoenzyme glutathione peroxidase at 0.2 nm resolution, *Eur. J. Biochem.,* 133, 51, 1983.
20. **Oldendorf, W. H., Sisson, W. B., Mehta, A. C., and Treciokas, L.,** Brain uptake of selenomethionine Se 75. I. Effects of elevated blood levels of methionine and phenylalanine, *Arch. Neurol.,* 24, 423, 1971.
21. **Sunde, R. A. and Hoekstra, W. G.,** Structure, synthesis and function of glutathione peroxidase, *Nutr. Rev.,* 38, 265, 1980.
22. **Burton, G. W., Cheeseman, K. H., Doba, T., Ingold, K. U., and Slater, T. F.,** Vitamin E as an antioxidant in vitro and in vivo, in *Biology of Vitamin E,* Ciba Foundation Symp. 101, Pitman, London, 1983, 4.
23. **Diplock, A. T.,** The role of vitamin E in biological membranes, in *Biology of Vitamin E,* Ciba Foundation Symp. 101, Pitman, London, 1983, 45.
24. **Muller, D. P. R., Lloyd, J. K., and Wolff, O. H.,** Vitamin E and neurological function: abetalipoproteinemia and other disorders of fat absorption, in *Biology of Vitamin E,* Ciba Foundation Symp. 101, Pitman, London, 1983, 106.
25. **Hayes, K. C.,** Retinal degeneration in monkeys induced by deficiency of vitamin E or A, *Invest. Ophthalmol.,* 13, 499, 1974.
26. **Westermarck, T.,** Selenium content of tissues in Finnish infants and adults with various diseases, and studies on the effects of selenium supplementation in neuronal ceroid lipofuscinosis patients, *Acta Pharmacol. Toxicol.,* 41, 121, 1977.
27. **Würtz-Jørgensen, A.,** *Spielmeyer-Vogt's Syndrome, 2nd symposium,* Dianalund, 1975, 30.
28. **Chow, C. K. and Tappel, A. L.,** Effect of dietary vitamin E on the activities of the glutathione peroxidase system in rat tissues, *J. Nutr.,* 103, 618, 1973.
29. **Yang, N. Y. J., McDonald, I. B., and Desai, I. D.,** Vitamin E supplementation and glutathione peroxidase activity, *Proc. Soc. Exp. Biol. Med.,* 151, 770, 1976.
30. **Machlin, L. J. and Gabriel, E.,** Kinetics of tissue α-tocopherol uptake and depletion following administration of high levels of vitamin E, in *Ann. N.Y. Acad. Sci.,* p. 48, 1982.
31. **Levander, O. A. and Morris, V. C.,** Interactions of methionine, vitamin E and antioxidants in selenium toxicity in the rat, *J. Nutr.,* 100, 1111, 1970.
32. **Griffith, O. W., Bridges, R. J., and Meister, A.,** Evidence that the gammaglutamyl cycle functions in vivo using intracellular glutathione: effects of amino acids and selective inhibition of enzymes, *Proc. Natl. Acad. Sci. U.S.A.,* 75, 5405, 1978.
33. **Swenseid, M. E. and Wang, M.,** Sulphur requirements in humans, in *Sulphur in Nutrition, Symposium,* Muth, O. H. and Oldfield, J. E., Eds., AVI Publishing, Westport, Conn., 1970, 209.
34. **Mudd, S. H.,** Diseases of sulphur metabolism: implications for the methionine-homocysteine cycle and vitamin responsiveness, in *Sulphur in Biology,* Elliot, K. and Whelan, J., Eds., Excerpta Medica, Amsterdam, 1980, 239.
35. **Tallgren, L. G.,** Inorganic Sulphates in Relation to the Serum Thyroxine Level and in Renal Failure, *Dissertation, University of Helsinki,* Helsinki, 1980, 18.
36. **Brady, P. S., Brady, L. J., Parsons, M. J., Ullrey, D. E., and Miller, E. R.,** Effects of riboflavin deficiency on growth and glutathione peroxidase activity system enzymes in the baby pig, *J. Nutr.,* 109, 1615, 1979.

37. **Saito, Y., Ohidi, N., and Yagi, K.,** Protective effect of riboflavin on suppression of growth caused by oxidized oil, *J. Nutr. Sci. Vitaminol.,* 27, 17, 1981.
38. **Komindr, S. and Nichols, G. E.,** Clinical significance of riboflavin deficiency, in *Nutritional Elements and Clinical Biochemistry,* Brewster, M. A. and Naito, H. K., Eds., Plenum Press, New York, 1980, 15.
39. **Westermarck, T.,** Antioxidant Therapy with Special Reference to Selenium, Rep. Ser. in Medical Physics, No. 4, University of Helsinki, Helsinki, 1982, 69.
40. **Westermarck, T. and Santavuori, P.,** Principles of antioxidant therapy in neuronal ceroid lipofuscinosis, *Med. Biol.,* 62, 148, 1984.
41. **Glatzle, D., Vuillenmier, J. P., Weber, F., and Deckeck, K.,** Glutathione reductase test with whole blood, a convenient procedure for the assessment of the riboflavin status in humans, *Experientia,* 30, 665, 1974.
42. **Raica, N. and Sauberlich, H. E.,** Blood cell transaminase activity in human vitamin B_6 deficiency, *Am. J. Clin. Nutr.,* 15, 67, 1964.
43. **Bayomi, R. A. and Rosalki, S. B.,** Evaluation of methods of coenzyme activation of erythrocyte enzymes for detection of deficiency of vitamins B_1, B_2 and B_6, *Clin. Chem.,* 22, 327, 1976.
44. **Vuonnala, U.,** Vitamin B_2 and B_6 treatment in Spielmeyer-Sjögren patients, in *Student Seminars in Children's Hospital,* University of Helsinki, Helsinki, 1984.
45. **Sun, G. Y. and Sun, A. Y.,** Dietary antioxidants, membrane lipids, and aging, in *Nutritional Intervention in the Aging Process,* Armbrecht, H. J., Prendergast, J. M., and Coe, R. M., Eds., Springer-Verlag, New York, 1984, 139.
46. **Gutteridge, J. M. C.,** Age pigments and free radicals: fluorescent lipid complexes formed by iron- and copper-containing proteins, *Biochim. Biophys. Acta,* 834, 144, 1985.
47. **Zeman, W.,** The dissection of a degenerative disease, in *4th Round Table Conf. on Batten's Disease,* Zeman, W. and Rider, J. A., Eds., American Elsevier, New York, 1975, 124.
48. **Johansson, E., Lindh, U., Alanen, T., Westermarck, T., Heiskala, H., Santavuori, P., and Elovaara, I.,** Elemental profiles of blood cells in certain neurological disorders, *Med. Biol.,* 62, 139, 1984.
49. **Heiskala, H., Westermarck, T., Alanen, T., and Santavuori, P.,** unpublished data, 1985.
50. **Westermarck, T., Gutteridge, J. M. C., Heiskala, H., Santavuori, P., Rahola, T., and Suomela, M.,** Involvement of iron in the pathogenesis of neuronal ceroid lipofuscinosis, *Life Chem. Rep. Suppl. Ser.,* Suppl. 2, 325, 1984.
51. **Gutteridge, J. M. C., Quinlain, G. J., Clark, I., and Halliwell, B.,** Aluminum salts accelerate peroxidation of membrane lipids stimulated by iron salts, *Biochim. Biophys. Acta,* 835, 441, 1985.
52. **Heiskala, H. and Santavuori, P.,** unpublished data, 1985.
53. **Santavuori, P., Westermarck, T., Rapola, J., Pohja, P., Moren, R., Lappi, M., and Vuonnala, U.,** Antioxidant treatment in Spielmeyer-Sjögren's disease, *Acta Neurol. Scand.,* 71, 136, 1985.
54. **Santavuori, P. and Moren, R.,** Experience of antioxidants in neuronal ceroid lipofuscinosis of Spielmeyer-Sjögren type, *Neuropaediatrie,* 8, 333, 1977.
55. **Norppa, H., Westermarck, T., Laasonen, M., Knuutila, L., and Knuutila, S.,** Chromosomal effects of sodium selenite in vivo. I. Aberrations and sister chromatid exchanges in human lymphocytes, *Hereditas,* 93, 93, 1980.
54. **Vladimirov, Y. U., Olenev, V. I., Suslova, T. B., and Cheremisina, Z. P.,** Lipid peroxidation in mitochondrial membrane, *Adv. Lipid Res.,* 17, 173, 1980.
57. **Santavuori, P., Rapola, J., Sainio, K., and Raitta, C.,** A variant of Jansky-Bielschowsky disease, *Neuropediatrics,* 13, 135, 1982.
58. **Lake, B. D. and Canavagh, P. C.,** Early juvenile Batten's disease — a recognizable subgroup distinct from other forms of Batten's disease, *J. Neurol. Sci.,* 36, 265, 1978.
59. **Libert, J., Martin, J. J., and Ceuterick, C.,** Protracted and atypical forms of ceroid-lipofuscinosis, in *Ceroid Lipofuscinosis (Batten's Disease),* Armstrong, D., Koppang, N., and Rider, J. A., Eds., Elsevier, Amsterdam, 1982, 45.
60. **Siakotos, A. N.,** personal communication, 1985.

CONSEQUENCES OF IMPAIRED ANTIOXIDANT PROTECTION OF THE RETINA

Martin L. Katz and W. Gerald Robison, Jr.

INTRODUCTION

The biological roles of chemical reactions involving free radical intermediates are currently subjects of extensive research. Particular interest has been focused on oxygen-based radicals and their potential roles in human pathologies. Oxygen-based radicals, generated as metabolic intermediates, are capable of initiating autoxidation chain reactions which can damage molecular components of cells and tissues.[1] In this chapter, evidence that autoxidation may be involved in certain pathologies of the mammalian retina will be reviewed.

The retina consists of a thin sheet of neural tissue lining the posterior two thirds of the eye. The nuclei of retinal cells are located in three distinct layers, which are separated from one another by plexiform or synaptic zones (Figure 1). Photoreceptor cells, which mediate the transduction of light into neural impulses, make up the outermost cell layer of the retina. Signals generated by the photoreceptors are transmitted to neurons with nuclei in the inner nuclear layer and then to ganglion cells, whose nuclei lie along the inner edge of the retina. The axons of the ganglion cells form the optic nerve, which passes out the back of the eye and ends for the most part at the lateral geniculate in the brain. In humans and most other mammals, two distinct capillary networks supply blood to the retina: one lying within the neural retina and the other, more extensive system, forming the choriocapillaris just exterior to the retina. The photoreceptor cells appear to depend mainly on the choriocapillaris for nutrition and waste removal, while cells with nuclei in the inner layers of the retina receive similar support from the retinal capillaries.[2] Between the photoreceptors and the choriocapillaris is a single layer of epithelial cells, the retinal pigment epithelium (RPE) (Figure 1), which forms a blood-retinal barrier. Exchange of most nutrients and metabolites between the choriocapillaris and the photoreceptors is mediated by the RPE. The RPE also plays a crucial role in the turnover of the photosensitive membranes of the photoreceptor cells. These membranes consist of flattened sacs or disks, in which are embedded the pigment molecules responsible for visible light absorbtion. Each photoreceptor cell has many disks, stacked in close apposition to one another in a specialized region called the outer segment (Figure 2). New disks are synthesized continuously and are added to the bases of the outer segments, while packets of the oldest disks are shed periodically from the outer segment tips. The shed outer segment material is phagocytized and digested by the RPE. Impairment of phagocytosis by the RPE has been found to lead to photoreceptor cell degeneration,[3-5] a fact which illustrates the importance of RPE function to photoreceptor cell maintenance. The RPE plays a number of other roles, including vitamin A transport and storage, which are crucial in maintaining photoreceptor cell integrity and function.

Most research on the potential roles of oxygen-based free radicals in retinal pathologies have been focused on the photoreceptor cells and RPE. Both of these cell types are subject to a variety of pathological alterations which can result in serious visual impairment. Investigations into the effects of impaired antioxidant protection on the photoreceptors and RPE will be the major focus of this chapter.

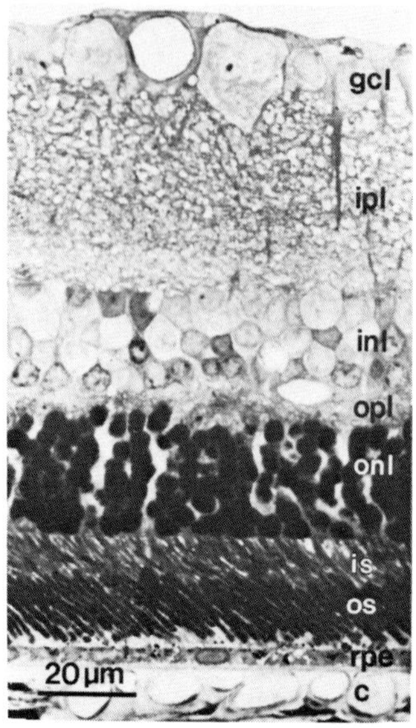

FIGURE 1. Light micrograph of a rat retina shown in cross section. The retina is organized into a number of distinct layers which consist of the ganglion cell layer (gcl), inner plexiform layer (ipl), inner nuclear layer (inl), outer plexiform layer (opl), outer nuclear layer (onl), and photoreceptor inner segments (is) and outer segments (os). The retinal pigment epithelium (rpe) separates the neural retina from the choriocapillaris (c).

FACTORS FAVORING POTENTIAL AUTOXIDATIVE DAMAGE TO THE RETINA

Oxygen Consumption by the Retina

One of the chief characteristics of the retina which suggests that this tissue may be particularly susceptible to autoxidative damage is its apparent high oxygen consumption. Both in vivo and in vitro measurements have indicated that oxygen utilization by the retina is substantial.[6-8] A variety of experimental evidence indicates that there is a considerable enhancement in retinal oxygen consumption in dark-adapted eyes relative to that which occurs in the light.[7-9] This increased oxygen consumption is probably due to the increased metabolic activity of the photoreceptors in the dark-adapted state.[9] The close coupling of retinal oxygen consumption and metabolic activity of the photoreceptor cells suggests that a large proportion of retinal oxidative metabolism occurs in the photoreceptors. Further evidence that these cells are responsible for a major fraction of retinal oxygen utilization was obtained by Adler and colleagues,[10] who measured oxygen partial pressures in the various layers of the cat retina *in situ*. It was found that a substantial drop in oxygen partial pressure occurred across the RPE and photoreceptor cells, suggesting that much of the oxygen derived from the choroidal circulation is consumed by these cell types. Cohen and Noell[6] also suggested that the photoreceptor cells are responsible for a major fraction of retinal oxidative metabolism, based on the finding that such metabolism increases dramatically as the photoreceptor inner and outer segments develop. The photoreceptor inner segments are mitochondrion-rich regions of these cells that lie just interior to the outer segments (Figure 2) and are where the energy necessary for critical steps in visual transduction is apparently generated.

Oxidative phosphorylation in mitochondria involves a number of one-electron transfers

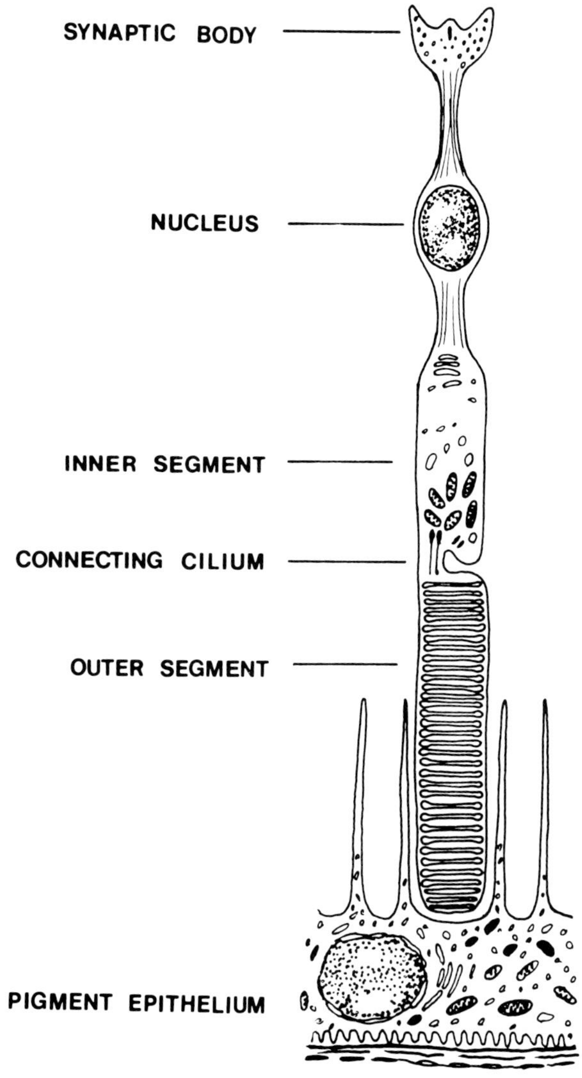

FIGURE 2. Diagram illustrating the structure of a rod photoreceptor cell. Photosensitive pigment molecules are located in the disk membranes of the outer segment.

in which molecular oxygen is ultimately reduced to water (Figure 3). The complete reduction of oxygen requires four electrons. Partially reduced forms of oxygen generated during mitochondrial electron transport include the two highly reactive radical species, superoxide anion (O_2^-) and the hydroxyl radical (OH·). Hydrogen peroxide (H_2O_2) is also generated as an intermediate during the metabolic reduction of oxygen. Hydrogen peroxide can in turn be broken down in the presence of reducing agents and trace amounts of metal ions to produce hydroxyl radicals.[1,11] Normally, the sequential steps involved in mitochondrial electron transport are tightly coupled, so that partially reduced forms of oxygen are not released. However, a certain amount of leakage of these intermediates apparently does occur,[12] resulting in the potential for damage to cellular macromolecules through reactions with the free radical species of partially reduced oxygen. The high metabolic rate of the photoreceptor cells, mediated by their abundant mitochondria, suggests that such oxygen-based radicals could be produced at a relatively high rate in these cells.

$$O_2 \xrightarrow{e^-} O_2^{\cdot -} \xrightarrow{e^- + 2H^+} H_2O_2 \xrightarrow{e^- + H^+} OH^\cdot \xrightarrow{e^- + H^+} H_2O$$
$$(H_2O_2 \rightarrow H_2O \text{ branch})$$

FIGURE 3. Metabolic reduction of molecular oxygen to water occurs via a series of one-electron transfers in which the intermediates superoxide ($O_2^{\cdot -}$), hydrogen peroxide (H_2O_2), and hydroxyl radical ($OH\cdot$) are produced.

PUFA Autoxidation

Initiation:

$$RH + OH^\cdot \rightarrow R^\cdot + H_2O$$
$$RH + O_2^{\cdot -} + H^+ \rightarrow R^\cdot + H_2O_2$$
$$RH + X^\cdot \rightarrow R^\cdot + X^\cdot$$

Propagation:

$$R^\cdot + O_2 \rightarrow ROO^\cdot$$
$$ROO^\cdot + RH \rightarrow ROOH + R^\cdot$$

Chain Branching:

$$ROOH + M^{n+} \rightarrow RO^\cdot + OH^- + M$$

Chain Termination:

$$R^\cdot + ROO^\cdot \rightarrow ROOR$$
$$R^\cdot + RO^\cdot \rightarrow ROR$$
$$R^\cdot + R^\cdot \rightarrow RR$$
$$X^\cdot + AH \rightarrow XH + A^\cdot$$

FIGURE 4. Autoxidation of PUFAs is initiated when a PUFA molecule reacts with a free radical to generate a PUFA radical. PUFA radicals can generate additional radicals via propagation reactions, and hydroperoxide intermediates can break down in the presence of metal ion catalysts to generate additional radicals. Free radical chain reactions can be terminated when two radical species react to form a nonradical product or when an antioxidant (AH) reduces the free radical (X·) to generate a nonradical species and a stable antioxidant radical (A·).

Lipid Composition of Photoreceptor Outer Segment Membranes

Polyunsaturated fatty acids, particularly those of membrane phospholipids, appear to be primary targets for damage by oxygen-based radicals. This is because reaction of a single free radical species with a polyunsaturated fatty acid (PUFA) moiety can trigger a chain reaction in which many PUFA moieties become involved (Figure 4), resulting in a process called autoxidation. On the basis of their lipid composition, the disk membranes of the photoreceptor outer segments would be expected to be particularly susceptible to autoxidative damage. In general, 50 to 60% of rod outer segment disk membrane fatty acids are polyunsaturates in mammalian retinas, with approximately 80 mol % of the total polyunsaturates consisting of docosahexaenoic acid (22:6w3),[13] a 22-carbon fatty acid with six double bonds. The more double bonds a fatty acid has, the more susceptible it is to autoxidative degradation.[14] Because of their high PUFA content and their close proximity to the actively metabolizing mitochondria of the inner segments, the disk membranes of the rod outer segments are quite likely to undergo autoxidative changes under normal conditions.

Vitamin A and Autoxidation

Derivatives of the retinol form of vitamin A play central roles in the process of visual transduction. Retinol obtained from the diet is delivered by a plasma carrier protein (retinol binding protein, RBP) to the tissues in which it is utilized.[15] Vitamin A transport from the plasma to the photoreceptors is mediated by cells of the RPE, which have receptors for plasma RBP along their basal and lateral plasma membranes,[16] In the photoreceptor cells, the 11-*cis*-retinaldehyde form of vitamin A is conjugated via a Schiff-base linkage to specific disk membrane proteins (the opsins) to form the visual pigments. Upon exposure to light, the visual pigments are bleached, which involves isomerization of the 11-*cis*-retinaldehyde chromophoric group to the all-*trans* configuration and its release from the opsin protein. The retinaldehyde is reduced to the retinol form and stored in the RPE as fatty acid esters until it is utilized for visual pigment regeneration. Particularly in the light, but apparently even under conditions of dark adaptation,[17] there is a rapid flux of retinol derivatives through the various compartments in the photoreceptor-RPE complex.

Vitamin A is an unsaturated lipid with a long series of conjugated double bonds. This feature of its chemical structure makes vitamin A particularly susceptible to autoxidative degradation. In the photoreceptor disk membranes, vitamin A compounds are in close proximity to the PUFAs of outer segment membrane phospholipids. Reactive intermediates generated during PUFA autoxidation in these membranes are likely to attack and involve vitamin A in the autoxidation process. Evidence for the susceptibility of vitamin A to autoxidative destruction in vivo has been provided by a variety of experiments which have shown that tissue levels of vitamin A are reduced in animals fed diets deficient in antioxidant nutrients, such as vitamin E.[18-20]

Not only can vitamin A participate as a co-oxidant with PUFAs in lipid autoxidation reactions, but under certain conditions some forms of vitamin A may also be involved in the initiation step of autoxidation. Delmelle[21] has shown that in an in vitro system, retinol can act as a photosensitizer for the production of singlet oxygen, an excited-state form of molecular oxygen capable of initiating autoxidation reactions. Light-activated production of singlet oxygen mediated by retinol is quite possible in the retina, where photon fluxes and retinol concentrations can be quite high. Evidence that vitamin A in some form may mediate light-induced damage to the retina was presented by Noell and Albrecht,[22] who found that light damage to the retinas of albino rats could be partially prevented by feeding the animals vitamin A-deficient diets. Further experimentation will be necessary, however, to determine whether the damaging effects of the interaction between light and vitamin A in the retina occur via an autoxidative mechanism.

Light

Evidence that light exposure may promote autoxidative damage to the retina has been obtained by Wiegand and colleagues,[23] who reported that exposure of albino rats to 100 to 125 fc of fluorescent light for 24 to 72 hr resulted in an elevation of lipid autoxidation products in the photoreceptor outer segments. Organisciak and colleagues[24] found that lipid autoxidation products were present in the retina in higher concentrations in rats reared in cyclic light than in dark-reared rats, but that 1 hr of intense light exposure did not result in significant elevation in lipid autoxidation products in the retina. These results indicate that light apparently promotes lipid autoxidation in the retina only under certain circumstances, suggesting that the mechanism of the light effect is probably quite complex.

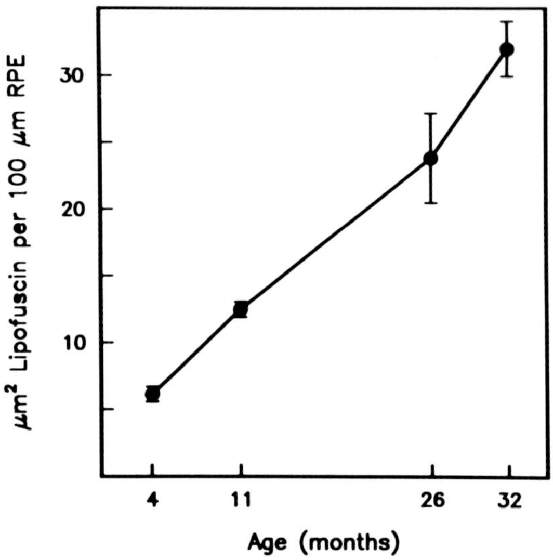

FIGURE 5. Relationship between age and RPE lipofuscin content in pigmented rats. There is an almost linear increase in RPE lipofuscin content during aging. (Data are from Katz, M. L. and Robison, W. G., *Exp. Eye Res.*, 38, 137, 1984.)

Recently, Organisciak and colleagues[25] reported that ascorbate exerted a significant protective effect against light damage to the retinas of albino rats. There is some evidence that under certain physiological conditions, ascorbate can act as an antioxidant in vivo.[26] Thus, it is possible that the protective effect of ascorbate against retinal light damage is due to the antioxidant activity of this compound.

While it is possible that light may promote autoxidation in the retina, some experimental evidence has been obtained that is inconsistent with the hypothesis that light damage to the retina acts via an autoxidative mechanism. If light damage occurs by promoting autoxidation, one would expect that antioxidant nutrient deficiency would increase the sensitivity of the retina to light damage. Stone and colleagues[27] subjected rats which had been maintained on an antioxidant-deficient diet to acute light stress and found that damage to the retina was no more severe than in rats which had received adequate levels of antioxidants in their diets. Likewise, Sykes and co-workers[28] found that vitamin E deficiency did not increase the sensitivity of albino rats to retinal light damage by high-intensity cyclic light. These findings are inconsistent with the hypothesis that damage to the retina caused by light exposure is the result of increased autoxidation in this tissue. Further experimentation will be necessary to establish the basis for the apparent association between light exposure and autoxidation in the retina.

EVIDENCE SUGGESTING ROLES FOR AUTOXIDATION IN AGING AND RETINAL PATHOLOGIES

Antioxidant Deficiencies and Lipofuscin

One of the most prominent age-related alterations known to occur in the mammalian retina is the progressive accumulation of lipofuscin, an autofluorescent pigment, within cells of the RPE (Figure 5).[29-31] Impairment of the antioxidant defense mechanisms of the retina results in a dramatic acceleration of RPE lipofuscin deposition (Figure 6),[18,19,32-34] suggesting that lipofuscin inclusions in the RPE may represent products of autoxidative damage. Support for this hypothesis is provided by the demonstration that the fluorescence emission spectra

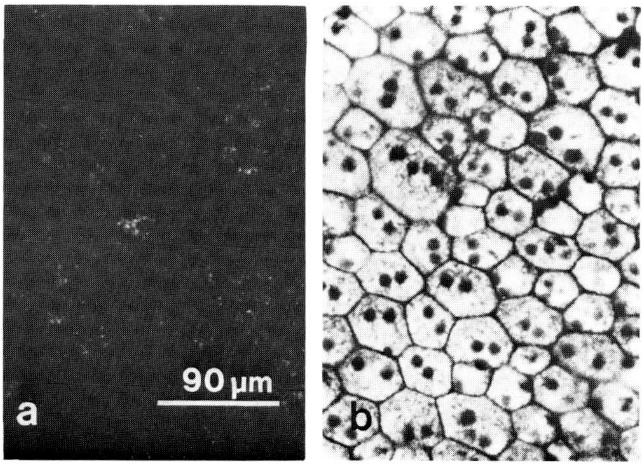

FIGURE 6. Fluorescence micrographs of whole mounts of unfixed RPE from an albino rat fed a vitamin E-adequate diet (a) and an animal fed a vitamin E-deficient diet (b). Vitamin E deficiency greatly accelerates RPE lipofuscin deposition. (Adapted from Katz, M. L., Robison, W. G., Jr., Herrmann, R. K., Groome, A. B., and Bieri, J. G., *Mech. Ageing Dev.*, 25, 149, 1984.)

of the age-related pigment and the pigment generated as a result of antioxidant deficiency are essentially identical.[35]

Likely precursors for RPE lipofuscin fluorophores are the polyunsaturated lipids of the photoreceptor outer segments. These lipids, and undoubtedly their autoxidation products, are taken up by the RPE during the normal process of outer segment membrane turnover.[36] Feeding animals diets deficient in antioxidant nutrients results in disruption of outer segment disk membrane morphology,[18,33] suggesting that autoxidative damage is occurring within these membranes. Deposition of the products of outer segment autoxidation in the RPE apparently underlies the accelerated rate of lipofuscin accumulation in animals fed antioxidant-deficient diets. In animals fed normal diets, autoxidative damage to the outer segment disk membranes apparently occurs at too slow a rate to become morphologically evident; however, over a long period of time enough autoxidative damage may occur to account for the age-related accumulation of lipofuscin observed in the RPE.

Recently definitive evidence has been obtained that photoreceptor outer segment membrane components can be converted directly into lipofuscin-like fluorophores.[37] In Royal College of Surgeons (RCS) rats with hereditary retinal dystrophy, the photoreceptor cells develop normally, but then gradually degenerate and disappear.[5] The degeneration apparently results from an inability of the RPE to ingest outer segment disk membranes.[3-5] As a consequence of this defect, photoreceptor outer segment material accumulates in the retina just interior to the RPE. This material develops autofluorescent characteristics that are quite similar to those of RPE lipofuscin (Figure 7),[37] suggesting that age-pigment in the RPE may be formed from components of the outer segments. While it has not been demonstrated that the outer segment material in the RCS rats develops its autofluorescent properties as a result of autoxidation, Zigler and Hess[38] have reported that the concentration of polyunsaturated fatty acid autoxidation products in the eyes of these rats reaches a peak when the amount of degenerating outer segment material is maximal. This finding suggests that the development of autofluorescence in the outer segment remnants may be related to the autoxidation process.

Vitamin A and Lipofuscin

Among the retinal lipids likely to be converted into RPE lipofuscin fluorophores via

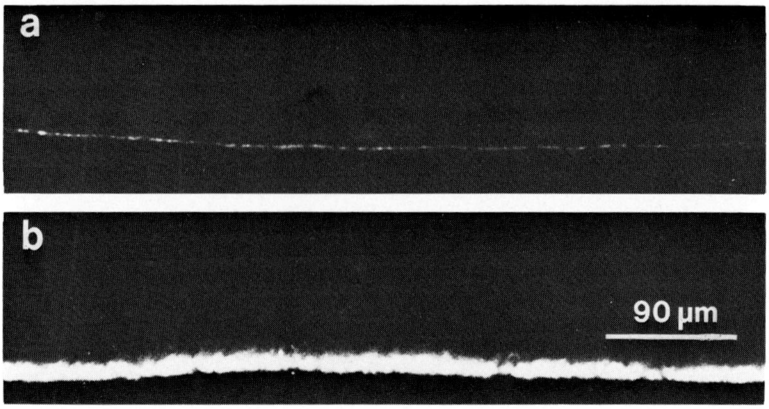

FIGURE 7. Fluorescence micrographs of the retina from a 66-day-old RCS rat with a normal retina (a) and from an RCS rat of the same age with a dystrophic retina (b). The material from the degenerating photoreceptor cells in the dystrophic rat retina is strongly fluorescent. The normal retina has only a small amount of autofluorescence in the RPE.

autoxidation are derivatives of retinol (a form of vitamin A), which are relatively abundant in the retina and RPE. In addition to being potential direct precursors of lipofuscin, retinol derivatives may also promote the autoxidation of other retinal components, as indicated earlier. In order to evaluate the possible roles of vitamin A compounds in RPE lipofuscin formation, the relationship between dietary retinol intake and RPE lipofuscin deposition has been studied.

Animals fed diets deficient in retinol precursors have been found to have substantially less RPE lipofuscin than do animals of the same age fed diets containing adequate levels of retinol.[19,39,40] This finding suggests that retinol derivatives play an important role in lipofuscin formation. Excessive dietary intake of retinol, on the other hand, failed to enhance RPE lipofuscin deposition over that seen in animals fed normal amounts of this vitamin.[39] The latter finding is probably related to the failure of high dietary retinol intake to produce elevated retinol levels in retinal tissues.[20]

Evidence that vitamin A compounds in the eye may be converted directly into lipofuscin fluorophores is provided by recent experiments in which RCS rats with hereditary photoreceptor degeneration were utilized.[41]. As indicated earlier, photoreceptor outer segment material accumulates at the apical surface of the RPE during the degeneration process. This material develops autofluorescent properties quite similar to those of RPE lipofuscin. When dystrophic rats were placed on a diet lacking retinol precursors, the amount of autofluorescence which developed in this outer segment material was substantially reduced, relative to that present in rats fed diets containing adequate amounts of retinol (Figure 8). This finding supports the conclusion that retinol derivatives play central roles in RPE lipofuscin fluorophore formation.

The question of whether the contribution of retinol to RPE lipofuscin deposition occurred via an autoxidative mechanism was addressed by determining whether deficiency in the lipid antioxidant vitamin E enhanced this effect. In animals fed vitamin E-adequate diets, retinol deficiency reduced RPE lipofuscin deposition by about 50%.[39] In contrast, retinol deficiency resulted in a reduction in RPE lipofuscin content of only about 22% in rats fed a vitamin E-deficient diet.[39] One possible interpretation of these findings is that retinol does not promote RPE lipofuscin deposition via an autoxidative mechanism. It was found, however, that vitamin E deficiency slowed the rate of vitamin A depletion from the retina in rats fed diets lacking retinol precursors. Thus, in vitamin E-deficient animals, more vitamin A compounds were available in the retina for enhancing RPE lipofuscin formation than were available in

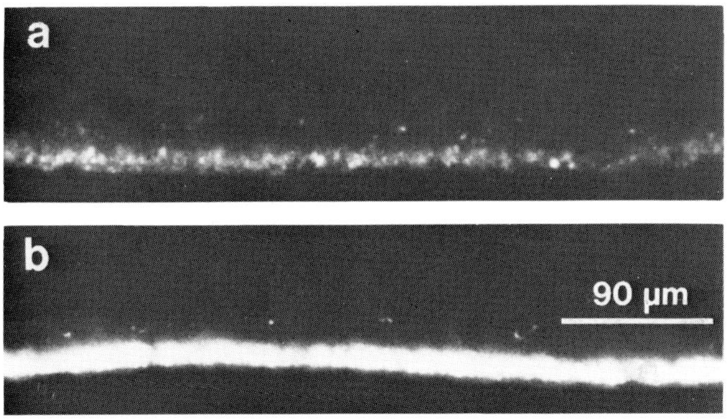

FIGURE 8. Fluorescence micrographs of retinas from 66-day-old dystrophic rats which had been fed either a retinol-deficient (a) or a retinol-containing diet (b) since weaning. The fluorescence intensity of the degenerating photoreceptor outer segment material is strongly dependent on dietary retinol content.

animals receiving dietary vitamin E. Because of the complexity of vitamin E and vitamin A interactions in vivo, no firm conclusion can be drawn from these experiments regarding the mechanism of the vitamin A effect on RPE lipofuscin accumulation.

Influence of Photoreceptor Cells on RPE Lipofuscin

The photoreceptor cells contain the most likely substrates for RPE lipofuscin formation: polyunsaturated fatty acids and vitamin A compounds. In addition, the photoreceptor cells appear to be responsible for a substantial proportion of retinal oxygen consumption. In light of these facts, the photoreceptor cells are likely to play significant roles in RPE lipofuscin formation. By comparing rates of RPE lipofuscin deposition in normal retinas and in retinas lacking photoreceptor cells, the contribution of the photoreceptor cells to accumulation of lipofuscin in the RPE has been evaluated.[37,42]

RCS rats with hereditary photoreceptor cell degeneration lack photoreceptor cells for most of their lives, while congenic control animals have a normal complement of these cells. Lack of photoreceptor cells in the RCS rat has been found to result in a substantial reduction in the age-related accumulation of lipofuscin in the RPE. At 16 months of age, rats with retinal dystrophy have about 45% less RPE lipofuscin than do congenic controls.[37] These findings indicate that the photoreceptor cells make major contributions to RPE lipofuscin deposition.

Further support for this conclusion comes from experiments in which RPE lipofuscin deposition has been compared in normal albino rats and in rats of the same strain whose photoreceptor cells have been destroyed at an early age by light damage. Rats lacking photoreceptor cells for most of their lives due to light stress demonstrated substantially less age-related RPE lipofuscin accumulation than did control animals with normal retinas.[40,42] Again, this finding suggests that the photoreceptor cells participate in the process that leads to RPE lipofuscin formation.

Autoxidation and Human Retinal Pathologies

One of the most prevalent causes of serious visual impairment in developed countries is senile macular degeneration (SMD),[43] an age-related pathology involving the central retina. While the underlying causes of this disease are unknown, primary age-related changes in the RPE are likely to play a role in the etiology of SMD.[44] The most well characterized of the age-related changes of the RPE is the progressive accumulation of lipofuscin in these cells.[29-31,34] While lipofuscin deposition per se has not been shown to be detrimental, the autoxidative reactions that apparently lead to its formation are quite capable of damaging important cellular molecules and organelles. Thus, the amount of lipofuscin within a cell may be indicative of the cumulative autoxidative damage to which it has been subjected. There is some evidence that autoxidation and SMD may be associated. Weiter and colleagues[45] have reported that human subjects with SMD have lower than normal levels of the selenium-dependent enzyme glutathione peroxidase in their plasmas. This enzyme plays an important antioxidant role, so the fact that it is depressed in patients with SMD suggests that autoxidation may be involved in the development of this disease. Further research will be necessary, however, to establish definitively a relationship between impaired antioxidant protection and SMD.

There is more extensive evidence linking another retinal disease, retinopathy of prematurity (ROP), and autoxidative damage.[1] As its name suggests, ROP is a disorder affecting premature infants. It involves abnormal development of the retinal vasculature, apparently as a result of exposure of infants to a higher oxygen tension than that present *in utero* at a time prior to completion of retinal vascularization. Protection against the development of ROP by vitamin E, as demonstrated in a number of clinical trials,[46-48] suggests that oxygen promotes ROP by inducing autoxidative damage. Apparently, normal protective mechanisms against autoxidative damage are not fully developed in the premature infant,[49-51] so that damage can occur to an extent great enough to affect normal developmental processes.

A number of other retinal pathologies may result from impaired antioxidant protection. Kolb and Gouras[52] reported that retinitis pigmentosa, a hereditary retinal degenerative disease, involved excessive lipofuscin deposition in the RPE, suggesting that abnormal amounts of autoxidative damage were occurring. Another hereditary disorder, ceroid lipofuscinosis or Batten's disease, involves deposition of material with fluorescence properties similar to lipofuscin in a variety of tissues, including the retina.[53] Not only the retina, but also other tissues in which this pigment is deposited undergo degenerative changes. A variety of evidence suggests that impaired antioxidant protection may underlie this disease,[54] but this has not yet been conclusively demonstrated. Perhaps the most definitive association between impaired antioxidant protection and human retinal pathology has been established for abetalipoproteinemia, a malabsorption syndrome, in which fat-soluble vitamins are not taken up effectively from the intestine. In patients with this defect, massive oral doses of vitamin E were found to be effective in preventing the chronic retinal degeneration that accompanies this disease.[55,56] This finding suggests a clear role for autoxidative damage in the retinal pathology that accompanies abetalipoproteinemia.

When considered as a whole, the human clinical evidence suggests that the retina is particularly sensitive to autoxidative damage. Impairment of normal antioxidant mechanisms, or exposure of the retina to oxidative stress before these mechanisms are fully developed, can result in serious damage to this tissue. A slow steady-state rate of autoxidative damage, despite the existence of optimal antioxidant protective mechanisms, may underlie the development of age-related retinopathies.

SUMMARY AND CONCLUSIONS

Various features of retinal metabolism, biochemical composition, and morphology suggest that this tissue may be quite susceptible to autoxidative damage. The retina, and particularly the photoreceptor cells, are very active in oxidative metabolism, the side products of which probably include oxygen-based radical species. These free radicals can initiate autoxidation chain reactions in the photoreceptor outer segment membranes, which are rich in polyunsaturated lipids, including vitamin A. There is some evidence that under certain conditions light may potentiate autoxidation in the retina by sensitizing singlet oxygen production.

Products of autoxidation generated from photoreceptor outer segment components apparently are deposited in the RPE in the form of autofluorescent pigment granules (lipofuscin). A definitive association between autoxidation and RPE lipofuscin deposition has been established by the finding that antioxidant nutrient deficiency greatly accelerates lipofuscin accumulation in the RPE. Dietary experiments also suggest that vitamin A compounds may be major retinal substrates for RPE lipofuscin formation. Since lipofuscin pigments accumulate progressively in the RPE during the aging process, it is possible that the cumulative effects of autoxidation play a role in retinal senescence.

Not only may autoxidation be involved in age-related retinal pathologies, but a variety of other human retinopathies have been linked to inadequate antioxidant protection as well. These include ROP, retinitis pigmentosa, and the retinopathies associated with Batten's disease and abetalipoproteinemia. In the cases of the latter disease and ROP, antioxidant supplementation has been shown to be effective in retarding the development of the retinal pathologies.

While a great deal remains to be learned about the antioxidant protective mechanisms of the retina, it is clear that antioxidant protection is essential for maintaining the integrity of retinal structure and function. Impairment of these protective mechanisms can have serious detrimental effects on the retina. In light of the fact that RPE lipofuscin deposition appears to be associated with the aging process even under optimal dietary conditions, it is likely that the antioxidant defense systems in the retina are imperfect, allowing autoxidative reactions to occur at a slow rate throughout the life span. The cumulative effects of these reactions may underlie age-related retinopathies and age-related pathological alterations in other tissues as well. Perhaps by gaining a better understanding of how to improve the antioxidant defenses of the retina and other tissues, it may be possible to develop a rational approach to retarding the development of pathologies associated with senescence.

ACKNOWLEDGMENTS

The work by M. L. K. reviewed in this chapter was supported in part by Grant EY 06458 from the National Institutes of Health. The authors would like to acknowledge the expert technical assistance of Anne B. Groome and Christine M. Drea.

REFERENCES

1. **Katz, M. L. and Robison, W. G., Jr.,** Autoxidative damage to the retina: potential role in retinopathy of prematurity, in *Retinopathy of Prematurity: Problem and Challenge,* Flynn, J. T., Ed., Alan R. Liss, New York, 1986, in press.
2. **Brown, K. T.,** The electroretinogram: its components and their origins, in *The Retina: Morphology, Function, and Clinical Characteristics,* Straatsma, B. R., Hall, M. O., Allen, A. A., and Crescitelli, F., Eds., University of California Press, Berkeley, 1969, 319.
3. **Herron, W. L., Riegel, B. W., Myers, O. E., and Rubin, M. L.,** Retinal dystrophy in the rat — a pigment epithelial disease, *Invest. Ophthalmol.,* 8, 595, 1969.
4. **Mullen, R. J. and LaVail, M. M.,** Inherited retinal dystrophy: primary defect in pigment epithelium determined with experimental rat chimeras, *Science,* 192, 799, 1976.
5. **LaVail, M. M.,** The retinal pigment epiethelium in mice and rats with inherited retinal degeneration, in *The Retinal Pigment Epithelium,* Zinn, K. M. and Marmor, M. F., Eds., Harvard University Press, Cambridge, 1979, 357.
6. **Cohen, L. H. and Noell, W. K.,** Relationships between visual function and metabolism, in *Biochemistry of the Retina,* Graymore, C. N., Ed., Academic Press, New York, 1965, 36.
7. **Sickel, W.,** Retinal metabolism in light and dark, in *Physiology of Photoreceptor Organs,* Fuortes, M. G. F., Ed., Springer-Verlag, Berlin, 1972, 667.

8. **Stefansson, E., Wolbarsht, M. L., and Landers, M. B.,** In vivo O_2 consumption in rhesus monkeys in light and dark, *Exp. Eye Res.*, 37, 251, 1983.
9. **Zuckerman, R. and Weiter, J. J.,** Oxygen transport in the bullfrog retina, *Exp. Eye Res.*, 30, 117, 1980.
10. **Adler, V. A., Cringle, S. J., and Constable, I. J.,** The retinal oxygen profile in cats, *Invest. Ophthalmol. Visual Sci.*, 24, 30, 1983.
11. **Pryor, W. A.,** Free radicals in autoxidation and aging, in *Free Radicals in Molecular Biology, Aging, and Disease,* Armstrong, D., Sohal, R. S., Cutler, R. G., and Slater, T. F., Eds., Raven Press, New York, 1984, 13.
12. **Chance, B., Sies, H., and Boveris, A.,** Hydroperoxide metabolism in mammalian organs, *Physiol. Rev.*, 59, 527, 1979.
13. **Fliesler, S. J. and Anderson, R. E.,** Chemistry and metabolism of lipids in the vertebrate retina, *Prog. Lipid Res.*, 22, 79, 1983.
14. **Holman, R. T. and Elmer, O. C.,** The rate of oxidation of unsaturated fatty acids and esters, *J. Am. Oil Chem. Soc.*, 24, 127, 1947.
15. **Ong, D. E.,** Vitamin A-binding proteins, *Nutr. Rev.*, 43, 225, 1985.
16. **Bok, D. and Heller, J.,** Transport of retinol from the blood to the retina: an autoradiographic study of the pigment epithelial cell surface receptor for plasma retinol-binding protein, *Exp. Eye Res.*, 22, 395, 1976.
17. **Defoe, D. M. and Bok, D.,** Rhodopsin chromophore exchange among opsin molecules in the dark, *Invest. Ophthalmol. Visual Sci.*, 24, 1211, 1983.
18. **Robison, W. G., Jr., Kuwabara, T., and Bieri, J. G.,** Vitamin E and the retina: photoreceptor and pigment epithelial changes, *Invest. Ophthalmol. Visual Sci.*, 18, 683, 1979.
19. **Robison, W. G., Jr., Kuwabara, T., and Bieri, J. G.,** Deficiencies in vitamins E and A in the rat: retinal damage and lipofuscin accumulation, *Invest. Ophthalmol. Visual Sci.*, 19, 1030, 1980.
20. **Katz, M. L., Drea, C. M., and Robison, W. G., Jr.,** Dietary vitamins A and E influence retinyl ester content and composition in the retinal pigment epithelium, *Biochim. Biophys. Acta,* 924, 432, 1987.
21. **Delmelle, M.,** An investigation of retinol as a source of singlet oxygen, *Photochem. Photobiol.*, 27, 731, 1978.
22. **Noell, W. K. and Albrecht, R.,** Irreversible effects of visible light on the retina: role of vitamin A, *Science,* 172, 76, 1971.
23. **Wiegand, R. D., Giusto, N. M., Rapp, L. M., and Anderson, R. E.,** Evidence of rod outer segment lipid peroxidation following constant illumination of the rat retina, *Invest. Ophthalmol. Visual Sci.*, 24, 1433, 1983.
24. **Organisciak, D. T., Favreau, P., and Wang, H. M.,** The enzymatic estimation of organic hydroperoxides in the rat retina, *Exp. Eye Res.*, 36, 337, 1983.
25. **Organisciak, D. T., Wang, H. M., Li, Z., and Tso, M. O. M.,** The protective effect of ascorbate in retinal light damage of rats, *Invest. Ophthalmol. Visual Sci.*, 26, 1580, 1985.
26. **Katz, M. L. and Robison, W. G.,** Nutritional influences on autoxidation, lipofuscin accumulation, and aging, in *Free Radicals, Aging, and Degenerative Diseases,* Johnson, J. E., Walford, R., Harman, D., and Miquel, J., Eds., Alan R. Liss, New York, 1987, 111.
27. **Stone, W. L., Katz, M. L., Lurie, M., Marmor, M. F., and Dratz, E. A.,** Effects of vitamin E and selenium deficiency on light damage to the rat retina, *Photochem. Photobiol.*, 29, 725, 1979.
28. **Sykes, S. M., Robison, W. G., Jr., and Bieri, J. G.,** Retinal damage by cyclic light and the effect of vitamin E, in Symposium on Biological Effects and Measurement of Light Sources, Health and Human Services Publ. FDA 81-8156, Rockville, Md., 1981, 86.
29. **Wing, G. L., Blanchard, G. C., and Weiter, J. J.,** The topography and age relationship of lipofuscin in the retinal pigment epithelium, *Invest. Ophthalmol. Visual Sci.*, 17, 601, 1978.
30. **Feeney-Burns, L., Hilderbrand, E. S., and Eldridge, S.,** Aging human RPE: morphometric analysis of macular, equitorial, and peripheral cells, *Invest. Ophthalmol. Visual Sci.*, 25, 195, 1984.
31. **Katz, M. L. and Robison, W. G., Jr.,** Age-related changes in the retinal pigment epithelium of pigmented rats, *Exp. Eye Res.*, 38, 137, 1984.
32. **Katz, M. L., Stone, W. L., and Dratz, E. A.,** Fluorescent pigment accumulation in retinal pigment epithelium of antioxidant-deficient rats, *Invest. Ophthalmol. Visual Sci.*, 17, 1049, 1978.
33. **Katz, M. L., Parker, K. R., Handelman, G. J., and Dratz, E. A.,** Effects of antioxidant nutrient deficiency on the retina and retinal pigment epithelium of albino rats: a light and electron microscopic study, *Exp. Eye Res.*, 34, 339, 1982.
34. **Katz, M. L., Robison, W. G., Jr., and Dratz, E. A.,** Potential role of autoxidation in age changes of the retina and retinal pigment epithelium of the eye, in *Free Radicals in Molecular Biology, Aging, and Disease,* Armstrong, D., Sohal, R. S., Cutler, R. G., and Slater, T. S., Eds., Raven Press, New York, 1984, 163.
35. **Katz, M. L., Robison, W. G., Jr., Herrmann, R. K., Groome, A. B., and Bieri, J. G.,** Lipofuscin accumulation resulting from senescence and vitamin E deficiency: spectral properties and tissue distribution, *Mech. Ageing Dev.*, 25, 149, 1984.

36. **Young, R. W. and Bok, D.,** Participation of the retinal pigment epithelium in the rod renewal process, *J. Cell Biol.*, 42, 392, 1969.
37. **Katz, M. L., Drea, C. M., Eldred, G. E., Hess, H. H., and Robison, W. G., Jr.,** Influence of early photoreceptor degeneration on lipofuscin in the retinal pigment epithelium, *Exp. Eye Res.*, 43, 561, 1986.
38. **Zigler, J. S. and Hess, H. H.,** Cataracts in the Royal College of Surgeons rat: evidence for initiation by lipid peroxidation products, *Exp. Eye Res.*, 41, 67, 1985.
39. **Katz, M. L., Drea, C. M., and Robison, W. G., Jr.,** Relationship between dietary retinol and lipofuscin in the retinal pigment epithelium, *Mech. Ageing Dev.*, 35, 291, 1986.
40. **Robison, W. G., Jr. and Katz, M. L.,** Vitamin A and lipofuscin, in *Cell and Developmental Biology of the Eye: The Microenvironment and Vision,* Sheffield, J. B. and Hilfer, S. R., Eds., Springer-Verlag, New York, 1987, 95.
41. **Katz, M. L., Eldred, G. E., and Robison, W. G., Jr.,** Lipofuscin autofluorescence: evidence for vitamin A involvement in the retina, *Mech. Ageing Dev.*, 39, 81, 1987.
42. **Robison, W. G., Jr., Kuwabara, T., and Bieri, J. G.,** The roles of vitamin E and unsaturated fatty acids in the visual process, *Retina,* 2, 263, 1982.
43. **Kini, M. M., Liebowitz, H. M., Colton, T., Nickerson, R. J., Ganley, J., and Dawber, T. R.,** Prevalence of senile cataract, diabetic retinopathy, senile macular degeneration, and open-angle glaucoma in the Framingham eye study, *Am. J. Ophthalmol.*, 85, 28, 1978.
44. **Hogan, M. J.,** Role of the retinal pigment epithelium in macular disease, *Trans. Am. Acad. Ophthalmol. Otolaryngol.*, 76, 64, 1972.
45. **Weiter, J. J., Dratz, E. A., Fitch, K., and Handelman, G. J.,** Role of selenium nutrition in senile macular degeneration, *Invest. Ophthalmol. Vis. Sci. (ARVO Suppl.),* 26, 58, 1985.
46. **Hittner, H. M., Godio, L. B., Rudolph, R. J., Adams, J. M., Garcia-Prats, J. A., Friedman, Z., Kautz, J. A., and Monaco, W. A.,** Retrolental fibroplasia: efficacy of vitamin E in a double-blind clinical study of preterm infants, *N. Engl. J. Med.*, 305, 1365, 1981.
47. **Finer, N. N., Schindler, R. F., Grant, G. D., Hill, G. B., and Peters, K. L.,** Effect of intramuscular vitamin E on frequency and severity of retrolental fibroplasia: a controlled trial, *Lancet,* 1, 1087, 1982.
48. **Finer, N. N., Schindler, R. F., Peters, K. L., and Grant, G. D.,** Vitamin E and retrolental fibroplasia: improved outcome with early vitamin E, *Invest. Ophthalmol. Visual Sci.*, 90, 428, 1983.
49. **Gallo-Torres, H. E.,** Transport and metabolism of vitamin E, in *Vitamin E: a Comprehensive Treatise,* Machlin, L. J., Ed., Marcel Dekker, New York, 1980, 193.
50. **Frank, L. and Groseclose, E. E.,** Preparation for birth into an O_2-rich environment: the antioxidant enzymes in the developing rabbit lung, *Pediatr. Res.*, 18, 240, 1984.
51. **Wispe, J. R., Bell, E. F., and Roberts, R. J.,** Assessment of lipid peroxidation in newborn infants and rabbits by measurements of expired ethane and pentane: influence of parenteral lipid infusion, *Pediatr. Res.*, 19, 374, 1985.
52. **Kolb, H. and Gouras, P.,** Electron microscopic observations of human retinitis pigmentosa, dominantly inherited, *Invest. Ophthalmol.*, 13, 487, 1974.
53. **Armstrong, D. and Koppang, N.,** Ceroid-lipofuscinosis, a model for aging, in *Age Pigments,* Sohal, R. S., Eds., Elsevier/North-Holland, Amsterdam, 1981, 355.
54. **Handelman, G. J. and Dratz, E. A.,** The role of antioxidants in the retina and retinal pigment epithelium and the nature of prooxidant-induced damage, *Adv. Free Radical Biol. Med.*, 2, 1, 1986.
55. **Bishara, S., Merin, S., Cooper, M., Azizi, E., and Deckelbaum, R. J.,** Combined vitamin A and E therapy prevents retinal electrophysiological deterioration in abetalipoproteinemia, *Br. J. Ophthalmol.*, 66, 767, 1982.
56. **Muller, D. P. R., Lloyd, J. K., and Wolff, O. H.,** Vitamin E and neurological function: abetalipoproteinemia and other diseases of fat absorption, in *Biology of Vitamin E,* Porter, R. and Whelan, J., Eds., Pitman, London, 1983, 106.

VITAMIN E AND RETINOPATHY OF PREMATURITY: HISTORICAL AND CLINICAL ASPECTS

Lois Johnson, Soraya Abbasi, and Graham Quinn

OXIDANT RADICALS AND PREMATURITY

Retinopathy of prematurity (ROP) is an oxidant disease of the incompletely vascularized retina. This chapter will discuss ROP in the context of current understanding of oxidative metabolism and the role of the antioxidant, vitamin E, in its prevention and treatment.

Almost all energy production by respiring cells is accomplished by the cytochrome enzymes of mitochondria through the four-electron reduction of molecular oxygen to water. This electron transport system, which appears to have evolved in response to the increasing oxygen content of the atmosphere of the earth, greatly minimizes the production of free radical intermediates.[1] A number of biologic oxidations, however, proceed by univalent or divalent reduction of oxygen and result in the production of superoxide, hydrogen peroxide, and the highly reactive hydroxyl radical. These include both enzymatic and autoxidations. The latter are increased by hyperoxia and oxidant metals and most frequently involve the unsaturated fatty acids of lipid membranes, oxyhemoglobin, and reduced electron-transferring components of cells. The enzymatic oxidations which produce significant amounts of superoxide include those catalyzed by xanthine oxidase, aldehyde oxidase, dihydro-orotic dehydrogenase, and, to a lesser extent, the flavoenzyme dehydrogenases.[1,2]

There are instances when oxidant radicals are produced for specific purposes. For example, during the respiratory burst of activated phagocytic cells, a membrane-bound primary oxidase initiates a sequence of precisely controlled reactions by which all three of these toxic species are generated and used to kill bacteria entrapped within the phagocytic vacuole. This essential component of antibacterial defenses usually proceeds without significant damage to the host.[3]

On the other hand, the excessive and/or uncontrolled free radical production which occurs during sustained hyperoxia, severe or chronic infection, severe allergic or autoimmune disease, systemic shock, and local hypoxic/ischemic events is destructive to the host. One important mechanism by which damage occurs involves the conversion of xanthine dehydrogenase to xanthine oxidase by proteases activated during hypoxia.[4] Hypoxanthine, breakdown product of the AMP which accumulates during hypoxia, and molecular oxygen which becomes readily available during reoxygenation provide the necessary substrates.

Defense systems have evolved to protect the organism from oxidant damage. These include the enzymes superoxide dismutase (SOD), catalase, glutathione peroxidase, and glutathione reductase, the lipid membrane constituent vitamin E, intracellular substances such as ascorbate, glutathione, cysteine, and cysteamine, and proteins such as serum albumin.[1] In ordinary circumstances, these defenses, which operate in a mutually supportive and interrelated fashion, are adequate to scavenge oxidant radicals before significant damage to the host occurs. As pointed out above, excessive production of these reactive species can result in major tissue injury. This is most apt to happen if the host is deficient in one or more antioxidant defense systems on the basis of immaturity, genetics, or nutritional lack.[2-5]

The relative hypoxia of the intrauterine environment requires minimal antioxidant defenses. In preparation for birth into an oxygen-rich atmosphere, a sharp increase occurs in activity of the major antioxidant enzymes SOD, catalase, and glutathione peroxidase in several animal species including man. This increase, which continues after birth, parallels maturation of the pulmonary surfactant system.[2] When birth occurs at term and without complications, antioxidant defenses are sufficient to withstand the oxidant challenge resulting from the inevitable transient asphyxia and sharp increase in oxygen tension which accompanies birth and the transition from fetal to neonatal circulation.

The risk of free radical damage to the premature infant is increased in direct proportion to degree of immaturity of antioxidant defenses and of metabolic and physiologic systems essential to extrauterine life (e.g., pulmonary, digestive, circulatory, neurologic, and musculoskeletal). It is also increased by the ability of modern medicine to sustain life without necessarily being able to prevent the wide swings from physiologic equilibrium which must occur in unstable infants who require extensive support systems for survival. Advances in perinatal and neonatal care are making possible not only increased survival, but also decreased morbidity through prevention and treatment of its major causes: birth asphyxia, intraventricular hemorrhage (IVH), respiratory distress, bronchopulmonary dysplasia (BPD), sepsis, necrotizing enterocolitis (NEC), and apnea. With each advance, however, survival is extended to include more and more immature infants, "micropremies", in whom the risk of morbidity is greatly increased.

These concepts will be illustrated by a discussion of ROP. In recent years, ROP has commanded almost as much attention from researchers and clinicians as it did during the epidemic of blindness among premature infants in the late 1940s and early 1950s. This discussion will include a brief description of the disease itself followed by a review of our work and that of others investigating the effect of vitamin E on incidence, severity, and long-term outcome of ROP. The chapter will end with mention of the results of vitamin E therapy for IVH, another common complication of prematurity in which oxidant damage plays an important role.

RETINAL VASCULARIZATION AND ROP

Until a gestational age of 4 months, metabolic needs of the human retina are adequately met by diffusion of oxygen from the underlying choroid. Differentiation, thickening, and stratification of the retina, which proceed centrifugally from the optic disk, then begin to position the inner parapapillary retina far enough from its oxygen source to limit the energy available for growth. This subtle metabolic compromise is believed to provide the stimulus for development of the intraretinal vasculature. In the human, vasculogenesis proceeds by a preliminary invasion of the nerve fiber layer by vasoformative spindle cells[6] rather than by the process of endothelial budding from the hyaloid artery seen in the kitten. Spindle cells reach the ora serrata at about 29 weeks of gestation. Most[6,7] but not all[8] investigators believe that spindle cells differentiate into endothelial cells, which then form a primitive capillary network from which mature retinal vessels evolve. Vascularization is not complete until term. Remodeling of vascular architecture continues for several months after birth.

Spindle cells and developing capillaries are highly sensitive to changes in oxygen tension. It is this sensitivity which, when infants are born prematurely, can result in the abnormal development of retinal vessels known as ROP.[6,7]

In the past, ROP was called retrolental fibroplasia (RLF). It was recognized only in its more severe forms, which could be diagnosed with the direct ophthalmoscope. It frequently led to blindness. With the use of the indirect ophthalmoscope, which allows visualization of the peripheral retina, the mild, far more common forms of ROP could be seen and studied. These include grades 1 and 2 ROP[9,10] which are confined to the peripheral retina and usually resolve spontaneously. ROP which progresses to significant extension of vessels into the vitreous (grade 3) or which is associated with obvious dilatation and tortuosity of posterior pole vessels (plus disease) usually results in retinal scarring and serious visual residua.[9,10]

In the late 1940s and early 1950s, before the toxic potential of oxygen was recognized, unmonitored quantities of this life-essential element were routinely given to all premature infants. The epidemic of blinding RLF which occurred was the subject of great concern and intensive study. Several workers reported the occurrence of severe persistent constriction of posterior pole retinal arteries and veins among infants who went on to develop the disease.

This was followed by vasodilatation of retinal vessels and overgrowth of vessels into the vitreous. The fibrovascular scarring and retinal detachment which occurred in the more severe cases led Terry, in 1942, to propose the name retrolental fibroplasia.

Experimental studies by Patz and Ashton on the effect of exposing normal 3-day-old kittens to high concentrations of oxygen demonstrated a similar pattern of pathology except that the end stages of scarring and retinal detachment did not occur. Controlled clinical trials in human infants of the effect of curtailment of oxygen therapy on incidence and severity of RLF followed these animal studies. The work of Gordon had already shown that providing oxygen only when there was obvious clinical need would not necessarily result in an increase in mortality. Second to prematurity, these clinical trials incriminated oxygen as the major risk factor for RLF. A recommendation for rigid limitation of oxygen use was made and resulted in a remarkable decrease in RLF. It was later recognized also to be associated with an increase in mortality and spastic diplegia. References as well as further details are available in the Review of ROP published by the American Academy of Pediatrics.[11]

With the development of micro blood-gas technology, it became increasingly possible to administer oxygen on the basis of need, even up to concentrations of 100%. Survival of premature infants again began to increase. Infant ventilators were developed and more sophisticated fluid and nutritional support became available. Even infants with birth weights under 1250 g began to survive. The incidence of severe RLF, which had never fallen to zero, began to increase. In contrast to the earlier epidemic, however, severe vasoconstriction of retinal vessels was not part of the disease process, hyperoxia documented on the basis of arterial oxygen measurements, could not be causally implicated, and degree of illness and hypoxic insult, both more common the more immature the infant, seemed to be major risk factors.[12] A mild and variable vasoconstriction in association with transient episodes of elevated arterial oxygen tensions could only be documented by serial fundus photographs.[13] However, in retinal digest preparation of eyes of infants who died, excessive capillary retraction in the nascent capillary bed could be demonstrated in association with oxygen treatment over a wide range of concentrations and duration.[14] In general, the extent of this obliterative process correlated with degree of oxygen exposure. It is proposed that hyperoxic damage to the primitive endothelial cell results in obliteration of newly formed capillaries and hypoxia of the adjacent retina. It is this presumed hypoxia which many believe to be central to the pathogenesis of ROP. Elsewhere in this volume (Pathogenesis and Antioxidant Suppression of Prematurity Retinopathy), Kretzer and Hittner present evidence that an oxidant insult to the spindle cell represents the essential initiating insult. In their hypothesis of ROP, retinal hypoxia does not occur and does not govern the sequence of pathology.[7] In any event, oxidant damage, either to the primitive endothelium or to the spindle cell, appears to be the initiator of ROP.

VITAMIN E DEFICIENCY AND ROP

Newborn infants are deficient in the membrane antioxidant vitamin E on the basis of body stores[15] and functional tests.[16] Body stores of vitamin E in the term infant are far below those of a healthy child or adult. They are nevertheless larger than in a premature infant (about 20 mg with a birth weight of 3600 g vs. 3 mg with a birth weight of 1000 g). The hypothesis of Owens and Owens that deficiency of this natural antioxidant predisposed premature infants to RLF was born out by the results of their first clinical trial. However, subsequent use of prophylactic oral vitamin E in their own nursery did not seem to affect the incidence of RLF as seen at that time. Interest in the vitamin was lost after oxygen was incriminated as a major risk factor for RLF.[11]

In 1972 we began retesting this theory under conditions of controlled oxygen therapy. Our studies with vitamin E in Philadelphia are still continuing. They were preceded by a 4-

Table 1
MEAN SERUM VITAMIN E LEVELS IN UNTREATED OR PLACEBO INFANTS, PENNSYLVANIA HOSPITAL — BIRTH WEIGHTS ≤2000 g[a]

	Mean serum vitamin E level		Standard nursery formula		
Years	Days 0—2 (mg/dℓ)	Days 7—28 (mg/dℓ)	Iron (mg/ℓ)	Vitamin E (IU/ℓ)	% Fatty acid as linoleic
1968—1971	0.30	0.25	12	4	32
1972—1975	0.31	0.40	Trace	9—12	32—26
1976—1978	0.57	0.74	Trace	15	23
1979—1981	0.58	0.93	Trace	15	23
1981—1984	0.55	1.16[b]	Trace	15	23

[a] This table does not stratify for birth weights below 2000 g because mean serum vitamin E levels, in contrast to body stores, are not significantly lower with increasing prematurity.
[b] Additional vitamin E was given as needed during these years to maintain serum vitamin E levels between 1 and 3 mg/dℓ.

Table 2
NEONATAL SURVIVAL RATE (%) AT PENNSYLVANIA HOSPITAL

	Birth weight			
Years	501—1000 g	1001—1250 g	1251—1500 g	1501—2000 g
1972—1975	15	48	75	87
1976—1978	23	59	83	94
1979—1981	39	77	93	95
1982—1984	67	92	99	99

year period of surveillance to define the incidence and natural history of RLF/ROP at Pennsylvania Hospital and to develop a classification system to include mild disease. The latter was amended in 1978 to include the phenomenon of plus disease and is in all essentials equivalent to the now accepted International Classification system for ROP (ICROP).[9,10]

A vitamin E-deficiency hemolytic anemia of premature infants had been found to be associated with use of formulas containing high levels of polyunsaturated fatty acids (PUFAs) and iron.[17] This information led to development of formulas with significantly higher vitamin E/PUFA ratios and the practice of delaying iron supplementation until hospital discharge. It also led, in about 1976, to the introduction of vitamin E to standard infant multivitamin drops. Vitamin E-deficiency anemia virtually disappeared and, as shown in Table 1, there was a progressive increase in the mean serum vitamin E levels of low-birth weight infants.

During this same time period there was a striking increase in survival of low-birth weight infants. Table 2 shows this, by birth weight group, for premature infants admitted to the intensive care nursery at Pennsylvania Hospital from 1972 to 1984. The quality of survival was also improving. This is illustrated with respect to ROP by the decreasing incidence of retinopathy over the same time period (Table 3). This decrease was most striking in infants with birth weight over 1250 g, and especially over 1500 g, in whom survival rates were beginning to stabilize. Other morbidities were also decreasing as the ability to prevent and/or treat the complications of prematurity improved. It is against this background of decreasing morbidity and improving nutrition that our studies and those of others on the effect of prophylactic vitamin E and ROP must be viewed.

Table 3
INCIDENCE OF ROP (%) AT PENNSYLVANIA HOSPITAL

	Birth weight			
Years	**501—1000 g**	**1001—1250 g**	**1251—1500 g**	**1501—2000 g**
1972—1975	100	75	48	16
1976—1978	75	63	15	8
1979—1981	88	58	18	5
1981—1984	83	49	13	3

CLINICAL TRIALS OF VITAMIN E AND ROP

In our first randomized clinical trial among infants with birth weights under 2000 g (1972 to 1974), a serum vitamin E level in the physiologic range (1.5 mg/dℓ) was targeted and achieved by age 1 day. It was maintained throughout the period of retinal vascular immaturity, i.m. vitamin E acetate, supplied by Hoffmann La Roche (HLR) as an investigational new drug (IND), was administered as needed to supplement oral intake. The formulas used were poor in iron, but rich in PUFAs, and had a low vitamin E/PUFA ratio. This contributed to the consistently low serum vitamin E levels (0.33 mg/dℓ) seen in placebo (P)- treated infants (Table 1). The mean age for initiating study medication was 11.3 hr ± 10.4 SD. The mean birth weight of study infants was 1662 g. A significant decrease in incidence and severity of ROP was associated with vitamin E treatment ($p<0.02$, n = 136). Cicatricial residua were also reduced.[18]

This work was continued from 1974 to 1976 using a higher target serum vitamin E level (3 mg/dℓ), a nonrandomized protocol, and the IND-HLR preparation of vitamin E-free alcohol, which is the active form of the vitamin. The duration of oxygen treatment doubled in comparison to the earlier study period and the survival rate in infants with birth weight under 1500 g improved; so also did the serum vitamin E levels of untreated infants. Only a trend toward a vitamin E-associated benefit was seen.[18]

In 1977 and 1978, Milner et al. in Ontario enrolled 268 infants in a randomized double-blind trial of oral vitamin E prophylaxis and ROP among infants with birth weight of ≤1500 g and likely to survive 5 days.[19] Study medication was used for 6 weeks. Mean serum vitamin E levels in P-treated infants were 0.9 and 1.0 mg/dℓ at ages 1 and 2 weeks, respectively. Vitamin E-treated infants, who received only 25 IU of vitamin E per day, had mean serum levels between 2 and 3 mg/dℓ by age 1 week and thereafter. Results of the initial eye exams (those between age 6 weeks and 4 months) were lost in a number of infants because of illness of the ophthalmologist. Only incidence of severe disease (5/114 P, 3/111 E) could be reported with confidence. The authors concluded, with respect to severe ROP, that " . . . of all the variables tested only length of time ventilated, birth weight, gestational age, vitamin E and health of the baby appear to have a significant contribution to make. Vitamin E administration, while perhaps ameliorating the development of RLF does not significantly reduce its incidence."

The randomized double-blind trial of Puklin et al. at New Haven was conducted during 1978.[20] It was designed to study BPD and included infants with birth weight above 1500 g. It used i.m. vitamin E acetate (IND-HLR), or placebo, within 24 hr of birth and through age 1 to 2 weeks if oxygen therapy was required. Vitamin E-treated infants had serum levels between 4 and 5 mg during the first week and 2 and 3 mg/dℓ during the second week. P-treated infants, as in the case of Milner, had higher mean serum vitamin E levels than infants in our first study (0.8 and 1.1 mg/dℓ for weeks 1 and 2). All infants were started on a standard vitamin E supplement of 25 to 50 IU/day once enteral feeds were tolerated. No significant difference between treatment groups was found for the 74 infants in whom ROP data were available.

In 1981 Hittner et al.[21] reported a significant vitamin E-associated decrease ($p < 0.01$) in severe ROP (grade 3+ or worse) in a double-blind randomized clinical trial of vitamin E prophylaxis at physiologic serum levels (50 vitamin E-treated, 51 P-treated infants, all with birth weights of ≤1500 g). Oral study medication at a dose of 100 mg/kg/day was instituted within 24 hr of delivery and continued throughout the period of retinal vascular immaturity. When this trial was conducted (1979 to 1980), vitamin E deficiency anemia was still present in the population served by the Texas Children's Hospital and was diagnosed in 10% of P-treated infants. In vitamin E-treated infants, the mean serum vitamin E level for the first 4 weeks after birth was only between 1 and 1.2 mg/dℓ in spite of the 100 mg/kg/day dosage. It was between 0.3 (at birth) and 0.6 mg/ℓ from age 1 to 4 weeks in P-treated infants. Kretzer's important ultrastructural findings[7,21] in the eyes of infants who died during this and subsequent studies are reported elsewhere in this volume.

In the randomized clinical trial of Phelps et al.[22] in Los Angeles (1980 to 1981), vitamin E-free alcohol (IND-HLR) or placebo by rapid i.v. infusion was initiated by age 1 day and continued as needed, to supplement oral intake, throughout the period of retinal vascular immaturity. The target serum vitamin E level was 3 mg/dℓ as measured prior to infusion. The mean serum vitamin E levels in P-treated infants were comparable to those of Puklin and Milner; 287 ≤1500-g infants birth weight, were enrolled. No difference between treatment groups with respect to incidence of ROP was noted (25/97 E vs. 28/99 P). One infant in each treatment group was blind.

The first study designed to test the ROP effect of prophylaxis at pharmacologic serum vitamin E levels (5 mg/dℓ) was conducted by Finer et al.[23] in Alberta between September 1987 and April 1981; 126 infants (≤1500-g birth weight, were randomized and enrolled: i.m. vitamin E acetate (IND-HLR) was initiated within 12 hr of birth and continued throughout the hospital stay if retinal vascularization was incomplete. In 48 vitamin E- and 51 untreated infants, a strong vitamin E-associated trend toward decreased severity of acute stage disease was found. There was a significant decrease ($p < 0.01$) in cicatricial sequelae among those who returned for follow-up at age 6 to 32 months. Mean serum vitamin E in control infants ranged from 0.5 at birth to 1.0 mg/dℓ by age 1 to 2 weeks.

Our own randomized, double-masked clinical trial of vitamin E prophylaxis at pharmacologic serum levels (5 mg/dℓ) enrolled 914 infants, ≤2000-g birth weight from January 1979 through April 1981. Follow-up continued until age 3 years post-term birth date. Because of its large size (755 infants with complete acute stage eye data, 424 with birth weight of ≤1500 g) and because its findings with respect to ROP as influenced by risk factors and age at enrollment are not yet in print, this trial will be reported on in more detail here.

Like the Los Angeles group, we used the IND parenteral vitamin E-free alcohol preparation of HLR, but by slow (usually 8-hr) infusion. This was followed by oral medication once enteral feedings were established. As in the case of Finer, the target serum vitamin E level was 5 mg/dℓ during the period of vascular immaturity or active ROP. It was then maintained at 3 mg/dℓ until healing was complete or until a postconceptual age of 1 year. All infants were followed ophthalmologically until age 3 years; 67.9% of infants with birth weight of ≤1500 g were started on study medication by age 1 day. The upper age limit for enrollment to the study was 5 days.

As in the clinical trials of Finer and Hittner in which cryopexy was used to treat any infant who developed severe ROP, our protocol required intervention if sight-threatening disease developed. We elected treatment with vitamin E at pharmacologic serum levels because we had treated ten infants in this way, during 1976 to 1978, with favorable results. In comparison to 14 untreated infants cared for during the same time period, visual morbidity in these ten vitamin E-treated infants was significantly reduced ($p < 0.02$). It was also

Table 4
NATURAL HISTORY OF SEVERE ROP[a]

	Grade 1 cic		≥Grade 2 cic	
Studies	**Number**	**%**	**Number**	**%**
Keith et al., Buffalo, 1977—81	1	6.3	15	93.8
Zak, Sydney, 1981—82	1	6.7	14	93.3
Johnson et al., Phila., 1968—78	1	4.3	22	95.7

[a] All infants had at least grade 3+ ROP with 360° ridge and extraretinal neovascularization in at least two quadrants. None of these studies reported an infant with absence of cicatricial residua (cic) after severe disease.

reduced in comparison to expected outcome[18,24-26] reported in the literature (Table 4). Therefore in this clinical trial, P infants with advanced grade 3+ ROP extra retinal neovascularization ([ERNV] in all four quadrants, Schaffer/Quinn and ICROP classification) were taken off study medication and treated with vitamin E to achieve a serum level of 5 mg/dℓ.

Characterization of our study population for degree of exposure to a large number of illness variables and suspected risk factors for ROP showed placebo and vitamin treatment groups to be very similar, as randomized within each of the four birth weight strata. This was true for all infants and for "early" and "late" enrolled infants. Measures of disequilibrium reflecting physiologic opposites, such as hypoxia and hyperoxia (or hypercapnia and hypocapnia), were highly correlated in both duration and extent with increasing severity of ROP regardless of treatment group ($p = 0.05$ multiple chi square). Similar findings have been reported by others.[12]

Among the 755 infants with complete eye data, a beneficial effect of vitamin E was seen on ROP categorized as none, mild (grade 1 to 2 ROP), moderate (grade 2+ to two quadrant grade 3+ ROP), or severe (four quadrant grade 3+ ROP or worse). Infants with birth weights of ≤1500 g had a decrease in ROP incidence ($p = 0.04$ chi square $p < 0.01$ logistic regression) and severity ($p = 0.05$) only if vitamin E prophylaxis was started by age 1 day. Nine placebo- vs. three vitamin-treated infants developed severe ROP. All had birth weights of ≤1500 g. Of these, ten infants weighed ≤1000 g (7 P, 3 E). No infant above 1500-g birth weight developed more than mild disease. The incidence of ROP in placebo-treated infants of this birth weight, while less than in previous years, was significantly higher than in vitamin E-treated infants (15/169 vs. 4/162, $p \leq 0.01$).

Of the 424 ≤1500-g birth weight infants, 50 developed ROP of moderate severity or worse (25 P, 25 E). Progression from moderate to severe disease was decreased in association with vitamin E treatment, regardless of enrollment age (9/25 P vs. 3/25 E, $p = 0.05$). At long-term follow-up only 26 infants (13 P, 13 E) were found to have developed the retinal sequelae of ROP. Among these, vitamin E therapy was associated with a decreased severity of cicatricial disease. Nine P vs. four vitamin E-treated infants showed cicatricial sequelae of more than grade 1 severity ($p < 0.025$). Five P vs. one vitamin E-treated infant had at least garde 3 cicatricial ROP by worse eye. This effect on cicatricial residua was seen even though the P-treated infants with severe ROP, eight of whom received late treatment with pharmacologic serum levels of E, had a better visual outcome than expected on the basis of reports in the literature.[24-26] One had no cicatricial sequelae; two had only grade 1 cicatricial sequelae. Thus, a favorable effect on incidence, severity, progression, and cicatricial sequelae of ROP was seen in association with high-dose vitamin E therapy.

The effect of vitamin E in decreasing the progression of moderate and severe ROP and in decreasing the extent of cicatricial sequelae was seen even if infants were not started on therapy very soon after birth. This supports the hypothesis that the pathogenesis of ROP, especially moderate and severe ROP, involves neonatal as well as perinatal events.

Pathologic evidence has been presented by Foos[27] that more than one insult, separated in time, can result in the vascular changes of ROP as shown by the finding of two demarcation lines, one more peripheral than the other. It is logical to expect that progression of abnormal vascularization would also be affected by the succession of hypoxic/acidotic/hyperoxic insults which can occur during periods of metabolic disequilibrium. This is supported by the high correlation between increasing severity of ROP and increasing degrees of illness found in our data.

It has been known for many years that vascularization is favored by decreased arterial oxygen tensions.[28] Recent studies have shown that hydroxy acids (HETEs) can modulate angiogenesis.[29] Several arachidonic acid isomers are produced by lipoxygenase and cyclooxygenase pathways in various tissues. The former include 12-HETE (12-hydroxy-eicosatetranoic acid) from platelets and 15-HETE (15-hydroxy-eicosatetroanoic acid) from blood vessel walls. 15-HETEs are also produced by free radical-mediated autoxidations, especially in the presence of metal catalysts and vitamin E deficiency. In human vessel culture systems, Stuart[29] has shown that hypoxia and, to a greater extent, hypoxia followed by normoxia increase the production of vascular 15-HETEs, while platelet production of antiangiogenic 12-HETEs are inhibitory to angiogenesis. Since vitamin E effectively blocks HETE formation through its singlet oxygen and hydroxy radical-scavenging ability,[30] it is reasonable to expect that it would decrease hypoxia-related production of proangiogenic substances in critically ill human babies. It would also be expected to decrease oxidant radical damage secondary to the xanthine oxidase/hypoxanthine mechanism. The optimal dose and serum level of vitamin E during periods of excessive free radical production might well exceed the physiological range.

RISK BENEFIT CONSIDERATIONS

In spite of the theoretical considerations outlined above and the findings in our clinical trial and that of Finer,[23] prophylactic therapy at pharmacologic serum vitamin E levels cannot be recommended. This is because, among vitamin E-treated infants with birth weights of ≤1500 g,[31] we found a significant increase in sepsis and late-onset NEC (an infectious disease). We believe this decreased resistance to infection was due to excessive scavenging of free radicals produced by phagocytic cells for bacterial killing.[32] Such a potential compromise of host defenses is unacceptable for a prophylactic medication for ROP since the vast majority of ROP is mild in degree and resolves completely. This is true even though a vitamin E-associated increase in sepsis or NEC was not found in the other studies using pharmacologic serum vitamin E levels, one for 6 to 10 weeks[23] and the other for 1 to 2 weeks,[20] perhaps because of the small size of their study populations. Finally it is important to point out that this side effect was not seen in any of the controlled clinical trials which targeted physiologic serum vitamin E levels for prophylaxis and it was not seen in our study among infants with birth weights of >1500 g.

In contrast to mild ROP, significant visual sequelae are often associated with moderate and severe retinopathy. Therefore treatment with pharmacologic serum vitamin E levels may be appropriate once this diagnosis has been made. ROP of this severity is not usually seen before age 2 months. It is reassuring that we found no vitamin E-associated change in incidence of sepsis or NEC after age 70 days.[31] This is perhaps because of the increase in myeloperoxidase content of white blood cells which occurs with advancing gestational and postnatal age[32a] which should make infants more comparable to adults in hydrogen peroxide potentiated killing mechanisms. In adults high-dose vitamin E does not significantly decrease intraphagocytic killing.[33]

No other significant vitamin E-associated side effects were recognized in any of the other controlled clinical trials, except in the Los Angeles study in which an increased incidence of severe IVH was found in ≤1000-g birth weight infants ($p < 0.01$). In contrast, the Philadelphia study showed a trend ($p < 0.1$) in the opposite direction. These differences may be related to characteristics of the study populations and/or techniques of vitamin E administration in the first few days after birth when most IVH occurs. More probably they relate to the fact that neither study was designed to assess the effect of vitamin E on IVH and that cranial ultrasound was not routinely available or required for its diagnosis.

IVH AND VITAMIN E

IVH is a serious cause of mortality and morbidity in premature infants, especially in those with birth weights of ≤1500 g. Although multifactorial in nature, IVH is generally considered to result from an initial ischemic/hypoxic insult. Oxidant damage must be involved in the pathogenesis. It is therefore not surprising that three studies[34,35] have reported a significant reduction in incidence of IVH in association with early administration of i.m. vitamin E so as to achieve serum vitamin E levels in the 3-mg/dℓ range by age 24 hr. In the study by Hittner et al.,[34] i.m. (15 mg/kg) vs. oral vitamin E-free alcohol administered within 24 hr of birth was studied with respect to effect on ROP. All infants had cranial ultrasound at age 5 to 7 days. A significant decrease in mortality and IVH was found in association with the more rapid increase in serum vitamin E levels achieved by the i.m. route. The two studies of Chiswick et al.[35] were specifically designed to study the effect of early administration of vitamin E on IVH. Both initiated therapy with 20 mg/kg of the slowly hydrolyzed parenteral vitamin E acetate within 12 hr of birth. The first randomized study did not require a double-blind protocol, but did include an examination by cranial ultrasound within 48 hr after birth and frequently thereafter. The second study, first reported before the British Pediatric Society in April 1986, was randomized, double-blind, and required cranial ultrasound on admission to the nursery and daily thereafter.[35a]

E-FEROL (PARENTERAL VITAMIN E ACETATE — O'NEAL JONES)

Serious toxicity occurred during use of a parenteral vitamin E acetate preparation known as E-Ferol which was marketed without FDA approval and has since been withdrawn. It was used in very high daily doses of 50 to 100 IU by i.v. infusion, as recommended in the package insert. Toxicity, which was dose related and most prominent in infants ≤1000-g birth weight, appears to have been related to the polysorbates present in the vehicle, in combination with the inert E-acetate form of the vitamin.[36,36a]

PARENTERAL VITAMIN E ALCOHOL (HOFFMANN LA ROCHE)

In nature, vitamin E exists as several isomers in the biologically active free alcohol form. Of these, α-tocopherol has the greatest antioxidant activity. Acetate or succinate esters of α-tocopherol are usually used in pharmaceutical preparations because of their greater stability. However, ester forms are inactive until after hydrolysis. Hydrolysis occurs rapidly in the gut after oral ingestion, but is accomplished only slowly after parenteral administration.[37] A parenteral free alcohol preparation of α-tocopherol in a nonpolysorbate emulsifier has been developed by Hoffmann La Roche, but is available only as an IND. It has been extensively studied for use in malabsorption syndromes such as abetalipoproteinemia and cholestatic syndromes. It has also been studied extensively in premature infants. However, therapeutic recommendations for this population have not been worked out. If the randomized, multicenter, ultrasound-controlled study of vitamin E and IVH now being organized in the U.S. confirms the British work, the task of writing guidelines for the prophylactic use of vitamin E will be greatly clarified.

In the meantime, several points need emphasizing. No undesirable side effects have ever been associated with prophylactic use of vitamin E resulting in serum levels in the physiologic range. Two controlled clinical trials of E and ROP using physiologic serum vitamin E levels showed significant treatment associated benefit.[18,21] Both were characterized by initiation of study medication before age 24 hr and control populations clearly deficient in vitamin E. A third study initiated vitamin E prophylaxis at a later age in infants with better background vitamin E nutrition and a trend toward less severe disease.[19] No study has ever found a vitamin E-associated increase in ROP associated with vitamin E treatment.[38]

CONCLUSIONS

Early supplementation with vitamin E appears to offer therapeutic advantages. We believe that supplementation with the active vitamin E-free alcohol should start either before birth or very soon thereafter. It should include an initial loading dose or doses followed by small daily supplements in a standard multivitamin preparation, with the aim of achieving rapid and sustained serum vitamin E levels near the midpoint of the physiologic range (0.8 to 3.0 mg/dℓ). This must be documented by serial laboratory measurements. Such a regime is not possible at present since the only available non-IND parenteral vitamin E is in the form of tocopherol acetate in a fixed multivitamin preparation (Pediatric Multivitamin Infusate, MVI-Armour).

It is important to remember that infants with birth weights of <1000 g and <27 weeks of gestation will not be completely protected from clinically significant ROP by vitamin E supplementation alone. The retinas of these clinically unstable infants concentrate the vitamin poorly because of low retinal levels of its carrier protein (see Chapter by Kretzer). Therefore adequate antioxidant therapy for these very premature infants has yet to be worked out.

REFERENCES

1. **McCord, J. M. and Fridovich, I.,** The biology and pathology of oxygen radicals, *Ann. Intern. Med.*, 89, 122, 1978.
2. **Massaro, D.,** Oxygen: toxicity and tolerance, *Hosp. Pract.*, July 15, 1986.
3. **Klebanoff, S. J.,** Oxygen metabolism and the toxic properties of phagocytes, *Ann. Intern. Med.*, 93, 480, 1980.
4. **McCord, J. M.,** Oxygen-derived free radicals in postischemic tissue injury, *N. Engl. J. Med.*, 312, 159, 1985.
5. **Boxer, L., Harris, R., and Baehner, R.,** Regulation of membrane peroxidation in health and disease, *Pediatrics*, 64, 713, 1979.
6. **Ashton, N.,** Retinal angiogenesis in the human embryo, *Br. Med. Bull.*, 26, 103, 1970.
7. **Kretzer, F. L., Hittner, H. M., Johnson, A. T., Mehta, R. S., and Godio, L. B.,** Vitamin E and retrolental fibroplasia: ultra-structural support of clinical efficacy, *Ann. N.Y. Acad. Sci.*, 393, 145, 1982.
8. **Cogan, D. G. and Kuwabara, T.,** Accessory cells in vessels of the paranatal human retina, *Arch. Ophthalmol.*, 4, 747, 1986.
9. **Quinn, G. E., Schaffer, D. B., and Johnson, L.,** A revised classification of retinopathy of prematurity, *Am. J. Ophthalmol.*, 94, 744, 1982.
10. The Committee for the Classification of ROP, An international classification of retinopathy of prematurity, *Arch. Ophthalmol.*, 102, 1130, 1984.
11. **James, L. S. and Lanman, J. T.,** History of oxygen therapy and retrolental fibroplasia, *Pediatrics., Suppl.* 57, 591, 1976.

12. **Lucey, J. F. and Dangman, B.,** A reexamination of the role of oxygen in retrolental fibroplasia, *Pediatrics,* 73, 82, 1984.
13. **Baum, J. D. and Bulpitt, C. J.,** Retinal vasoconstriction in premature infants with increased arterial oxygen tensions, *Arch. Dis. Child.,* 45, 350, 1970.
14. **Garner, A. and Ashton, N.,** Vaso-obliteration and retrolental fibroplasia, *Proc. R. Soc. Med.,* 64, 774, 1971.
15. **Dju, M. Y., Mason, K. E., and Filer, L. J., Jr.,** Vitamin E (tocopherol) in human fetuses and placentae, *Etud. Neo Natalis,* 1, 49, 1952.
16. **Gutcher, G. R., Raynor, W. J., and Farrell, P. M.,** An evaluation of vitamin E status in premature infants, *Am. J. Clin. Nutr.,* 40, 1078, 1984.
17. **Williams, M. L., Scott, R. J., O'Neil, P. L., and Oski, F. A.,** Role of dietary iron and fat on vitamin E deficiency anemia of infancy, *N. Engl. J. Med.,* 292, 887, 1975.
18. **Johnson, L., Schaffer, D. B., Quinn, G. E., Goldstein, D., Mathis, M. J., Otis, C., and Boggs, T. R., Jr.,** Vitamin E supplementation and the retinopathy of prematurity, *Ann. N.Y. Acad. Sci.,* 393, 473, 1982.
19. **Milner, R. A., Watts, J. L., Paes, B., and Zipursky, A.,** RLF in 1500 gram neonates: part of a randomized clinical trial of the effectiveness of vitamin E, in Retinopathy of Prematurity Conference Syllabus, Washington, D.C., December 4 to 6, 1981, 703.
20. **Puklin, J. E., Simon, R. M., and Ehrenkranz, R. A.,** Influence on retrolental fibroplasia of intramuscular vitamin E administration during respiratory distress syndrome, *Ophthalmologica,* 89, 96, 1982.
21. **Hittner, H. M., Godio, L. B., Rudolph, A. J., Adams, J., Gracia-Prats, J., Friedman, Z., Kautz, J., and Monoco, W.,** Retrolental fibroplasia: efficacy of vitamin E in a double-blind clinical study of preterm infants, *N. Engl. J. Med.,* 305, 1365, 1982.
22. **Phelps, D. L., Rosenbaum, A. L., Isenberg, S. J., et al.,** Tocopherol efficacy and safety for preventing retinopathy of prematurity: a randomized, controlled, double-masked trial, *Pediatrics,* 79, 489, 1987.
23. **Finer, N. N., Schindler, R. F., Grant, G., and Hill, G., and Peters, K.,** Effect of intramuscular vitamin E on frequency and severity of retrolental fibroplasia; a controlled trial, *Lancet,* 1, 1087, 1982.
24. **Keith, C. G.,** Visual outcome and effect of treatment in Stage III developing retrolental fibroplasia, *Br. J. Ophthalmol.,* 66, 446, 1982.
25. **Zak, T. A.,** Discussion — natural history of the active retinopathy of prematurity, *Ophthalmologica,* 91, 1502, 1984.
26. **Palmer, E. A. and Phelps, D.,** Multicenter trial of cryotherapy for retinopathy of prematurity, commentary, *Pediatrics,* 77, 428, 1986.
27. **Foos, R. Y.,** Chronic retinopathy of prematurity, *Ophthalmologica,* 92, 563, 1985.
28. **Miller, A. T., Jr. and Hale, D. M.,** Increased vascularity of brain, heart, and skeletal muscle of polycythemic rats, *Am. J. Physiol.,* 219, 702, 1970.
29. **Stuart, M. J., Ganley, C., and Setty, Y.,** Hydroxyacids and angiogenesis: potential role in the retinopathy of prematurity (ROP), *Pediatr. Res.,* 19, 163a, 1985.
30. **Panganamala, R. and Cornwell, D. G.,** The effects of vitamin E on arachidonic acid metabolism, *Ann. N.Y. Acad. Sci.,* 393, 376, 1982.
31. **Johnson, L., Bowen, F. W., Abbasi, S., Herrmann, N., Weston, M., Sacks, L., Porat, R., Stahl, G., Peckham, G., Delivoria-Papadopoulos, M., Quinn, G., and Schaffer, D.,** Relationship of prolonged pharmacologic serum levels of vitamin E to incidence of sepsis and necrotizing enterocolitis in infants with birth weight 1500 grams or less, *Pediatrics,* 75, 619, 1985.
32. **Engle, W. A., Yoder, M. C., Baruley, J. L., et al.,** Vitamin E decreases superoxide anion production by polymorphonuclear leukocytes, *Pediatr. Res.,* 23(3), 345, 1988.

32a. **Christensen, R. D. and Rothstein, G.,** Neutrophil myeloperoxidase concentration: changes with development and during bacterial infection, *Pediatr. Res.,* 19, 278, 1985.

33. **Baehner, R. L., Boxer, L. A., Ingraham, L. M., et al.,** The influence of vitamin E on human polymorphonuclear cell metabolism and function, *Ann. N.Y. Acad. Sci.,* 393, 237, 1982.
34. **Hittner, H. M., Speer, M. E., Rudolph, A. J., Blifeld, C., Chadda, P., Holbein, M. E. B., Godio, L. B., and Kretzer, F. L.,** Retrolental fibroplasia and vitamin E in the preterm infant — comparison of oral versus intramuscular: oral administration, *Pediatrics,* 73, 238, 1984.
35. **Chiswick, M. L., Johnson, M., Woodhall, C., Gowland, M., Davies, J., Toner, N., and Sims, D. G.,** Protective effect of vitamin E (DL-alpha-tocopherol) against intraventricular haemorrhage in premature babies, *Br. Med. J.,* 287, 81, 1983.

35a. **Sinha, S., Davies, J., Toner, N., et al.,** Vitamin E supplementation reduces frequency of periventricular hemorrhage in very preterm babies, *Lancet,* 1, 466, 1987.

36. **Alade, S. L., Brown, R. E., and Paquet A., Jr.,** Polysorbate 80 and E-ferol toxicity, *Pediatrics,* 77, 593, 1986.

36a. **Rivera, A., Abdo, K. M., Bucher, J., et al.,** Toxicity of IV vitamin E vs. polysorbate vehicle in newborn rabbits, *Pediatr. Res.,* 23, 263, 1988.

37. **Newmark, H. L., Pool, W., Bauernfeind, J. C., and DeRitter, E.,** Biopharmaceutic factors in parenteral administration of vitamin E, *J. Pharm. Sci.,* 64, 655, 1975.
38. **Johnson, L., Quinn, G., Abbasi, S., et al.,** Vitamin E and Retinopathy of Prematurity—Letter to the Editor, *Pediatrics,* 81, 329, 1988.

PATHOGENESIS AND ANTIOXIDANT SUPPRESSION OF RETINOPATHY OF PREMATURITY

Frank L. Kretzer and Helen M. Hittner

INTRODUCTION

Retinopathy of prematurity (ROP) is a potentially blinding disease of the developing preterm retina that is triggered by free radical damage to spindle cells.[1-4] The subsequent cascade of events may include inner retinal neovascularization and serous and/or tractional retinal detachment.

Historically, two epidemics of ROP have occurred. The first followed the introduction of oxygen into neonatal intensive care units in the early 1940s. This epidemic lasted until the early 1950s, when it was recognized that excessive oxygen administration was etiologically related to ROP. This first epidemic consisted primarily of larger infants (≥1000-g birth weight), since survival of 750 to 1000-g birth weight infants was <8% at that time. The second epidemic followed the technologic advances in respiratory therapy of the early 1970s and the recognition that judiciously curtailed oxygen was essential to prevent an inordinately high mortality and morbidity. This second epidemic is currently rising with continued neonatal advances which have increased the survival of 750 to 1000-g birth weight infants to >85% in many centers. Termination of the current epidemic requires an understanding of retinal development as it relates to antioxidant systems (selenium-dependent glutathione peroxidase and vitamin E) and to the pathogenesis of ROP.

RETINAL DEVELOPMENT

Photoreceptor Maturation (Outer Retina)

The gradient of photoreceptor maturation from the optic disk to the ora serrata has been simplified into four stages[3,4] which are shown and defined in Figure 1.

Correlation with Selenium-Dependent Glutathione Peroxidase

At approximately 18 weeks of gestational age, the neuroblastic retina transforms into Stage I photoreceptors with increasing metabolism and nascent synapse formation. Concomitant with the development of Stage I photoreceptors, there is an increase in endogenous retinal selenium-dependent glutathione peroxidase activity (SeGSH-Px) (Figure 2). Each mole of this rate-limiting enzyme requires four mol of selenium attached to four cysteine residues at the active sites. Therefore, selenium supplementation (maternal prepartum or infant postpartum) can maintain and/or induce this retinal antioxidant system in the human fetus as early as 18 weeks. The feasibility of pre- and postpartum supplementation is related to the known 3:2 maternal to fetal plasma selenium ratio and the water solubility of selenium.[5,6]

Correlation with Vitamin E

As Stage I photoreceptors mature into Stage II through IV photoreceptors, they secrete increasing amounts of interstitial retinol binding protein (IRBP) (Figures 1, 3, and 4)[7-9] that bind vitamins A and E.[10,11] Endogenous retinal vitamin E levels are low until near term primarily because of a 4:1 maternal to fetal plasma vitamin E ratio. Other contributing factors responsible for low endogenous vitamin E levels in low-birth weight infants are the small amounts of s.c. fat required for storage, low levels of plasma polyunsaturated fatty acids (PUFAs), and minimal IRBP-dependent retinal uptake. Exogenous vitamin E elevates retinal

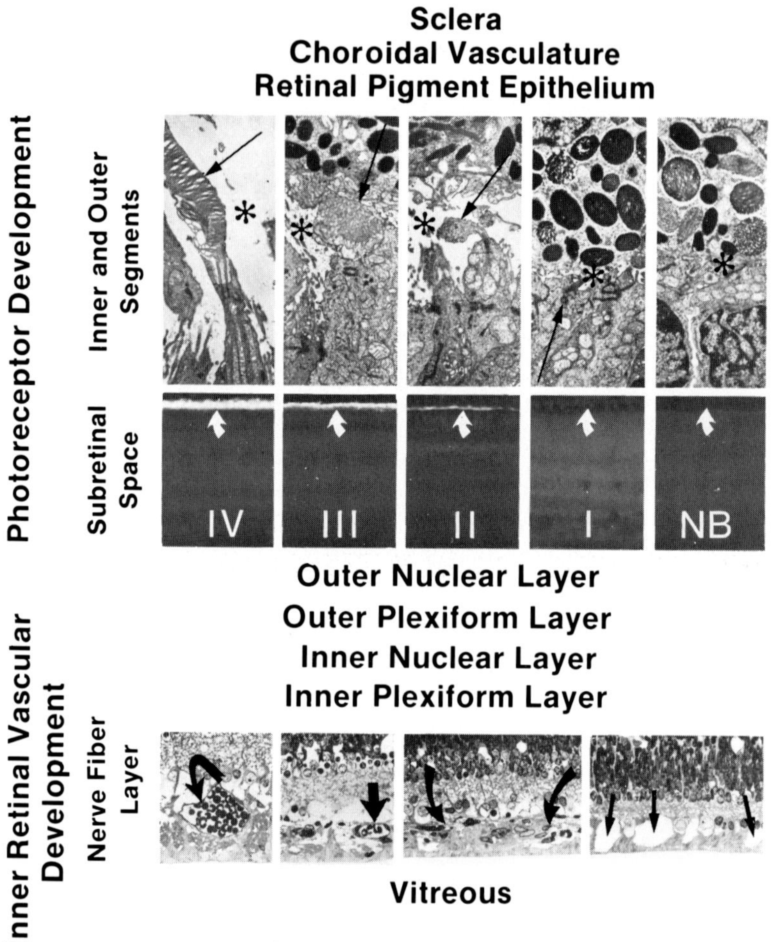

FIGURE 1. This montage defines photoreceptor development, from mature (left) to immature (right), as related to IRBP and inner retinal vessels. Each of the five stages of photoreceptor development along the optic disk (left) to ora serrata (right) axis is depicted as a vertical column. The first horizontal row defines the ultrastructure of photoreceptor outer and inner segments. (Magnification × 6000.) In Stage IV (IV), long outer segments with uniformly stacked disks (indicated by arrow) are in a large subretinal space (*). Inner segments with densely packed mitochondria are not visible. In Stage III (III), short bulbous outer segments with random disks (indicated by arrow) are in an enlarging subretinal space (*). Inner segments have abundant mitochondria. In Stage II (II), small balloon-shaped outer segments with random, tubular profiles (indicated by arrow) are in a small subretinal space (*). Inner segments contain few mitochondria. In Stage I (I), immature photoreceptors with apical centrioles (indicated by arrow) abut a tiny subretinal space (*). In neuroblastic retina (NB), the retinal pigment epithelium and neural retina are in tight apposition along their common apical surfaces (*). The second horizontal row defines the specific fluorescence pattern of the subretinal space (white arrows) by immunocytochemical localization of IRBP in fluorescence micrographs. (Magnification × 100.) In Stage IV, a wide bright band surrounds photoreceptor outer and inner segments and terminates at the outer limiting membrane. In Stage III, a narrower band surrounds outer and inner segments and terminates at the outer limiting membrane. In Stage II, a thin homogeneous band appears. In Stage I and neuroblastic retina, there is no immunospecific fluorescence, indicating that immature photoreceptors without outer segments do not secrete IRBP. The third horizontal row defines the extent of vascularization in the nerve fiber layer that corresponds to various stages of photoreceptor development as demonstrated by light micrographs. (Magnification × 100.) The transretinal relationship is emphasized by the large space representing retinal parenchyma that contains the stratified outer and inner nuclear and plexiform layers. At all gestational ages, there is an absolute relationship between Stage IV and inner retinal vessels (indicated by curved arrow) and Stage III and inner retinal arterioles (indicated by arrow). Therefore, these two columns are precisely aligned. In contrast, the migrating spindle cell apron (indicated by arrows) is shown to be transretinal to both Stage II and Stage I photoreceptors. Cystoid spaces (indicated by arrows) devoid of any endothelial precursors occur transretinal to both Stage I and neuroblastic retina.

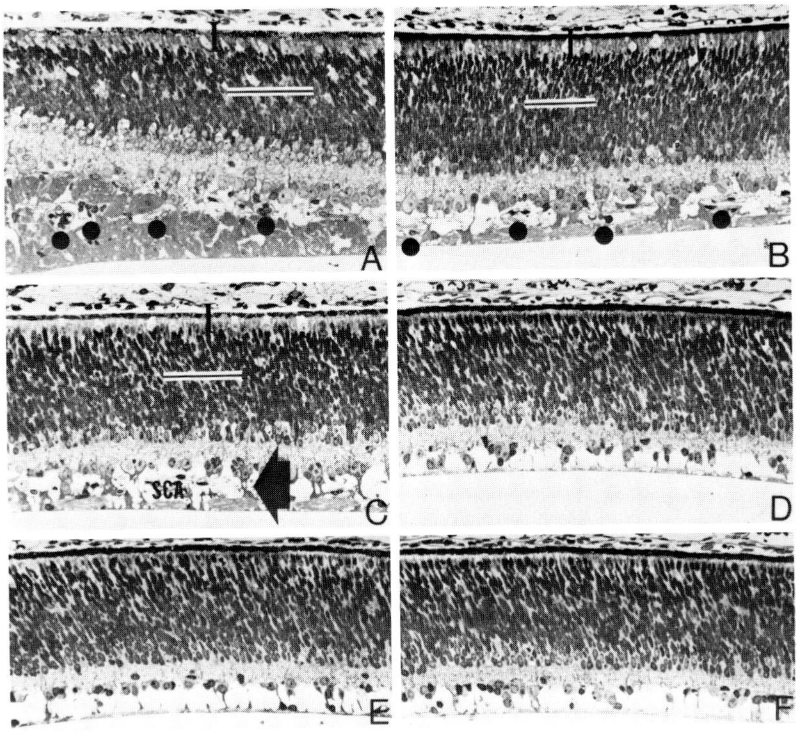

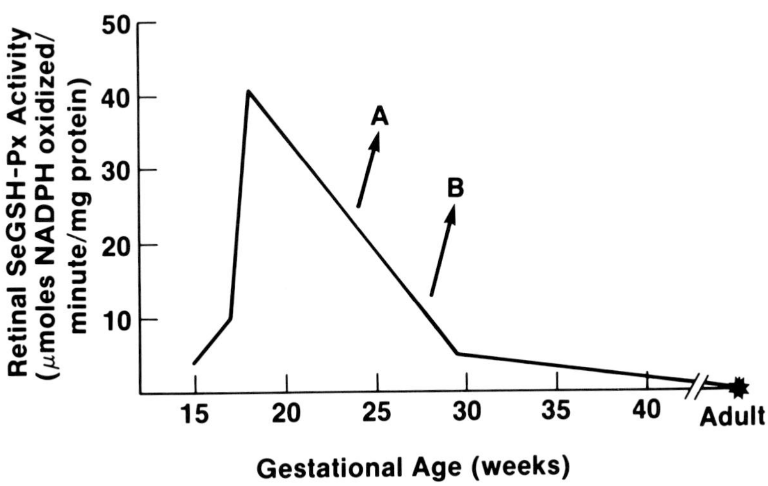

FIGURE 2. Light micrographs showing retinal development (each micrograph is one sixth of the distance from the optic disk to the ora serrata) in a 20-week gestational age infant (300-g birth weight) who survived for 3 hr. Notice Stage I photoreceptors (I) in A, B, and C; neuroblastic retina in D, E, and F; and the differentiation of the outer and inner nuclear layers (▬) in A, B, and C. Within the nerve fiber layer, there are tiny inner retinal capillaries (●) in A and B; a spindle cell apron (SCA) in C; the anterior edge of the spindle cell apron (←) in C; and no vasoformation or endothelial precursors in D, E, and F. (Magnification × 430.) This degree of retinal maturation correlates with high selenium-glutathione peroxidase (SeGSH-Px) activity and minimal potential for retinal uptake of exogenous vitamin E due to the very small radial distribution of IRBP. The graph plots endogenous retinal activity of SeGSH-Px as a function of gestational age. A and B represent the similar kinetics of maintained and/or induced retinal SeGSH-Px activity with exogenous selenium supplementation as a function of gestational age. The asterisk marks mean adult activity.

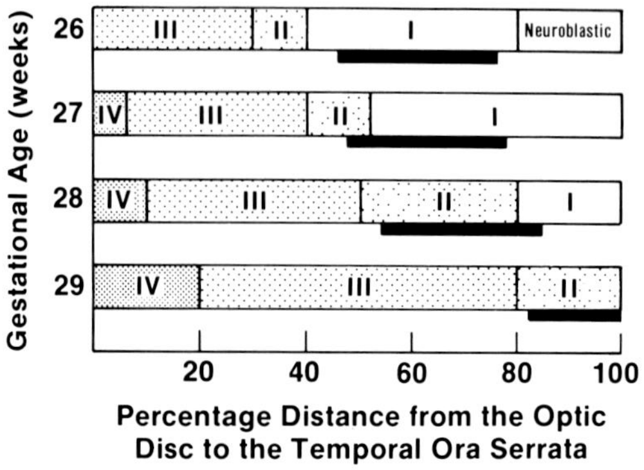

FIGURE 3. This bar graph depicts the gradient of photoreceptor maturation expressed as a percentage distance of each stage along the total optic disk- ora serrata axis. The data are generated from continuous montages of retinas of four preterm infants who survived <4 hr. Roman numerals indicate the stage of photoreceptor development; the stippled area represents the distribution and density of IRBP within the subretinal space; and solid black bars mark the transretinal location of the spindle cell apron.

levels maximally only after 28 weeks of gestational age,[12-14] when IRBP is present for >70% of the distance from the optic disk to the ora serrata (Figure 4). These biochemical facts predict that vitamin E supplementation to postpartum preterms provides maximal retinal antioxidant protection only after 28 weeks of gestational age. Therefore, the potential for prepartum supplementation is minimal. Megadose vitamin E therapy is inadvisable, since vitamin E can compete with clotting factors[15] yet provide no additional retinal protection.

Retinal Vasoformation (Inner Retina)

Spindle Cell Migration

Spindle cells invade the cystoid spaces of the inner retina as early as 16 weeks of gestational age. They progress toward the ora serrata over the next 13 weeks. Stage II photoreceptors first significantly appear around the optic disk at 22 weeks of gestational age and reach the ora serrata 8 weeks later. The transretinal geometry between the slowly migrating spindle cells and the faster-developing Stage II photoreceptors changes with time. In retinas ≤27 weeks of gestational age, the majority of the spindle cell apron is transretinal to Stage I photoreceptors, while in retinas ≥28 weeks of gestational age, the majority of the spindle cell apron is transretinal to Stage II photoreceptors (Figures 1 and 3). Furthermore, the low metabolic requirements of Stage I and II photoreceptors indicate that after premature birth, spindle cells are in a hyperoxic environment. These photoreceptors create a minimal barrier to the transretinal flux of oxygen from the nonconstricting choroidal vasculature.

Canalization

With maturation of photoreceptors to Stages III and IV, there is increased photoreceptor metabolism and oxygen consumption as evidenced by the mitochondrial aggregation within inner segments. Stage III and IV photoreceptors (Figures 1 and 3) create a formidable barrier to the transretinal flux of oxygen. Oxygen is then supplied to the inner retina via development of an inner retinal vasculature. This occurs by transformation of spindle cells into endothelium through the process of canalization. At all gestational ages, inner retinal vessels are transretinal to Stage IV photoreceptors, and inner retinal arterioles are transretinal to Stage III.

Muller Cell Function (Link Between Outer and Inner Retina)

Muller cells are retinal glia. These transretinal cells have their apical processes within the

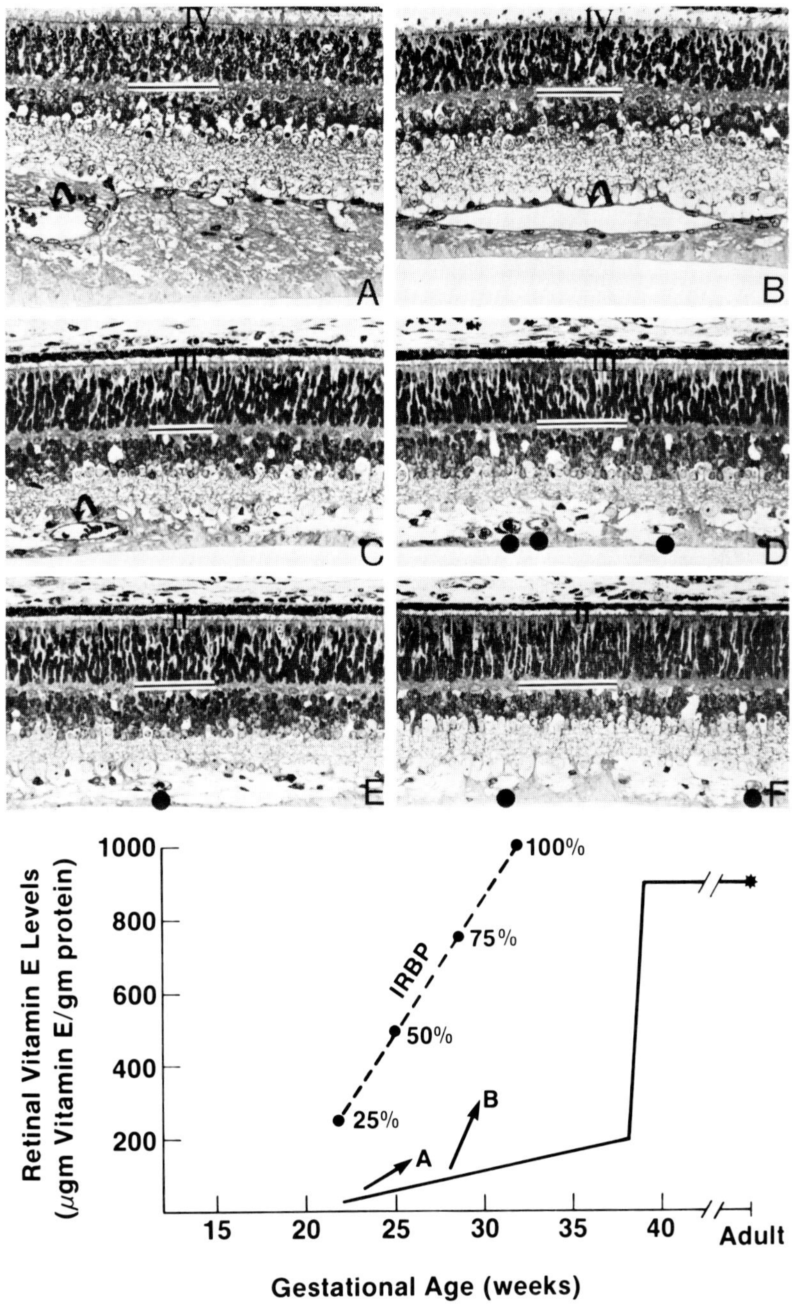

FIGURE 4. Light micrographs showing retinal development (each micrograph is one sixth of the distance from the optic disk to the ora serrata in a 30-week gestational age infant) (1330-g birth weight) who survived for 32 hr. Notice Stage IV photoreceptors (IV) in A and B; Stage III photoreceptors (III) in C and D; the Stage II photoreceptors (II) in E and F; and pronounced stratification of the outer and inner nuclear layers (▬) in all six micrographs. Within the nerve fiber layer, there are large inner retinal vessels (indicated by curved arrow) in A, B, and C; and tiny inner retinal capillaries (●) in D, E, and F. (Magnification × 430.) This degree of retinal maturation correlates with minimal SeGSH-Px activity and excellent potential for retinal uptake of exogenous vitamin E due to the very large radial distribution of IRBP. The graph plots two parameters as a function of gestational age: endogenous retinal levels of vitamin E (solid line) and the percentage of the distance from the optic disk to the ora serrata of IRBP (dashed line). A and B represent the different kinetics of increased retinal vitamin E levels with exogenous vitamin E supplementation as a function of gestational age. The asterisk marks mean adult levels.

subretinal space and their basal processes at the retina-vitreous interface. Muller cells enshroud retinal neurons and have large surface contacts with spindle cells. Fat-soluble vitamin E is shuttled across the polar subretinal space in nonspecific, hydrophobic pockets of IRBP. Vitamin E is then absorbed by Muller cell membranes and diffuses to the inner retina. The antioxidant is subsequently transferred to inner retinal membranes in contact with these glia. Because Muller cells are differentiated in the neuroblastic retina, these transport functions are potentially operant at very low gestational ages.

PATHOGENESIS OF ROP

Biochemical Events

The normal processes of photoreceptor maturation and inner retinal vasoformation occur in a hypoxic, intrauterine environment. Relative hypoxia with subsequent reperfusion or hyperoxia creates an adverse antioxidant retinal environment with production of free radicals which insult membranes within hours.[16]

Ultrastructural Events and Assayable Factors

In the Periphery (Altered Spindle Cell Migration)

Gap junctions — The normal cellular response to an adverse environment is the formation of extensive gap junctions between adjacent cells. This process has been induced experimentally in numerous cell types by relative hyperoxia, temperature extremes, hormone induction, pH changes, and cellular destruction.[17-20] Similarly, in the development of ROP, the first morphologic event following hypoxic or hyperoxic insults is the formation of extensive gap junctions between adjacent spindle cells within 4 to 20 days.[1-4] With formation of these gap junctions (activation), canalization no longer continues. However, spindle cells remain as a thin, anastomosing apron within the cystoid spaces of the inner retina. Spindle cells can remain gap junction-linked for as long as 8 to 12 weeks. If gap junctions disappear (down modulate) very soon after activation, normal canalization from spindle cells can resume. If gap junction down modulation is delayed, spindle cells stack, and abnormal inner retinal vasoformation from metaplastic cells can occur. The kinetics of these cellular events are summarized in Figure 5.

Rough endoplasmic reticulum — Extensive protein synthesis correlates ultrastructurally with a large cytoplasmic volume of rough endoplasmic reticulum (RER). Subsequent to gap junction formation, there is a unique increase in the cytoplasmic volume of RER within spindle cells. Conversely, following the down modulation of gap junctions, there is a subsequent unique decrease in the cytoplasmic volume of RER within spindle cells. The kinetics of these intracellular events are summarized in Figure 5.

Angiogenic factors — Utilizing the chorioallantoic membrane assay,[21] angiogenic factors are detectable only from extracts of the peripheral, avascular retina when it contains a population of gap junction-linked spindle cells with a large cytoplasmic volume of RER.[3,4] Thus, angiogenic factors are synthesized and secreted for a finite period of time and from a unique location (Figure 5). At no time during the pathogenesis of ROP can angiogenic factors be extracted from the vascularized retina. This occurs in the absence of any ultrastructural evidence of endothelial necrosis or retraction. If gap junction formation between adjacent spindle cells could be *totally prevented* by sequestering free radicals, then the incidence and severity of ROP would be reduced. However, since gap junction formation can only be *partially reduced*, then only the severity of ROP will be decreased without affecting incidence.

At the Interface (Altered Canalization)

The interface corresponds to the junction between immature inner retinal vessels and the spindle cell apron. With development of ROP, the interface becomes a line (Stage 1), then

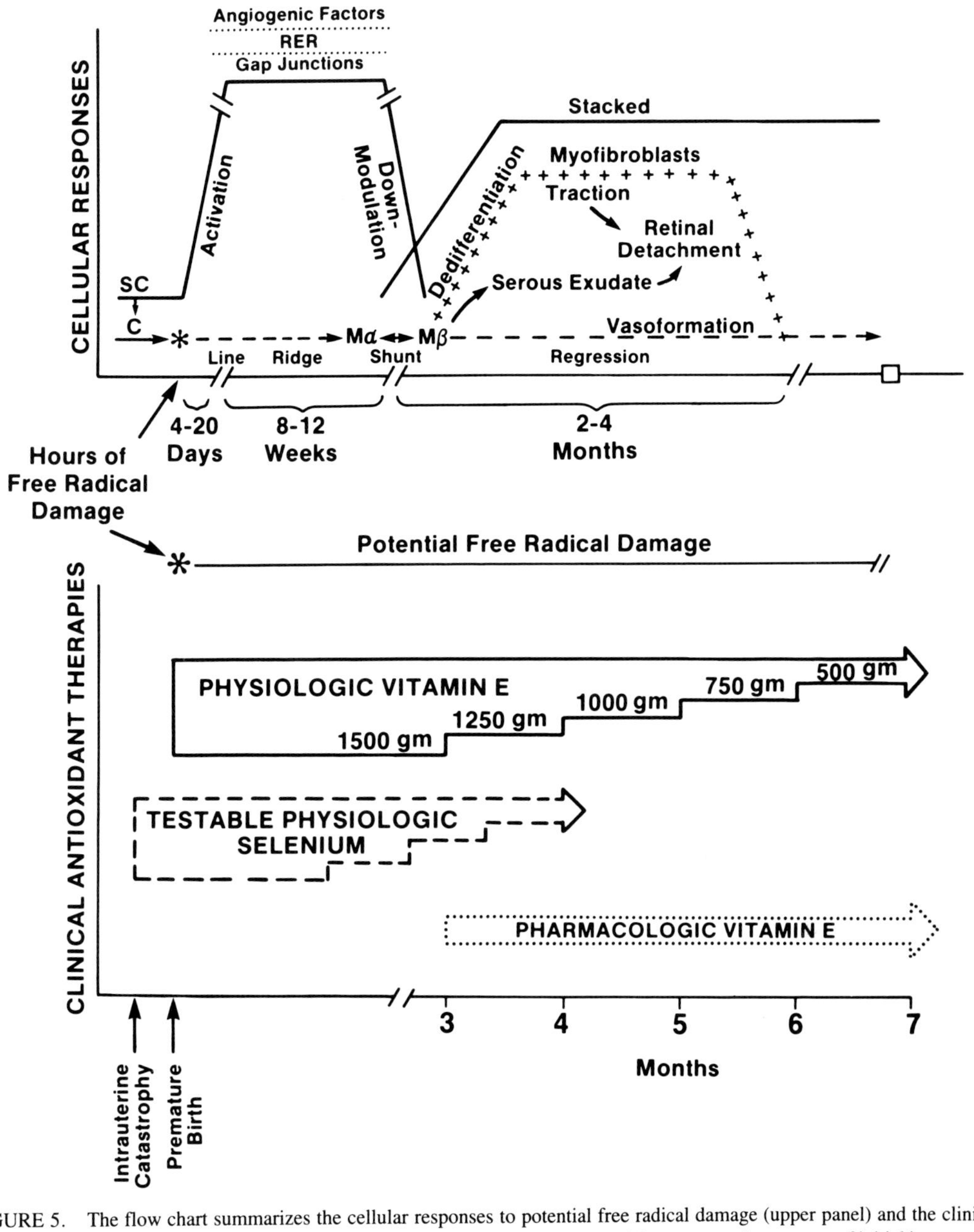

FIGURE 5. The flow chart summarizes the cellular responses to potential free radical damage (upper panel) and the clinical antioxidant therapies (lower panel) as a function of time. In the upper panel, asterisks represent the moment of initial intrauterine or extrauterine free radical damage. Prior to this damage, spindle cells (SC) normally differentiate into canalizing cells (C). After this damage, cellular responses are divided into four time intervals: hours (biochemical events), days (spindle cell activation), weeks (secretion of angiogenic factors and formation of a line, ridge, and shunt), and months (spindle cell stacking and determination of metaplastic cell differentiation within the shunt). Solid lines represent pathologic spindle cell events. Dotted lines denote pathologic intracellular spindle cell events. Dashed lines plot pathologic canalizing cellular events. Hatched lines represent myofibroblast proliferation. The box marks the time when inner retinal vessels reach the ora serrata and spindle cells have disappeared from the retina. In the lower panel, following premature birth, physiologic vitamin E supplementation (1.2 to 3.3 mg %) (solid line box) should begin immediately and continue uninterruptedly until inner retinal vessels reach the temporal ora serrata (approximate durations of physiologic vitamin E supplementation are shown as a function of birth weight). Prepartum supplementation is not feasible. Following an intrauterine catastrophe or the onset of premature labor, physiologic selenium supplementation (dashed line box) pre- or postpartum may provide antioxidant protection to spindle cells. Selenium supplementation need to be given only until retinal maturation develops adequate IRBP for sufficient vitamin E uptake. Following development of severe ROP, pharmacologic vitamin E supplementation (5.0 mg %) (dotted line box) could minimize myofibroblast formation and/or contraction.

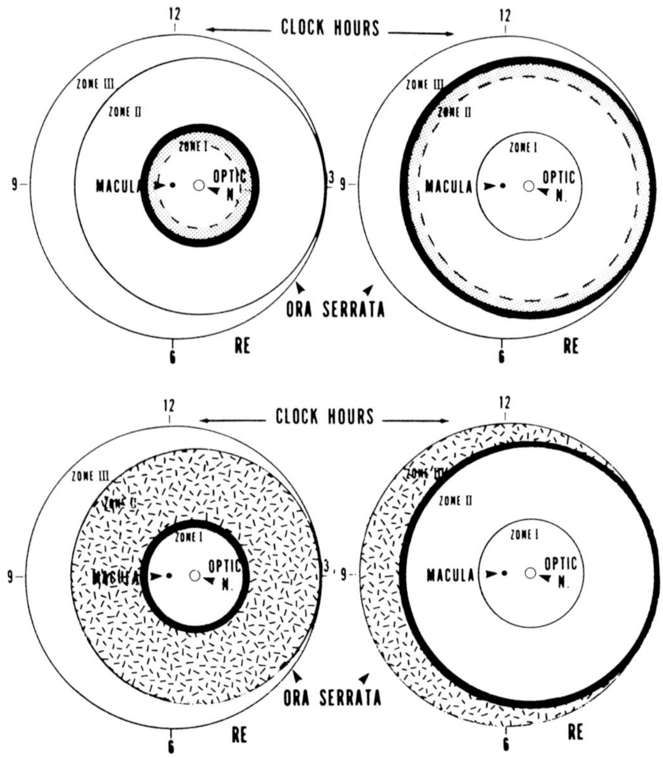

FIGURE 6. This figure contrasts the surface areas of potential endothelial necrosis or retraction (dotted areas) with that of the surface area of potential spindle cell free radical damage (dashed areas). The dense circle represents the extent of retinal vascularization when this damage occurred. The retinas on the left represent damage when vessels extend through Zone I (lower gestational age). Those on the right represent damage when vessels extend through Zone II (greater gestational age).

a ridge (Stage 2), and subsequently a ridge with extraretinal fibrovascular proliferation (Stage 3). The location of the interface as a circumference whose radius begins at the optic disk relates to the zones of the International Classification for ROP.[22] This advancing inner retinal vascularization correlates with gestational age. At lower gestational ages, the radius of the vascularized zone is usually smaller, and the area of the avascular retina is generally greater.

Figure 6 contrasts the surface area of potential endothelial necrosis or retraction with the surface area of spindle cells in infants with inner retinal vascular development extending through Zone I (lower gestational age) and extending through Zone II (greater gestational age). Should endothelial necrosis[23,24] or retraction[25] be a major factor in the development of ROP, the potential area of endothelial necrosis or retraction would be greater in older as compared to younger preterms. Development of severe ROP would then erroneously equate with greater gestational age. Should the area of the spindle cell apron which can experience oxidative damage be the major factor in the development of ROP,[1-4] then it would correctly predict the increased potential sites of synthesis and secretion of angiogenic factors in smaller rather than in larger preterms.

Metaplastic Cells — In the embryonic retina (Figure 7), canalization occurs at the interface between the vascular and avascular retina where a shunt can form when ROP develops. During the finite period of synthesis and secretion of angiogenic factors and with development of severe ROP, cells composing the shunt are increasingly separated from the spindle cell

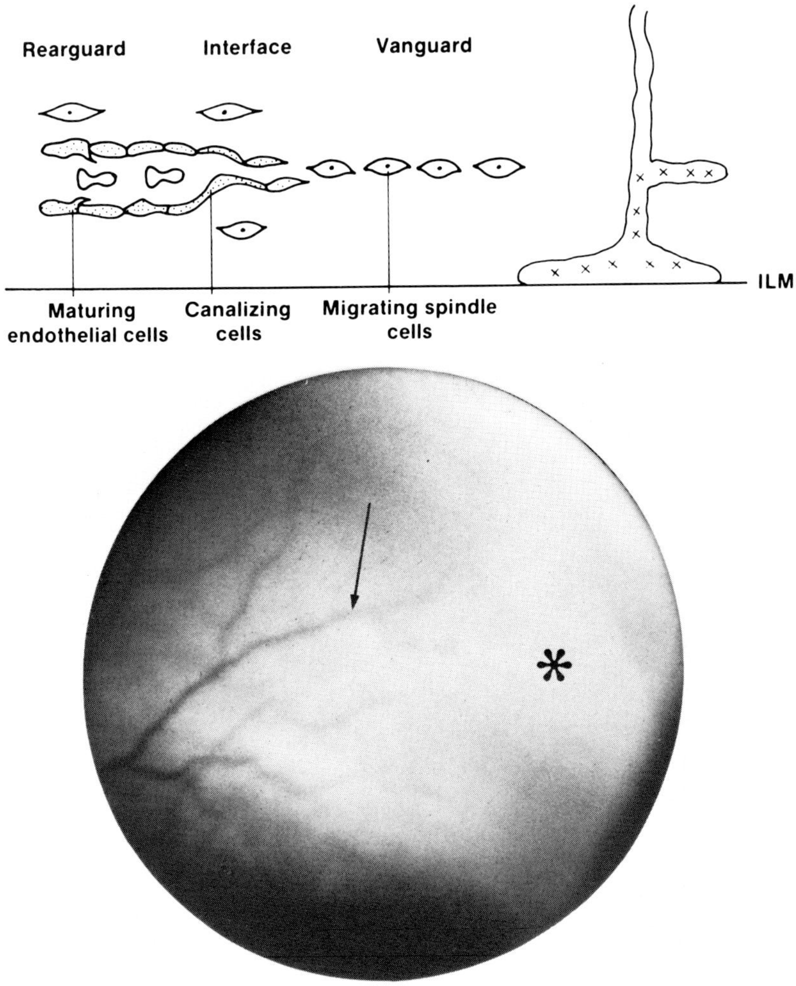

FIGURE 7. The diagram depicts the vascularized rearguard, interface, and avascular vanguard retina in embryonic vasoformation. Multiple-dot cells represent maturing endothelial cells; single-dot cells equate with migrating spindle cells; X areas denote basal processes of transretinal Muller cells (glia) which form the inner limiting membrane (ILM). The fundus photograph shows the interface between the vascular and avascular retina at premature birth. Clinically, the undamaged preterm retina at birth shows a gradual diminution in the caliber of inner retinal vessels (indicated by arrow) which taper into a gray, avascular retina (*) with no distinct border.

apron by a hypocellular region.[26] This hypocellular region is seen clinically as a dark line that separates the shunt from the avascular retina (Figure 8) and represents the visualization of choroidal pigment. This region is most clearly seen in dark races. Two populations of metaplastic cells compose the shunt (Figure 8). The first population is the anterior, α-metaplastic cells that are highly undifferentiated. These cells appear as a white line that is posterior to the hypocellular region. The second population is the posterior, β-metaplastic cells that are committed to forming primative capillaries in the presence of angiogenic factors. These β-metaplastic cells appear as a red line that is posterior to the white line. For unknown reasons, following the cessation of angiogenic factor secretion, one cell type (α- or β-metaplastic cells) predominates and one of two clinical situations develops (Figure 5). If α-metaplastic cells predominate, regression ensues. An aberrant vasoformative process (not

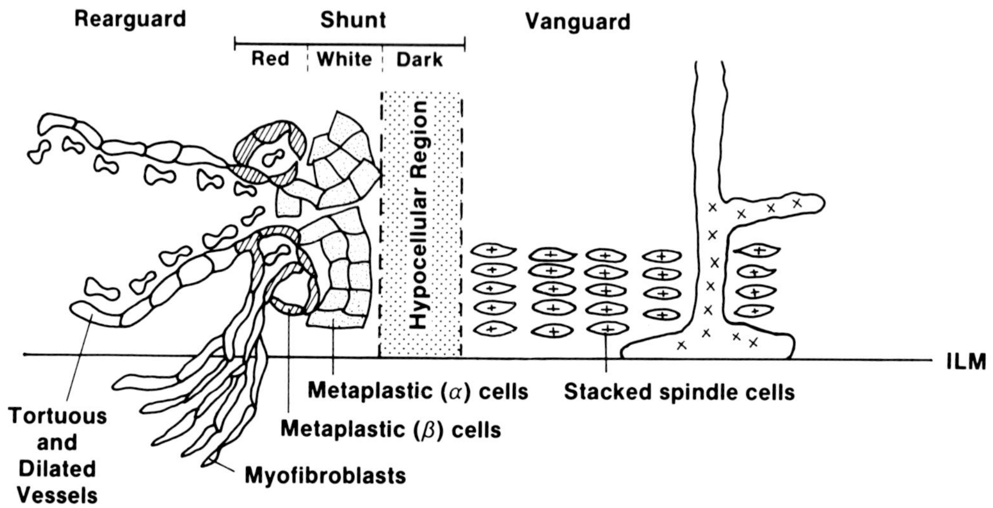

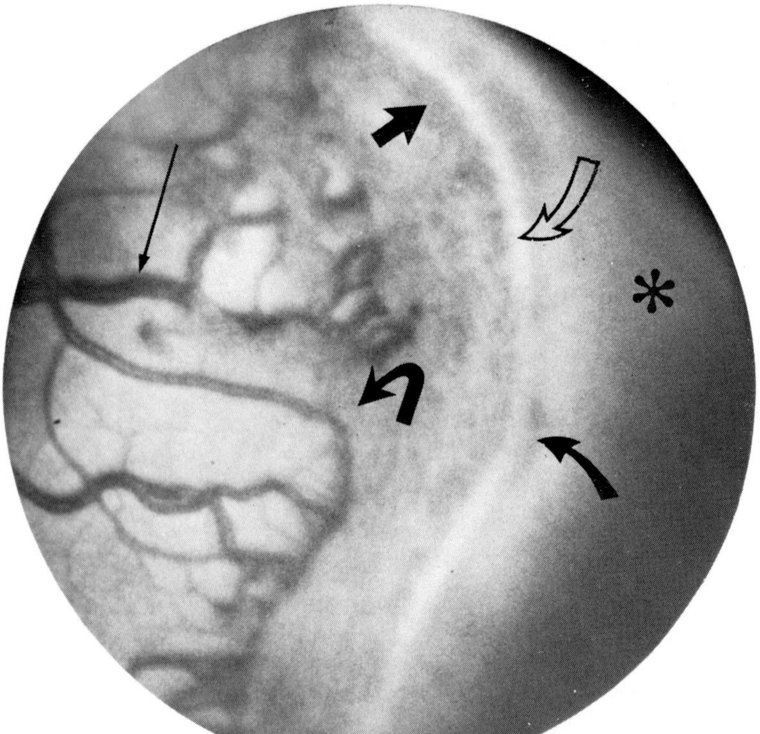

FIGURE 8. The diagram depicts the vascularized rearguard, interface, and avascular vanguard retina in advancing ROP Stage 3b(+). Plus cells designate stacked spindle cells, and X areas denote basal processes of transretinal Muller cells (glial) which form the inner limiting membrane (ILM). The interface of advancing ROP contains a hypocellular region and α- (dotted) and β- (striped) metaplastic cells. Myofibroblasts are dedifferentiated cells from β-metaplastic cells. Myofibroblasts are depicted as streaming into the vitreous and overlaying the dilated and tortuous retinal vessels. The fundus photograph shows the interface between the vascular and avascular retina at Stage 3b(+). Clinically, the severely damaged preterm retina demonstrates dilated and tortuous retinal vessels (indicated by thin arrow) whose terminal arborations are covered by myofibroblast sheets (indicated by curved arrow). The shunt has two components, a posterior red line (indicated by arrow head) composed of β-metaplastic cells and an anterior white line (indicated by open arrow) composed of α-metaplastic cells. Anterior to the shunt is a dark line (indicated by black arrow) which is a hypocellular region separating metaplastic cells from stacked spindle cells in the avascular retina (*).

from spindle cell canalization) becomes operant. α-Metaplastic cells give rise to β-metaplastic cells which form inner retinal vessels with regression of ROP. Clinically, this is seen as an anterior advance of the shunt toward the ora serrata. This often leaves one concentric ring of remnant myofibroblasts in the vitreous of the peripheral, vascularized retina. If subsequent hypoxic or hyperoxic insults are experienced during regression, a second cessation of the shunt advance can occur. Clinically, this can be seen as a second concentric ring in the peripheral, vascularized retina. If β-metaplastic cells predominate, plus disease (dilatation and tortuosity) develops via a shunting mechanism. Serous retinal detachment can be the result of leakage from primitive endothelial cells that are differentiating from β-metaplastic cells. A tractional process ensues when β-metaplastic cells in extensive numbers dedifferentiate into myofibroblasts which invade the vitreous.[27] This creates tractional forces which can trigger retinal detachment.[28]

Myofibroblasts — Myofibroblasts are fusiform cells which form organized sheets within the vitreous. They invade the vitreous from the posterior edge of the shunt concomitant with stacking of the spindle cells. They initially overlay the vascularized retina and remain as the dominant cellular component of the vitreous for about 3 months (Figure 5).[29] Ultrastructurally, myofibroblasts contain numerous actin microfilaments which provide the contractile apparatus of these cells. With their subsequent contraction, these sheets which are growing posteriorly become a central vector force. This central contraction is similar to tightening of a purse string with production of a tractional retinal detachment.

CLINICAL RISK FACTORS

Free radical damage to spindle cell plasma membranes is proposed to be the causative event in the development of ROP.[1-4] By understanding the possible sources of free radicals that can impinge on spindle cells in premature retinas, the multiple reported ROP risk factors[3,4] can be appreciated as a unified group of clinical parameters. Figure 9 summarizes the multifactoral nature of ROP. The variable expressivity of ROP severity is a function of the number of spindle cells experiencing various degrees of free radical damage.

Immaturity

The most consistently identified ROP risk factor is immaturity. The unifying principle is that spindle cells in the avascular, hyperoxygenated, peripheral retina are present in large numbers when inner retinal vessels are restricted to Zone I. These spindle cells are highly vulnerable to free radical damage. They are transretinal to Stages I and II photoreceptors and experience minimal vitamin E antioxidant stabilization due to the restricted central domain of IRBP. Therefore, low gestational age and low birth weight correlate with maximal risk for development of severe ROP.

Despite judicious curtailment of oxygen, there has been an increase in the incidence of severe ROP due to the increased survival of immature infants (≤1000-g birth weight). These infants have large spindle cell aprons projecting equally into both the nasal and temporal hemispheres. This large spindle cell apron can be induced to secrete maximal levels of angiogenic factors by multiple sources of free radical formation in the sickest, smallest infants.

Free Radicals

The unifying principle is that damage to spindle cell plasma membranes can be initiated by multiple clinical parameters. Each parameter increases the probability that free radicals generated by enhanced cellular oxygenation can overwhelm the sequestering ability of endogenous antioxidant systems.

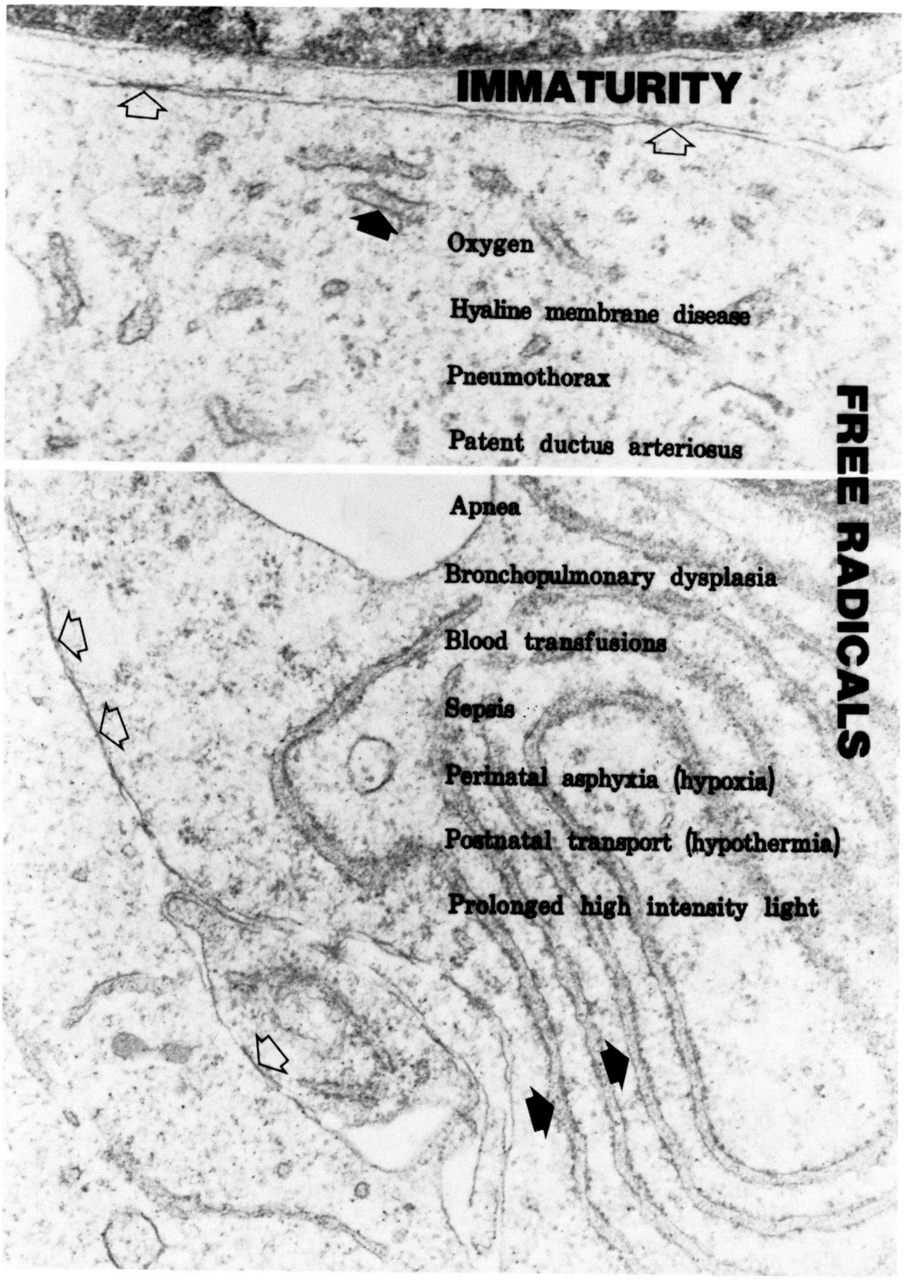

FIGURE 9. Electron micrographs demonstrating the transformation of undamaged spindle cells (upper figure) with minimal gap junctions (indicated by open arrow) and a small cytoplasmic volume of rough endoplasmic reticulum (indicated by black arrow) to free radical-damaged spindle cells (lower figure) with extensive gap junctions (open arrows) and a large cytoplasmic volume of rough endoplasmic reticulum (black arrows). Superimposed on this ultrastructural transformation are two risk factors: IMMATURITY and FREE RADICALS. (Magnification × 54,000.)

Oxygen

One consistent ROP risk factor is the cumulative amount of oxygen administered. The pathologic response of spindle cells is increased by oxygen tension greater than the hypoxic, intrauterine environment. Risk factors such as hyaline membrane disease, pneumothorax, patent ductus arteriosus, apnea, and bronchopulmonary dysplasia are associated with the therapeutic use of supplemental oxygen. They enhance the degree of free radical damage which potentiates gap junction formation between adjacent spindle cells.

The amount of oxygen administered is not the sole cause of ROP, but it is one unavoidable factor if these infants are to survive with intact central nervous systems. There is no safe, minimal level of oxygen that can be administered. All extrauterine levels are hyperoxic challenges to the peripheral retina.

Transfusions

Transfusions with adult, packed red blood cells change the fetal oxygen dissociation curve. Adult hemoglobin has less affinity for oxygen than fetal hemoglobin. As a result, more free radical damage is imposed on the developing retina. Very sick preterms receive relatively large quantities of adult blood in replacement transfusions. Transfusions are necessary due to blood loss related to multiple blood tests, to intraventricular and pulmonary hemorrhage, and to exchange transfusions related to hyperbilirubinemia.

Sepsis

Sepsis is also identified repeatedly as a risk factor for ROP. Macrophages release a large quantity of free radicals concomitant with the ingestion of bacteria. The burst of free radicals overloads the endogenous antioxidant systems of the preterm retina and potentiates gap junction formation between adjacent spindle cells.

Reperfusion

Difficult stabilization of preterms, such as perinatal asphyxia with a resultant low pH or postnatal transport with a resultant low temperature, is another ROP risk factor. These factors cause increased tissue utilization of oxygen with increased free radical formation.

Light

Prolonged high-intensity light has been suggested to increase the risk of ROP. Light is known to damage photoreceptors, resulting in decreased utilization of oxygen by even the most immature photoreceptors. Thus, the minimal oxygen barrier created by photoreceptors is further reduced and permits a greater oxygen flux across the retina from choroidal vessels. This increases the degree of free radical damage to spindle cell plasma membranes (Figure 10).

POTENTIAL ANTIOXIDANT CLINICAL THERAPIES

Of the multiple cellular antioxidant systems, the combination of selenium-glutathione peroxidase activity and vitamin E has the potential to create a favorable oxidant-antioxidant balance in the inner retina of all viable preterms. Selenium-glutathione peroxidase activity is present in significant quantities as early as 18 weeks. Selenium uptake into the retina to maintain these high enzymatic activities following birth is not age-dependent because selenium is water soluble (Figure 2). In contrast, vitamin E is only present in significant quantities after 28 weeks because its retinal uptake is dependent on age-related photoreceptor secretion of IRBP (Figures 3 and 4).

Of these two systems, only the efficacy of vitamin E has been appraised in four, double-masked, controlled, randomized clinical trials in which vitamin E was administered early

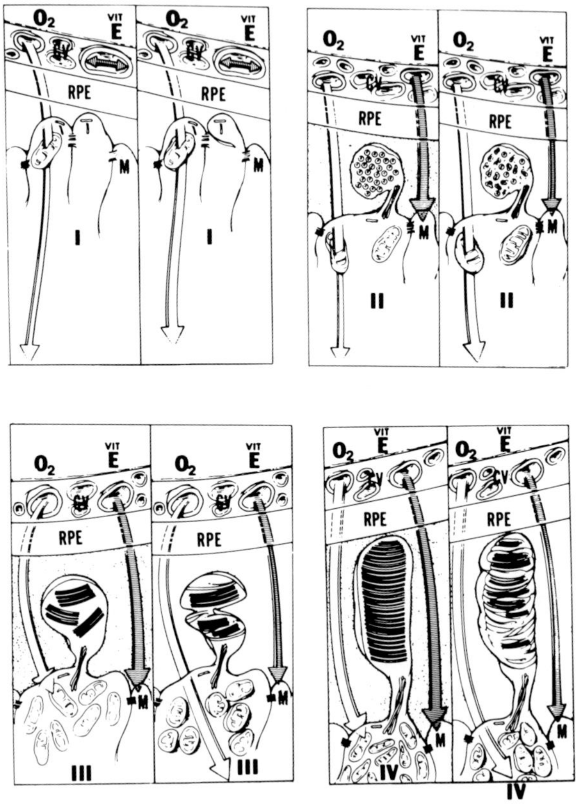

FIGURE 10. These drawings represent the theoretical effect of prolonged, high light intensity on Stage I to IV photoreceptors. Open arrows represent oxygen flux from choroidal vessels (CV) across the retinal pigment epithelium (RPE) into the neural retina. Hatched arrows depict potential vitamin E uptake from choroidal vessels (CV), through the retinal pigment epithelium (RPE), across the subretinal space within IRBP (stippled material), and into Muller cells (M). For each stage of photoreceptor maturation, normal photoreceptors are on the left, and light-damaged photoreceptors are on the right. Note the difference in thickness and length of the arrows and in stippling density which indicate quantitative changes in oxygen flux, vitamin E uptake, and IRBP, respectively. Increased oxygen flux across light-damaged Stage IV and III photoreceptors would relate to physiologic vasoconstriction of inner retinal vessels. Changes affecting Stage II and I photoreceptors would increase the probability of free radical damage to spindle cells in the nerve fiber layer.

and continuously until the inner retinal vasculature reached the temporal ora serrata.[30-34] With increasing survival of lower gestational age infants (≤27 weeks), results of these clinical trials have failed to produce a noncontroversial, all-or-none effect demonstrable by univariate analysis. In reality, vitamin E affords insufficient protection to the lower gestational age infants of the current epidemic.

Vitamin E Supplementation

The four clinical trials each employed a different route of administration without an appreciation of retinal uptake kinetics related to efficacy and without consideration of plasma kinetics related to toxicities. If the pathogenesis of human ROP involved endothelial necrosis[23,24] or retraction[25] as in various animal models, vitamin E in the plasma alone should be sufficient to minimize the development of human ROP. However, with acquisition of preterm whole eye donations, the focus of the pathogenesis of ROP shifted from endothelial necrosis or retraction to spindle cells. The concept of vitamin E uptake into remote, retinal parenchyma became paramount.[1-4]

If vitamin E were water soluble, all routes of administration should be efficacious for all viable infants. However, only oral, i.m., and slow i.v. administrations of vitamin E allow for the timely association with low-density lipoproteins[35] and for rapid retinal uptake by IRBP present within the subretinal space. Rapid i.v. administration apparently has delayed retinal uptake for unknown reasons.

Toxicity can be introduced by excessive dose via any parenteral route.[36] Plasma vitamin E levels ≥5 mg % can suppress macrophage activity and result in sepsis and necrotizing enterocolitis.[37] Rapid i.v. administration produces extreme plasma peaks of vitamin E which can compete with vitamin K and result in a bleeding diathesis.[38] Parenteral administration of vitamin E must be carefully monitored, and all i.v. solutions containing even "small" quantities of vitamin E must be appreciated as potent exogenous sources. Oral administration of vitamin E (breast milk or oral supplements) cannot achieve toxic levels or produce toxic peaks.[36] Other toxicities related to vitamin E in preterms are examples of inappropriate vehicles.[36]

Early Physiologic Vitamin E

Physiologic supplementation results in plasma sufficiency in the range of 1.2 to 3.3 mg %. Megadose therapies do not provide greater protection to remote spindle cells that are suspended in peripheral, avascular cystoid spaces. A critical prerequisite for potential efficacy is that supplementation must begin in the early hours following birth.[39] Furthermore, vitamin E sufficiency must be maintained without interruption. Free radical insult to spindle cells at any time prior to completion of retinal vascularization can initiate spindle cell activation and subsequent development of ROP.

With appropriate early and continuous supplementation, vitamin E can affect incidence and severity in infants ≥28 weeks of gestational age. Rarely, exceptions will occur that result from genetic conditions which simulate ROP or from intrauterine catastrophies, asphyxia, and hypothermia.[40] In these situations, free radical insults to spindle cells have already triggered a cascade of events which subsequent vitamin E supplementation cannot reverse.

Late Pharmacologic Vitamin E

Pharmacologic supplementation targets plasma levels ≥5 mg %. After plus disease has developed, initiation of pharmacologic supplementation has been reported to decrease the severity of cicatricial RLF.[41] This prolonged pharmacologic therapy must minimize contraction of formed myofibroblast sheets or favorably tip the balance of metaplastic cells within the shunt.

Current Recommendations

Target a plasma vitamin E level of 2.0 to 3.0 mg %. This can be accomplished by giving three early i.m. injections (alcohol in water) (15, 10, and 10 mg/kg on the first, second, and fifth days of life, respectively) or by giving early, slow i.v. infusions (alcohol in water) (3 mg/kg/day). This early parenteral vitamin E administration within the first hours of life decreases infant mortality and the development of severe intraventricular hemorrhage.[42,43]

Maintain physiological supplementation with continuous oral vitamin E (alcohol or acetate in medium chain triglycerides) with low osmolality (100 mg/kg/day)[34] until inner retinal vessels reach the temporal ora serrata. Inner retinal vasoformation halts in the immediate postpartum period in these immature infants. Thus, postpartum age plus gestational age at premature birth does not equate to gestational age in uninterrupted pregnancies. The clinical consequence is that vitamin E supplementation must continue until ophthalmologic examination confirms inner retinal vascularization to the ora serrata (Figure 5). When oral feedings must be interrupted, utilize parenteral vitamin E (alcohol in water) (3 mg/kg/day of slow i.v. or 10 mg/kg i.m. every third day). With use of parenteral vitamin E, plasma monitoring is mandatory.

Selenium Supplementation

No clinical trials have been performed to date which capitalize on the early presence of selenium-glutathione peroxidase activity in the human retina (Figure 2). Maintenance and/or induction of this high enzymatic activity following birth with selenium supplementation may suppress the development of severe ROP in infants ≤27 weeks of gestational age who do not have genetic conditions which simulate ROP.

The current epidemic can end only after spindle cells are protected by antioxidants whose retinal uptake is not age dependent. Thus, the concept of an antioxidant cocktail designed for infants of specific gestational ages must evolve. For infants ≤27 weeks, the cocktail initially would be weighted toward selenium. With increasing postnatal age, the cocktail gradually would change and be weighted toward vitamin E.

ACKNOWLEDGMENTS

This research was supported by grants from the Retina Research Foundation, Hoffmann-LaRoche, Cullen Foundation, Research to Prevent Blindness, and the Alabama Agricultural Experiment Station (AAES) of Auburn University. The clinical trials were only possible through the unending support of Dr. Arnold J. Rudolph and the neonatal staff, pediatric house staff, and nurses in the Neonatal Intensive Care Units of Texas Children's and Jefferson Davis Hospitals. The authors thank the administrators and technicians of the Lions Eyes of Texas Eye Bank for sensitively obtaining these difficult whole eye donations. The morphologic analyses include data generated by Rekha S. Mehta, Dr. A. Tim Johnson, Dr. David G. Hunter, Evelyn S. Brown, Dr. Donna Goad, Dr. Richard Fish, Dr. Chaula Rana, Dr. Cordell Adams, Dr. Hien L. Nguyen-Ngo, Dr. James Chodosh, and Dr. Michele Sprechman Curtis. The IRBP immunocytochemical studies were performed in collaboration with Dr. C. D. B. Bridges and Dr. Dominic M. K. Lam, with rabbit antibovine IRBP antibody provided by Dr. Shao-Ling Fong and Dr. Gregory I. Liou and cryosections produced by Pat A. Glazebrook. The selenium studies include data generated by Dr. Helen W. Lane with the technical assistance of Sharon P. Barron and Janet Johnston. The authors acknowledge the photographic expertise of Alexander Kogan and Gilma Miranda, the secretarial talents of Dorothy Carr, the drawings and graphs by Anne Sparks, and the manuscript critique of Dr. Michael Osato.

REFERENCES

1. **Kretzer, F. L., Mehta, R. S., Johnson, A. T., Hunter, D. G., Brown, E. S., and Hittner, H. M.,** Vitamin E protects against retinopathy of prematurity through action on spindle cells, *Nature (London),* 309, 793, 1984.
2. **Hittner, H. M., Rudolph, A. J., and Kretzer, F. L.,** Suppression of severe retinopathy of prematurity with vitamin E supplementation: ultrastructural mechanism of clinical efficacy, *Ophthalmology,* 91, 1512, 1984.
3. **Kretzer, F. L. and Hittner, H. M.,** Human retinal development: relationship to the pathogenesis of retinopathy of prematurity, in *Retinopathy of Prematurity: Current Concepts and Controversies,* McPherson, A. R., Hittner, H. M., and Kretzer, F. L., Eds., B. C. Decker, Toronto, 1986, 27.

4. **Kretzer, F. L., McPherson, A. R., and Hittner, H. M.,** An interpretation of retinopathy of prematurity in terms of spindle cells: relationship to vitamin E prophylaxis and cryotherapy, *Graefe's Arch. Clin. Exp. Ophthalmol.*, 224, 205, 1986.
5. **Combs, C. F. and Combs, S. B.,** The nutritional biochemistry of selenium, *Annu. Rev. Nutr.*, 4, 257, 1984.
6. **Rudolph, N. and Wong, S. L.,** Selenium and glutathione peroxidase activity in maternal and cord plasma and red cells, *Pediatr. Res.*, 12, 789, 1978.
7. **Johnson, A. T., Kretzer, F. L., Hittner, H. M., Glazebrook, P. A., Bridges, C. D. B., and Lam, D. M. K.,** Development of the subretinal space in the preterm human eye: ultrastructural and immunocytochemical studies, *J. Comp. Neurol.*, 233, 505, 1985.
8. **Schneider, B. G., Papermaster, D. S., Liou, G. I., Fong, S. L., and Bridges, C. D.,** Electron microscopic immunocytochemistry of interstitial retinol-binding protein in vertebrate retinas, *Invest. Ophthalmol. Visual Sci.*, 27, 679, 1986.
9. **Wiggert, B., Lee, L., Rodrigues, M., Hess, H., Redmond, T. M., and Chader, G. J.,** Immunochemical distribution of interphotoreceptor retinoid-binding protein in selected species, *Invest. Ophthalmol. Visual Sci.*, 27, 1041, 1986.
10. **Fong, S. L., Liou, G. I., Landers, R. A., Alvarez, R. A., and Bridges, C. D. B.,** Purification and characterization of a retinol-binding glycoprotein synthesized and secreted by bovine neural retina, *J. Biol. Chem.*, 259, 6534, 1984.
11. **Liou, G. I., Bridges, C. D. B., Alvarez, R. A., and Fong, S. L.,** Binding specificity of interstitial retinol-binding protein — possible physiological consequences of competition between vitamins A and E, *Invest. Ophthalmol. Visual Sci.*, 24(Suppl. 3), 42, 1983.
12. **Bhat, R., Raju, T., Barrada, A., and Evans, M.,** Disposition of vitamin E in the eye, *Pediatr. Res.*, 22, 16, 1987.
13. **Nishida, A. and Togari, H.,** Effect of vitamin E administration on alpha-tocopherol concentrations in the retina, choroid, and vitreous body of human neonates, *J. Pediatr.*, 108, 150, 1986.
14. **Nielson, J. C., Naash, M. I., and Anderson, R. E.,** The regional distribution of vitamins E and C in mature and premature human retinas, *Invest. Ophthalmol. Visual Sci.*, 29, 22, 1988.
15. **Helson, L.,** The effect of intravenous vitamin E and menadiol sodium diphosphate on vitamin K dependent clotting factors, *Thromb. Res.*, 35, 11, 1984.
16. **Halliwell, B. and Gutteridge, J. M. C.,** Lipid peroxidation, oxygen radicals, cell damage, and antioxidant therapy, *Lancet*, 1, 1396, 1985.
17. **Picard-Schneider, G., Cartentier, J. L., and Girardier, L.,** Quantitative evaluation of gap junctions in rat brown adipose tissue after cold acclimation, *J. Memb. Biol.*, 78, 85, 1984.
18. **Decker, R. S.,** Hormonal regulation of gap junction differentiation, *J. Cell Biol.*, 69, 669, 1976.
19. **Turin, L. and Warner, A. E.,** Intracellular pH in early *Xenopus* embryos: current flow between blastomeres, *J. Physiol. (London)*, 300, 489, 1980.
20. **Yee, A. G. and Revel, J. P.,** Loss and reappearance of gap junctions in regenerating liver, *J. Cell Biol.*, 78, 554, 1978.
21. **Glaser, B. M., D'Amore, P. A., Michels, R. G., Brunson, S. K., Fenselau, A. H., Rice, T., and Patz, A.,** The demonstration of angiogenic activity from ocular tissues, *Ophthalmology*, 87, 440, 1980.
22. The Committee for the Classification of Retinopathy of Prematurity, An international classification of retinopathy of prematurity, *Pediatrics*, 74, 127, 1984.
23. **Ashton, N., Tripathi, B., and Knight, G.,** Effects of oxygen on the developing retinal vessels of the rabbit. I. Anatomy and development of the retinal vessels of the rabbit, *Exp. Eye Res.*, 14, 214, 1972.
24. **Patz, A.,** Currrent concepts of the effects of oxygen on the developing retina, *Curr. Eye Res.*, 3, 159, 1984.
25. **Kushner, B. J., Essner, D., Cohen, I. J., and Flynn, J. T.,** Retrolental fibroplasia. II. Pathologic correlation, *Arch. Ophthalmol.*, 95, 29, 1977.
26. **Cogan, D. G. and Kuwabara, T.,** Accessory cells in vessels of the paranatal human retina, *Arch. Ophthalmol.*, 104, 747, 1986.
27. **Garner, A.,** The pathology of retinopathy of prematurity, in *Retinopathy of Prematurity,* Silverman, W. A. and Flynn, J. T., Eds., Blackwell Scientific, Boston, 1985, 19.
28. **Soong, H. K., Eller, A. W., Hirose, T., Hanninen, L., and Kenyon, K. R.,** *In situ* actin distribution in excised retrolental membranes in retinopathy of prematurity, *Arch. Ophthalmol.*, 103, 1553, 1985.
29. **McPherson, A. R., Hittner, H. M., Moura, R. A., and Kretzer, F. L.,** Treatment of retrolental fibroplasia with open-sky vitrectomy, in *Retinopathy of Prematurity: Current Concepts and Controversies,* McPherson, A. R., Hittner, H. M., and Kretzer, F. L., Eds., B. C. Decker, Toronto, 1986, 193.
30. **Hittner, H. M., Godio, L. B., Rudolph, A. J., Adams, J. M., Garcia-Prats, J. A., Friedman, Z., Kautz, J. A., and Monaco, W. A.,** Retrolental fibroplasia: efficacy of vitamin E in a double-blind clinical study of preterm infants, *N. Engl. J. Med.*, 305, 1365, 1981.

31. **Finer, N. N., Grant, G., Schindler, R. F., Hill, G. B., and Peters, K. L.,** Effect of intramuscular vitamin E on frequency and severity of retrolental fibroplasia: a controlled trial, *Lancet,* 1, 1087, 1982.
32. **Johnson, L., Quinn, G. E., Abbasi, S., Otis, C., and Bowen, F. W.,** Vitamin E and retinopathy of prematurity, *Pediatrics,* 81, 329, 1988.
33. **Phelps, D. L., Rosenbaum, A., Isenberg, S. J., Leake, R. D., and Dorey, F. J.,** Tocopherol efficacy and safety for preventing retinopathy of prematurity: a randomized, controlled, double-masked trial, *Pediatrics,* 79, 489, 1987.
34. **Hittner, H. M. and Kretzer, F. L.,** Efficacy of vitamin E in retinopathy of prematurity, in *Retinopathy of Prematurity: Current Concepts and Controversies,* McPherson, A. R., Hittner, H. M., and Kretzer, F. L., Eds., B. C. Decker, Toronto, 1986, 89.
35. **Traber, M. G. and Kayden, H. J.,** Vitamin E is delivered to cells via the high affinity receptor for low-density lipoprotein, *Am. J. Clin. Nutr.,* 40, 747, 1984.
36. **Hittner, H. M. and Kretzer, F. L.,** Toxicity of vitamin E in preterm infants, in *Retinopathy of Prematurity: Current Concepts and Controversies,* McPherson, A. R., Hittner, H. M., and Kretzer, F. L., Eds., B. C. Decker, Toronto, 1986, 111.
37. **Johnson, L., Bowen, F. W., Abbasi, S., Herrmann, N., Weston, M., Sacks, L., Parat, R., Stahl, G., Peckham, G., Delivoria-Papadopoulos, M., Quinn, G., and Schaffer, D.,** Relationship of prolonged pharmocologic serum levels of vitamin E to incidence of sepsis and necrotizing enterocolitis in infants with birth weight 1,500 grams or less, *Pediatrics,* 74, 619, 1985.
38. **Rosenbaum, A. L., Phelps, D. L., Isenberg, S. J., Leake, R. D., and Dorey, F.,** Retinal hemorrhage in retinopathy of prematurity associated with tocopherol treatment, *Ophthalmology,* 92, 1012, 1985.
39. **Finer, N. N., Schindler, R. F., Peters, K. L., and Grant, G. D.,** Vitamin E and retrolental fibroplasia: improved visual outcome with early vitamin E, *Ophthalmology,* 90, 428, 1983.
40. **Hittner, H. M. and Kretzer, F. L.,** Differential diagnosis of retinopathy of prematurity, in *Retinopathy of Prematurity: Current Concepts and Controversies,* McPherson, A. R., Hittner, H. M., and Kretzer, F. L., Eds., B. C. Decker, Toronto, 1986, 53.
41. **Johnson, L., Schaffer, D., Quinn, G., Goldstein, D., Mathis, M. J., Otis, C., and Boggs, T. R.,** Vitamin E supplementation and the retinopathy of prematurity, *Ann. N.Y. Acad. Sci.,* 393, 473, 1982.
42. **Speer, M. E., Blifeld, C., Rudolph, A. J., Chadda, P., Holbein, M. E. B., and Hittner, H. M.,** Intraventricular hemorrhage and vitamin E in the very low birth weight infant: evidence for efficacy of early intramuscular administration, *Pediatrics,* 74, 1107, 1984.
43. **Sinha, S., Davies, J., Toner, N., Bogle, S., and Chiswick, M.,** Vitamin E supplementation reduces frequency of periventricular haemorrhage in very preterm babies, *Lancet,* 1, 466, 1987.

PHARMACOLOGICAL PROTECTION AGAINST FREE RADICAL-INDUCED RETROLENTAL FIBROPLASIA

László Karmazsin and György Balla

INTRODUCTION

As a result of advances in neonatology, neonatal mortality has decreased significantly, although this decreasing rate is not the same for different countries.[1] Further, the incidence of disease characteristic for this population is increasing in the survivors, as a consequence of their immaturity. Among these disorders are the neonatal syndromes caused by oxygen toxicity, the most important of which are retrolental fibroplasia (RLF), bronchopulmonary dysplasia, and central nervous system (CNS) neuronal necrosis.[2] RLF is being investigated very intensively because it may result in blindness.[3]

RLF is a multifactorial disease, but it can be attributed primarily to neonatal immaturity and oxygen toxicity, although these two constituents are closely related to each other.[4] In these two fields, great knowledge has been accumulated, but the final solution is still in the future.[5] The most effective way may be a decrease in the rate of premature babies, but until this is accomplished, the pathophysiological problems have to be better understood, as even moderate success can save the patient's eyesight.

The mechanisms of oxygen toxicity and the cellular and extracellular protective antioxidants of immature babies are well known from many points of view. The natural defense mechanism against oxygen radicals is ineffective in premature babies, and therefore substitution with exogenous antioxidants is necessary.[2] The formation of free radicals cannot be decreased beyond a certain limit because the extrauterine environment is hyperoxic as compared to intrauterine life. This is especially true in oxygen therapy, although increased oxygen toxicity can be observed during hypoxic states.[6]

Since the general developmental and pathogenetic aspects of RLF have been dealt with in previous chapters of this Handbook and elsewhere,[6-16] the specific role played by light in triggering immaturity retinal degeneration and on the effects of antioxidant therapy is focused on below.

RETINOPATHY OF PREMATURITY (ROP)-LIGHT RELATION

For a premature baby, besides exposure to increased oxygen tension, light also represents a dramatic change. Clinical studies have demonstrated the role of light in the pathogenesis of ROP, which is the reversible stage of RLF.[7]

After birth, all parts of the retina, first of all the spindle cells, come under a significant photon effect. A cascade of photochemcial and biochemical reactions is initiated by electron excitation, and uncontrolled free radical formation is unavoidable, especially in the immature retina.[17] Calcium plays a central role in the phototransduction linked to the cyclic nucleotide cascade.[21] During the process of photoisomerization, there is a possibility for free radicals to be formed, but the basic mechanisms of photochemical damage are unknown.[22,23]

In the photoreceptors, there is a large quantity of retinal lipids which are rich in functionally important polyunsaturated fatty acids. These molecules are the source of energy (through their β-oxidation) for the photoreceptors with help of the mitochondria, and they take part in the repair mechanisms of cellular membranes. As the exposure to light increases, there is a progressive decrease in the relative percent of those phospholipid classes containing two polyunsaturated fatty acids per molecule, and lipid hydroperoxides accumulate, which causes photoreceptor degeneration.[22,24] The photolyzed rhodopsin seems to be the initiator of retinal destruction due to constant illumination.

Constant high-intensity illumination results in vitamin E loss and larger amounts of lipid hydroperoxides form in vitamin E-deficient patients such as premature infants. These changes result in spindle cell transformation releasing vasoproliferative agents, so the effect of light can stimulate the neovascularization indirectly.

During development, the levels of ascorbate and glutathione per retina increase. Light exposure influences this process as well. Intense light exposure results in a significant decrease in retinal ascorbate and an increase in retinal glutathione.[25] In the retina, glutathione is maintained in reduced form by the hexose monophosphate shunt and therefore GSH metabolism is not damaged seriously.

Ascorbate supplementation may prove useful in the prevention of ocular light damage. The photoreceptors, having a high rate of metabolism, derive their energy from mitochondria. Damage to these organelles by hyperoxia, hypoxia, and photons results in an increase in free radicals, and this process leads to oxygen toxicity and cellular destruction. During normal function of the mitochondria, there is a low level of free radical formation, and the endogenous antioxidant enzymes can scavenge them.[26] After light exposure, especially in the premature retina, a formation of oxygen radicals is to be expected.

Light can modify the metabolism of the catecholamine systems of the retina.[27] The retina contains three catecholamine neurotransmitters: dopamine, epinephrine, and norepinephrine. The predominant catecholamine is the dopamine in the amacrine cells; it represents 85 to 90%. Hydrophilic compounds do not readily enter the retina because of the blood-retina barrier. Thus, dopamine and its metabolites are synthesized in the retina. During maturation, treatment with dexamethasone increases the retinal concentration of dopamine. Light results in activation of tyrosine hydroxylase and this photic stimulation modulates the synthesis of retinal dopamine. Activation of the enzyme apparently takes place within seconds after light exposure. This is probably the consequence of a cyclic-AMP-dependent protein kinase-mediated phosphorylation of the enzyme. Epinephrine also can be found in the retina, but its concentration is low.[28] Exposure to light increases retinal epinephrine in the presence or absence of sympathetic nerves. Norepinephrine is associated primarily with sympathetic nerves, as superior cervical ganglionectomy results in almost complete loss of this neurotransmitter. The retinal content of norepinephrine is decreased when the environment changes from light to dark, and it can modulate the retinal circulation directly. Light has a direct effect on catecholamines, increasing their autoxidation. This process might be a significant source of free radicals in the retina.[29] When administering drugs which interact with catecholamines and their metabolism, the effects of light must be considered.

Macrophages can be observed in the subretinal space during development, and constant light results in a rapid increase in their number. Similar cells appear within the vascularized part of the inner retina during vascular growth.[30] The function of these cells is to clean away the degraded cell elements, thus the macrophages forming free radicals can contribute to oxygen toxicity of the retina.

Damage to the retina by irradiation is well known.[31] Hyperbaric oxygen increases the injury.[23] In adults, hyperbaric oxygen causes a toxic inactivation of respiratory enzymes, selective rod cell death, and depression of the electroretinogram.

Recently, cryotherapy and photocoagulation have been performed on infants, who have developed ROP stage 3 or 4.[32] This can induce free radicals and therefore these treatments are not recommended for infants with acute retinopathy stage 1 or 2. By contrast, our procedures did not result in severe cicatricial disease or blindness.[33]

ANTIOXIDANT THERAPY

The eyes of premature babies are in danger both before the 28th week of gestation and after it. This danger cannot be prevented by only monitoring the blood oxygen concentration and, discussed below, substitution with exogenous antioxidants may be necessary.

D-Penicillamine

D-Penicillamine is used in adults with rheumatoid arthritis, Wilson's disease, various forms of heavy metal intoxication, and pulmonary interstitial fibrosis.[36] Because of its well-known side effects, D-penicillamine should be used under close control. Some of the side effects are proteinuria, leukopenia, thrombocytopenia, aplastic anemia, glomerulopathy, myasthenia gravis, and allergic reactions. These facts prevent frequent application in the field of neonatology. Nevertheless during the last decade we have not observed any side effect in term and preterm infants, probably because of the short-term administration, though D-penicillamine has been given in high doses. The other reason for its apparent innocuousness may be the different effects of D-penicillamine on newborns as compared to adults.[37]

The administration of D-penicillamine for prevention and cure of ROP is indicated by its mechanisms of action and by our clinical observations. D-Penicillamine as other drugs containing sulfhydryl groups, can scavenge free radicals.[38] The copper-D-penicillamine complex has a superoxide dismutase-like activity.[39] In vivo, it decreases the concentration of superoxide dismutase induced by free radicals.[40] The effect of this drug might result from its chelating properties, especially for copper and iron, which are potent catalysts of free radical formation. D-Penicillamine can reduce plasma bilirubin concentration and, for this reason, photodynamic bilirubin becomes less dangerous.[41] It influences heme metabolism, decreases heme oxygenase, and increases Δ-amino-levulinate synthetase activity. This prevents oxygen toxicity. D-Penicillamine can enter into the corpus vitreum, thereby reducing vitreous proliferation.[42] The formation of insoluble collagen is inhibited by D-penicillamine.[36]

In our clinic all infants weighing less than 1500 g at birth and requiring supplemental oxygen receive D-penicillamine therapy. D-Penicillamine, 300 mg/kg/day, is administered intravenously during the first 3 days and, after this period, 50 mg/kg/day is administered intravenously for 2 weeks. If any sign of ROP occurs, we continue the treatment, 50 mg/kg/day, for another 2 weeks, under close control.[37] We found that no infant under 1500 g treated with D-penicillamine developed cicatricial disease or blindness. Therefore we can recommend its administration for the prevention of RLF.[43]

Methyldihydroquinoline-Type Antioxidants

Two novel nontoxic, synthetic radical scavengers have recently been developed in Hungary,[44-47] namely, dihydroquinoline-type antioxidant, MTDQ (6,6′-methylene-bis-2,2,4-trimethyl-1,2-dihydroquinoline), and its water-soluble disulfonic derivate, MTDQ-DA. Their protective effect against oxygen toxicity is complex. Their molecular weight is more than twice that of the hitherto known dihydroquinoline-type antioxidants, while the ratio of molecular weight to functional groups remains the same. It has been assumed that the significant decrease in toxicity is due to the higher molecular weight. The NH groups of the dihydroquinoline moieties are mobile and readily form stable nitroxyl radicals, which are demonstrable for several hours by means of electron spin resonance spectroscopy.[48] The NH groups can react with alkyl peroxide and hydrogen peroxide radicals, delivering molecular oxygen. This mechanism is important not only against free radicals, but, in a hypoxic environment, it becomes a source of oxygen. Presumably, the methylene bridge at position 6 correlates with the peroxide decomposing action, while hydrol formation can be observed. The reaction is reversible, the low pH is advantageous to form hydrol, and the alkaline pH pushes the balance into the opposite direction.[49] This quality may be useful for newborns suffering from acidosis. MTDQ is a lipid-soluble scavenger and MTDQ-DA is a water-soluble one. MTDQ-DA can be desulfonated in vivo and it can accumulate into the cell membranes protecting them from the harmful chain reactions of oxygen.

The beneficial effect of MTDQ was also observed in the course of experimental studies. MTDQ was administered to newborn rats kept in an oxygen-rich environment. It was found to decrease liver and lung malondialdehyde content. It also decreased superoxide dismutase activity in the lungs.[50] Orally administered MTDQ prevented hyperoxia-induced retinal injuries in newborn rats.[51]

REFERENCES

1. **Barson, A. J., Tasker, M., Lieberman, B. A., and Hillier, V. F.,** Impact of improved perinatal care on the causes of death, *Arch. Dis. Child,* 59, 199, 1984.
2. **Phelps, D. L.,** Neonatal oxygen toxicity — is it preventable?, *Pediatr. Clin. North Am.,* 29, 1233, 1982.
3. **Purohit, D. M., Ellison, R. C., Zierler, S., Mictinnen, O. S., and Nadas, A. S.,** Risk factors for retrolental fibroplasia: experience with 3025 premature infants, *Pediatrics,* 76, 339, 1985.
4. **Silverman, W. A.,** *Retrolental Fibroplasia: A Modern Parable,* Grune & Stratton, New York, 1980.
5. **Weiter, J. J.,** Retrolental fibroplasia: an unsolved problem, *N. Engl. J. Med.,* 305, 1404, 1981.
6. **Fridovich, I.,** Hypoxia and oxygen toxicity, *Adv. Neurol.,* 26, 255, 1979.
7. **Glass, P., Avery, G. B., Subramanian, K. N. S., Keys, M. P., Sostek, A. M., and Friendly, D. S.,** Effect of bright light in the hospital nursery on the incidence of retinopathy of prematurity, *N. Engl. J. Med.,* 313, 401, 1985.
8. **Kretzer, F. L., Mehta, R. S., Johnson, A. T., Hunter, D. G., Brown, E. S., and Hittner, H. M.,** Vitamin E protects against retinopathy of prematurity through action on spindle cells, *Nature (London),* 309, 793, 1984.
9. **Johnson, A. T. and Kretzer, F. L.,** Interstitial retinol binding protein in the developing human retina: a proposed explanation for vitamin E suppression of retinopathy of prematurity, in *The Interphotoreceptor Matrix in Health and Disease,* Bridges, C. D., Ed., Alan R. Liss, New York, 1985, 251.
10. **Hittner, H. M., Rudolph, A. J., and Kretzer, F. L.,** Suppression of severe retinopathy of prematurity with vitamin E supplementation: ultrastructural mechanism of clinical efficacy, *Ophthalmology,* 91, 1512, 1984.
11. **Milley, J. R., Rosenberg, A. A., and Jones, M. D.,** Retinal and choroidal blood flows in hypoxic and hypercarbic newborn lambs, *Pediatr. Res.,* 18, 410, 1984.
12. **Patz, A.,** Current concepts of the effect of oxygen on the developing retina, *Curr. Eye Res.,* 3, 159, 1984.
13. **Tripathi, B. J. and Tripathi, R.C.,** Cellular and subcellular events in retinopathy of oxygen toxicity with a preliminary report on the preventive role of vitamin E and gamma aminobutyric acid: a study in vitro, *Curr. Eye Res.,* 3, 193, 1984.
14. **Flower, R. W., Blake, D. A., Wajer, S. D., Egner, P. G., Mcleod, D. S., and Pitts, S. M.,** Retrolental fibroplasia: evidence for a role of the prostaglandin cascade in the pathogenesis of oxygen-induced retinopathy in the newborn beagle, *Pediatr. Res.,* 15, 1293, 1981.
15. **Ernest, J. T. and Goldstick, T. K.,** Retinal oxygen tension and oxygen reactivity in retinopathy of prematurity in kittens, *Invest. Ophthalmol. Visual Sci.,* 25, 1129, 1984.
16. **Phelps, D. L. and Rosenbaum, A. L.,** Effects of marginal hypoxemia on recovery from oxygen-induced retinopathy in the kitten model, *Pediatrics,* 73, 1, 1984.
17. **Halliwell, B. and Gutteridge, J. M. C.,** Oxygen free radicals and iron in relation to biology and medicine: some problems and concepts, *Arch. Biochem. Biophys.,* 246, 501, 1986.
18. **Noell, W. K.,** Effects of environmental lighting and dietary vitamin A on the vulnerability of the retina to light damage, *Photochem. Photobiol.,* 29, 717, 1979.
19. **Bridge, R. R., Murray, L. P., Pierce, B. M., Akita, H., Balogh-Nair, V., Findsen, L. A., and Nakanishi, K.,** Two-photon spectroscopy of locked-11-cis-rhodopsin: evidence for a protonated Schiff base in a neutral protein binding site, *Proc. Natl. Acad. Sci. U.S.A.,* 82, 4117, 1985.
20. **Stryer, L.,** Cyclic GMP cascade of vision, *Annu. Rev. Neurosci.,* 9, 87, 1986.
21. **Lamb, T. D.,** Transduction in vertebrate photoreceptors: The roles of cyclic GMP and calcium, *TINS,* May, 224, 1986.
22. **Kagan, V. E., Shvedova, A. A., Novikov, K. N., and Kozlov, Y. P.,** Light-induced free radical oxidation of membrane lipids in photoreceptors of frog retina, *Biochim. Biophys. Acta,* 330, 76, 1973.
23. **Ruffolo, J. J., Ham, W. T., Mueller, H. A., and Millen, J. E.,** Photochemical lesions in the primate retina under conditions of elevated blood oxygen, *Invest. Ophthalmol. Visual Sci.,* 25, 893, 1984.
24. **Anderson, R. E., Rapp, L. M., and Wiegand, R. D.,** Lipid peroxidation and retinal degeneration, *Curr. Eye Res.,* 3, 223, 1984.
25. **Organisciak, D. T., Wang, H. M., and Kon, A. L.,** Ascorbate and glutathione levels in the developing normal and dystrophic rat retina: effect of intense light exposure, *Curr. Eye Res.,* 3, 257, 1984.

26. **Boverins, A.**, Mitochondrial production of superoxide radical and hydrogen peroxide, *Adv. Exp. Med. Biol.*, 78, 67, 1977.
27. **Hadjiconstantinou, M. and Neff, N. H.**, Catecholamine systems of retina: a model for studying synaptic mechanism, *Life Sci.*, 35, 1135, 1984.
28. **Hadjiconstantinou, M., Cohen, J., and Neff, N. H.**, Epinephrine: a potential neurotransmitter in retina, *J. Neurochem.*, 41, 1440, 1983.
29. **Misra, H. P. and Fridovich, I.**, The role of superoxide anion in the autoxidation of epinephrine and a simple assay for superoxide dismutase, *J. Biol. Chem.*, 247, 3170, 1972.
30. **Sanyal, S., De Ruiter, A., and Dees, Ch.**, Light dependent accumulation of macrophages at the photoreceptor-pigment epithelial interface in the retina of albino mice, *Experientia*, 40, 852, 1984.
31. **Stanford, M. R.**, Retinopathy after irradiation and hyperbaric oxygen, *J. R. Soc. Med.*, 77, 1041, 1984.
32. **Reisner, S. H., Amir, J., Shohat, M., Krikler, R., Nissenkorn, I., and Ben-Sira, I.**, Retinopathy of prematurity: incidence and treatment, *Arch. Dis. Child*, 60, 698, 1985.
33. **Hittner, H. M., Kretzer, F. L., and Rudolph, A. J.**, Prevention and management of retrolental fibroplasia, *Hosp. Pract.*, 19, 85, 1984.
34. **Condell, R. A. and Tappel, A. L.**, Evidence for suitability of glutathione peroxidase as a protective enzyme: studies of oxidative damage, renaturation, and proteolysis, *Arch. Biochem. Biophys.*, 223, 407, 1983.
35. **Fridovich, I.**, Biological effects of the superoxide radical, *Arch. Biochem. Biophys.*, 247, 1, 1986.
36. **Chapela, R., Zuniga, G., and Selman, M.**, D-Penicillamine in the therapy of fibrotic lung diseases, *Int. J. Clin. Pharmacol. Ther. Toxicol.*, 24, 16, 1986.
37. **Lakatos, L., Hatvani, I., Oroszlán, Gy., Karmazsin, L., and Lakatos, Zs.**, Clinical observations in the prevention of retrolental fibroplasia with D-penicillamine, in *Physiologic Foundations of Perinatal Care*, Stern, L., Ed., Praeger, New York, 1984, 293.
38. **Lal, M., Lin, W. S., Graucher, G. M., and Armstrong, D. A.**, The repair protection and sensitization of papin with respect to inactivation by H_2O_2 and OH: effects of dithiothreitol, penicillamine, cysteine, and penicillamine disulphide, *Int. J. Radiat. Biol.*, 28, 549, 1975.
39. **Lengfelder, E. and Elstner, E. F.**, Determination of the superoxide dismutating activity of D-penicillamine copper, *Hoppe Seylers Z. Physiol. Chem.*, 359, 751, 1978.
40. **Karmazsin, L., Lakatos, L., Balla, Gy., Makay, A., and Hatvani, I.**, Experimental data on the prevention of retrolental fibroplasia by D-penicillamine, *Acta Paediatr. Acad. Sci. Hung.*, 21, 131, 1981.
41. **Cohen, A. N. and Ostrow, J. D.**, New concepts in phototherapy: photoisomerization of bilirubin IX and potential toxic effects of light, *Pediatrics*, 65, 740, 1980.
42. **Weiss, J. F. and Belkin, M.**, The effect of penicillamine on posttraumatic vitreous proliferation, *Am. J. Ophthalmol.*, 92, 625, 1981.
43. **Lakatos, L., Hatvani, I., Oroszlán, Gy., Balla, Gy., Karmazsin, L., Olubunmi, A., Kincses, É., Szabó, I., and Vekerdy, Zs.**, A koraszülöttek retinopathiájának megelózése D-penicillaminnal, *Orv. Hetil.*, 126, 1391, 1985.
44. **Bär, V., Böszörményi, J., Richter, P., Rozsnyai, T., and Mercz, J.**, Method for the Development of Water Soluble Dihydroquinoline Type Antioxidants (in Hungarian), Hungarian Patent 8081, 1979.
45. **Bär, V., Böszörményi, J., and Richter, P.**, Water Soluble Dihydroquinoline-Type Antioxidant Derivatives, U.S. Patent 4,356,306, 1980.
46. **Bär, V., Mercz, J., Szvoboda, J., Pollák, Zs., and Mátyás, J.**, Method for the Development of Dihydroquinoline Type Antioxidant (in Hungarian), Hungarian Patent 162, 1975.
47. **Bär, V., Mercz, J., Szvoboda, J., Pollák, Zs., and Mátyás, J.**, Dihydroquinoline Derivatives of Antioxidant Activity, U.S. Patent 4,025,631, 1977.
48. **Belágyi, J.**, Spin Trapping and Labelling of Dihydroquinoline Type Radical Scavengers: MTDQ and MTDQ-DA, Department of Biophysics, University Medical School unpubl. rep., Pécs, Hungary, 1985.
49. **Pollák, Zs., Rodé, I., and Bär, V.**, Beitrag zu den bio- und radiochemischen Wirkungen der Verbindung MTDQ, *Strahlentherapie*, 154, 499, 1978.
50. **Balla, G., Makay, A., Pollák, Zs., Matkovics, B., Lakatos, L., and Karmazsin, L.**, Damaging effect of free radicals liberated during the reduction of oxygen: its influencing by drugs, *Acta Paediatr. Acad. Sci. Hung.*, 23, 319, 1982.
51. **Balla, Gy., Makay, A., Hatvani, I., Boros, A., Bär, V., Pollák, Zs., Lakatos, L., and Karmazsin, L.**, Nem toxikus dihidrokinolintípusú antioxidáns hatása a retinális érburjánzásra hiperoxigenizált újszülött patkányokon, *Gyermekgyogyaszat*, 32, 357, 1981.

Index

INDEX

A

ABAP, see 2,2′-Azobis(2-amidinopropane hydrochloride)
Abetalipoproteinemia, 245, 248—249, 298—299
Acetaminophen, 121
Acetaminophen hepatotoxicity, 182
Acetylation, 215
Acetylcholine, 87, 128
N-Acetylcysteine, 121, 274
Acivicin, 128
Acquired intestinal malabsorption, 249
Actin, 325
Acute hepatitis, 173—174
Acute myocardial infarction, 194, 227—236
Adenosine triphosphate (ATP), 45, 168
Adenylate cyclase, 203
Adjuvant arthritis, 77, 80
Adrenal cortex, role of antioxidants in function of
  cytochrome P450, pseudosubstrate effects on, 3—5
  in vivo role of cytochrome P450/pseudosubstrate effects, 12—13
  protection of cytochrome $P450_{11\beta}$ by antioxidants, inhibitors, and lowered oxygen
    ascorbic acid, 8—12
    comparison of protection, 5—6
    mode of action of phenols and sulfoxides, 7—8
    vitamin E, selenium, and other biological antioxidants, 12
Adrenal medulla, 12
Adrenal vascular system, 12—13
Adrenaline, 82
Adrenochrome, 178, 180
Adrenochrome-adrenaline system, 180
Adrenochrome-induced toxicity, 181
Adrenocortical cells, 269
Adult hemoglobin, 327
Age, exercise and, 138
Age-pigment, 295
Aging, 18, 55—56
  autoxidation, role for, 294—298
  centrophenoxine, effects of, 90
  effects of
    concept of synaptic plasticity, 255—256
    effect on synaptic plasticity in the cerebellar cortex, 256—260
    effect of vitamin E deficiency, 260—263
  free radical theory, 125
  free radicals and, 264
  glutathione, role of, 125
  membrane hypothesis of, 88, 90—91, 255, 259—260
  oxygen consumption, 133
  oxygen radical-mitochondrial injury theory, 125
  synaptic membranes, 256—260
AIBN, see Azobis (-isobutyronitrile)
Air pollutant intoxication, 100
Albumin, 30, 41
Alcohol, 18
Alcoholic hepatitis, 169, 173
Alcoholism, 19—20
Aldehyde oxidase, 180, 303
Aldosterone, 3
Alkoxyl radicals, 30, 136
Amino acid translocation, 128—129
Amino acid transport, 128—129
Amino acid uptake, 128—129
Aminochrome, 178, 180
Analgesic actions, 80
Angiogenesis, 310
Anthocyanins, 77
Anthglutin, 128
Anthracycline antibiotics
  carcinogenic action of, protective effect of α-tocopherol against, 275
  cardiotoxicity induced by, modulation by α-tocopherol of, 274
  cardiotoxicity of, 269, 273—274
  cytotoxic activity of, 273
  free radical generation in cardiac tissue and cardiac defense system, 273—274
  lipid peroxidation in cardiac tissue induced by, 274
  mammary gland, carcinogenic action in, 274—275
  mechanism of action of, 271—273
  membrane-active antitumor agents, 272—273
  metabolism of, 271—272
  physiological function of α-tocopherol, 269—271
  redox cycling, 273
  side effects, 269
Antiaging drug, 87
Antibodies, 156
Anticancer drugs, 271
Anticonvulsant, 109
Antiedematous activity, 77—79
Antihistaminic effects, 82
Anti-inflammatory corticosteroids, 81
Anti-inflammatory drugs, 197—203
  depolymerization of hyaluronic acid in "in vitro" systems, 197—201
  depolymerization of hyaluronic acid in "in vivo" systems, 201—203
  oxygen free radicals, proteoglycans, and collagen, 203
Anti-inflammatory effects, 77—80, see also specific topics
  catechol-*O*-methyltransferase, 82
  connective tissue, 82
  inflammatory mediators, 80—81
  leukocyte functions, 81—82
  mechanisms of action, 80—83
Antimycin, 49—50
Antioxidant capacity, 39
Antioxidant coenzyme Q pool, 57
Antioxidant defense system, 180, 269, 303

Antioxidant effect, 38
Antioxidant systems, 122, 315
Antioxidant therapy, 333—336
  D-penicillamine, 335
  methyldihydroquinoline-type antioxidants, 335—336
  neuronal ceroid lipofuscinosis, 281—287
  retinopathy of prematurity, 327—330
Antioxidants, 39, 41, 57, 294, 299, see also specific types
  adrenal cortex, role in function of
    cytochrome P450, pseudosubstrate effects on, 3—5
    in vivo role of cytochrome P450/pseudosubstrate effects, 12—13
    protection of cytochrome $P450_{11\beta}$ by antioxidants, inhibitors, and lowered oxygen
      ascorbic acid, 8—12
      comparison of protection, 5—6
      mode of action of phenols and sulfoxides, 7—8
      vitamin E, selenium, and other biological antioxidants, 12
  amount present, 32
  atherosclerosis, prevention of, 220
  coenzyme Q, role of
    antioxidant properties of, 50—56
    lipid peroxidation and free radical formation, 45—50
    theoretical discussion: mechanisms, 56—58
  deficiencies, 294—295
  glutathione peroxidase, 298
  11β-hydroxylase, protection of
    ascorbic acid, 8—12
    comparison of protection, 5—6
    mode of action of phenols and sulfoxides, 7—8
    vitamin E, selenium, and other biological antioxidants, 12
  lipid-soluble, 33—37
  mechanisms of action
    chain-breaking
      general, 31—33
      lipid-soluble, 33—37
      water-soluble, 38—40
    preventive, 30
  modulation of zonation by, 13
  phenolic, 33
  retina, protection of
    factors favoring potential autoxidative damage, 290—294
    general, 289—290, 298—299
    roles for autoxidation, 294—298
  total, 33
  tumors, activity in, 21
  vitamins C and E, see Vitamin C; Vitamin E
  water-soluble, 38—40
Antipyretic actions, 80
Antisterility vitamin factor, 168
Antitumor agents, 272—273
Apnea, 304, 327
Apoferritin, 199
Apolipoproteins, 211—212
Apoprotein A-I, 193
Apoprotein B, 248
Aqueous radical-trapping antioxidants, 38
Arachidonic acid, 18, 50, 126, 220
Arachidonic acid cascade, 81
Arachidonic acid metabolites, 80
Arachidonic acid metabolization, 82
Aromatic amines, 31
Arterial plasma GSSG content, 146
Arterial wall
  derangement of enzyme activities, 218
  LDL, 218—223
  lipid metabolism, 215—217
Arteriosclerosis, 238—239
Arteriovenous differences of amino acids, 129
Arthritis, see also Rheumatoid arthritis
  lipid peroxidation and, 202
  oxygen free radicals and, 201—202
Ascorbate, 3, 31, 38—41, 45, 101, 294, see also Ascorbic acid
Ascorbate radical, 39, 95—96
Ascorbic acid, 77, 95—96, 99, 180, 283
  deficiency, 101
  degradation of hyaluronic acid, 198—199
  mode of action, 8—12
  protection by, 5, 8, 11, 181—182
Ascorbyl radical, 40
Atherogenesis, 187, 189
Atheroma, 212
Atherosclerosis, 187—188, 191, 194, 207, 209—210, 212—213, 215
  changes of enzyme activities in arterial wall, 216
  lipid peroxides associated with, 227
  prevention by antioxidants, 220
  vitamin C deficiency, 192—193
Atopic eczema, 18—19
ATP, see Adenosine triphosphate
Autofluorescence, 296, 299
Autofluorescent pigment, 294
Autoxidation, 32, 38, 40—41, 289, 298, 303
  aging, role in, 294—298
  catecholamines, 178
  free radical-mediated, 310
  glutathione, 124
  initiation, 29
  light, 294
  mechanism of, 29
  photo-induced, 30
  photoreceptor outer segment membrane, 299
  polyunsaturated fatty acids, 292, 295
  propagation, 29
  retinal components, 296
  retinal pathologies, human, 297—298
  retinal pathologies, role in, 294—298
  termination, 29
  thermally initiated, 32—33, 35, 38
  vitamin A, 293
Autoxidative damage, 290—294
Avascular retina, 323—324
Axonal degeneration, 251
Axons, 243—245, 248—251

Azo dyes, 72
2,2′-Azobis(2-amidinopropane hydrochloride) (ABAP), 32, 38, 40
Azobis (-isobutyronitrile) (AIBN), 32

B

B-lymphocytes, 153, 156—157
Bacteriophage T_2, 200
Barbiturate intoxications, 87
Bassen-Kornzweig syndrome, 248—249
Batten's disease, 298—299
Behavioral performance, 90
BHA, see Butylated hydroxyanisole
BHA analogs, 69—71
BHT, see Butylated hydroxytoluene
Bilayers, 35
Bile acids, 189—190
Bioassays, 35, see also specific types
Bioflavonoids, 77
Biological antioxidants, 36
Biotransformation enzymes, 99, 102
Biotransformation reactions, 100
Biotransformation systems, 100—102
Bipotencies, 36
Birth asphyxia, 304
Blackcurrant, 18
Bleomycin-dependent degradation of DNA, 199
Blindness, 333
Blood-brain barrier, 36, 87, 281
Blood components, 155
Blood flow, 129
Blood lipid levels, 207, see also α-Tocopherol
Blood plasma, 36
Blood-retinal barrier, 289
Blood vessel walls, 310
Body stores of vitamin E, 305
Borage, 18
Bovine serum albumin (BSA), 88
Bradykinin, 80—81
Brain cells, 87
Brain cortex, 90
Brain injury, 87
Breast cancer, human, 22
Breast milk, 329
Bronchopulmonary dysplasia, 304, 327
BSA, see Bovine serum albumin
Buthionine sulfoximine, 123—124
Butylated hydroxyanisole (BHA), 23, 63—66
Butylated hydroxytoluene (BHT), 23, 34, 63, 66, 218, 222, 283

C

Calcium, 114
Calcium fluxes, 115
Calcium homeostasis, 126, 134, 177, 183
Calcium paradox, 113
Calcium translocation, 273
Calcium uptake, 181
Canalization, 318
Cancer, 19—23
Capillary permeability, 80—82
Carbon-centered radical, 29
Carbon-centered, resonance-stabilized radical, 29, 37
Carbon dioxide production, 87
Carbon tetrachloride, 56
Carbon tetrachloride-induced hepatotoxicity, 168, 170—171
Carcinogenesis, 101
Carcinogenic action of anthracycline antibiotics
  mammary gland, 274—275
  protective effect of α-tocopherol against, 275
Carcinogenic chemicals, 20
Carcinogenicity, 67, 269
Carcinogens, 69
Cardiac cells
  oxidation of catecholamines, 178—181
  proposed mechanisms of iso-induced cardiotoxicity, 177—178
  protective effects of ascorbic acid and sodium bisulfate, 181—182
Cardiolipin, 272, 274
Cardiotoxicity, 177—178, 182, 273—274
Cardiovascular actions of taurine, 109
Cardiovascular disease, 19, 209, 212, 237—242
Carnitine biosynthesis, 189
β-Carotene, 30—31, 36—37, 41
β-Caroteneperoxyl radical, 37
Carotenoids, 36—37
β-Carotenyl radical, 37
Catalase, 30, 41, 122, 180, 218, 273, 303
Cataract formation, 122
Catechol-*O*-methyltransferase, 82
Catecholamine, 18, 177—181, 334
Catecholamine-induced cardiotoxicity, 177—178
Catecholamine-induced toxicity, 183
Catecholamine oxidase, 180
Catechols, 71
Cell-cell interactions, 221, 223
Cell cycling, 57
Cell death, see Cellular necrosis
Cell membrane, 82, 91, 127
Cellular colloids, 91
Cellular necrosis, 23, 113—114, 121
Central nervous system endothelial cells, 244
Centrophenoxine (CPH)
  aging, effects on, 90
  chemical methods for superoxide radical reactions, 89
  effects of, 87—88
  electron spin resonance spectroscopy in spin trapping experiments, 88—89
  nootropic effects, 91
  OH· radical scavenging ability of, 89
  oxy radical scavenger properties, 88—89
  protein cross-linking model, 88
  superoxide radical scavenger properties of, 89
Cerebellar glomeruli
  concept of synaptic plasticity in, 255—256
  effect of aging on synaptic plasticity, 256—260
  effect of vitamin E deficiency, 260—263

Cerebral atrophy, 87
Cerebral vascular accident, 227—236
Cerebrospinal fluid, 90
Cerebrovascular mortality, 187
Ceroid lipofuscinosis, 298
Ceruloplasmin, 30, 41, 199, 201
Chain-breaking antioxidants, 30—33, 41, 100
Chain-carrying peroxyl radicals, 29
Chain initiation, 30, 32
Chain oxidation, 32
Chain reaction, 29—31, 39
Chelating agents, 222
Chemiluminescence, 67
Chemotactic factors, 83
Chemotaxis, 155—156
Chiral discrimination, 36
Chloramine derivative, 117
p-Chlorophenoxyacetic acid (PCPA), 87—89
Cholestane-3β, 5α, 6β-triol, 218
Cholestatic liver disease, 247
Cholesterol, 18, 57, 187—188
  relationship with α-tocopherol, 169, 173
  transformation into bile acids, 189—190
Cholesterol catabolism, 161—162
Cholesterol ester, 215—216
Cholesterol esterifying enzyme, 216
Cholesterol homeostasis, 161
Cholesterol 7α-hydroxylase, 190
  cholesterol catabolism, 161
  circadian rhythm, 161—162
  plasma cholesterol level, 162
  protein kinase, 162
  protein phosphatase, 162
  vitamin E status, effect of, 161—165
Cholesterol hydroxylation, 161
Cholesterol synthesis, 161
Cholestyramin, 193
Choline, 87—88
Cholinergic vesicles, 128
Chorioallantoic membrane assay, 320
Choroidal vasculature, 318
Chroman heterocyclic ring, 34
Chromosome aberrations, 99
Chronic alcoholism, 87
Chronic hepatitis, 173—174
Chronic inflammation, 80
Chronic inflammatory diseases, 203
Chronic necrotizing myopathy, 244
Chronic vitamin E deficiency, see Vitamin E, at deficiency
Chrysin, 80
Cigarette smoking, 188
Circadian effects, 138
Circadian rhythm, 161—162
Clarke's column, 243
Clotting factors, 318
CMO, see Corticosterone methyl oxidase
Coenzyme, glutathione as, 126
Coenzyme Q
  acute myocardial damage, 227—236
  aging, 55—56
  antioxidant properties of, 50—56
  electron transfer, 46
  endurance training, 55
  intact animals, 54—56
  mechanisms, 56—58
  mitochondrial electron transfer chain, 46
  protonmotive reactions, 46
  redox state, role of, 52—53
  stroke, 227—236
  structure, 46
  subcellular systems, 50—54
Cold acclimation, 54
Collagen, 203
Collagen biosynthesis, 188—189
Collagen cross-linking, 82
Compactin, 57
Compartmentalization of cell, 98
Compensative synaptogenesis, 259
Congenital biliary atresia, 245—248, 250—251
Conjugation, 121
Connective tissue, 82
Copper, 199, 222
Coronary artery disease, 207
Coronary artery occlusion, 209
Coronary atherosclerosis, 194
Coronary mortality, 187, 190
Corticosterone methyl oxidase (CMO), 3—5
  loss of, 4
  protection by antioxidants, inhibitors, and lowered oxygen
    ascorbic acid, 8—12
    comparison of protection, 5—6
    mode of action of phenols and sulfoxides, 7—8
    vitamin E, selenium, and other biological antioxidants, 12
Cortisol, 3
Cotton pellet granuloma, 77, 79
CPH, see Centrophenoxine
Creatine kinase, 133
Cultured heart cells, 181
Cuneate nuclei, 243, 245, 248
Cyclooxygenase, 220
Cyclooxygenase inhibitors, 23
Cyclooxygenase products, 80—81
Cystathionine pathway, 124
Cysteamine, 123, 274
Cysteine, 123—124
Cysteine requirements, 138
Cysteine sulfinate, 105—107
Cysteinyl-glycine, 123, 126
Cystic fibrosis, 18, 251
Cytochrome *b*, 47—49
Cytochrome b-c_1, 45
Cytochrome-indophenol oxidase system, 180
Cytochrome P450, 63, 66, 190, 193
  ascorbic acid, protection by and mode of action of, 8—12
  cholesterol hydroxylation, 161
  interactions with vitamins C and E, 96—97
    effect of vitamins C and E on detoxication processes, 97—101

redox systems, 95—96
metyrapone, protection by, 5—6
peroxidase activity, 5
phenols, protection by and mode of action of, 5—8
pseudosubstrate effects on, 3—5, 12—13
reduction by lipid peroxidation, 162
sulfoxides, protection by and mode of action of, 5—8
Cytochrome P450 cycle, 97—99, 102
Cytochrome P450-dependent microsomal hydroxylation, 193
Cytochrome P450 metabolism, 72
Cytosol, 56, 91
Cytotoxicity, 218, 273

D

Damaged cells, 137
Daunorubicin, 271—272, see also Anthracycline antibiotics
De novo synthesis, 68
Dehydroascorbate, 39
Deiodination, 125—126
Density, 91
Depolymerization of hyaluronic acid, see also Hyaluronic acid
in vitro systems, 197—201
in vivo systems, 201—202
Desaturation, 21
Δ-5-Desaturation, 17
Δ-6-Desaturation, 17—18
Desferrioxamine, 283
Detoxication
enzyme induction, mechanism of, 71—72
glutathione, function of, 146
glutathione, role of, 121—122
Detoxification, see Detoxication
Detoxification enzymes, 63, 65—71, see also specific types
Developing capillaries, 304
DGLA, see Dihomogammalinolenic acid
DHA, see Docosahexaenoic acid
DHBF, see Dihydrobenzofuran
Diabetes, 19—20
Diabetes mellitus, 18
Diabetic neuropathy, 20
Diabetic retinopathy, 77
Diamide, 128
Dicoumarol, 67
Dietary fiber, 187
Diethylmaleate, 124
Dihomogammalinolenic acid (DGLA), 17—18
Dihydrobenzofuran (DHBF), 35
Dihydro-orotic dehydrogenase, 303
Dimethylaminoethanol (DMAE), 87, 89, 91
Diol-epoxides, 66
Diosmin, 77
Distal axonopathy, 251
Disulfide bonds in protein, 127
Disulfide bridges, 127
Disulfiram, 66
Dithiothreitol, 123
Divalent cations, 222
DMAE, see Dimethylaminoethanol
DMPO, 89
DNA synthesis, 123
Docosahexaenoic acid (DHA), 19, 293
Dolichols, 57
Dorsal columns of spinal cord, 243—245, 251
Dorsal horns, 243
Doxorubicin, 271—273, see also Anthracycline antibiotics
Drug detoxication, see Detoxication
Drugs, 121
DT diaphorase, 56, 272
Dystrophic axons, 243, 245—246, 248, 251
Dystrophic retina, 296

E

Edema, 77, 81
EDTA, 218
EFA, see Essential fatty acids
Efamol, 18—20
Eggs, 18
EH, see Epoxide hydrolase
Eicosanoids, 17, 220
Eicosapentaenoic acid (EPA), 18—19
Electron spin resonance spectroscopy (ESR), 88—89
Electron transfer reactions, 46, 48
Electron-transferring components of cells, 303
Electron transport, 52
Electron transport chain, 45
Electron transport system, 303
Electrophilic compounds, protection against, 72
Electrophoretic mobility, 223
Endocytosis, 222
Endogenous antioxidant mechanisms, 181
Endogenous metabolic peroxides, 146
Endoplasmic reticulum, 100, 134—135, 167, 180
Endothelial cell derived growth factor, 221
Endothelial cells, 218—221
Endothelial injury, 188
Endothelial necrosis, 320, 322, 328
Endothelial retraction, 320, 322, 328
Endotoxins, 188
End-plate potentials, 128
Endurance capacity, 135—136
Endurance training, 55, 136, 138—141
Energy production, 303
Enzymatic oxidations, 303
Enzymes, 121
EPA, see Eicosapentaenoic acid
Epidermal growth factor, 221
EPO, see Evening primrose oil
Epoxide hydrolase (EH), 64, 66—67, 72
E-PTA, see Ethanolic phosphotungstic acid technique
Erythrocyte disorders, 148
Erythrocyte membrane, 203
Erythrocytes, 271
binding sites for vitamin E, 269

glutathione concentration in, 125
oxidative stress inducing efflux of GSSG from, 146
Escherichia coli, 123
Eskimos, 19
ESR, see Electron spin resonance
Essential fatty acids (EFA)
alcoholism, 19—20
cancer, 20—23
diabetes, 19—20
foods associated with, 18
general, 17—19
illnesses associated with, 18
inflammatory disorders, 19—20
metabolism, 18
therapeutic administration of, 19—20
Essential nutrients, 157
Ethanol, 168
Ethanolic phosphotungstic acid (E-PTA) technique, 256—257
Ethoxyquin, 66
Ethylene, 198
n-Ethylmaleimide, 49
Evening primrose oil (EPO), 18
Excitable tissues, 113—115
Exercise
age, 138
caloric demands, 138
cysteine requirements, 138
effects of, 133
endurance capacity, 135—136
endurance training, 136, 138—141
to exhaustion, 135—136
glucose requirement, 138
glutathione studies, 136—138
lipid peroxidation during, 135—136
muscle injury, 133
oxidative stress, effects on, 133
oxygen utilization, 133
physical fitness, 138, 141
tissue injury, 133
vasopressin, 137
vitamin E depletion during, 135—136
Exogenous antioxidants, 333
Extrahepatic tissues, 68—69
Extraretinal fibrovascular proliferation, 322
Extraretinal neovascularization, 309

F

Fasting, 138
Fatigue, 133, 189
Fatty acid peroxidation, 192
Fe^{2+} overload, 90
Fenton reaction, 88, 198, 259
Ferritin, 199
Fetal hamsters, 245
Fetal to neonatal circulation, transition from, 303
Fisetin, 80
Fish oils, 19, 21
Flavanones, 78—79
Flavanonols, 78—79
Flavin semiquinone, 48
Flavoenzyme dehydrogenases, 303
Flavones, 78—79
Flavonoids, 77, 80—83
Flavonols, 78—79
Flavoproteins, 48, 180
Fluorescence, 294—295, 297—298
Fluorescence anisotropy, 90
Fluorescence recovery after photobleaching (FRAP), 90
Fluorophores, 295—296
Foam cell, 215, 220, 222
Formaldehyde dehydrogenase, 126
Formaldehyde glyoxalase, 126
Four-electron reduction of molecular oxygen, 303
FRAP, see Fluorescence recovery after photobleaching
Free fatty acids, 178
Free radical chains, 32
Free radical damage
liver, 167—168
protection against, 153, 155
spindle cells, 315
Free radical formation, role of coenzyme Q
aging, 55—56
antioxidant properties, 50—56
electron transfer, 46
endurance training, 55
intact animals, 54—56
mechanisms, 56—58
mitochondrial electron transfer chain, 45—49
protonmotive reactions, 46
redox state, role of, 52—53
structure of, 46
subcellular systems, 50—54
vitamin E involvement, 147
Free radical generation, 113—117
Free radical-induced damages, 90—91
Free radical intermediates, 303
Free radical-mediated autoxidations, 310
Free radical pathology, 47
Free radical scavengers, 81, 101
synovial fluid degradation, action on
depolymerization of hyaluronic acid in "in vitro" systems, 197—201
depolymerization of hyaluronic acid in "in vivo" systems, 201—202
oxygen free radicals, proteoglycans, and collagen, 203
vitamin E, 148
Free radical theory of aging, 125
Free radicals, 17, 20—24, 36, 41, 83, 134, 180—181, 292, 298, 333
lipid peroxidation, see Lipid peroxidation
protection against, 121
retinal pathologies, 289
retinopathy of prematurity risk factor, 325—327
synovial fluid degradation induced by
depolymerization of hyaluronic acid in "in vitro" systems, 197—201
depolymerization of hyaluronic acid in "in vivo"

systems, 201—202
oxygen free radicals, proteoglycans, and collagen, 203
uncontrolled production, 303
Fruit, 187—190
Fulminant hepatitis, 169, 173
Functional tests for vitamin E, 305

G

G6PD, see Glucose 6-phosphate dehydrogenase
GABA, 108
D-Galactosamine hepatitis, 168—172
Gamma-linolenic acid (GLA), 18
Gap junctions, 320
Gene activity regulation, 271
Genetics, 303
GES, see Guanidinoethane sulfonate
Ghost membranes, 272
GI side effects, 80
GLA, see Gamma-linolenic acid
Gluconeogenesis, 121, 124, 137
Glucose, 146
D-Glucose, 127
Glucose 6-phosphate dehydrogenase (G6PD), 65, 124—125, 127, 146
Glucose uptake, 87
γ-Glutamyl amino acids, 128—129
γ-Glutamyl cycle, 123, 128—129
γ-Glutamyl-cysteine synthetase, 124
γ-Glutamyl glutamine, 129
γ-Glutamyl transpeptidase, 123—124, 126, 129
Glutaraldehyde, 223
Glutaredoxin, 123
Glutathione, 38, 40, 45, 180, 218, 274
aging of cells, 125
amino acid uptake, 128—129
autoxidation, 124
calcium homeostasis, 126
cataract formation, prevention of, 122
coenzyme, 126
cysteine, 123—124
damaged cells, 137
deficiency
hemoglobin changes caused by, 147—148
reduction in RBC survival, 147
deiodination of thyroid hormone, 125—126
depletion, 121, 124, 138, 282
detoxication, role in, 121—122
detoxification as main function of, 146
disulfide bonds in protein, role in formation of, 127
DNA synthesis, 123
exercise-induced oxidative stress, effects on
effects of endurance training, 138—140
general, 133—134, 140—141
lipid peroxidation and vitamin E depletion, 135—136
methods, 134—135
studies, 136—138
gluconeogenesis, 124
glucose to prevent oxidation of, 146
γ-glutamyl linkage, 121, 128—129
hemoglobin structure, modification of, 127
in tissues, 138
insulin and, 127
intracellular levels, 134
leukotriene synthesis, 126
macromolecular structures, role in maintenance of, 127
metabolic functions of, 123—126
methionine-sparing effect of, 124
neurotransmitter release, 128
oxidants, line of defense against, 146
oxidative stress, 122, 134, 136—138
parathyroid hormone and, 127—128
pentose-phosphate activity, control of, 146
pentose phosphate pathway, 124—125
physiological functions, see other subtopics hereunder
polypeptide hormones, secretion of, 127—128
protective roles, 121—123
protein synthesis, 123
proteolysis, 123
radiation, 122—123
recovery, 138
resting animals, 137
synergism between vitamin E and, 146—148
synthesis, 134
thiol status, maintenance of, 127
third messenger, 134
α-tocopherol regeneration, 270
transport processes, role in, 127—129
vitamin E regeneration by, 150
Glutathione disulfide, see GSSG
Glutathione levels, 63
Glutathione peroxidase, 30, 41, 56—57, 122, 125, 180, 281, 298, 303
Glutathione-peroxidase pathway, 146
Glutathione reductase (GR), 65, 122, 125, 146, 180, 273
Glutathione synthetase deficiency, 156
Glutathione system, 148
Glutathione *S*-transferase (GST), 64—66, 68—70, 72, 124
Glutathione-transferase pathway, 146
Glutathionuria, 128
Glycine receptors, 108
Glycogenolysis, 121
Glycosylation reaction, 57
Glycosides, 77
Glyoxalase, 126
GR, see Glutathione reductase
Gracile nuclei, 243—246, 248, 250
Granulocytosis, 194
GSH, see Glutathione
GSH pools, 124
GSH/GSSG ratio, 122, 124, 126
GSH status, 127
GSH turnover, 129
GSSG, 122—123, 125, 134, 146
GST, see Glutathione *S*-transferase
Guanidinoethane sulfonate (GES), 108

Guinea pig, 17

H

Haber-Weiss reaction, 115, 122, 198—199
Halothane anesthesia, 167
HDL, see High density lipoproteins
Heart, 140—141
Heart attacks, 19
HeLa cells, 272, 275
Hemoglobin, 147—148
Hemoglobin structure, 127
Hemolytic anemia, 306
Hepa 1c1c7 cells, 71—72
Hepatocytes, 124
Hepatoma cells, 71—72
Hepatotoxic mechanisms, 167
Hepatotoxicity, 100, see also Liver injury
Hepatotoxins, see Liver injury
12-HETE, 310
15-HETE, 310
High-density lipoproteins (HDL), 193
  cholesterol levels, 210—212
  low-density lipoprotein cytotoxicity, 218
High-intensity light, 327
High pressure liquid chromatography, 134
Histamine, 80—81
HMG-CoA reductase, 161, 190
Human plasma, 35, 37—40
Human serum, 40
Hyaline membrane disease, 327
Hyaluronic acid
  anti-inflammatory drugs, 199—200
  arthritis, 201—202
  depolymerization
    in vitro systems, 197—201
    in vivo systems, 201—202
  hyaluronidase, 200—201
  iron implication, 198—199
  lipid peroxidation, 202
  oxygen free radicals, implication of, 197—198, 201—202
  polymorphonuclear leukocytes, role of, 200
Hyaluronidase, 200—201
Hydrazine derivatives, 167
Hydrogen abstraction, 136
Hydrogen peroxide, 48, 101, 115, 122, 180—181, 218, 291—292, 303
Hydrolases, 180
Hydrolysis, 121, 311
Hydroperoxidation, 115
Hydroperoxide, 30, 122
Hydroperoxyl radical, 40
Hydroquinones, 71
Hydroxy acids, 310
Hydroxy fatty acids (HETEs), 81, see also 12-HETE; 15-HETE
3-Hydroxy-3-methylglutaryl-coenzyme-A reductase, see HMG-CoA reductase
Hydroxyl radical production, 198
Hydroxyl radical scavengers, 197, 199—200
Hydroxy radical scavenging, 310
β-Hydroxybutyrate, 50
25-Hydroxycholesterol, 218
12-Hydroxyeicosatetraenoic acid, 212
Hydroxyl radicals, 30, 101, 115—117, 122, 181, 291—292, 303
11β-hydroxylase, see also Cytochrome P450
  protection by antioxidants, inhibitors, and lowered oxygen
    ascorbic acid, 8—12
    comparison of protection, 5—6
    mode of action of phenols and sulfoxides, 7—8
    vitamin E, selenium, and other biological antioxidants, 12
  pseudosubstrate effects in loss of, 3—5
  selenium, effects of, 12
  α-tocopherol, effects of, 12
  vitamin E, effects of, 12
Hydroxylation, 121, 197
4-Hydroxynonenal, 23
Hydroxyproline, 89
Hyperactivity, 18
Hyperbilirubinemia, 327
Hypercholesterolemia, 190, 192—194, 216
Hyperglutathionemia, 128
Hyperlipidemia, 216
Hyperoxia, 303, 320, 334
Hyperoxic environment, 318
Hypertension, 18, 188, 194
Hypertriglyceridemia, 189, 210
Hypobetalipoproteinemia, 248
Hypocellular region, 323
Hypochlorous acid, 116
Hypotaurine
  antioxidant effect, 115—116
  biosynthesis, 105—107
  concentration in tissues, 105
  free radical generation, 113—117
  lipid peroxidation, 113—117
  membrane stabilizer, 109—113
  metabolic degradation, 107
  metabolic pathways, 105—107
  pharmacological effects, 108—109
  physiological effects, 108—109
  protective effect, 113
  quenching of hydroxyl radicals, 115—117
  regulatory mechanisms, 105—107
  sexual tissue, 115
  sources, 105—108
  synthesis, 107—108
  uptake, 107—108
Hypotaurine aminotransferase, 107
Hypoxia, 146, 303, 305, 310, 320, 334
Hypoxic insult, 305

I

Immaturity, 303, 325—326
Immune responses
  nonspecific, 154—156
  specific, 156—157

Immunoenhancement, 157
Increased survival, 304
Indomethacin, 23
Induction
BHA, studies with
analogs, 69—71
extrahepatic tissues, 68—69
liver, 67—68
detoxification enzymes
mechanism, 71—72
phase II, 65—67
in rodent tissues, 67—71
general, 63—65, 72
Induction period, 31—33, 39
Inflammation, 133
Inflammatory diseases, 19—20
Inflammatory mediators, 80—81
Inhibitors, protection of 11β-hydroxylase
ascorbic acid, 8—12
comparison of protection, 5—6
mode of action of phenols and sulfoxides, 7—8
vitamin E, selenium, and other biological antioxidants, 12
Initiating radicals, 29
Initiation, 29-30, 41, 293
Initiators, 32—33, 38
Insulin, 127
Intermediary metabolism, changes in, 178
International Classification for ROP, 306, 322
Interstitial retinol binding protein, 315, 318—319, 327
Intracellular acidity, 178
Intracellular density, 91
Intracellular enzymes, 149
Intracellular hydration, 91
Intracellular potassium, 90—91
Intracellular signal, 129
Intraretinal vasculature, 304
Intrauterine environment, 303
Intraventricular hemorrhage, 304, 311—312, 327
Investigational new drug, 307—308, 311—312
Ionic composition, 91
Iodothyronine 5′-deiodinase, 125
Ionizing radiation, 20
Iproniazid, 167
Iron, 23, 45, 198—199, 222
Iron-induced peroxidation, 199
Iron-sulfur flavoproteins, 48
Iron supplementation, 306
Irreversible mitochondrial swelling, 45
Ischemic heart disease, 212
Ischemic/hypoxic insult, 311
Isolated vitamin E deficiency, 249, 251
Isoniazid, 167
Isoproterenol
calcium homeostasis, alterations in, 177
cardiotoxicity induced by, 177—178
oxidation of catecholamines, 178—181
protective effects of ascorbic acid and sodium bisulfate in cardiac cells, 181—182
Isozymes, 67—68

K

Kaempferol, 80
KCN, 49
Ketoaldehydes, 126
2-Ketoisocaproate, 127
Ketone bodies, 129
Kidneys, 18
Kittens, 305

L

Lacrimal gland, 19
Lactating rat, 128
Lactation, 128—129
Large-caliber myelinated sensory axons, 251
Lateral diffusion constant membrane proteins, 90
LDL, see Low density lipoproteins
Learning ability, 87
Lectin receptors, 272
Lens cells, 146
L-Leucine, 127
Leukocyanidin, 77
Leukocyte functions, 81—82
Leukotriene synthesis, 126
Leukotrienes (LTs), 17, 81, 138
Light, 289—290, 293—294, 327
Light damage, 293—294, 297
cis-Linoleic acid, 17—18
cis-α-Linolenic acid, 17—18
Lipids, 207—210, 293
Lipid composition, 292—293
Lipid membranes, 303
Lipid metabolism, 161, 192—193
Lipid metabolism disorders
biochemical aspects, 187—192
epidemiology, 187
optimum intake of vitamin C, 193—194
vitamin C, lipids, and atherosclerosis, 192—193
Lipid peroxidation, 20—24, 29, 83, 113—117, 127, 157, 213, 218, 223, 273
arthritis, 202
in cardiac tissue, induced by anthracyclines, 274
cytochrome P450, reduction in, 162
exercise-induced oxidative stress, 135—136
experimental cardiovascular disease, 237—239
inhibition of, 50
liver injury, 167—174
mitochondrial electron transfer chain, 45—49
Lipid peroxides, 17, 113
Lipid peroxyls, 39
Lipid-soluble antioxidants, 33—37
Lipid-soluble, plasma antioxidant, 37
Lipid vesicles, 52
Lipofuscin, 87, 91, 125, 299
antioxidant deficiencies, 294—295
photoreceptor cells, influence of, 297
retinal pathologies, 294—297
senile macular degeneration, 297—298
vitamin A, 295—297
Lipoperoxidation, 100—101, 191—192

Lipoperoxidation reactions, 188
Lipopigment, 244, 247, 249, 251—252
Lipoproteins, 161, 210—211
β-Lipoprotein, 169, 173
β-Lipoproteinemia, 169
Liposomes, 38—40, 52—53, 271
Lipoxygenase (LO) products, 80—81
Liver, 18, 136—138, 140—141
  cholesterol transformation into bile acids, 189—190
  induction of detoxification enzymes by BHA, 67—68
  oxidative stress inducing efflux of GSSG from, 146
Liver cancer, 173
Liver cirrhosis, 168, 173—174
Liver glutathione, see Exercise; Glutathione
Liver injury
  carbon tetrachloride-induced, 168, 170—171
  ethanol, 168
  free radical damage, 167—168
  D-galactosamine as cause, 168, 172
  halothane anesthesia, 167
  hydrazine derivatives, 167
  iproniazid, 167
  isoniazid, 167
LO, see Lipoxygenase products
Local hypoxic/ischemic events, 303
Long-chain fatty acids, 56
Low-birth weight infants, 306, 315
Low-density lipoprotein-dextran complex, 222
Low-density lipoprotein pathway, 215
Low-density lipoprotein receptor, 190, 193
  acetylated, 223
  cholesterol homeostasis, 161
  down regulation of, 215
Low-density lipoproteins (LDL), 187—192
  acetylated, 222
  cholesterol levels, 210, 212
  denatured, 215
  endothelial cells, 218—221
  macrophages and, 222—223
  metabolism in arterial wall, effect of vitamin E on
    derangement of enzyme activities, 218
    endothelial cells, 218—220
    general, 215—217, 223
    macrophages, 222—223
    SMCs, 221—222
  modified, 222
  native, 215
  oxidized, 215, 221
  smooth muscle cells, 221—222
Low temperature, 327
LTs, see Leukotrienes
Lymphocytes, 99
Lymphokines, 156
Lysophosphatidylcholine-induced lens damage, 145
Lysosomal activity, 91
Lysosomal enzymes, 83
Lysosomal membrane, 203
Lysosomes, 81—82

M

Macromolecular structures, 127
Macrophage-derived growth factor, 220—221
Macrophages, see Mononuclear cells
Magic bullet treatment, 23
Major antioxidant enzymes, 303, see also specific types
Malabsorption syndrome, 298
Malonaldehyde, 23
Malondialdehyde (MDA), 49, 115, 122, 223
Mammalian organ, 129
Mammary cancer, 21
Mammary gland, 124, 128—129, 274—275
Manganese ion, 45
Mastocytoma cells, 272
Maturation period, 91
MDA, see Malondialdehyde
Meat, 18
Medium life span, 90
Membrane-active antitumor agents, 272—273
Membrane-associated enzymes, 149
Membrane-bound enzymes, 162
Membrane hypothesis of aging, 88, 90—91, 255, 259—260
Membrane lipid fluidity, 90
Membrane permeability, 91
Membrane phospholipids, 88
Membrane proteins, 90
Membrane stabilization, 162, 203, 270—271
Membrane stabilizers, 109—113
Menadione, 54
Mercapturic acid, 121
Metabolic disequilibrium, 310
Metabolic disorders, 194
Metabolic rate, 54
Metal catalysts, 310
Metal ion chelation, 40
Metal ions, 291
Metaplastic cells, 322—325
Methional, 198
Methionine, 121, 283
L-Methionine, 128, 282
Methionine-sparing effect of GSH, 124
Methylation, 215
3-Methyl-cholanthrene, 66
Methyldihydroquinoline-type antioxidants, 335—336
Methylglyoxal, 126
Metyrapone, 5—6
Mevinolin, 57
Mice, 17, 67
Microbicidal radicals, 48
Micropremies, 304
Microsomal monooxygenases, 98
Microsomes, 52, 56
β-Migrating very low-density lipoprotein, 222
Milk, human, 18
Mitochondria, 50, 53, 56—57, 168, 290
Mitochondrial electron transfer chain, 45—49
Mitochondrial injury, 181—182

Mitochondrial membrane permeability, 181—183
Mitochondrial oxidative phosphorylation, 273
Mitochondrial photo-oxidation, inhibition of, 51
Mitochondrial swelling, 157
Mitochondrion, 290
Monkey, vitamin E-deficient, 248—249
Monoamine oxidase, 180
Monobromobimane, 127
Monocytes, 220
Mononuclear cells, 153, 156, 201, 220, 327
  isolated, 200
  low-density lipoproteins and, 222—223
Monounsaturated fatty acids, 21
Morbidity, 207, 212, 304, 306
Mortality, 207, 212, 305
Mosquito, glutathione concentration, 125
mRNA synthesis, 88, 90
Muller cells, 318, 320
Muscle, 140
Muscle injury, 133
Muscle soreness, 133
Mussel hepatopancreas, 125
Mutagenicity, 67
Myelin sheaths, 243—244, 248, 250
Myocardial cell injury, 177, 183
Myocardial hypoxia, 178
Myocardial infarction, 18, 189, 212
Myocardial ischemia, 177
Myocardial necrosis, 227, 230
Myocardium, 87
Myofibroblasts, 325, 329
Myricetin, 80

N

NADH dehydrogenase, 45, 49, 272
NADH/NAD^+ ratio, 127
$NADP^+$, 125
NADP coenzyme, 146
NADPH, 122—123, 125, 134
NADPH-cytochrome *c* reductase, 180
NADPH cytochrome P450 reductase, 180
NADPH dehydrogenase, 272
NADPH/$NADP^+$ ratio, 127
NAD(P)H:quinone reductase, 67
Natural phenolic compounds, anti-inflammatory effects of, 77—83, see also Anti-inflammatory effects
NCL, see Neuronal ceroid lipofuscinosis
Necrotizing enterocolitis, 304
Neonatal care, 304
Neonatology, 333
Neovascularization, 315
Nerve cell membranes, 88—89
Neuroblastic retina, 315
Neuromuscular preparations, 128
Neuronal ceroid lipofuscinosis (NCL), 281—285
Neuronal perikarya, 244
Neuronal plasticity, 255
Neurons, 255
Neuropathology of chronic vitamin E deficiency
  experimental vitamin E deficiency, 243—245
  in humans, 245—251
Neurotransmitter release, 128
Neutrophil functions, 81
Neutrophils, 47, 116, 154—156
Newborn infants, vitamin E deficiency, 305
N-Nitroso compounds, formation of, 101—102
p-Nitrosodimethylalanine, 199
Nonheme iron-sulfur proteins, 48
Nootropic effects, 87, 91
Nootropica, 87
Noradrenaline, 82
NSAI drugs, 80
Nucleus of tractus solitarius, 243
Nutrition, 306
Nutritional lack, 303

O

Obliteration of capillaries, 305
Ocular lens, glutathione metabolism in, 122
OH· free radical scavenger centrophenoxine as, 87—91
OH· free radicals, 88—90
OH· radicals, 89—90
OH· radical scavenger, 89
OH· radical scavenger properties of DMAE, 91
OH· radical scavenging ability of CPH, 89
Oleic acid, 18
Opsin, 293
Organ contractions, 81
Organelle membranes, 83
Osteoarthritis, 202
Ovaries, 18
Oxidant damage, 304—305, 311
Oxidant disease, 303
Oxidant metals, 303
Oxidant radical damage, 310
Oxidant radical scavengers, 303
Oxidant stress, see Oxidative stress
Oxidants, glutathione as line of defense against, 146
Oxidation, 121
Oxidative damage, 140, 147—148, 181
Oxidative metabolism, 303
Oxidative metabolites, 180, 182
Oxidative phosphorylation, 47
Oxidative reactions, 108
Oxidative stress, 127, 222, see also Exercise
  efflux of GSSG from erythrocytes, liver, and lens cells, 146
  exercise-induced, 133
  glutathione as protection against, 122, 134
  vitamin E as system of defense against, 149
Oxidized sterols, 219
5-Oxoproline, 128—129
Oxy radical-protecting effect, 88
Oxy radical scavenger properties of CPH, 88—89
Oxygen, 292, 298, 327
  divalent reduction of, 303

protection of 11β-hydroxylase
ascorbic acid, 8—12
comparison of protection, 5—6
mode of action of phenols and sulfoxides, 7—8
vitamin E, selenium, and other biological antioxidants, 12
retinal pathologies, 289
retrolental fibroplasia, 305
toxic potential, 304
univalent reduction of, 303
unmonitored quantities, 304
Oxygen administration, 315
Oxygen affinity, 127
Oxygen consumption, 87, 290—292, 297
Oxygen free radicals, 192, 197—198, 201—203
Oxygen radical-mitochondrial injury theory of aging, 125
Oxygen radicals, 99
Oxygen tension, 304
Oxygen toxicity, 133, 333
Oxygen utilization, 133
Oxyhemoglobin, 303

P

Pain, 80
Pancreas, 18
Pancreatic B-cell, 127
Parabolic responses, 99
Paracetamol, 121
Parathyroid cells, 127—128
Parathyroid hormone, 127—128
Passive potassium permeability, 88, 91
Patent ductus arteriosus, 327
PBN, 89
PCB, see Polychlorinated biphenyls
PCPA, see *p*-Chlorophenoxyacetic acid
Pectin, 187, 190
D-Penicillamine, 335
Pennsylvania Hospital, 306
Pentose-phosphate activity, 146
Pentose-phosphate pathway, 124—125, 146
Perinatal care, 304
Peripheral arterial occlusive disease, 77
Peripheral vascular insufficiency, 77
Peroxidase, 5, 7, 122
Peroxidation, 30, see also Lipid peroxidation
Peroxides, 30, 121
Peroxyl radical-catalyzed autoxidative destruction, 39—40
Peroxyl radical scavenger, 35
Peroxyl radical-trapping capacity, 38
Peroxyl radicals, 136
antioxidants, mechanisms of
chain-breaking
general, 31—33
lipid-soluble, 33—37
water-soluble, 38—40
preventive, 30
autoxidation, mechanism of, 29
Phagocyte functions, 155—156
Phenobarbital, 66
Phenolic antioxidants, 33, 63—65
detoxification by enzyme induction, 71—72
induction of detoxification enzymes, 67—71
phase II detoxification enzymes, 65—67
Phenolic compounds, anti-inflammatory effects of, 77—83, see also Anti-inflammatory effects
Phenolic derivatives, 77
Phenols, 5—8, 31—33, 35
Phenoxyl radical, 31, 34—35
Phloretin, 80
Phorone, 124
Phosphate, 45
Phosphatidyl-DMAE, 88, 91
Phospholipase A_2, 222
Phospholipases, inhibition of, 82
Phospholipid liposomes, 35
Phospholipid vesicles, 57
Phospholipids, 169, 174, 292
Phosphoryl-DMAE, 87
Photochemical damage, 333
Photo-induced autoxidation, 30
Photoreceptor cells, 289—291, 293, 297—298
Photoreceptor maturation, 315—318
Photoreceptor outer segment membrane, 295, 298
Photoreceptors, 318, 327
Photosensitizer, 293
Physical exercise, see Exercise
Physical fitness, 138, 141
Physiologic equilibrium, 304
Placenta, 129
Plasma, 36, 38, 40, 136—138, 140—141
Plasma antioxidants, 38
Plasma cholesterol level, 162
Plasma lipids, 36, 39, see also α-Tocopherol
apolipoproteins, 211—212
clinical implications, 212—213
effect of α-tocopherol on lipoproteins, 210
effect of α-tocopherol on serum lipids, 207—210
Plasma membrane, 91
Plasma proteins, 38, 40
Platelet aggregation, 178, 191, 213
Platelet-derived growth factor, 218, 221
Platelets, 218
Plus disease, 306, 325, 329
Pneumothorax, 327
4-POBN, 89
Polychlorinated biphenyls (PCB), 66
Polycyclic aromatic hydrocarbons, 66, 72
Polymorphonuclear leukocytes, 200—201
Polypeptide hormones, 127—128
Polyphenols, 77
Polysaccharide chains, 82
Polysorbates, 311
Polyunsaturated fatty acids (PUFAs), 37, 39, 100—101, 126, 145
autoxidation, 292, 295
cancer, 20—23
EFAs, therapeutic administration of, 19—20

general, 17—19
groups of, 17
photoreceptor cells, 297
Polyunsaturated lipid, 32, 298
Post-thrombotic edema, 77
Postapoplectic disorders, 87
Posterior column of spinal cord, 248, 251
Posterior pole retinal arteries and veins, 304
Prawns, 18
Premature infants, 306, 309
Prematurity, 304, 306
Premenstrual syndrome, 18
Preventive antioxidants, 30, 40—43
Primary antioxidants, 41
Primary necrotizing myopathy, 243, 247
Primate plasma, 40
Primitive capillary network, 304
Primitive endothelial cell, 305
Probucol, 223
Progressive neurological disorder, 245—248, 250—251
Proline, 89
Proline hydroxylation, 82
Pro-oxidant, 37, 45, 147, 180
Propagation, 29—30
Propranolol, 181
Prostacyclin, 191, 220
Prostacyclin synthetase, 220
Prostaglandin biosynthesis, 191—192
Prostaglandin E_1, 191
Prostaglandin E_2, 203
Prostaglandins, 17, 80—81, 138
Proteases, 303
Protein component damage, 162
Protein cross-linking model, 88
Protein kinase, 162
Protein phosphatase, 162
Protein-SH groups, 38, 40, 146, 150
Protein synthesis, 91, 123, 145
Protein thiols, 41
Protein turnover, 91
Proteoglycans, 203
Proteolysis, 123
Protonmotive Q cycle, 45
Psychometric performance, 90
Pseudosubstrates
cytochrome P450, effects on 3—5, 12—13
11β-hydroxylase, effects in loss of, 3—5
PUFAs, see Polyunsaturated fatty acids
Pulmonary hemorrhage, 327
Pyridoxine, 283
Pyruvate kinase, 36

Q

QR, see Quinone reductase
Quality of life, 212
Quercetin, 80
Quinol, 56
Quinone reductase (QR), 64, 67, 71—72, 271
Quinones, 180

R

Rabbit, 17
Radiation, 121—123
Radical chain reaction, 31
Radical processes, 91
Radical propagation stage, 41
Radical-trapping capacity, 40
Radicals, 31, 38—39
Radiosensitivity, 123
Rat curative myopathy bioassay, 36
Rat liver, 35
Rat liver microsomes, 54
Rats, 17
chronic vitamin E deficiency, 243—245
induction by BHA, 67
vitamin E-deficient, 244—247
Raynaud's syndrome, 19
Reactive metabolite, 121
Red blood cell glutathione, 145—150, see also Glutathione
Red blood cells, 35
reduction in survival of, vitamin E and GSH association with, 145, 147
specific binding sites for α-tocopherol, 145
vitamin E to protect against hemolysis induced by oxidative agents, 145
Red cells, 36
Redox cycling, 178, 273
Redox systems of vitamins C and E, 95—96
Reduced glutathione (GSH), see Glutathione
Reduction, 121
Rehydration of intracellular mass, 90
Reperfusion, 320, 327
Reperfusion injury, 183
Residual heat production, 91
Respiratory burst, 154—155
Respiratory chain inhibitors, 49, 52
Respiratory disease, 304
Response to injury hypothesis, 218
Reticulocytes, 123
Retina
autoxidation, 292—294, 296
autoxidative damage to, 290—294
avascular, 323—324
canalization, 318
development of, 315—320
hypoxia, 305
incompletely vascularized, 303
inner, 318
light, 293—294
lipid composition, 292—293
Muller cells, 318, 320
oxygen consumption, 290—292, 297
photoreceptor maturation, 315—318
preterm, 315, see also Retinopathy of prematurity
spindle cells, 318
vitamin A, 293

Retinal degeneration, 298
Retinal detachment, 315
Retinal dystrophy, 295
Retinal pathologies, 294—298
Retinal pathologies, human, 297—298
Retinal pigment epithelium, 289—290, 293—294, 296, 299
Retinal vascularization, 304—305
Retinal vasoformation, 318
Retinal vessels, 304—305
Retinaldehyde, 293
Retinitis pigmentosa, 298—299
Retinol, 180, 293, 296—297
Retinol binding protein, 293
Retinol derivatives, 296
Retinopathy, 20, 299
Retinopathy of prematurity (ROP), 298—299, 333—334
  angiogenic factors, 320
  antioxidant therapies, 327—330
  biochemical events, 320
  categorization, 309
  cicatricial residua, reduction in, 307
  cicatricial sequelae, 309—310
  classification system to include mild disease, 306
  early enrolled infants, 309
  experimental studies, 305
  free radicals, 325—327
  gap junctions, 320
  high-dose vitamin E therapy, 309
  hypocellular region, 323
  hypoxic/acidotic/hyperoxic insults, 310
  immaturity, 325—326
  incidence, 304, 306—307, 309, 320
  initiator of, 305
  intervention, 308
  late enrolled infants, 309
  light, 327
  long-term outcome, 304
  metabolic disequilibrium, 310
  metaplastic cells, 322—325
  moderate, 310
  myofibroblasts, 325
  neonatal events, 310
  oxygen, 327
  perinatal events, 310
  progression from moderate to severe, 309—310
  reperfusion, 327
  risk factors, 305, 309, 325—327
  rough endoplasmic reticulum, 320
  selenium supplementation, 330
  sepsis, 327
  severe, 304, 307—310, 320
  standard vitamin E supplement, 308
  transfusions, 327
  ultrastructural events, 320—325
  visual sequelae, 310
  vitamin E acetate, 307
  vitamin E and
    clinical trials, 307—310
    deficiency, 305—307
    E-Ferol, 311
    IVH, 311
    oxidant radicals, 303—304
    parenteral, 311—312
    retinal vascularization, 304—305
    risk benefit considerations, 310—311
  vitamin E therapy, 327—330
Retroendocytosis, 222
Retrolental fibroplasia, 304—306, 333—337
Reversible mitochondrial swelling, 45
Rhesus monkeys, 244
Rheumatoid arthritis, 19, 201—202, see also Arthritis
Riboflavin, 283
Ribonucleotides, 123
RNA synthesis, 91
ROP, see Retinopathy of prematurity
Rotenone, 49, 52
Rough endoplasmic reticulum, 320
Rutin, 77
Rutin derivatives, 77

S

Salicylate hydroxylation, 197
Sarcoma 180 cells, 272
Sarcoplasmic reticulum, 134—135
Scanning electron microscopy, 189
Schiff-base linkage, 293
Schwann cells, 251
Seafoods, 18—19
Seizures, 109
Selenium, 237—242, 281, 298
  11β-hydroxylase activity, effects on, 12
  retinopathy of prematurity, supplementation in, 330
  vitamin E as redox controller of, 145, 149
Selenium-dependent glutathione peroxidase, 315
Selenium-glutathione peroxidase, 327
Selenium supplementation, 315
Semiquinone, 48, 67, 135, 180
Senescence, 3, 299
Senile macular degeneration, 297—298
Sensory nucleus of fifth nerve, 243
Sepsis, 304, 327
Serine-borate, 128
Serotonin, 80—81
Serous retinal detachment, 325
Serum topics, see specific compound
Sexual functioning, 108
Sexual tissue, 115
Singlet oxygen, 36, 48, 115, 293, 298, 310
Sjögren's syndrome, 18–19
Skeletal muscle, 133, 136–138, 140—141
Skeletal muscle fibers, 251
Skin, 17
Slow reacting substance, 126
Smooth endoplasmic reticulum, 190
Smooth muscle, 180
Smooth muscle cells, 221—222
SOD, see Superoxide dismutase
Sodium bisulfite, 181—182

Sodium selenite, 281—282
Spastic diplegia, 305
Spielmeyer-Sjögren disease, 281, 283, 285
Spin adducts, 88
Spin trapping experiments, 88—89
Spin traps, 88—89
Spinal cord, vitamin E deficiency in
  experimental, 243—245
  in humans, 245—251
Spindle cell, 304—305, 315, 318, 329
Spinocerebellar disorder, 245, 252
Starvation, 129
Stereoelectronic effect, 35
Stereoelectronic enhancement, 35
Steroid-related toxicity, 81
Steroids, 3—4
Stimulus-induced calcium influx, 82
Strenuous exercise, 133, see also Exercise
Streptozotocin, 20
Stroke, 227—236
Subependymal hemorrhages, 245
Submitochondrial electron transfer chain components, redox state, influence of, 54
Submitochondrial particles, 47, 49—50, 52—53
Substituted phenols of enzymes, induction by, 63—65
  BHA, 67—71
  detoxification, 71—72
  phase II detoxification enzymes, 65—67
Succinate, 51—52
Succinate dehydrogenase, 45
Succinate oxidation, 52
Suicidal attempts, 121
Sulfhydryl groups, 38, 40, 273—274
Sulfoxides, 5—8
Sulfur-containing amino acids, 125
Superoxide, 40, 115, 181, 303
Superoxide anion, 40, 47, 101, 178, 180, 218, 222, 291—292
Superoxide dismutase (SOD), 23, 40—41, 46, 56, 122, 180, 199, 201—202, 218, 281, 303
Superoxide radicals, 23, 45, 47, 49, 52, 67, 89, 122
Sural nerve, 249, 251
Survival of low-birth weight infants, 306
Synaptic density and aging, 263
Synaptic junctions, 255—256, 259
Synaptic membranes, 256—260
Synaptic morphometry, 263
  aging, effect of, 256—260
  concept of, 255—256
  vitamin E deficiency, effect of, 260—264
Synaptic plasticity, see Synaptic morphometry
Synaptic remodeling, 258—259
Synaptic turnover, 263
Synaptic ultrastructure, 259—261
Synaptosomal membranes, 90
Synaptosomes, 90
Synergism, 38, 190
Synergistic effect, 39—40
Synergistic interactions, 41
Synergistic mechanism, 40
Synovial fluid, degradation of
  depolymerization of hyaluronic acid
    in vitro systems, 197—201
    in vivo systems, 201—202
  oxygen free radicals, proteoglycans, and collagen, 203
  UV illumination of, 197
Systemic sclerosis, 19
Systemic shock, 303

T

TBA, see Thiobarbituric acid
TBARS, see Thiobarbituric acid reactive substances, 215, 217
TCDD, see 2,3,7,8-Tetrachlorodibenzo-*p*-dioxin
T-lymphocytes, 153, 156—157
Taurine
  antioxidant effect, 113, 116—117
  biosynthesis, 105—107
  calcium fluxes, 115
  cardiovascular actions of, 109
  concentration in tissues, 105—106
  deficiency, 108, 110—112
  excitable tissues, 113—115
  free radical generation, 113—117
  lipid peroxidation, 113—117
  metabolic degradation, 107
  metabolic pathways, 105—107
  membrane stabilizer, 109—113
  pharmacological effects, 108—109
  physiological effects, 108—109
  protective effects, 112—113
  regulatory mechanisms, 105—107
  sexual tissue, 115
  sources, 105—108
  synthesis of, 105, 107—108
  uptake, 107—108
Taxifolin, 80
Temperature control, 109
Termination, 29
tert-Butylhydroquinone, 70
Testes, 18
2,3,7,8-Tetrachlorodibenzo-*p*-dioxin (TCDD), 66
Thiobarbituric acid (TBA), 21, 23, 202
  acute myocardial damage, 227—236
  stroke, 227—236
Thiobarbituric acid reactive substances (TBARS), 215, 217
Thiols, 38, 121—122, 124, 127, see also Glutathione
Thioredoxin, 123
Third messenger, glutathione as, 134
Thrombosis, 220
Thyroid hormone, 54, 125—126
Thyroxine, 45, 125
Tissue injury, 133
Tocopherol, 101, 180, 217—218, 283
α-Tocopherol, 33—36, 45, 50, 95—96, 263, see also Vitamin E
  acute myocardial damage, 227—236
  anthracycline antibiotics, potential modifier of

toxicity of, 269—275
anthracycline-induced cardiotoxicity, modulation of, 274
antioxidant, 269—270
apolipoproteins, 211—212
carcinogenic action of anthracycline antibiotics, protection against, 275
clinical implications, 212—213
gene activity regulation, 271
11β-hydroxylase, effects on, 12
lipoproteins, effect on, 210—211
lysophosphatidylcholine-induced lens damage, effect on, 145
membrane-structure stabilizer, 270—271
physiological function of, 269—271
red blood cells, specific binding sites for, 145
relationship with cholesterol, 169, 174
relationship with β-lipoprotein, 169, 173
relationship with phospholipid, 169, 174
serum high-density lipoprotein cholesterol levels, 210—212
serum lipids, effect on, 207—210
serum low-density lipoprotein cholesterol levels, 210, 212
serum very low-density lipoprotein cholesterol levels, 210, 212
stroke, 227
Tocopheroxyl radical, 39
α-Tocopheroxyl radical, 39—40
Tocopheryl radicals, 101—102
Total antioxidant, 33
Total cholesterol levels, 207—209
Total protein content, 90
Total synthesis, 88, 90
Toxic electrophiles, protection against, 63
Toxicity, 67
Trans acids, 17
Transcription, 91
Transferrin 30, 41, 199
Transformed cell lines, 21
Transforming growth factor, 221
Transfusions, 327
Translation, 91
Transpeptidase deficiency, 128
Transpeptidation, 128
Transport processes, 127–129
Treadmill, 136
Triglyceride levels, 209—210
3,3′-5-Tri-iodothyronine, 125
Tumor formation, 65
Tumors, 123

U

Ubiquinone, 45
UDP-glucose dehydrogenase (UDPGD), 65
UDP-glucuronosyltransferase (UDPGT), 64; 66, 72
Unsaturated fatty acids, 303
Unstable infants, 303
Urate, 38, 40—41
Urate proteins, 38
Urea, 88
Uric acid, 40
Uridine diphosphate glucose, 168
Uridine triphosphate, 168
UV illumination, 197

V

Vascular damage, 213
Vascular diseases, 207
Vascular fragility, 82
Vascularization, 310
Vasodilatation, 80
Vasodilation, 177
Vasopressin, 137
Vegetables, 187—190
Venous plasma GSSG content, 146
Venous insufficiency, 77
Verapamil, 181
Very low-density lipoprotein cholesterol levels, 210—212
Viral hepatitis, 168
Viral infections, 18
Vitamin A, 289, 293—298, 315, 318
Vitamin B_2, 283—284
Vitamin B_6, 283—284
Vitamin C, 38—40, 82, 283
blood components, contents of, 155
cytochrome P450 cycle, 97—99
deficiency, 97—98, 102, 155—156
atherosclerosis, 192—193
biochemical aspects, 187—192
carnitine biosynthesis, 189
cholesterol transformation into bile acids, 189—190
collagen biosynthesis, 188—189
epidemiology, 187
lipid metabolism, 192—193
lipoperoxidation, 191—192
prostaglandin biosynthesis, 191—192
detoxication processes, effects on, 97—101
immune responses, effect of interaction with vitamin E on, 153—158
interaction with cytochrome P450, 95—101
interaction with vitamin E, 96—97
evidence of, 154
immune response, effect on, 153—158
megadoses, 102, 193
nonspecific immune responses, 154—156
optimum intake of, 193—194
optimum doses, 102
redox system of, 95—96
specific immune responses, 156—157
synergism between vitamin E and, 101
Vitamin C-Vitamin E interaction, 41
Vitamin deficiencies, 157, see also specific vitamins
Vitamin E, 19, 33—36, 38—40, 134, 222, 282, 284, 295, 315, 334, see also α-Tocopherol
antioxidant action of, 145—146

bioactivity, 35—36
biological membranes, active element in, 146
biological membranes, effects on, 147
biological membranes, incorporation into, 148
blood components, contents of, 155
cell death, 23
cholesterol catabolism, 161—162
chronic deficiency, see deficiency, hereunder
clinical trials, 327—330
deficiency, 97, 99, 102, 145, 156, 162
abetalipoproteinemia, 248—249
acquired intestinal malabsorption, 249
congenital biliary atresia, 245—248, 250—251
cystic fibrosis, 251
effects of
on plasticity of cerebellar synaptic junctions, 260—263
synaptic plasticity
in the cerebellar cortex, 256—260
concept of, 255—256
experimental, 243—244
15-HETE, 310
isolated, 249, 251
light exposure, 294
liver diseases, 168
protein component damage, 162
reduction in RBC survival, 145, 147
retinopathy of prematurity, 305—307
depletion during exercise, 135—136
detoxication processes, effects on, 97—101
exogenous sources, 329
free radical scavenger, 148
functions, 146
high doses, 311
11β-hydroxylase, effects on, 12
immune responses, effect of interaction with vitamin C on, 153—158
immunostimulation by, 157
in vivo antioxidant, 162
interaction with cytochrome P450, 95—101
interaction with vitamin C, 96—97
evidence of, 154
immune response, effect on, 153–158
interstitial retinol binding protein, 315, 318
intracellular enzymes, influence on, 149
IVH and, 311
lipid peroxides, inhibition of, 227
liver injury, 167—174, see also Liver injury
low-density lipoprotein metabolism in arterial wall
derangement of enzyme activities, cause of, 218
and endothelial cells, 218—220
and macrophages, 222—223
and SMCs, 221—222
and vitamin E, 215—217
mechanisms of action of, 149
megadoses, 102, 318, 329
membrane-associated enzymes, modulation of activity of, 149
membrane stabilization, 162
nonspecific immune responses, 154—156
optimum doses, 102
oxidative stress, system of defense against, 149
pharmacologic supplementation, 329
physiologic supplementation, 329
protein synthesis, regulatory role in, 145
red blood cell glutathione and, 145—150
red blood cells protected against hemolysis induced by oxidative agents, 145
redox controller of selenium, 145, 149
redox system of, 95—96
regeneration by glutathione, 150
retinal pathologies, 296, 298
retinopathy of prematurity, see Retinopathy of prematurity
route of administration, 328
serum total cholesterol levels, effect on, 207—209
serum triglyceride levels, effect on, 209—210
side effects, 311
specific immune responses, 156—157
synergism between glutathione and, 146—148
synergism between vitamin C and, 101
therapeutic implications, 148—150, 310—311
toxicity, 328—329
Vitamin E acetate, 307—308, 311
Vitamin E alcohol, 311—312
Vitamin E analog, 35
Vitamin E-free alcohol, 307—308
Vitamin E/PUFA ratios, 306
Vitamin E status, cholesterol 7α-hydroxylase, effect on, 161—165
Vitamin K, 329
Vitamin K_5, 54

W

Water content, 91
Water permeability, 88, 91
Water solubility of proteins, 90
Water-soluble antioxidants, 38–40
Westermarck-Santavuori therapy, 282, 284
Workloads, 140

X

Xanthine dehydrogenase, 303
Xanthine oxidase, 180, 271—272, 303
Xanthine oxidase/hypoxanthine mechanism, 310
Xanthine oxidase system, 200
Xanthine-xanthine oxidase system, 115, 197, 203
Xenobiotics, 67, 69, 167

Z

Zeman formula, 282
Zinc, 18
Zonation, 12—13